KV-646-566

LIVERPOOL JMU LIBRARY
3 1111 01471 5559

SYMBOLIC COMPUTATION

Computer Graphics

Springer Series
SYMBOLIC COMPUTATION – *Computer Graphics*

J. Encarnação, E. G. Schlechtendahl:
Computer Aided Design. Fundamentals and System Architectures. IX, 346 pages, 183 figs., 1983

G. Enderle, K. Kansy, G. Pfaff:
Computer Graphics Programming. GKS – The Graphics Standard. XVI, 542 pages, 93 figs., 1984

J. Encarnação, R. Schuster, E. Vöge (eds.):
Product Data Interfaces in CAD/CAM Applications. Design, Implementation and Experiences. XIV, 254 pages, 147 figs., 1986

U. Rembold, R. Dillmann (eds.): Computer-Aided Design and Manufacturing. Methods and Tools. 2nd, revised and enlarged edition. XIV, 458 pages, 304 figs., 1986

Computer-Aided Design and Manufacturing

Methods and Tools

Second, Revised and Enlarged Edition

Edited by
U. Rembold and R. Dillmann

With 304 Figures

Springer-Verlag
Berlin Heidelberg New York
London Paris Tokyo

Prof. Dr.-Ing. Ulrich Rembold
Dr.-Ing. Rüdiger Dillmann
Universität Karlsruhe
Institut für Informatik III
Postfach 69 80
D-7500 Karlsruhe 1

This book is a completely revised version of the Lecture Notes in Computer Science, Vol. 168 "Methods and Tools for Computer Integrated Manufacturing", Springer-Verlag Berlin Heidelberg New York Tokyo 1984

ISBN-13:978-3-642-82750-1 e-ISBN-13:978-3-642-82748-8
DOI: 10.1007/978-3-642-82748-8

Library of Congress Cataloging-in-Publication Data.
Computer-aided design and manufacturing.
(Symbolic computation. Computer graphics)
Completely rev. version of Methods and tools for computer integrated manufacturing.
Bibliography: p.
1. CAD/CAM systems – Congresses. I. Rembold, Ulrich. II. Dillmann, R., 1949–.
III. Advanced CREST Course on Computer Integrated Manufacturing (1983 : Karlsruhe, Germany). Methods and tools for computer integrated manufacturing. IV. Series.
TS155.6.C644 1986 670.42'7 86-19943 ISBN-13:978-3-642-82750-1 (U.S.)

2145/3140-5 4 3 2 1 0

Introduction

Manufacturing contributes to over 60 % of the gross national product of the highly industrialized nations of Europe. The advances in mechanization and automation in manufacturing of international competitors are seriously challenging the market position of the European countries in different areas. Thus it becomes necessary to increase significantly the productivity of European industry. This has prompted many governments to support the development of new automation resources. Good engineers are also needed to develop the required automation tools and to apply these to manufacturing. It is the purpose of this book to discuss new research results in manufacturing with engineers who face the challenge of building tomorrow's factories.

Early automation efforts were centered around mechanical gear-and-cam technology and hardwired electrical control circuits. Because of the decreasing life cycle of most new products and the enormous model diversification, factories cannot be automated efficiently any more by these conventional technologies. With the digital computer, its fast calculation speed and large memory capacity, a new tool was created which can substantially improve the productivity of manufacturing processes. The computer can directly control production and quality assurance functions and adapt itself quickly to changing customer orders and new products. It allows the automation of long and short range forecasting, product design, process planning, production scheduling and manufacturing control. The flow of information through a plant can be monitored and corrective action can easily be initiated if necessary. Probably the most important asset of the computer is its capacity to integrate the entire manufacturing system from product design to fabrication. Its ability to make decisions will contribute to the conception and design of flexible manufacturing systems. With the progress of artificial intelligence activities and with powerful low cost computing equipment, it will be possible to conceive expert systems for manufacturing. Upon entry of a product order they will be able to plan and supervise automatically the production process. This will be realized with the help of decision rules and a knowledge data base about the production process and the available manufacturing resources.

The object of this book is to discuss advanced and future CAD/CAM technologies. Of course, the vast scope of manufacturing cannot be satisfactorily presented in one volume. Thus only those topics which are of common interest to many manufacturing organisations and which appear to be very important in the future are selected. In addition, an attempt is made to present to the reader the computer activities in manufacturing as an entity. For this reason the contents of the book can be used for courses covering this subject.

The following topics are presented:

- Computer Control for Manufacturing Equipment
- Design of Production Control and Flexible Manufacturing Systems
- CAD Systems and their Interface with CAM
- Technological Planning for Manufacture
- Economic Analysis of Manufacturing Systems
- Quality Assurance and Robot Vision
- The Interface of Process Planning with Production and Facility Planning
- Product Design for Assembly
- Production Control and Information Systems
- Programming of Machine Tools and Industrial Robots

This book is based on the lecture notes of the Advanced Course on Computer Integrated Manufacturing (CIM '83), held at the University of Karlsruhe, September 5-16, 1983, Karlsruhe, Federal Republic of Germany. The course was financed by the Ministry of Research and Technology (BMFT) of the Federal Republic of Germany and by the Commission of the European Communities. The course was a continuation of a course on "Computer Aided Design, Modelling, Systems Engineering, CAD-Systems" held at the Technical University of Darmstadt, September 8-19, 1980. The lecture notes were published in 1980 by Springer-Verlag under the same title, and edited by J. Encarnação.

The principal contributors to this book are:

- R. Dillmann, University of Karlsruhe, Karlsruhe, Germany
- G. Doumeingts, University of Bordeaux, Bordeaux, France
- W. Epple, University of Karlsruhe, Karlsruhe, Germany
- M. Gini, Politecnico of Milan, Italy
- H. Grabowski, University of Karlsruhe, Karlsruhe, Germany
- F.-L. Krause, IPK-Berlin, Berlin, Germany
- F. Leimkuhler, Purdue University, West Lafayette, Indiana, USA
- P. Levi, University of Karlsruhe, Karlsruhe, Germany
- C.R. Liu, Purdue University, West Lafayette, Indiana, USA
- A. Redford, University of Salford, Salford, Great Britain
- U. Rembold, University of Karlsruhe, Karlsruhe, Germany
- A.-W. Scheer, University of Saarland, Saarbrücken, Germany
- G. Spur, IPK and Technical University of Berlin, Berlin, Germany

All these authors deserve words of sincere thanks for their excellent manuscripts. Mrs. D. Jung and Miss B. Seufert were responsible for typing the final manuscript, and Mr. S. Müller helped with the editing. The authors would like to express their gratitude to all the people who helped to contribute to the success of this book.

Karlsruhe, October 1986

Ulrich Rembold
Rüdiger Dillmann

Contents

4 Evolutionary Trends in Generative Process Planning

1 CAD-Systems and Their Interface with CAM

H. Grabowski and R. Anderl

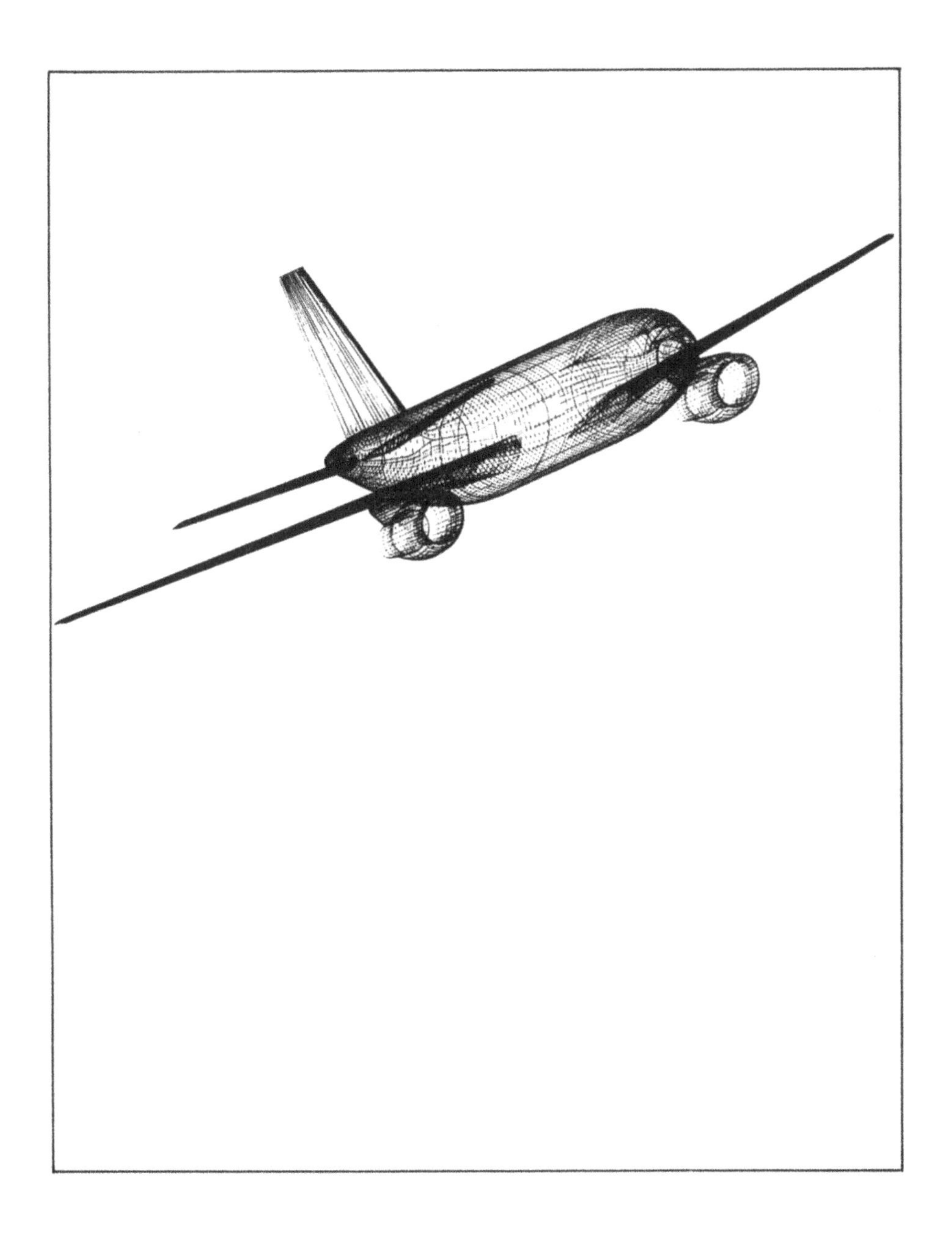

Design study of an airbus

(Courtesy of Messerschmidt-Bölkow-Blohm, Hamburg, FRG)

1.1 Introduction

Business and competition force companies to automate their production processes and to increase the product quality. Computers used for the control of the production process improve the productivity and increase the quality of the product. The computer can be applied to the following different production areas:

CAD	Computer Aided Design
CADR	Computer Aided Drafting
CAE	Computer Aided Engineering
CAP	Computer Aided Planning
CAM	Computer Aided Manufacturing
CAQ	Computer Aided Quality Assurance
CAM	Computer Integrated Manufacturing

Generally stated the computer supports the production processes and improves these via high performance hardware, algorithms, logic and the processing of information. During the last years hardware and software performance has been increasing steadily. This trend has helped to promote the computer as a tool for design, planning, manufacturing and quality assurance.

Today's developements aim to link and integrate computer aided processes. The question of the interface, as a key to data exchange between computer applications, becomes more and more important. In addition to the technical questions, organizational aspects relating to the information flow and operational organization have to be considered.

1.2 Philosophy of the Application of CAD Systems

Computer Aided Design (CAD) is a new technology in which the design processes are computer supported. A CAD-system contains hardware and software components, which are effectively integrated into one system. A typical CAD-System hardware configuration is shown in Fig. 1.1.

Several vendors of CAD-equipment offer so-called CAD Turnkey Systems which contain CAD hardware and CAD software as one product and which can readily support a specific design process [1]. In addition to the CAD-turnkey systems computer independent CAD-software packages are offered. These CAD software packages have to be installed on a hardware configuration which consists of the computer and the CAD workstations.

CAD systems can be defined by their hardware and software components or by the task for which they are used in a company. Figure 1.2 represents the characteristics of a CAD-system.

The use of the CAD-system technology for the design process increases productivity because it shortens the time for the development phases of a product. Some facts which illustrate the importance of CAD-Systems for design are summarized in Fig. 1.3.

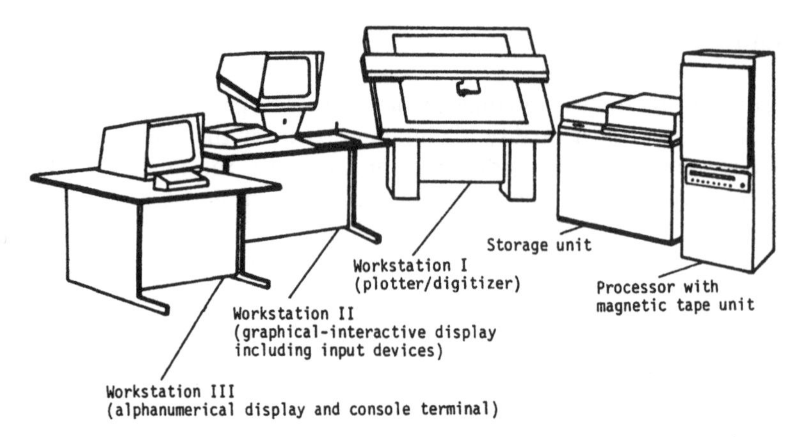

Fig. 1.1. Typical CAD-hardware configuration

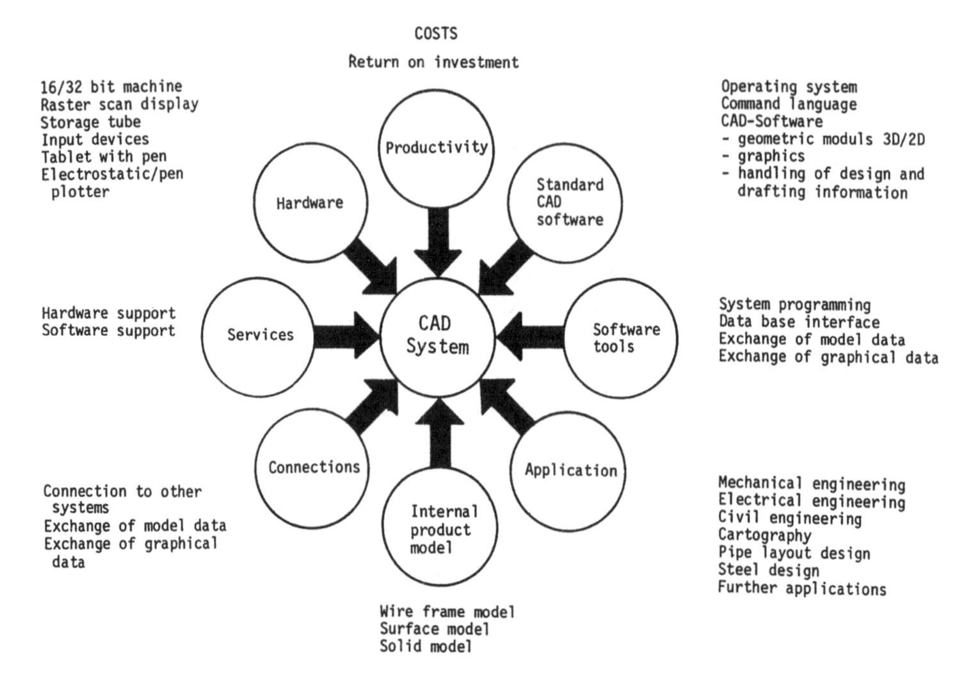

Fig. 1.2. Characteristics of CAD-systems

On the basis of these characteristics it becomes evident that the design department needs powerful new tools:

- To increase productivity
- To improve the efficiency of the design process
- To increase the design quality and product quality
- To reduce processing time of orders

Why the computer should be used to support the design process:

- the transit-time of an order through design is about 25-30% of its total processing time
- design is responsible for determining about 75% of the total manufacturing cost
- the design productivity has increased by about 20% since 1900, that of manufacturing by about 1000%
- design has an investment of about $ 1200 per workstation manufacturing about $ 20.000

Fig. 1.3. Characteristics of the design process

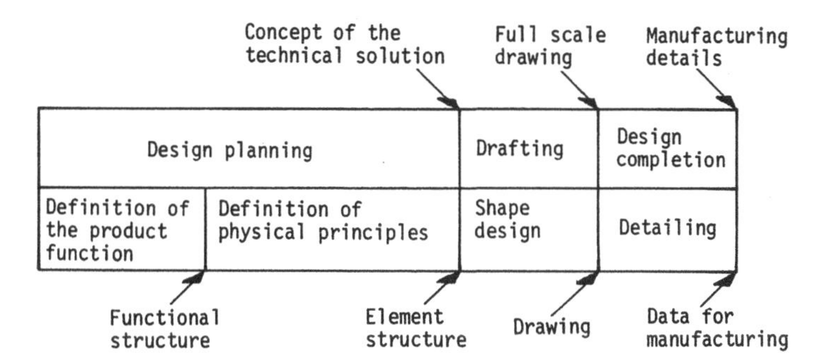

Fig. 1.4. Comparison of design activities and design phases

- To reduce development time
- To ensure the reliability of design data

To fulfill these requirements, a CAD-system has to support the entire design process, which can be structured in four different design phases (based on the VDI-directive 2210). The design phases, their activities and the results of each activity are summarized in Fig. 1.4 [2].

Today most CAD-systems only support the detailing process with the goal of generating a finished technical documentation of the technical solution which can be used for manufacturing and assembly. The activities in producing a finished and documented technical solution are summarized as follows:

1. Producing the manufacture and assembly oriented design of the product (process of designing)
2. Technological drafting (process of detailing)
   - Dimensioning and tolerancing
   - Shape and positional tolerancing
   - Definition of surface finishes
   - Layout of quality control procedures
   - Determination of material
   - Textual instructions for technical and organisational subjects

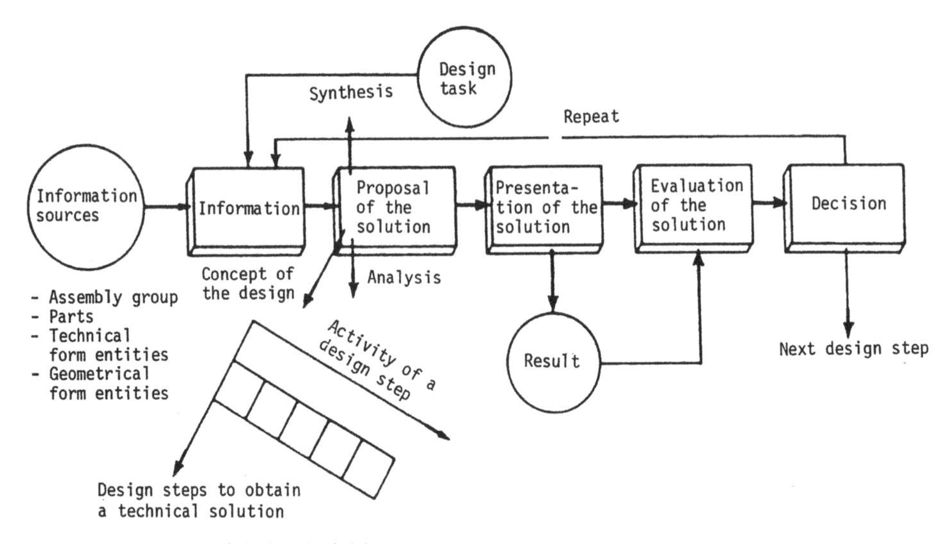

Fig. 1.5. Sequence of design activities

3. Generation of technical drawing
   - Manufacturing drawing
   - Assembly drawing
4. Optimization and control
   - Tolerance analysis
   - Control of completeness and correctness
   - Control of using standardized parts
   - Order release for production

In general, the technical solution conceived under the aspects of functionality and efficiency is optimized and laid out by design for economic manufacture and assembly. Thereby the full functionality and efficieny of the product is maintained. From the designers point of view his activities can be regarded as a repetitive activity of informing, calculating, drafting and deciding. Figure 1.5 shows the sequence of these activities [3].

In addition to information gathering and mathematical calculation methods, the graphical representation of the technical solution is an important basis for the designer's decision as to whether or not the constructive idea fulfils the requirements.

The CAD-system has to support the designer via a graphic-interactive communication technique. Data describing the technical solution are stored in a database as a computer internal product model.

The computer internal model contains the definition data of the product, their semantics and relations between corresponding data. Using this approach the integration of further production processes can be performed. To illustrate the capability of a CAD-system the software structure and the computer internal model of the product are outlined.

1.3 Software Structure of CAD Systems

The graphical-interactive communication capabilities of a CAD-system are supported by tools like geometry processing, graphics processing, etc. The results are stored in the computer internal model. With these characteristics CAD-software can be structured as illustrated in Fig. 1.6.

This logical structure divides CAD-software into four parts which communicate with each other:

1. *The communication module* is the interface between designer and CAD-system. It contains commands, controls, graphic-interactive input/output functions and initialization methods with assigned data. Figure 1.7 shows an example of a graphical-interactive communication pad.
2. *The methods module* provides a number of specific methods to solve design tasks. Methods contain algorithms including instructions about the performance of the algorithms, their scope and application. Typical methods are modules for geometry processing, calculation, dimensioning, tolerancing, etc. The integration of the methods into CAD-systems is shown in Fig. 1.8 [4].
3. *The database module* provides a functional interface to the database manipulation language (DML) to manipulate (to insert, delete, modify and get) product definition data. Product definition data are stored as a computer internal

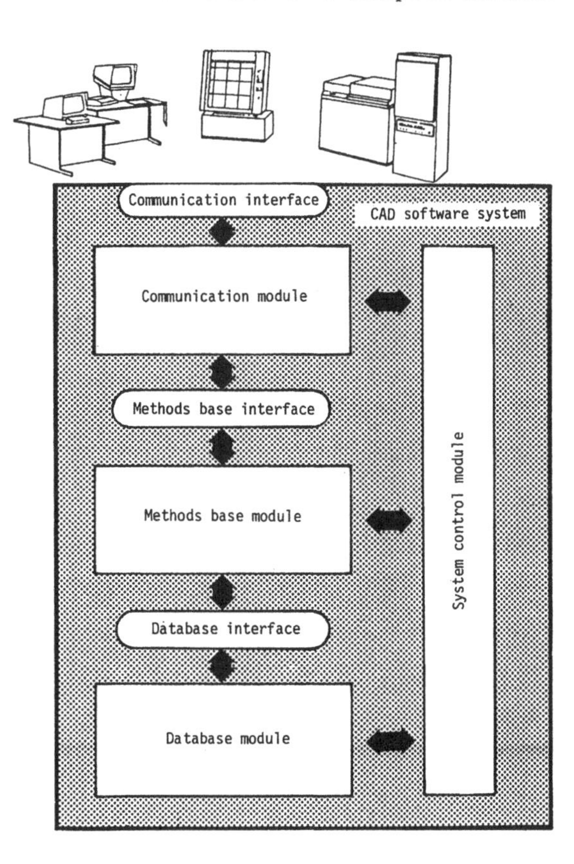

Fig. 1.6. Logic structure of CAD-systems

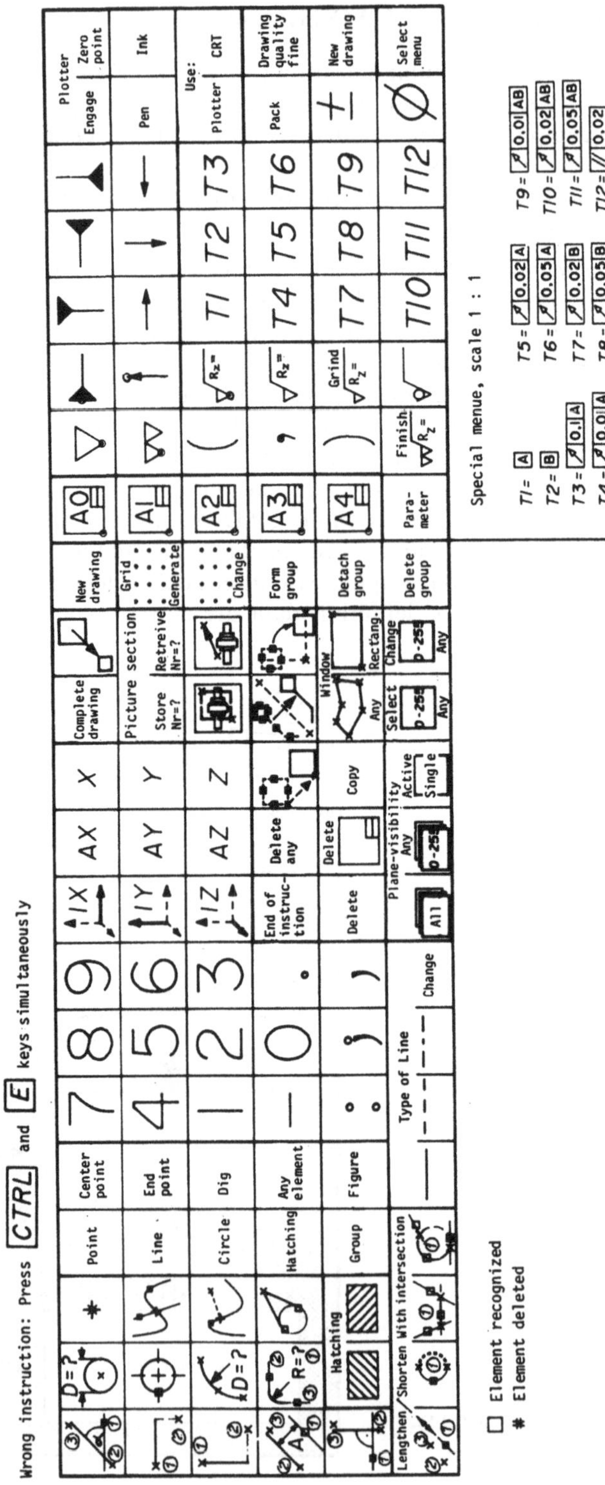

Wrong instruction: Press CTRL and E keys simultaneously

Point
Center point
7 8 9
AX X
Complete drawing
New drawing
A0
Plotter
Engage | Zero point
Line
End point
4 5 6
AY Y
Picture section
Store Nr=? | Retreive Nr=?
Grid Generate
A1
Pen
Ink
D=?
Circle
Dig
1 2 3
AZ Z
Change
A2
R_z=
T1 T2 T3
Use:
Plotter | CRT
R=?
Hatching
Any element
0 .
End of instruc-tion
Delete any
Form group
A3
R_z=
T4 T5 T6
Pack
Drawing quality fine
Hatching
Group
Figure
Delete
Delete
Copy
Window
Any | Rectang.
Detach group
A4
Grind R_z=
T7 T8 T9
New drawing
Lengthen/Shorten With intersection
Type of Line
Change
Plane-visibility
All
Any 0-255
Active Single
Select 0-255 Any
Change 0-255 Any
Delete group
Para-meter
Finish R_z=
T10 T11 T12
Select menu

☐ Element recognized
\# Element deleted

Special menue, scale 1 : 1

T1 = A
T2 = B
T3 = ↗ 0.1 A
T4 = ↗ 0.01 A
T5 = ↗ 0.02 A
T6 = ↗ 0.05 A
T7 = ↗ 0.02 B
T8 = ↗ 0.05 B
T9 = ↗ 0.01 AB
T10 = ↗ 0.02 AB
T11 = ↗ 0.05 AB
T12 = // 0.02

Fig. 1.7. Graphic-interactive communication Pad

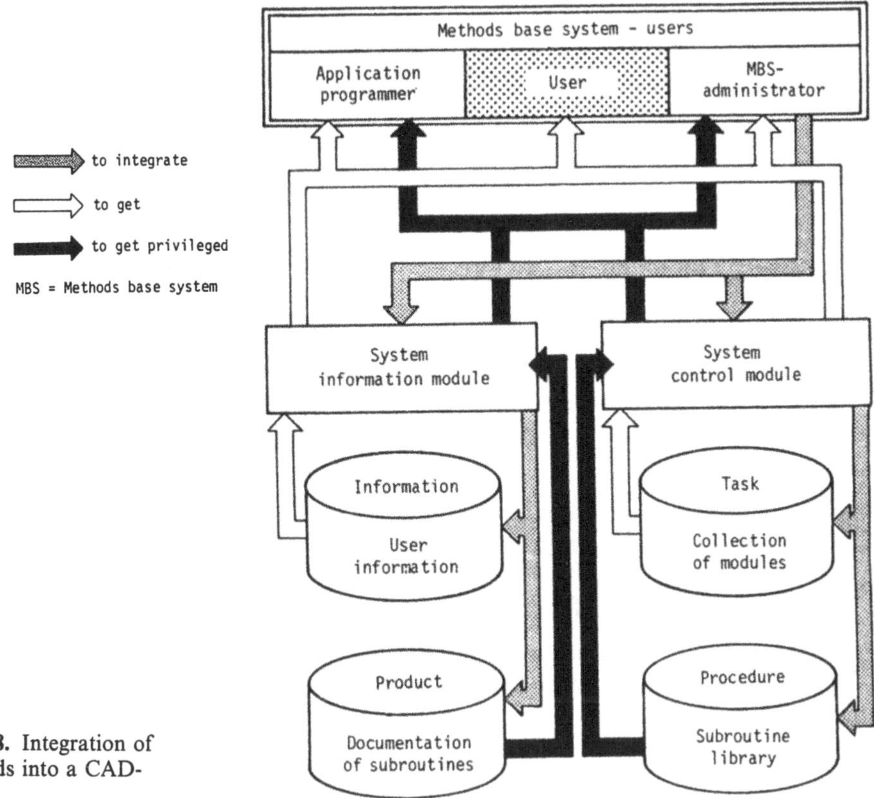

Fig. 1.8. Integration of methods into a CAD-system

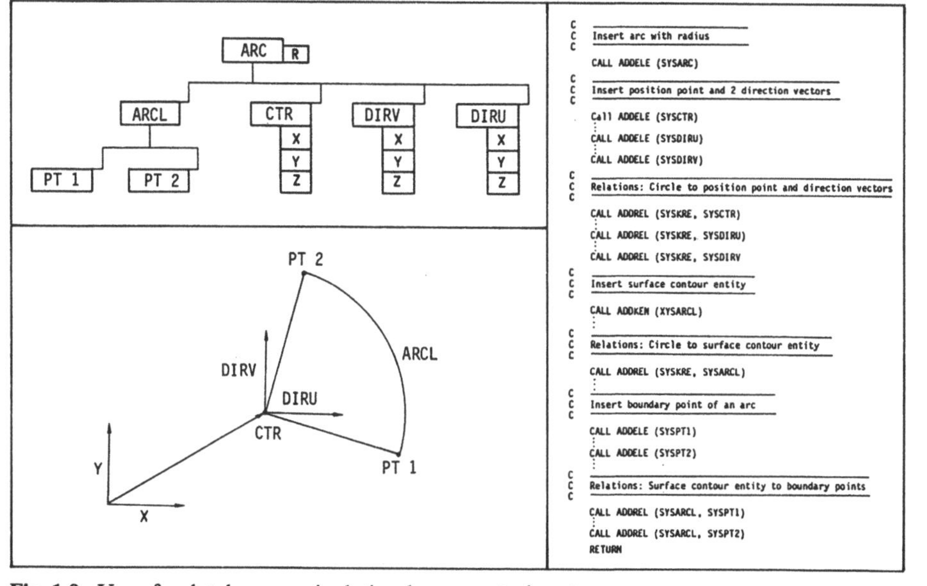

Fig. 1.9. Use of a database manipulation language to insert an arc

model [5]. An example of the use of the database manipulation language to insert an arc is represented in Fig. 1.9.

4. *The control module* controls and coordinates the different modes of the system.

This logical structure allows definition of seperate modules, which perform specific tasks as needed. The purpose of this structure is to make CAD software more flexible and efficient. These modules may also be realized with hardware.

1.4 Computer Internal Model

The technical solution of a product is represented by the computer internal model. The purpose of the computer internal model is to provide product definition data for the different design functions such as:

- Modifying, optimizing, finishing and testing of the technical solution of the designed product
- Generation of production schedules and part lists
- Generation of NC-programms for machine-tools, robots and measuring machines
- Simulation of manufacture and assembly operations

Computer internal models are usually stored in and controlled by a database management system. When the product definition data are stored separately from processing data the flexibility of the software is improved. Thus, data for process-

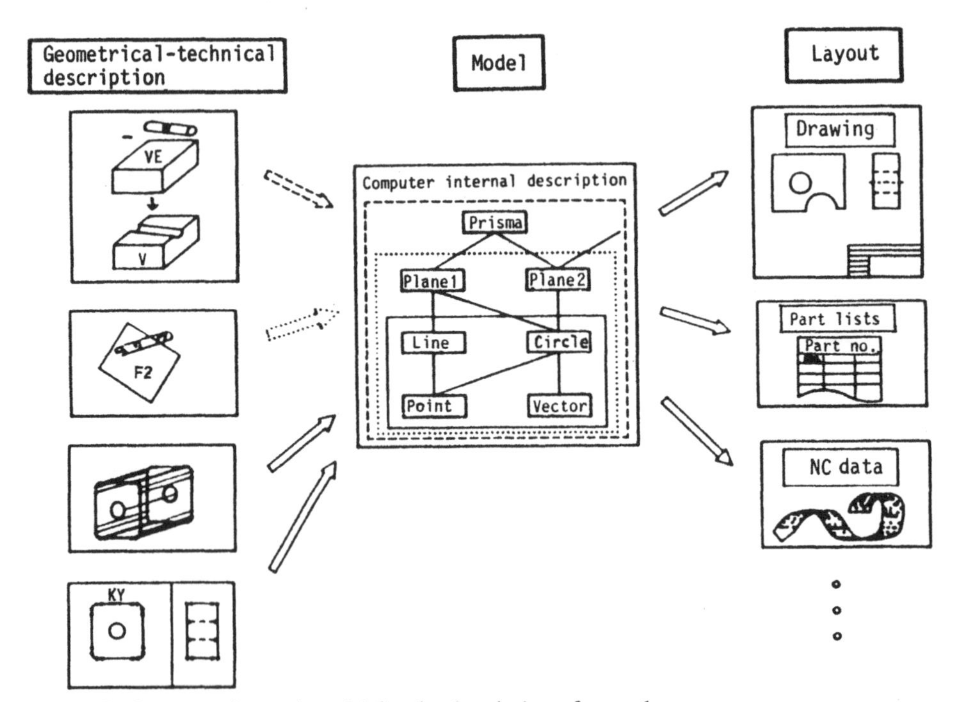

Fig. 1.10. Computer internal model for the description of a product

ing methods can be added and updated. Changes to these processing methods alter the computer internal model and make modifications or additions necessary. With the concept of a CAD-system which is based on the computer internal model further applications can be integrated.

The complexity and performance of the computer internal model also influence the efficacy of the communication interface. The efficiency increases if the number of commands needed to solve a special problem is kept to a minimum. This, however, necessitates that the man-machine communication is supported by algorithms supplied with data from the computer internal model. Figure 1.10 shows how the computer internal model is located between the description and processing functions of the product definition data [6].

The degree of integration or addition of other production planning and control activities depends on the complexity of the computer internal model used in the CAD-system.

1.4.1 Different Geometric Models for CAD

The geometric representation is one of the main components of the computer internal model. Computer internal models can be structured as a matrix where the types of generated model data (digital cell models, topological models) are listed in rows and the computer internal representation (procedure internal, data structure), in columns. The geometry of a product can be represented in steps of different complexity reaching from 2-dimensional line geometry up to a solid representation of the product. Another classification divides the geometric elements into analytical and numerical ones. Figure 1.11 shows geometric model elements of different complexity.

A feature of the computer internal model is its parametric capability which allows the change of dimensions for a defined geometry whereby the topology of

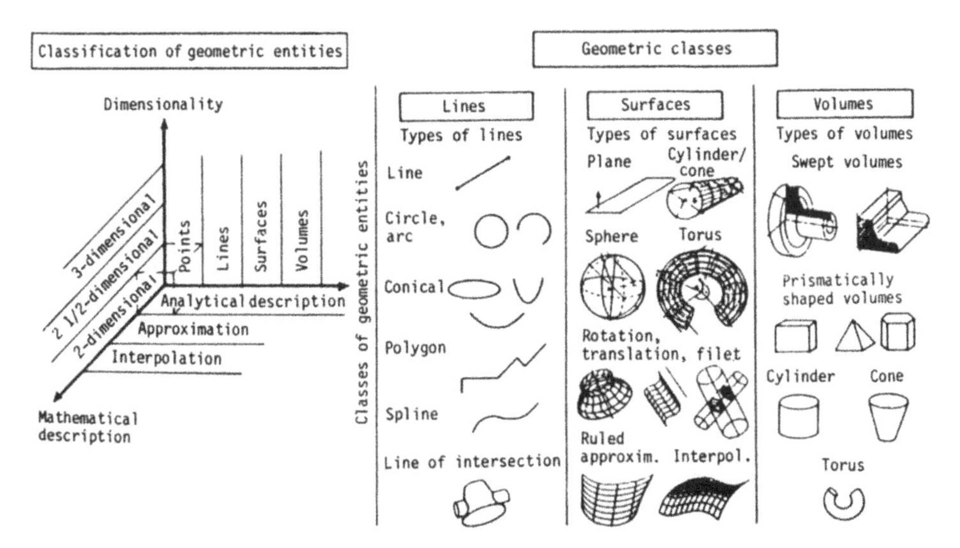

Fig. 1.11. Geometric model elements of different complexity

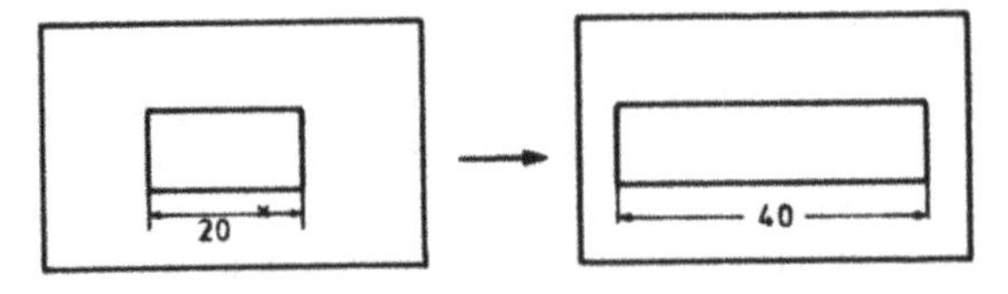

CHANGE LDIM ⟨DIG⟩; L = 40

Fig. 1.12. The use of a parametric model

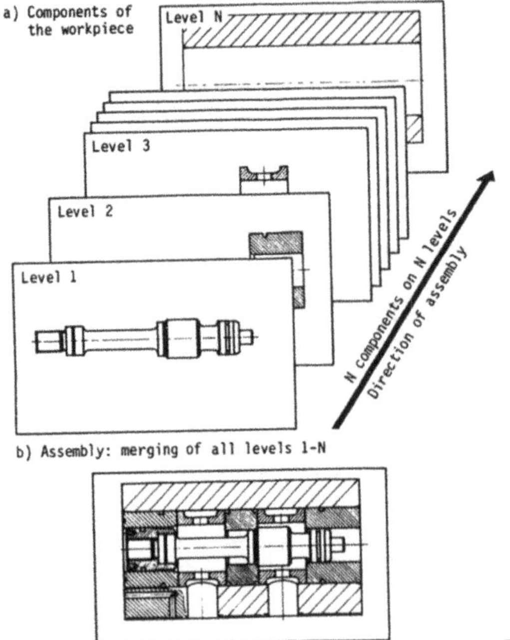

Fig. 1.13. Layer capability of CAD-systems

the product is maintained. With parametric models product variants can be generated easily. An application of the parametric capability is illustrated in Fig. 1.12.

To use parametric models efficiently, algorithms for automatic dimensioning and tolerancing of the referenced geometric element are required. With these algorithms a variant of a product can be generated, for example by modifying the dimensioning element of the referred geometry.

A model can be structured from different layers. The possibility of relating model information to the layers, is used to structure drawings and to generate assembly drawings. The layer technique allows overlaying of layers and drawing at all related information. An example of the use of the layer capability is shown in Fig. 1.13.

In addition to the geometric model, other important information is generated during the design process. One essential output is the technological description of

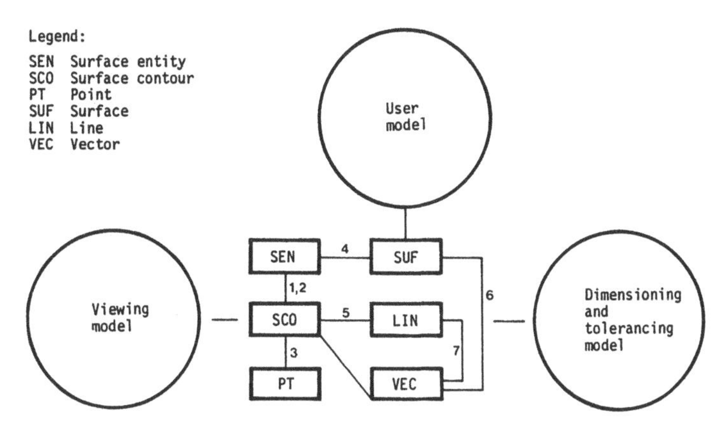

Fig. 1.14. Structure of the computer internal product model of the CAD-system DICAD (Dialog oriented integrated CAD-system)

the product. When the modelling process is considered as an entity a computer internal product model may be structured as shown in Fig. 1.14.

Based on this structure, four "submodels" are defined which are related to each other.

The user model contains elements of the user application and the history of the design process. The geometry model contains the product geometry represented by a volume model which is based on the description of the product boundary. The drawing model is generated from the geometry model and is the 2-dimensional representation of the product. The technology model provides technological information and supplements the geometric representation needed for the generation of manufacture and assembly data.

It is important that in addition to the model data the semantics of the data and the relations between the data are stored.

1.4.2 Importance of Technology-Oriented Model for CAD/CAM

The integration of the design process with the manufacturing process is mainly influenced by the geometric representation and the technological description of the workpiece which is also related to the geometry. Furthermore the semantics of the technological data have to be controlled by a seperate data structure [7]. An example of a technology model which is interconnected with the geometry and drawing models is presented by Fig. 1.15.

The technology model contains the following element classes:

- Dimension
- Surface finish
- Tolerance
- Text

Each element class includes different element types. The interconnection with the geometry and drawing models is necessary because of the following aspects:

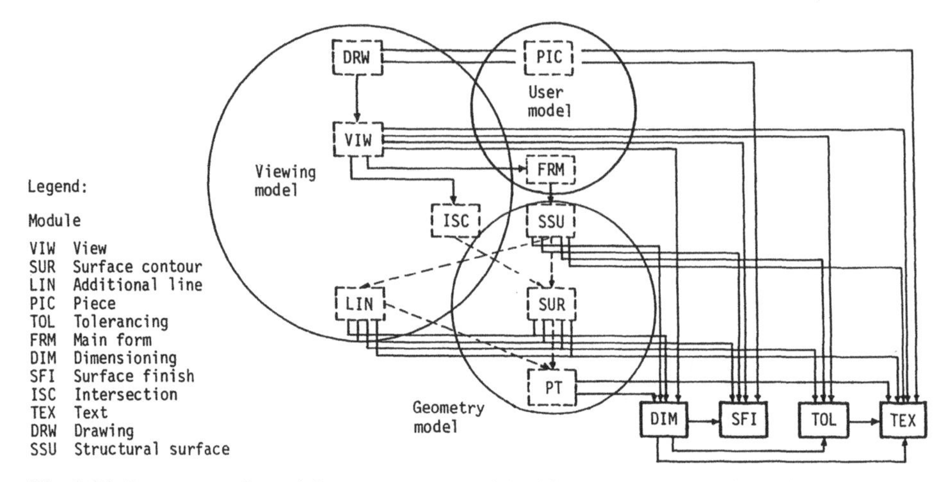

Fig. 1.15. Interconnection of the technology model with the geometry and drawing models

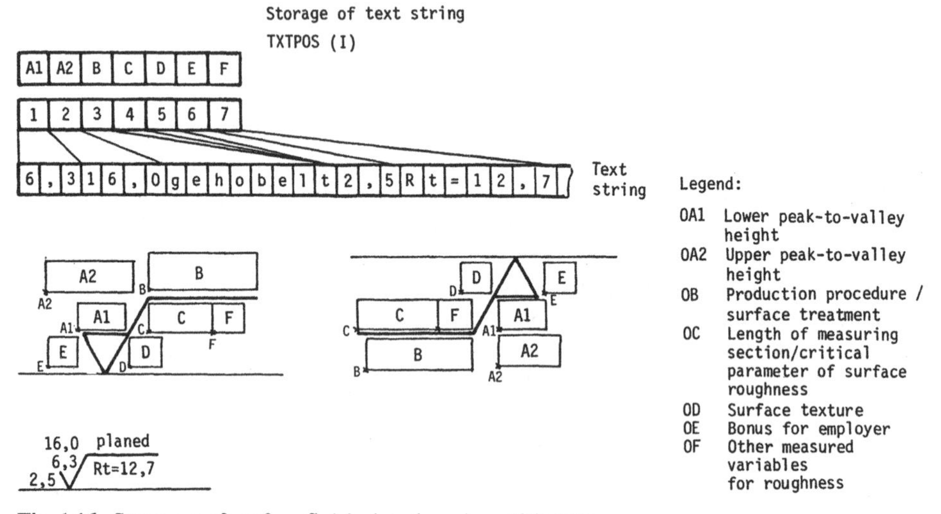

Fig. 1.16. Structure of surface finish data based on ISO 1302

1. References to the geometry model
   A modification of the geometry may cause a modification of the technology
   The model of the finished product can be tested with regard to the following questions:
   Does every surface have it's technological information?
   – Does it fit into an assembly part?
   The planning processes can be supported by the following features:
   The geometry is available with the description of
   Surfaces
   Contours

- The tolerances are available for every surface
- The surface finish description is available

2. References to the drawing model
   - The graphical representation of the geometry and the technological description make up the complete technical drawing.
   - The designer is able to formulate his technological description on the basis of the technical drawing whereas the CAD-system builds up the technology model with respect to the geometry.

These requirements can be met if the semantics of the technological information are available. Figure 1.16 illustrates how surface finish data can be structured and related to the geometry.

On the basis of this structure the geometry and technology are available and can be used by other algorithms to generate further information.

1.5 Interfaces of CAD Systems

CAD-systems consist of a number of hard- and software components which have to exchange data. The hardware contains different processors, storage units, alphanumerical and graphical input and output devices which are interconnected via a data and address channel. The software contains modules to process and

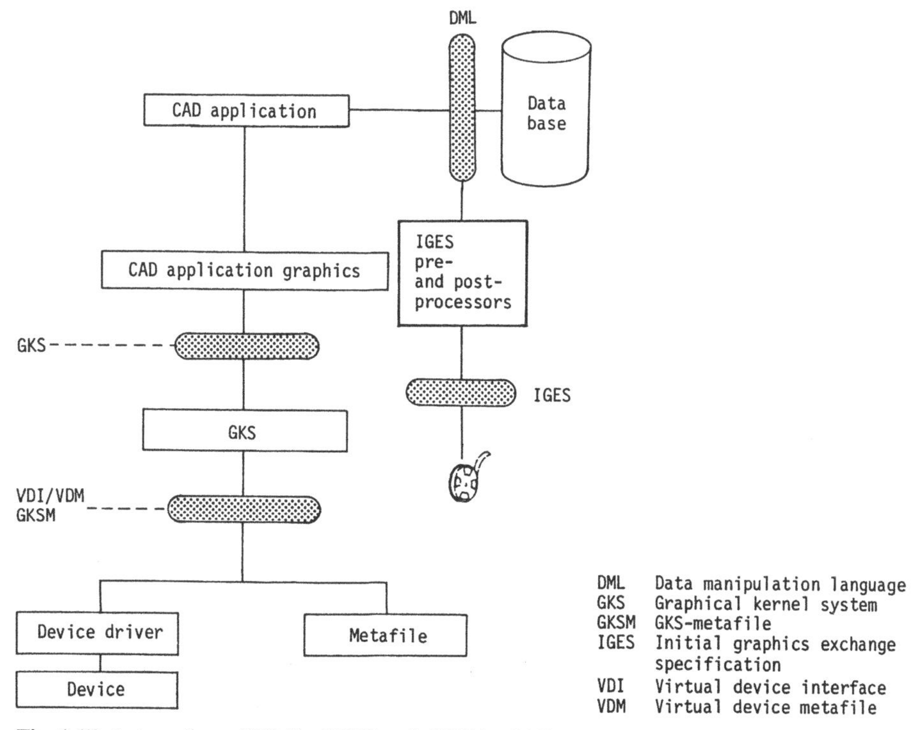

Fig. 1.17. Integration of DML, IGES and GKS in CAD-systems

store design data and to support the communication between the system and the designer. In order to be able to connect the CAD-modules standardized interfaces have to be defined. Industry-wide standardization work is required if such interfaces are to be of general interest. The goals of the standarization effort are:

- To define general interfaces
- To reduce dependence on a specific equipment manufacturer
- To avoid duplicating of work

In general it can be stated that increasing generalization of a standard reduces its efficiency. Another problem arises when different groups of specialists work out a standard. They may not necessarily reach the status of a real standard (such as an ISO-standard) but it may be very practical to use this standard. In the field of CAD several standards, recommendations and directives have been established. Some important interfaces such as the Database Manipulation Language (DML), the Initial Graphics Exchange Specification (IGES) and the Graphical Kernel System (GKS) will be discussed in more detail. Figure 1.17 shows, how these interfaces are integrated in a CAD-system.

It was shown in Fig. 1.6 that a CAD-system can be structured from three parts:

- The communication module
- The methods module
- The data base module containing the computer internal model

1.5.1 Database Manipulation Language (DML)

For the data base system module four components are necessary. They are shown in Fig. 1.18.

The components of the data base systems are:

1. The data query system; it allows retrieval of data from the computer internal model and uses attributes to describe the query task.
2. The data administration system; it includes operations to perform and control data base operations such as:
   Initialization of files
   Organization of files
   Storage of files
   Definition and identification of code for data protection
   Backup file management
3. The data manipulation system; it allows the user to operate on logical elements and relations of the data structure model.
4. The data base; it contains the information which has to be stored and manipulated.

The data manipulation language (DML) has been defined to allow the user to work with logical elements and relations of the data structure model. The operations of the data manipulation language can be classified in 4 groups:

– Insert	– Modify
– Get	– Delete

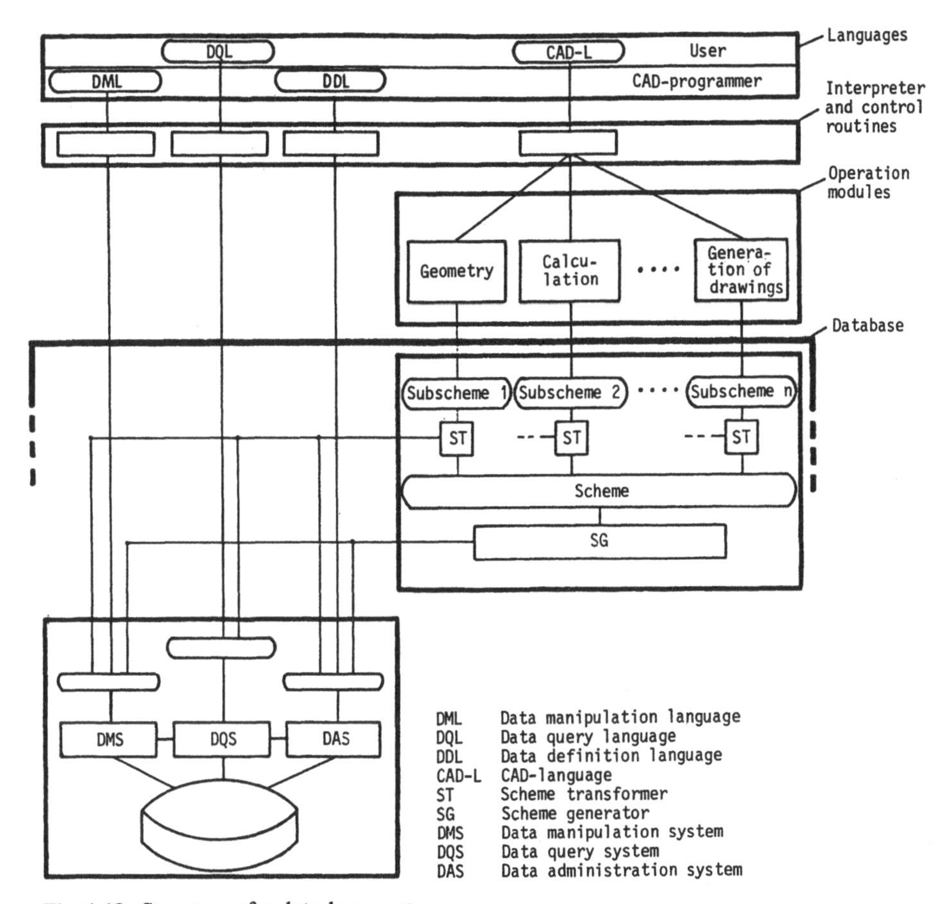

Fig. 1.18. Structure of a data base system

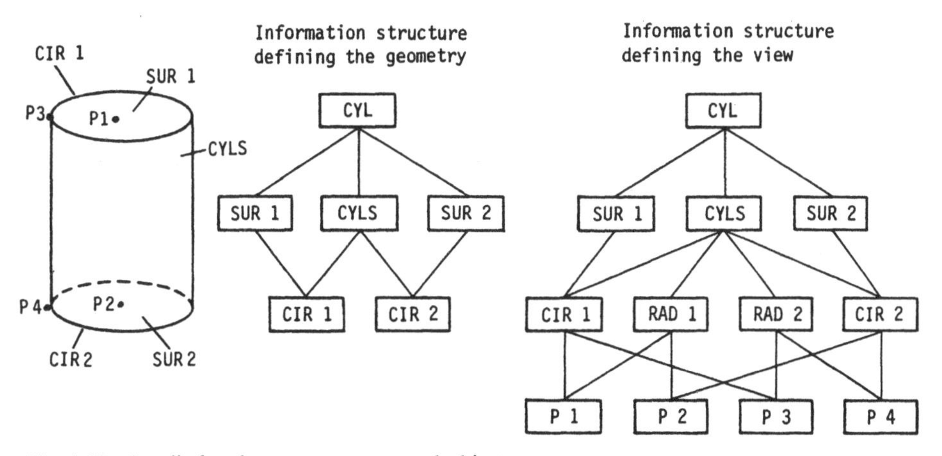

Fig. 1.19. A cylinder shown as a structured object

All of these operations affect the elements and their relation within the data structure model. In connection with the elements and their relations the operations allow manipulation of the database structure (schema).

For CAD-application a data manipulation language is required to operate on structured objects. A structured object is composed of geometric and technologic elements, having a hierarchical or network-like structure. Figure 1.19 shows the structured object of a cylinder.

With the description defined by the data definition language the data manipulation language offers all functions for describing structured objects.

1.5.2 Initial Graphics Exchange Specification (IGES)

The exchange of product definition data is based on technical documents such as technical drawings, functional descriptions, part lists, etc. When a product is described computer internally, by either a solid surface or a wire frame model, the exchange of the model with other workstations is possible. The Initial Graphics Exchange Specification (IGES), which is part of the ANSI standard Y 14.26 M specifies a data format for exchanging product definition data. This exchange of data may be necessary for the following purposes:

1. To perform computer aided design functions
   Exchange of data between CAD-systems
   Exchange of data between CAD-systems and other program modules, such as Finite Element Systems, NC-Systems etc.
2. To process organisational data of a company
   To exchange information between departments of a company
   To exchange information between cooperating companies
3. To process organisational data which refer to the technical application
   To assure the availability of standardized parts bought from suppliers
   To maintain system independent model archives

The exchange of product definition data as described by IGES is shown in Fig. 1.20.

The exchange process is done in 2 steps.

First the IGES preprocessor of the CAD-system 1 transforms its computer internal model into IGES format. It is set up in a sequential file. In the second step the IGES postprocessor of the CAD-system 2 transforms the IGES file into the

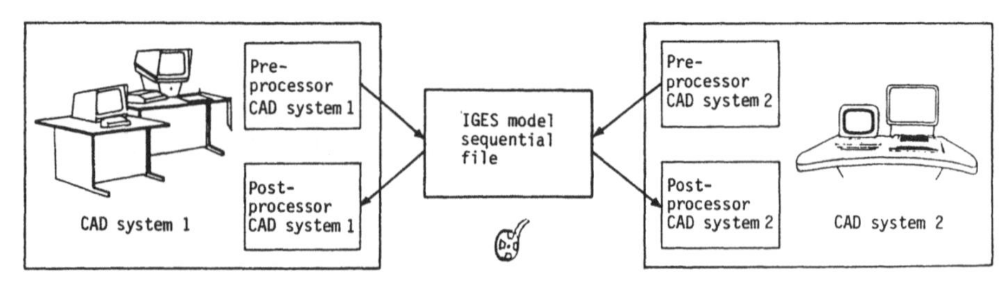

Fig. 1.20. The exchange of product definition data with the IGES system

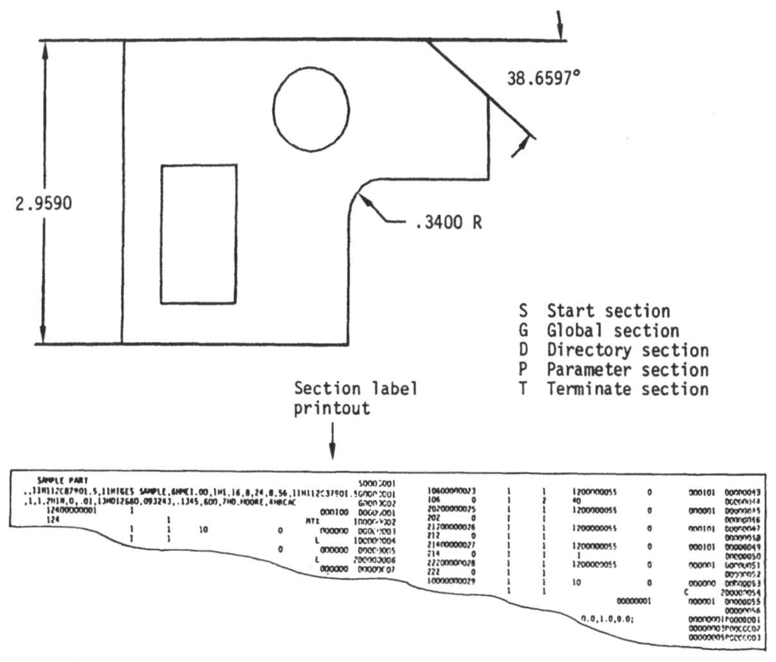

Fig. 1.21. Example of an IGES file

computer internal model of the CAD-system 2. The IGES file itself contains five sections:

1. The "Start Section" which contains comments
2. The "Global Section" which contains the description of the preprocessor and the information for the postprocessor to process the file. Information such as identification, data of origination, the name of the CAD-system, dimensions etc. are included in this section.
3. The "Directory Entry Section" which contains the directory data of IGES entities
4. The "Parameter Section" which contains the parameter data refered to entities of the "Directory Entry Section"
5. The "Terminate Section" which identifies the end of the IGES-file and controls the other sections.

Figure 1.21 gives an example of an IGES file.

Entities which are supported by IGES are shown in Fig. 1.22. IGES offers a Macro capability to exchange data on product variants.

During the exchange of product definition data, it is necessary that IGES preprocessors and postprocessors inform the user about errors, and that they issue warnings, which may indicate the following problems:

1. An IGES-entity cannot be transformed
2. A transformation has been performed with a loss of information
3. The IGES-entities are transformed in system specific entities.

IGES entities

Geometrical entities	Drawing entities	Structural entities	Associative element
Circle	Dimension of angle	Font pattern line	Allows definition of schemes
Composite entity	Centerline	Subfigure	
Cone	Dimension of diameter	Font text	
Copious data block	Flag note	Associativity instance	
Face	General label	Drawing	
Line	General note	Property	
Parametric spline	Leader (arrow)	Subfigure instance	
Parametric bicubic spline surface	Dimension of line	View	
Point	Dimension of point		
Ruled surface	Dimension of radius		
Surface of revolution	Section		
Tabulated cylinder	Center line		
Transformation matrix			

Fig. 1.22. Entities supported by IGES

The success of the exchange of product definition data depends on the CAD-system vendor and user. The CAD-system vendor has to offer efficient preprocessors and postprocessors and the CAD-system user has to prepare his designed parts such that the loss of information during the exchange process is minimized [8].

1.5.3 Graphical Kernel System (GKS)

The Graphical Kernel System (GKS) is a specification of functions for 2-dimensional line and raster graphics. GKS was developed to be independent of computers, programming languages, graphical peripherals and the user applications. To guarantee the consistency of the defined functions, a well balanced modular concept was worked out. Figure 1.23 shows the concept of GKS.

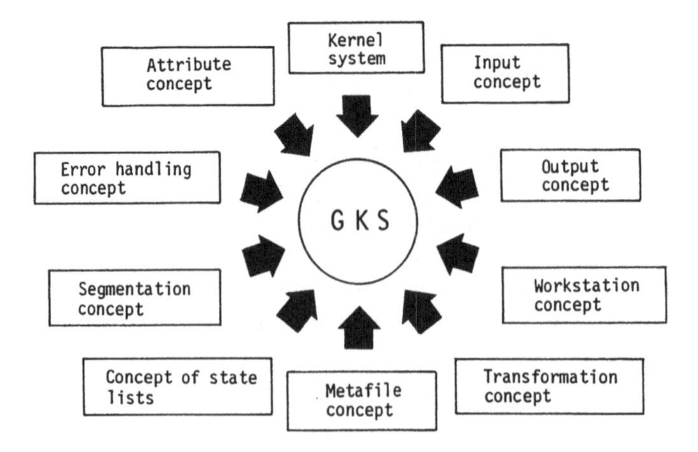

Fig. 1.23. Concept of GKS

One of the main ideas of GKS is the concept of abstract workstations. Based on this concept graphical workstations are defined as a connection of input and output devices. A workstation contains 1 to n input devices and 1 output device. The characteristics of the workstations are described in the workstation description table. Figure 1.24 shows different types of graphical workstations in GKS.

When GKS is embedded in a programming system, it can be regarded as a standardized interface between the application program and the graphical devices (Fig. 1.25).

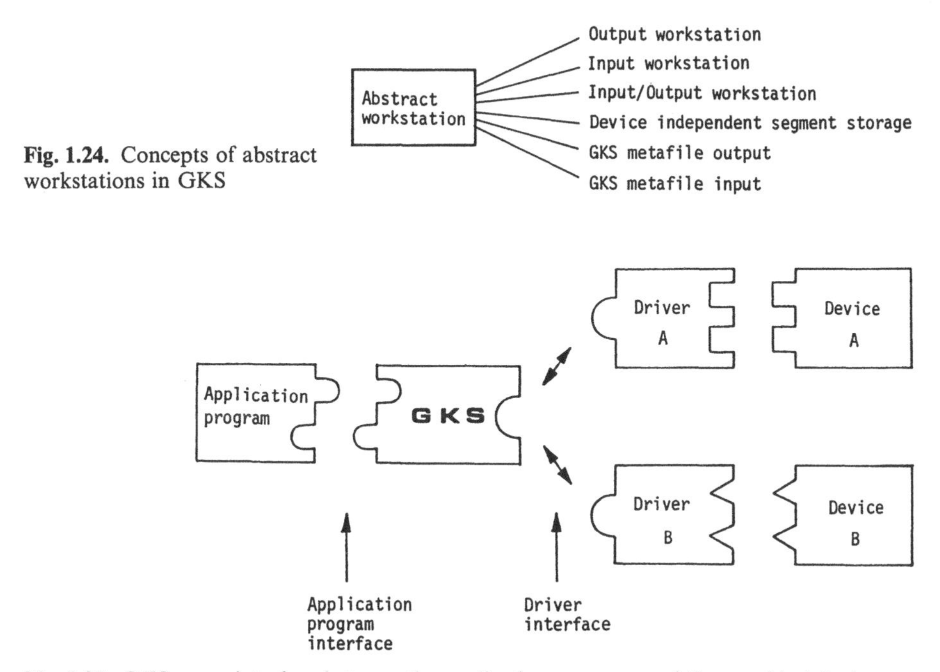

Fig. 1.24. Concepts of abstract workstations in GKS

Fig. 1.25. GKS as an interface between the application program and the graphical device

GKS functions are integrated in the application program, for example via FORTRAN subroutine calls. The connection of graphical devices is done by the DI/DD (device independent, device dependent) interface.

To offer different levels of graphic support, a GKS level structure was defined. This level structure offers levels of different capabilities ranging from plotting to dynamic motion and real time interaction. The levels are structured as a matrix, Fig. 1.26 [9].

The input capabilities are represented by the lines, and additional functions by the columns.

The application independent storage of graphical data is supported by the GKS metafile concept. When metafile data are generated the metafile is handled like a workstation, which means that GKS controls the metafile output. After the metafile has been entered, it is stored in GKS.

	a	b	c
0	No input, minimal control, only pre-defined bundles, multiple normalization transformation facilities but minimum set table required is 1, and all output functions with subset of attributes metafile functions optional	REQUEST input, mode setting and initialise functions for input devices, no pick input, and set viewport input priority	SAMPLE and EVENT input no PICK
1	Full output including full bundle concept, multiple workstation concept, basic segmentation (everything except device independent segment storage): metafile functions required	REQUEST PICK, mode setting and initialise for PICK	SAMPLE and EVENT input for PICK
2	Device independent segment storage		

Fig. 1.26. GKS-level structure

With the metafile concept GKS supports the following CAD capabilities:

1. The storage of graphical data, independent of the system
2. The exchange of graphical data between systems
3. The organization of the output of the graphical data

1.6 Integration of the Manufacture Planning Process

When production planning and control functions are automated they should be integrated with the CAD system to use existing product definition data. The connection of these functions should be done via a defined interface, which means that product definition data has to be transformed to a specified format. Any additional applications should be able to process the format of that interface. The integration is based on the common product definition data, for example as a computer internal model stored in a data base. Communication techniques and processing methods for the special process are integrated into the system and access data from the common definition (Fig. 1.27).

The connection and integration of further production planning and control functions to a CAD-system can be achieved by the following different tasks:

1. Preparation of design data for the manufacture process (for example NC-oriented dimensioning of a part)
2. Generation of standard interfaces (for example IGES)

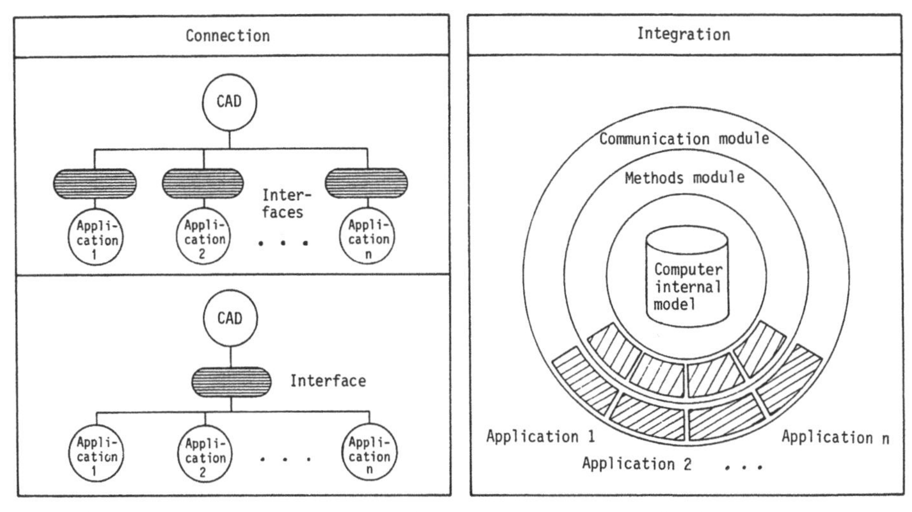

Fig. 1.27. The difference between connecting and integrating of production planning and control functions to a CAD-system

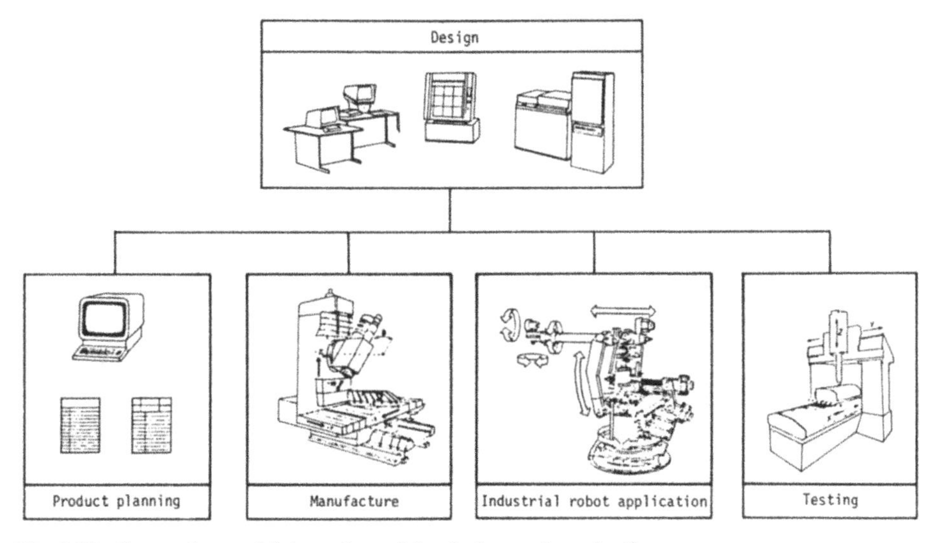

Fig. 1.28. Connection and integration of the design and production process

3. Generation of the NC-program, for example the part description in a NC-language format
4. Generation of complete production information

These tasks allow different degrees of automation to support manufacture planning and NC-programming. Additional work may have to be done if the support does not satisfy all manufacturing requirements.

The integration of production planning and control functions will result in the following advantages:

1. Exchange of complete product definition data by digital means
2. Computer aided preparation of technical data
3. Reduction of manual effort
4. Fast exchange of complex technical data
5. Avoidance of duplicated work (for example the repeated description of a product)
6. Reduction of errors
7. Computer support for the control of technical data
8. Reduction of test and control procedures

Figure 1.28 shows how the CAD process can be connected directly to production planning, manufacturing, industrial robot application and testing.

1.6.1 Planning Process Based on CAD Models

The generation of production schedules is one of the main tasks of the manufacture planning process. The process of generating the production schedule is based on the technical solution of the design process, the order requirements, the manufacturing know how and the financial means of the company. This planning process includes the selection of raw materials and the manufacture sequence, the definition of machine tools and the generation of manufacture times. The requirements for the generation of a production schedule are represented by Fig. 1.29.

A well designed CAD system should offer the following capabilities:

1. To make available product definition data which is represented by a computer internal model, including
   - Geometric data
   - Technological data
   - Organizational data

   in addition to the semantics and the relations of the workpiece attributes must be known
2. To provide a simulation capability to graphically model the manufacturing process including the tool path, machine tool workspace and fixture geometry

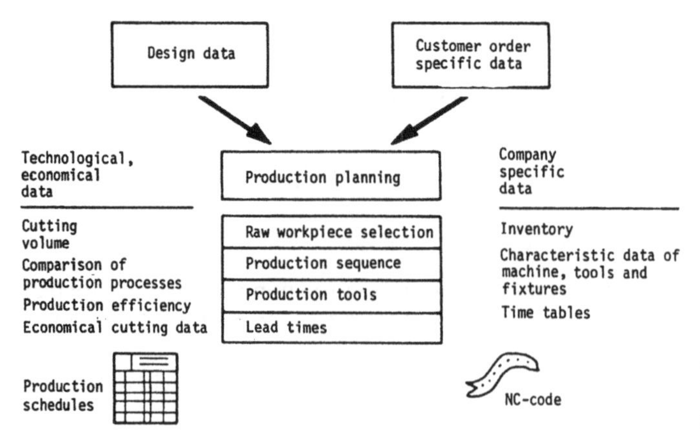

Fig. 1.29. Data needed for the generation of a production schedule

3. To offer technological and organizational information
4. To provide communication and efficient planning techniques.

With these capabilities the production schedule can be generated by graphical interactive communication techniques. The simulation of the planning process is done with the help of the computer internal model. The use of connection modules or common interfaces is another possibility to connect a CAD-system with computer aided planning system (CAP).

An overview of CAP-systems is given by Esch and Kohl [10]. With increasing complexity of the computer internal model the possibility of automatically generating a production schedule becomes increasingly difficult. An approach to simplifying planning is to set up digital discription tables and to use digital methods to generate a plan. These tables also facilitate a possible modification of the planning process.

1.6.2 NC-Machine Tool Programming Based on CAD Models

Programming of a NC machine tool by a CAD system can be divided into three phases:

1. The preparation of the technical solution for manufacturing
2. Programming of the NC-machine tool by graphical interactive communication with the CAD system
3. The completion of the NC-data

Figure 1.30 shows how a CAD system is used to program machine tools.

Different CAD-system offer different supports for NC-tool programming. The following example shows the generation of a NC-program with the help of a CAD system which contains a wire frame model.

The first programming step is to prepare the manufacturing process for the finished technical solution. The example of a rotational part shows that the part

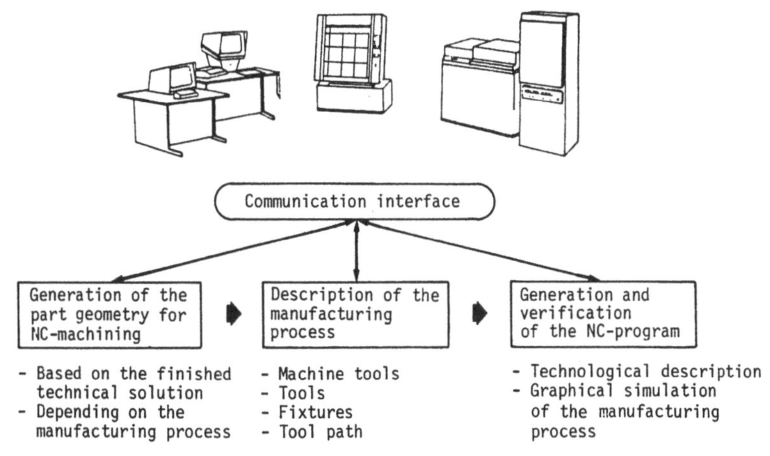

Fig. 1.30. NC-machine tool programming with a CAD-system

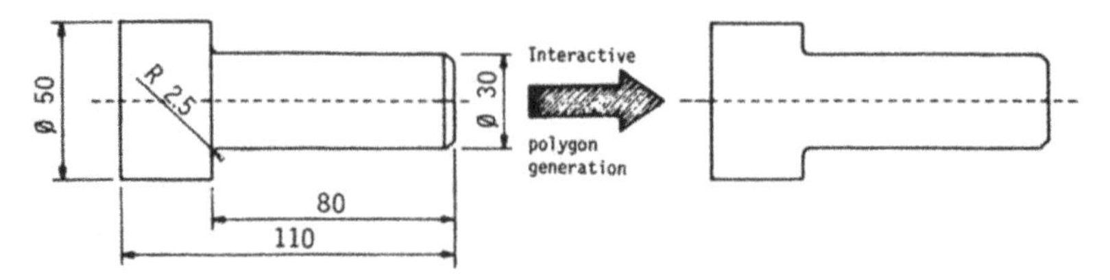

Fig. 1.31. Generation of a contour

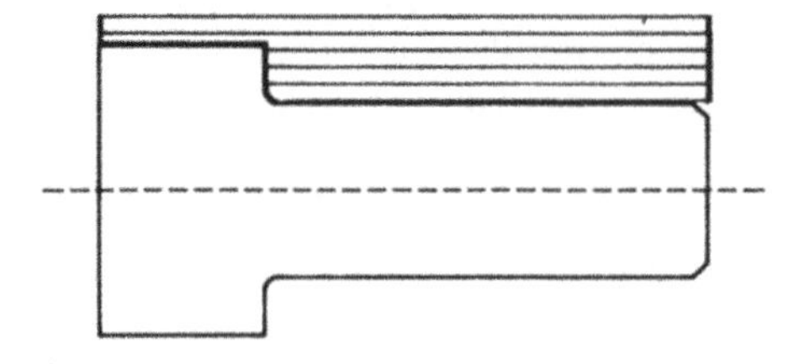

Fig. 1.32. Graphical representation of a machining process

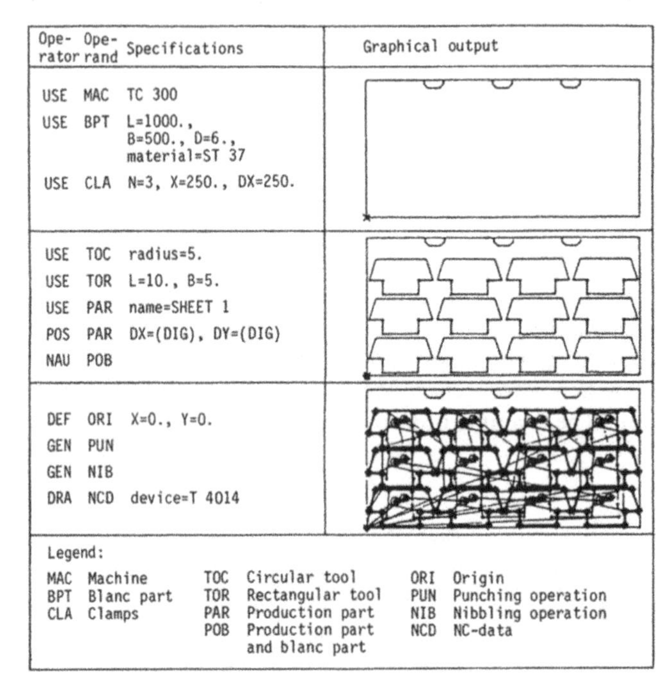

Fig. 1.33. The process of generating an NC-program with DICAD

geometry has to be modified from the drawing specific representation to a contour specific representation (Fig. 1.31).

The second step is to define the raw material, the machine tools, fixtures, tools and part contour to be machined. The machining process is determined and the result is represented graphically on the screen (Fig. 1.32).

In the third step the NC-program is generated. Depending on the capability of the CAD-system varions NC data may have to be added.

A CAD-system may support different NC-languages, such as APT, COMPACT II etc.

With a simple system, technological data has to be inserted via dialogue or an editor.

CAD-systems which contain computer internal models of higher complexity (technological oriented volume models) are also capable of processing technological data with the help of specific algorithms.

Figure 1.33 shows the generation of a NC program for a sheet metal part with the DICAD system, which uses a technological oriented volume model.

It is also possible to connect a CAD-system to the NC-system. An example is the interface module CADCPL of the EXAPT-Association (Fig. 1.34).

The connection is done with a standardized interface (IGES) which has to be generated by the CAD-system. The CADCPL module uses this interface to generate EXAPT-code.

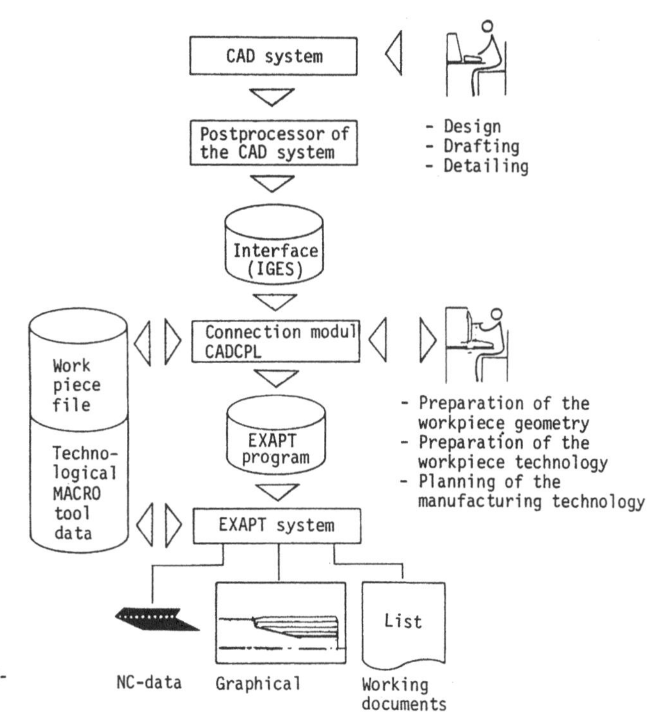

Fig. 1.34. Connecting a CAD-system with an NC-system via CADCPL

1.7 Economic Aspects

Economic considerations are an important aspect of the decision to purchase a CAD-system for a specific application. The profitability of the system has to be proven with a return on investment analysis. For each application the possible increase in productivity should be investigated. Depending on the application different methods of investment analysis can be used:

1. *Cost accounting method.* The purpose of this method is to calculate the cost savings yearly encountered by a substitution investment or the yearly costs of an alternative new investment. Using this method all cost items of the project have to be considered.

2. *Profit accounting method.* The purpose of this method is to compare the resulting profit of different investment alternatives. This method is mainly used for investment needed to expand a facility.

3. *Profitability calculation.* The purpose of this method is to calculate the yearly return on investment which can be expressed by following equation:

$$R = \frac{\text{annual profit of an alternative}}{\text{annual investment}} \cdot 100\,(\%)$$

or in the case of an investment for automation

$$R = \frac{\text{annual cost savings}}{\text{annual investment}} \cdot 100\,(\%)$$

4. *Payback method.* The purpose of this method is to determine the time when the proposed investment shows a profit. The payback time is calculated by the following equation:

$$A = \frac{\text{required investment}}{\text{annual profit}}\ \text{(years)}$$

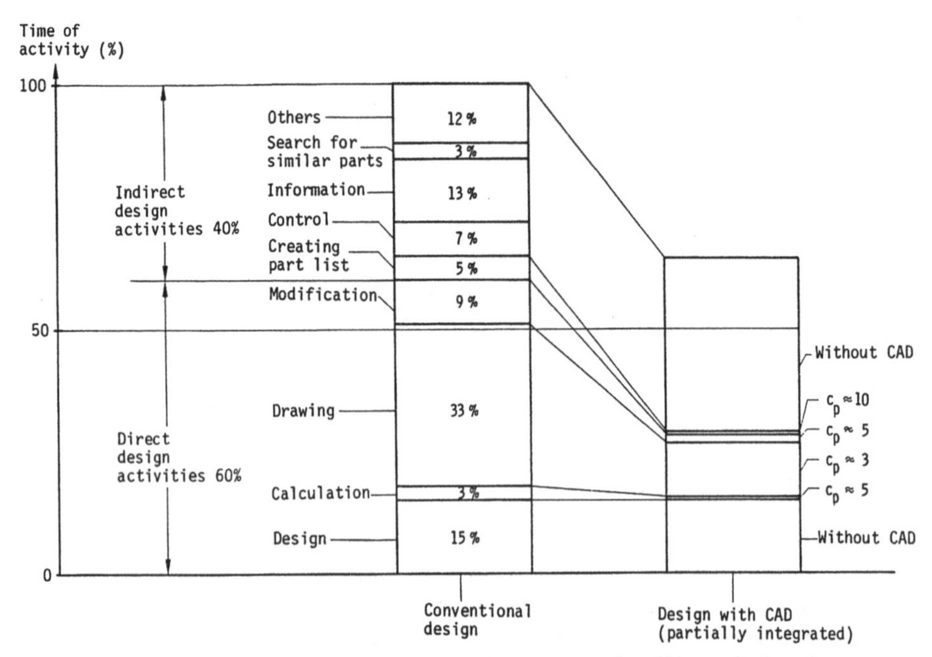

Fig. 1.35. Productivity increase obtained with a CAD-system for different design phases

or in the case of an investment for automation

$$A = \frac{\text{required investment}}{\text{annual savings}} \text{ (years)}$$

The problem of all methods is how to determine a characteristic index for the increased productivity. For the design process the indices shown in Fig. 1.35 can be used.

Table 1.1. Productivity factors for CAD systems

Author/application	Process	Index Cp
(Rapp)	Administration	4
Tool manufacture		1
	Looking for information	4
	Calculating	1.5
	Drafting	3
(Bakey)	Project definition	1
architecture	Project engineering	2
	Project design	3
	Project drafting	10
	Project estimating	13
	Project documentation	10
	Preplanning the erection	
	or Construction of a building	7
	"As-built" drawings	5
(Scott)	Design engineering	3
Tool manufacture	Design drafting	4
	NC-programming	10
	Production engineering	2
	Quality engineering	3
	Tool design	6

Table 1.2. Productivity factors for CAD systems

Author/application	Drawing	Index Cp
(Chasen)	Simple logic drawings	4.5–5.0
Mechanical-	Single-line drawings	3.5–4.0
engineering	Wiring diagrams	3.0–3.5
	Piping and instrumentation diagrams	3.0–3.5
	Raw part drawings	4.3
	Assembly/detail	3.7
	Sheet metal drawings	3.7
	Extrusion drawings	3.2
	Numerical control	2.7
	Detail aircraft drawings	2.4
	Layout drawings	1.7–2.2
	Structural steel	1.5–2.0
	Piping layout	1.25–1.75
(Müller)	New design	2.0
Car	Variant design	3.0
manufacture	Average value	3–4

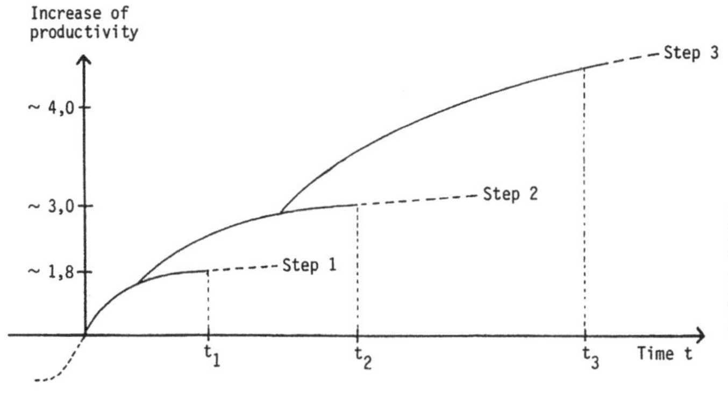

Fig. 1.36. Productivity increase obtained with a CAD-system as a function of its installation time

Further factors given by different authors are shown in Tables 1.1 and 1.2 [11].

The specified factors can be improved if a CAD system is conceived for a special application. The productivity increase against the time of the introduction of a system can be represented by curves shown in Fig. 1.36.

1.8 Conclusion

With the use of a CAD system it is possible to increase substantially the productivity of the design process. When CAD is integrated with planning and control functions, with the help of internal models, an additional potential for automation is opened up. Production schedules for manufacturing may be automatically generated and control data may be supplied to machine tools, robots and measuring machines.

1.9 References

1. Grabowski H, Anderl R (1983) CAD-Turnkey-Systems. Computer & Graphics. University of Karlsruhe
2. Seiler W (1983) Fundamentals for the Use of Computers for Product Development. Scriptum of the VDI Seminar CAD/CAM Computer Aided Design and Manufacturing. VDI Düsseldorf
3. Grabowski H, Maier H, Rausch W (1981) CAD-Seminar. Technische Akademie, Wuppertal
4. Maier H (1982) Administration of Methods and their Interfacing with Workpiece Models of an Integrated CAD-System. Dissertation at the University of Karlsruhe
5. Eigner M (1980) Semantic Data Models as an Aid to Information Handling in CAD-Systems and their Program Realization on Small Computers. Dissertation at the University of Karlsruhe
6. Seiler W (1983) Fundamentals for the Use of Computer for Product Development. Scriptum of the VDI Seminar CAD/CAM Computer Aided Design and Manufacturing. VDI Düsseldorf
7. Grabowski H, Anderl R, Seiler W (1983) DICAD – A Technical Modelling System. G.M.-Seminar, Berlin
8. Anderl R, Tröndle K (1983) Model Exchange – A Necessity for the Integration of CAD/CAM Applications. VDI-Z 125, no. 4, Febr.

9. N.N. (1982) DIN-ISO 7942, Graphic Kernel System – A Functional Description. DIN Standard
10. Esch H, Kohl HJ (1980) CAP/CAM Systems, Overview and Features, System Selection. Industrie-Anzeiger, no. 88
11. Grabowski H, Hettesheimer E (1983) Organisational Aspects with the Introduction of Computer Supported Design Methods. VDI-Z

1.10 Additional Literature

ANSI (American National Standards Institute) (Sept. 1981) Digital Representation of Product Definition DATA (Y 14.26M)
Anderl R, Rix J, Wetzel H (1983) The Use of GKS for CAD. Informatik-Spektrum, vol 6, no 2
Bradford M Smith, Wellington J (1983) IGES, A Key Interface Specification for CAD/CAM Systems Integration. National Bureau of Standards, January
Encarnação J, Enderle G (1983) An Overview of the Development of the Graphic Kernel System. Informatik-Spektrum, vol 6, no 2
Encarnação J, Schlechtendahl EG (1983) Computer Aided Design. In: Symbolic Computation (Computer Graphics). Springer, Berlin Heidelberg New York Tokyo
Grayer AR (1980) Alternative Approaches in Geometric Modelling. CAD, vol 12, no 4
Jones PF (1982) Four Principles of Man-Computer Dialogue. Business Press, no 3
Lewis JM (1981) Interchanging Spline Curves Using IGES. IPC Business Press, vol 13, Nov.
Liewald MH, Kennicott PhR (1982) Intersystem Data Transfer via IGES. IEEE CG A, May
National Bureau of Standards (1982) IGES-Newsletter, vol 2, no 1, March
Rembold U, Blume C, Dillmann R, Mörtel G (1981) Technical Requirements of Future Robots, part 1. VDI-Z, no 4, Feb.
Scott DJ (1980) Computed Aided Graphics, Determining System Size, Estimating Costs and Savings. In: Proceedings Autofact West., CAD/CAM VIII. Society of Manufacturing Engineers, Anaheim
Scowen RS (1982) IGES – A Critical Review and some Suggestions. Report IFIP Working Group on CAD-Systems – Framework Roros, June
Weissflog U (1983) Experience in Design and Implementation of an IGES Translator. Lecture Notes, CAMP '83, Berlin

2 Design for Assembly

A. H. Redford

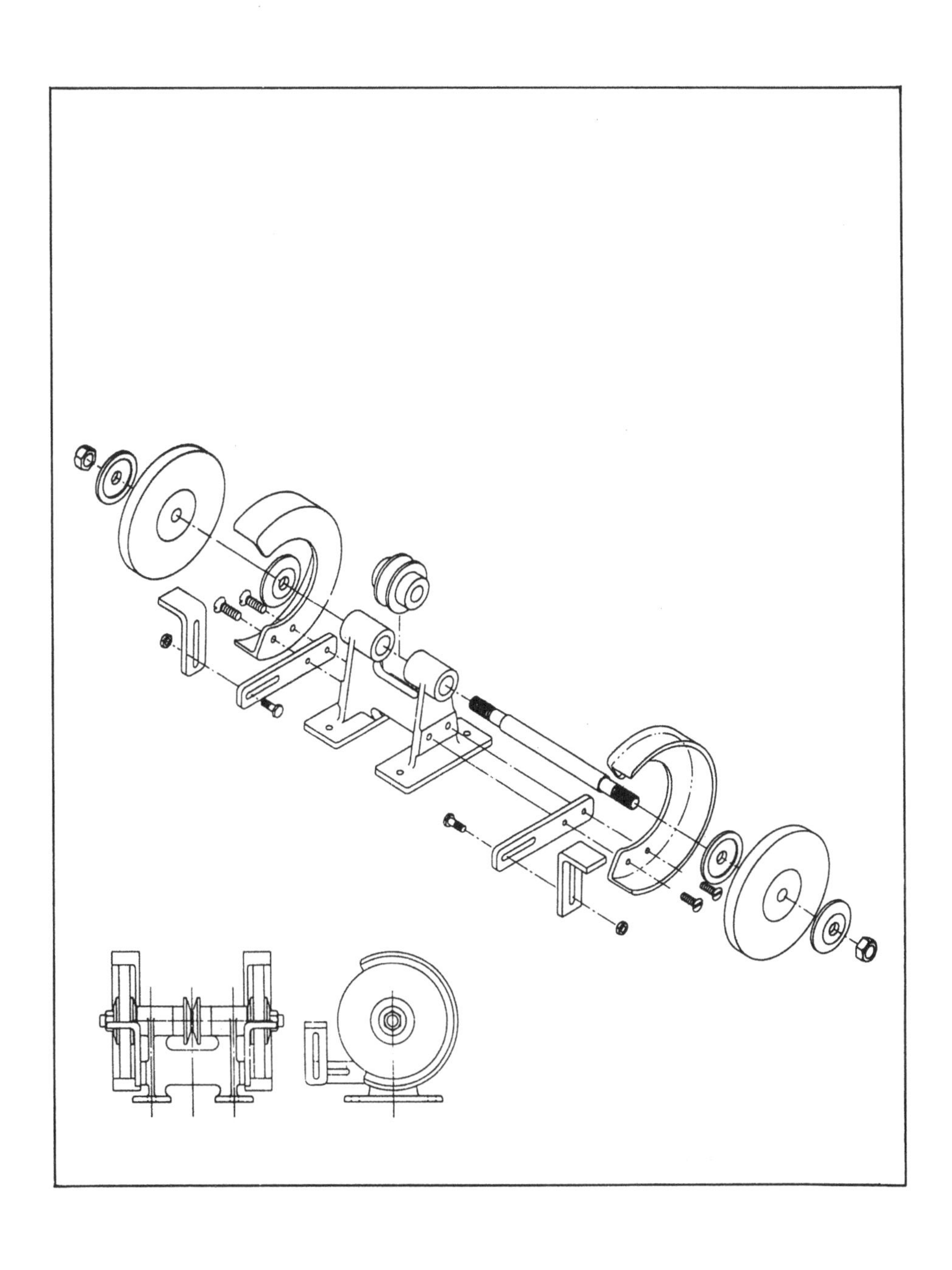

Layout of a grinder

(Courtesy of International Business Machine Corporation, White Plains, New York, USA)

2.1 Introduction

Over the last two hundred years, much effort has been devoted to devising, developing and improving the equipment which manufacturers use to produce individual parts. This has resulted in a manufacturing capability in which there is little scope for improvement, and in general the main efforts in part manufacture are rightly concerned with work organisation and job scheduling in the small to medium batch size production industries. Further, in parts manufacture, there is a wealth of literature on the technology of manufacture whereas in assembly, by comparison, there is virtually none.

Currently, much is being said about the benefits of improving productivity in all types of industry, but before this can be explored the term productivity, which is used freely to emphasise particular points of view, needs to be qualified if it is to have any valid meaning.

The most common way of defining and measuring productivity is output per man hour and this is usually referred to as labour productivity.
Another more realistic way of defining productivity is the relationship between output and total input and this is usually referred to as total productivity.

Because labour productivity is the most commonly used measure of productivity, it is important to recognise that an improvement in labour productivity can reflect any one or any combination of the following:

1. Improved efficiency in the use of labour and capital
2. An increase in the amount of capital associated with each man hour of labour
3. An increase in the average quality of labour.

Thus, whilst (2) and (3) above are desirable aims, they do not necessarily help to increase total productivity as defined and it is quite possible to increase labour productivity whilst at the same time reducing total productivity. Taking a hypothetical example, if a company is persuaded to install a machine costing $ 100,000 and which effectively does a job equivalent to one worker, the effect will clearly be to improve labour productivity. However, since it is unlikely to be economic to spend $ 100,000 to replace one worker, the capital is not being used efficiently and total productivity will fall. Similarly, if the average quality of labour improves and this results in an increased output per man hour but at the same time there is an excessive increase in wage rates, total productivity may fall.

Clearly, improving labour productivity is important but interpretation of the significance of improvements must be carefully considered.

The above discussion refers to productivity in all industries that produce goods and services. Manufacturing productivity should be considered to be particularly important because of the relatively high impact that this sector of industry has on the generation of national wealth. In the advanced manufacturing nations, the manufacturing industries absorb a large proportion of the nations workforce and are usually the largest contributors to the gross national product. Within the manufacturing industries, the discrete parts/durable goods industries form a major target for productivity improvements since these, in particular, are under direct attack from economically priced, high quality items from other sources. These industries are highly significant in international trade but they are not the most efficient and highly automated mass production units one might expect. The great bulk of their products are not mass produced but are produced in small to medium size batches in inefficient factories using out-dated machines, tools and methods. These industries are very dependent on manual labour for the handling and assembly of parts and this labour is often provided with tools no more sophisticated than screwdrivers, spanners and hammers. It is not surprising, therefore, that for a wide variety of manufacturing industries, assembly accounts for more than 50 per cent of the total manufacturing cost of a product, and more than 40 per cent of the labour force. Since, in general, very little thought is given to optimising product assembly, it is likely that improvements in this area will yield the biggest reduction in manufacturing costs. Further, it can also be stated that improvements in product design leading to greater economy of manufacture of parts and assembly of products will always lead to improvements in both labour productivity and total productivity. To design products for manual assembly that can be assembled more easily or more quickly needs little expenditure on capital equipment, and yet significant improvements in productivity can often be achieved. Similarly, designing products for automated assembly which results in easier and cheaper handling and assembly will also lead to reduced costs and improvements in productivity.

The design principles described in this chapter can be used with conventional table look-up methods. However, in the future they can be integrated into expert systems, allowing the automation of the design for assembly.

2.2 Design for Assembly Philosophy

The main objective of systematic design for assembly is to help the designer, who is under considerable pressure to meet other important design criteria, to consider design for assembly in an organised manner, such that each aspect of the activity is considered in a fixed sequence and that the implication of decisions made are both known and consistent with decisions which might have been made had someone else been carrying out the study.

The procedures and recommendations are in no way meant to inhibit the designer in his basic task of conceptual design. Rather, the information presented should be used to complement a designer's wide knowledge of the more traditional aspects of design to ensure that, at least, before products are finalised, the implication of design on assembly costs have been fully considered.

A designer might argue that, instinctively, he always has in mind how the product is to be assembled and that design rules for handling and assembly are common sense. This is to some extent true, but in the absence of some formalised procedure for considering design for assembly, many factors tend to be overlooked or ignored by an individual designer, and certainly designers do not usually consider the special requirements of machanised or automatic assembly at the design stage. More importantly, if a group of designers were given the same problem they would inevitably produce a variety of designs, and each would consider their own way to assemble.

Because in the design for assembly methodology proposed, parts handling and assembly operations are classified using a simple yet comprehensive classification system, it is also possible for the designer to examine other products or sub-assemblies produced by the company to identify those which have the same or similar handling and/or assembly classification numbers. The benefits of this are:

1) If enough identical handling and assembly processes exist, it is possible that each would have production rates which would be too low for mechanisation but which collectively would have the volume required to make mechanisation feasible.
2) If several identical handling or assembly processes exist then:
   (a) The method of carrying out the process would be common to them all and planning time would be saved.
   (b) It is conceivable that wider use could be made of standardised parts, particularly items such as fasteners.
3) If similar handling or assembly processes exist, it might be possible to alter the designs so that these become identical processes.

Such a classification system would be of great benefit if used with the aid of a computerized process planning system.

2.3 Determination of the Most Appropriate Process

There are three basic types of assembly: manual, dedicated and flexible. Manual assembly, as the name implies, makes use of human capabilities with or without the aid of sophisticated jigs, fixtures and power tools to perform the assembly task. Manual assembly costs remain relatively constant and independent of the quality of parts being used and the required production volume. Further, manual assembly is versatile in that it can cope with a wide variety of products manufactured in a wide range of batch sizes. Two types of manual assembly are considered; that without the aid of sophisticated jigs and fixtures etc. (MA) and that with these aids (MM).

Dedicated assembly equipment (special purpose assembly machines) is expensive and requires considerable engineering development. Its performance is very susceptible to the quality of the parts being assembled and, as the name implies, is not easily adapted to assemble more than a single or at the most a small 'family' of products. It cannot be justified for low annual production volumes and, in general, the cost of assembly falls as the annual production volume increases. Two types of dedicated assembly equipment are considered, indexing machines (AI) and free transfer machines (AF).

Flexible assembly equipment, sometimes referred to as either programmable or robotic systems, provides for considerable flexibility in production volume and greater adaptability to design changes and different product styles than dedicated assembly equipment. Two types of flexible assembly equipment are considered, programmable assembly (AP) and robotic assembly (AR).

The system takes into account one hundred different assembly situations based on a 10 × 10 matrix. The questions to be answered to determine the row

			single product has a market life of 3 years or more without significant fluctuations in demand, the manual fitting or adaption of parts is not required and the parts are of sufficiently high quality								variety of different but similar products, no manual fitting, high quality parts and investment in automation encouraged (6)	variety of products, manual fitting, low quality parts, fluctuations in demand or automation discouraged (6)
			investment in automation encouraged SQ/W $\geq$ 3 (3)				investment in automation discouraged SQ/W $<$ 3 (3)					
			few product styles $y \leq 1.5$ (4)		several product styles $y > 1.5$ (4)		few product styles $y \leq 1.5$ (4)		several product styles $y > 1.5$ (4)			
			few design changes $n_d \leq 0.5$ (5)	several design changes $n_d > 0.5$ (5)	few design changes $n_d \leq 0.5$ (5)	several design changes $n_d > 0.5$ (5)	few design changes $n_d \leq 0.5$ (5)	several design changes $n_d > 0.5$ (5)	few design changes $n_d \leq 0.5$ (5)	several design changes $n_d > 0.5$ (5)		
			0	**1**	**2**	**3**	**4**	**5**	**6**	**7**	**8**	**9**
annual production volume per shift greater than 0.7 million assemblies $V_{as} > 0.7$	7 or more parts in the assembly $n \geq 7$	**0**	AF	AF	AP	AP	AF	AF	AP	AP	AP	MA
	less than 7 parts in the assembly $n < 7$	**1**	AI	AI	AI	AI	AI	AI	AI MM	MM AI	AP	MA
annual production volume per shift greater than 0.5 million assemblies $0.5 < V_{as} \leq 0.7$	25 or more parts in the assembly $n \geq 25$	**2**	AF	AP	AP	AP	AF AP	AP	AP	AP	AP	MA
	15 or more parts in the assembly $15 \leq n < 25$	**3**	AF AP	AP AF	AP	AP	AF MM	AP MM	AP MM	AP MM	AP	MA
	10 or more parts in the assembly $10 \leq n < 15$	**4**	AI	AI	AI	AP AI	AI	AI	MM AI	MM AP	AP	MA
	7 or more parts in the assembly $7 \leq n < 10$	**5**	AI	AI	AI	AI AP	AI	MM AI	MM	MM	AP MM	MA
	less than 7 parts in the assembly $n < 7$	**6**	AI	AI	AI	AI MM	AI	MM AI	MM	MM	MM AR	MA
annual production volume per shift greater than 0.1 million assemblies $0.1 < V_{as} \leq 0.5$	10 or more parts in the assembly $n \geq 10$	**7**	AI AP	AP	AP	AP	MM AI	AP MM	AP MM	AP	AP	MA
	less than 10 parts in the assembly $n < 10$	**8**	MM	MM	MM	MM	MM	MM	MM	MM MA	AR MA	MA
annual production volume per shift less than or equal to 0.1 million assemblies $V_{as} \leq 0.1$		**9**	MM	MM MA	MM MA	MA	MM MA	MA	MA	MA	MA	MA

Fig. 2.1. Classification system for products and assemblies

and column numbers of the matrix, together with the matrix are shown in Fig. 2.1. It is of interest to note that in general moving from the bottom right to the top left of the picture indicates a decrease in assembly cost by a factor usually between 5 and 10.

The cost data for the algorithms used to develop Fig. 2.1 are 'typical' for industries in the U.S. and Great Britain. In the computer model which has been developed, data appropriate to a particular company can be used and these, if significant, will obviously modify the results shown in Fig. 2.1. However, experience has indicated that for reasonable variations in cost parameters the trends of Fig. 2.1 are typical. It can also be seen from Fig. 2.1 that, of the two flexible assembly systems, programmable assembly is far more relevant than robotic assembly, and it is anticipated that in the near future, with an increasing emphasis on the assembly of families of products in relatively small batches, there will be considerable scope for line assembly systems using single arm assembly robots in the place of dedicated workheads, whereas the 'single station' multi-arm robot assembling in series is less likely to be cost effective for the majority of products at present.

2.4 Re-design for Manual Assembly

If manual assembly is found to be the most appropriate assembly process then the classification and coding system developed for handling and insertion are both two digit codes which have been designed to accomodate the majority of handling and insertion problems. Associated with handling and insertion codes are time penalties which have been determined experimentally and which represent the additional time necessary to handle and/or insert 'non-standard' parts.

2.4.1 Classification and Coding for Handling and Insertion

Factors which affect manual handling are considered to be:

1. The symmetry of the part in the direction of insertion
2. The symmetry of the part perpendicular to the direction of insertion
3. The need for grasping tools
4. The interaction of random parts (do they tangle and/or nest)
5. The need for two handed operation
6. The handling difficulty due to factors other than geometry (fragile, slippery etc.)
7. The part thickness
8. The part size

Part of the coding sheet for manual handling is shown in Fig. 2.2.

Factors which affect manual insertion are considered to be:

1. Whether or not the part is a fastener
2. The difficulty associated with restricted view or obstructed access
3. The stability of the part after insertion (does the part need holding down for subsequent operations)

Choice of second digit

Choice of first digit

First digits 0–3: parts can be grasped and manipulated by one hand without the use of grasping tools(1); they do not severely nest or tangle, and are not flexible(2)

First digit		parts are easy to grasp and manipulate (4), thickness > 2 mm (5): size > 15 mm (6)	parts are easy to grasp and manipulate (4), thickness > 2 mm (5): 6 mm ≤ size ≤ 15 mm (6)	parts are easy to grasp and manipulate (4), thickness > 2 mm (5): size < 6 mm (6)	parts are easy to grasp and manipulate (4), thickness ≤ 2 mm (5): size > 6 mm (6)	parts are easy to grasp and manipulate (4), thickness ≤ 2 mm (5): size ≤ 6 mm (6)	parts are not easy to grasp or manipulate (4), thickness > 2 mm (5): size > 15 mm (6)	parts are not easy to grasp or manipulate (4), thickness > 2 mm (5): 6 mm ≤ size ≤ 15 mm (6)	parts are not easy to grasp or manipulate (4), thickness > 2 mm (5): size < 6 mm (6)	parts are not easy to grasp or manipulate (4), thickness ≤ 2 mm (5): size > 6 mm (6)	parts are not easy to grasp or manipulate (4), thickness ≤ 2 mm (5): size ≤ 6 mm (6)
		0	1	2	3	4	5	6	7	8	9
$(\alpha+\beta) < 360°$ (3)	0	0	0	0	0						
$360° \leq (\alpha+\beta) < 540°$ (3)	1	0	0	1							
$540° \leq (\alpha+\beta) < 720°$ (3)	2	0	1								
$(\alpha+\beta) = 720°$ (3)	3	0									

First digits 4–7: parts can be grasped and manipulated by one hand with the use of grasping tools(1); they do not severely nest or tangle, and are not flexible(2)

Second-digit columns: parts need standard tools (1) for grasping and manipulation; parts need tweezers for grasping and manipulation.

First digit		parts do not require optical magnification for manipulation; parts are easy to grasp and manipulate (4): thickness > 0.25 mm (5)	parts do not require optical magnification for manipulation; parts are easy to grasp and manipulate (4): thickness ≤ 0.25 mm (5)	parts do not require optical magnification for manipulation; parts are not easy to grasp or manipulate (4): thickness > 0.25 mm (5)	parts do not require optical magnification for manipulation; parts are not easy to grasp or manipulate (4): thickness ≤ 0.25 mm (5)	parts require optical magnification for manipulation; parts are easy to grasp and manipulate (4): thickness > 0.25 mm (5)	parts require optical magnification for manipulation; parts are easy to grasp and manipulate (4): thickness ≤ 0.25 mm (5)	parts require optical magnification for manipulation; parts are not easy to grasp or manipulate (4): thickness > 0.25 mm (5)	parts require optical magnification for manipulation; parts are not easy to grasp or manipulate (4): thickness ≤ 0.25 mm (5)	parts need standard tools(1) other than tweezers	parts need special tools(1) for grasping and manipulation
		0	1	2	3	4	5	6	7	8	9
$\alpha \leq 180°$ (3), $0° \leq \beta \leq 180°$ (3)	4	2	5	3	6						
$\alpha \leq 180°$ (3), $\beta = 360°$ (3)	5	3	6	3							
$\alpha = 360°$ (3), $0° \leq \beta \leq 180°$ (3)	6	3	7								
$\alpha = 360°$ (3), $\beta = 360°$ (3)	7	4									

First digit 8: parts can be grasped and lifted by one hand with the use of grasping tools(1) if necessary; they severely nest or tangle, or are flexible(2) (two hands may be used for manipulation of parts)

First digit	parts are easy to grasp and manipulate (4), $\alpha \leq 180°$ (3): size > 15 mm (6)	parts are easy to grasp and manipulate (4), $\alpha \leq 180°$ (3): 6 mm ≤ size ≤ 15 mm (6)	parts are easy to grasp and manipulate (4), $\alpha \leq 180°$ (3): size < 6 mm (6)	parts are easy to grasp and manipulate (4), $\alpha = 360°$ (3): size > 6 mm (6)	parts are easy to grasp and manipulate (4), $\alpha = 360°$ (3): size ≤ 6 mm (6)	parts are not easy to grasp or manipulate (4), $\alpha \leq 180°$ (3): size > 15 mm (6)	parts are not easy to grasp or manipulate (4), $\alpha \leq 180°$ (3): 6 mm ≤ size ≤ 15 mm (6)	parts are not easy to grasp or manipulate (4), $\alpha \leq 180°$ (3): size < 6 mm (6)	parts are not easy to grasp or manipulate (4), $\alpha = 360°$ (3): size > 15 mm (6)	parts are not easy to grasp or manipulate (4), $\alpha = 360°$ (3): size ≤ 6 mm (6)
	0	1	2	3	4	5	6	7	8	9
8	3									

First digit 9: two hands, two persons or mechanical assistance required for grasping and manipulation of parts

Second-digit columns 0–7: parts require two hands for grasping and manipulation; parts do not severely nest or tangle and are not flexible (2).

First digit	parts are not heavy (7); parts are easy to grasp and manipulate (4): $\alpha \leq 180°$ (3)	parts are not heavy (7); parts are easy to grasp and manipulate (4): $\alpha = 360°$ (3)	parts are not heavy (7); parts are not easy to grasp or manipulate (4): $\alpha \leq 180°$ (3)	parts are not heavy (7); parts are not easy to grasp or manipulate (4): $\alpha = 360°$ (3)	parts are heavy (7); parts are easy to grasp and manipulate (4): $\alpha \leq 180°$ (3)	parts are heavy (7); parts are easy to grasp and manipulate (4): $\alpha = 360°$ (3)	parts are heavy (7); parts are not easy to grasp or manipulate (4): $\alpha \leq 180°$ (3)	parts are heavy (7); parts are not easy to grasp or manipulate (4): $\alpha = 360°$ (3)	parts severely nest or tangle, or are flexible (2)	two persons or mechanical assistance required for parts manipulation
	0	1	2	3	4	5	6	7	8	9
9	1									

Fig. 2.2. Difficulty levels for manual handling processes

Choice of second digit

Choice of first digit

Choice of first digit				after assembly no holding down required to maintain orientation and location (3)						holding down required during subsequent process(es) to maintain orientation and location (3)			
				not self-locking				self-locking					
				easy to align or position during assembly (4)		not easy to align or position during assembly (no features facilitating alignment or positioning)				easy to align or position during assembly (4)		not easy to align or position during assembly (no features facilitating alignment or positioning)	
				no resistance to insertion (5)	resistance to insertion (5)	no resistance to insertion (5)	resistance to insertion (5)	only part inserted is located (3)	part inserted and other part(s) located (3)	no resistance to insertion (5)	resistance to insertion (5)	no resistance to insertion (5)	resistance to insertion (5)
addition of any part[(1)] where neither the part provided nor any other part is being finally secured[(7)] immediately	part and the associated tool (including hands) can easily reach the desired location and the tool can be operated easily			**0**	**1**	**2**	**3**	**4**	**5**	**6**	**7**	**8**	**9**
	part and the associated tool (including hands) cannot easily reach the desired location or the tool cannot be operated easily	due to obstructed access or restricted vision (6)	0	0	1	2							
			1	3	4								
		due to obstructed access and restricted vision (6)	2	4									

Choice of first digit					no screwing operation or plastic deformation immediately after insertion (snap/press fits, circlips, spire nuts, etc)		plastic deformation immediately after insertion						screwing immediately after initial insertion	
							plastic bending			rivetting or similar plastic deformation				
								not easy to align or position during assembly (no features facilitating alignment or positioning)			not easy to align or position during assembly (no features facilitating alignment or positioning)			
					easy to align and position with no resistance to insertion (4)	not easy to align or position (no features facilitating alignment or positioning) and/or resistance to insertion	easy to align and position during assembly (4)	no resistance to insertion (5)	resistance to insertion (5)	easy to align and position during assembly (4)	no resistance to insertion (5)	resistance to insertion (5)	easy to align and position with no resistance to screwing (4)	not easy to align or position (no features facilitating alignment or positioning) and/or resistance to insertion
addition of any part[(2)] where the part itself and/or any other parts are being finally secured[(7)] immediately	part secures itself only	part and the associated tool (including hands) can easily reach the desired location and the tool can be operated easily			**0**	**1**	**2**	**3**	**4**	**5**	**6**	**7**	**8**	**9**
		part and the associated tool (including hands) cannot easily reach the desired location or the tool cannot be operated easily	due to obstructed access or restricted vision (6)	3	2	4	3	4	5	6				
				4	4	6	5	6	7					
			due to obstructed access and restricted vision (6)	5	6	8	7	8						
	part is securing other parts immediately	part and the associated tool (including hands) can easily reach the desired location and the tool can be operated easily		6	2	4	3							
		part and the associated tool (including hands) cannot easily reach the desired location or the tool cannot be operated easily	due to obstructed access or restricted vision (6)	7	4	6								
			due to obstructed access and restricted vision (6)	8	6									

Choice of first digit		mechanical fastening processes (part(s) already in place but not secured immediately after insertion)				non-mechanical fastening processes (part(s) already in place but not secured immediately after insertion)				non-fastening processes	
		none or localized plastic deformation				metallurgical processes					
							additional material required				
		bending or similar processes	rivetting or similar processes	screwing or other processes	bulk plastic deformation (more than 75% of the part material is plastically deformed during fastening)	no additional material required (e.g. resistance welding, etc.)	soldering processes	weld/braze processes	chemical processes (e.g. adhesive bonding, etc.)	manipulation of part(s) or sub-assembly (e.g. orienting, fitting or adjustment of part(s) or sub-assemblies, etc.)	other processes (e.g. liquid insertion, etc.)
assembly process where: all solid parts are in place or non-solids are added or part(s) or sub-assemblies are manipulated		**0**	**1**	**2**	**3**	**4**	**5**	**6**	**7**	**8**	**9**
	9	3									

Fig. 2.3. Difficulty levels for manual assembly processes

4. The ease with which a part can be aligned and positioned
5. The degree of resistence to insertion
6. The type of fastening process

Part of the coding sheet for manual insertion is shown in Fig. 2.3.

With the appropriate classification and coding data, it is possible to analyse a product design systematically for ease of assembly.

2.4.2 Sequence of Design Analysis

The various stages of the analysis are:

1. Determine the sequence of assembly
2. Complete the design for manual assembly worksheet
3. Determine the assembly efficiency
4. Consider the elimination of potentially redundant parts
5. Consider the re-design of high cost handling and/or insertion parts
6. After re-design, determine the new design efficiency.

2.4.3 Determination of the Sequence of Assembly

The simplest method of determining the sequence of assembly is to disassemble the product in a 'reasonable' manner. This would not generally be the only order of dis-assembly but the results of analysis carried out to date indicate that the effectiveness of the exercise is not significantly changed if other dis-assembly sequences were considered, provided these were practical.

2.4.4 Compilation of the Worksheet

A simple product is shown in Fig. 2.4 and the worksheet associated with this product is shown in Fig. 2.5. Before the assembly efficiency can be determined it is necessary to estimate the theoretical minimum number of parts for the product; this is identified by column 15 of the worksheet.

Three criteria are applied to determine if a part needs to be separate. These are:

1. The part has to be capable of moving relative to its mating part(s)
2. The part has to be manufactured from a different material to its mating part(s)
3. Either assembly or dis-assembly, if required, would not be possible if the part were not separate.

2.4.5 Determination of Assembly Efficiency

Assembly efficiency is defined as the standard time to assemble the theoretical minimum number of parts divided by the actual assembly time. Tests have indicated that the time taken to handle and insert a 'simple' part is two seconds and this figure has been used to determine the assembly efficiency for the product analysed in this paper. However, the standard assembly time is subject to the specific problems of individual industries and this and the other basic data can be changed accordingly to suit the requirements.

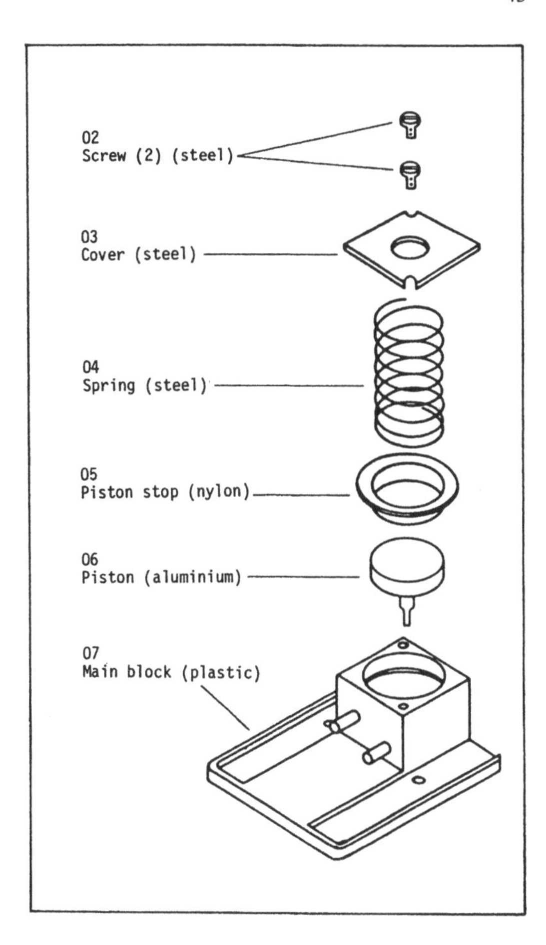

Fig. 2.4. Original design of piston assembly

It should be noted that whilst improving assembly efficiency absolutely, it is more important to relate the assembly efficiency of the new design to that of the old design, i.e. An improvement in assembly efficiency from 10 per cent to 20 per cent represents a halving of assembly cost even though the final efficiency, due to conflicting factors which prevent re-design for assembly, is not particularly high.

2.4.6 Elimination of Potentially Redundant Parts

In this particular example three parts have been considered to be potentially redundant, the cover, because it can be combined with the piston stop since it does not move relative to the piston stop and can be made of the same material, and the cover screws since it should be possible, under appropriate circumstances, to eliminate all separate fasteners. In this particular example the combined cover and piston stop and the main block could be designed such that the cover is a snap-fit on the main block and it was accepted that functional, manufacturing and aesthetic requirements would still be met by the new design. Clearly, if any other essential design requirement were to be impaired by a re-design for assembly exercise, this would not be acceptable.

1	2	3	4	5	6	7	8	9	10	11	12	13	14	15	
Part I.D. No.	Number of Times the Operation is Carried Out Consecutively	End-To-End Symmetry 180° or 360°	Rotational Symmetry About Insertion Axis	Handling Process Code	Handling Difficulty Level	Basic Handling Time, t_h	Assembly Process Code	Assembly Difficulty Level	Basic Assembly Time, t_a	I.D.'s of Parts Added to	I.D.'s of Other Parts Located or Secured	I.D. of New Sub-assembly or Group of Parts	Operation Time in Seconds 2(6+7+9+10)	Figures for Estimation of Theoretical Minimum Number of Parts	Remarks
		α	β												
07	1	360	360	30	0	1	00	0	1	91		90	2	1	main block
06	1	360	0	10	0	1	02	2	1	07		90	4	1	piston
05	1	360	0	10	0	1	00	0	1	07		90	2	1	piston stop
04	1	180	0	80	3	1	00	0	1	07		90	5	1	spring
03	1	180	180	13	1	1	06	4	1	07		90	7	0	cover
02	2	360	0	11	0	1	69	7	1	07	03,04 05,06	01	18	0	screws
													38	4	
													t_{ma}	n_{min}	

DESIGN EFFICIENCY

$$E_{ma} = \frac{n_{min}(t_h + t_a)}{t_{ma}} \qquad \frac{4(2)}{38} = 0.21$$

UNLESS SPECIFIC DATA IS AVAILABLE

$t_h = t_a = 1s$

I.D. No's:
01-89 Parts of the Final Assembly
90 Unsecured Group of Parts
91-99 Tools, Fixtures, etc.

Fig. 2.5. Design for manual assembly worksheet

2.4.7 Re-design of High-Cost Handling or Insertion Parts

From the worksheet shown in Fig. 2.5 it can be seen that, of the parts that remain after the elimination of appropriate parts, the spring is a high cost handling part (5 s to handle) whilst the piston is a high cost insertion part (4 s to insert). It was thought that nothing could be done to improve the design of the spring to improve its handling properties, but that the piston could be re-designed to improve its insertion properties. The problem associated with the piston is that, on insertion, the piston has to be released before the piston spindle enters the main block and this sometimes results in the piston tilting and the spindle not aligning with the hole in the main block. Two reasonable re-design alternatives exist: Firstly, the spindle could be extended such that it entered the hole in the main block before the piston had to be released; this was discounted on the grounds that it would impair the function of the product. Secondly, a small stub spindle could be added to the top of the piston which could be gripped by the operator and which would then facilitate easy insertion. This was considered acceptable and if the volume above the piston was considered to be critical this could be accomodated by slight re-design of the piston stop.

The re-designed product is shown in Fig. 2.6 and its worksheet is shown in Fig. 2.7 where it can be seen that the assembly efficiency has increased from 21 per cent to 62 per cent, an increase in assembly efficiency of 41 per cent.

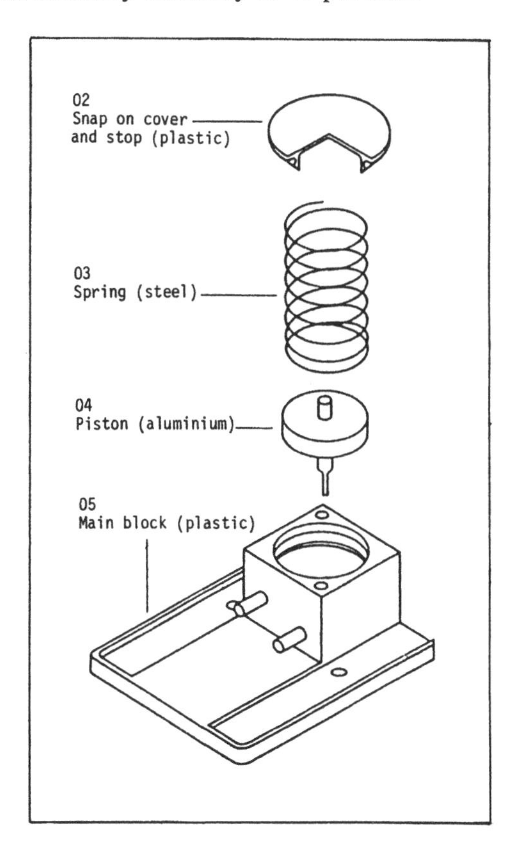

Fig. 2.6. Re-designed piston assembly

1	2	3	4	5	6	7	8	9	10	11	12	13	14	15	
Part I.D. No.	Number of Times the Operation is Carried Out Consecutively	End-To-End Symmetry 180° or 360°	Rotational Symmetry About Insertion Axis	Handling Process Code	Handling Difficulty Level	Basic Handling Time, t_h	Assembly Process Code	Assembly Difficulty Level	Basic Assembly Time, t_a	I.D.'s of Parts Added to	I.D.'s of Other Parts Located or Secured	I.D. of New Sub-assembly or Group of Parts	Operation Time in Seconds 2[6+7+9+10]	Figures for Estimation of Theoretical Minimum Number of Parts	Remarks
		α	β												
05	1	360	360	30	0	1	00	0	1	91		90	2	1	main block
04	1	360	0	10	0	1	00	0	1	05		90	2	1	piston
03	1	180	0	80	3	1	00	0	1	05		90	5	1	spring
02	1	360	90	10	0	1	60	2	1	05	04 03	01	4	1	cover & stop
													13	4	
													t_{ma}	n_{min}	

DESIGN EFFICIENCY

$$E_{ma} = \frac{n_{min}(t_h + t_a)}{t_{ma}} \qquad \frac{4(2)}{13} = 0.62$$

UNLESS SPECIFIC DATA IS AVAILABLE

$t_h = t_a = 1s$

I.D. No's:
01-89 Parts of the Final Assembly
90 Unsecured Group of Parts
91-99 Tools, Fixtures, etc.

Fig. 2.7. Design for manual asembly worksheet

2.5 Re-design for Automatic Assembly

2.5.1 Classification and Coding for Automatic Handling

The techniques used for re-design for automatic assembly are similar to those used for manual assembly, the only basic difference being that the coding system for automatic handling consists of a three digit code. This code deals sequentially with basic shape, basic symmetry and properties and orientations of asymmetric features. The two digit code for insertion deals with the type of insertion (fastener or non-fastener) and the direction of insertion for the first digit and the quality of insertion and/or the type of fastener for the second digit.

The basic geometry of the part is defined as being one of three shapes, rotational (first digit 0, 1 or 2), square or triangular prismatic (first digit 3, 4 or 5) and rectangular prismatic (first digit 6, 7 or 8). The sub-divisions between each basic shape are as shown in Fig. 2.8 where it can be seen that parts are basically short (flat), cubic or long. In practice, whilst the basic categorisations may appear to be

Parts can easily be fed (but not necessarily oriented) using conventional hopper feeders	rotational parts	$L/D<0.8$	discs	**0**
		$0.8 \leq L/D \leq 1.5$	short cylinders	**1**
		$L/D>1.5$	long cylinders	**2**
	triangular or square prismatic parts	$L/D<0.8$	flat parts	**3**
		$0.8 \leq L/D \leq 1.5$	cubic parts	**4**
		$L/D>1.5$	long parts	**5**
	rectangular parts	$A/B \leq 3$, $A/C>4$ (see note 6)	flat parts	**6**
		$A/B>3$	long parts	**7**
		$A/B \leq 3$, $A/C \leq 4$	cubic parts	**8**
Parts are difficult to feed using conventional hopper feeders				**9**

Fig. 2.8. Choice of first digit of automatic handling code

CHOICE OF THIRD DIGIT / CHOICE OF SECOND DIGIT

parts are small and non-abrasive				parts will not tangle or nest: not light, not sticky	not light, sticky	light, not sticky	light, sticky	parts will tangle or nest but not severely: not light, not sticky	not light, sticky	light, not sticky	light, sticky	parts will severely nest but not severely tangle	parts will severely tangle
				0	**1**	**2**	**3**	**4**	**5**	**6**	**7**	**8**	**9**
parts do not tend to overlap during feeding	not delicate	non-flexible	**0**	0	1	2	3	2	3	3	4		
		flexible	**1**	2	3	4	5	4	5	5	6		
	delicate	non-flexible	**2**	1	2	3	4	3	4	4	5		
		flexible	**3**	3	4	5	6	5	6	6	7		
parts tend to overlap during feeding	not delicate	non-flexible	**4**	2	3	3	4	4	5	4	5		
		flexible	**5**	4	5	5	6	6	7	6	7		
	delicate	non-flexible	**6**	3	4	4	5	5	6	5	6		
		flexible	**7**	5	6	6	7	7	8	7	8		

	very small parts: rotational parts: discs or short cylinders $L/D \le 1.5$	long cylinders $L/D > 1.5$	non-rotational parts: flat parts $A/B \le 3$, $A/C > 4$	long parts $A/B > 3$	cubic parts $A/B \le 3$, $A/C \le 4$	large parts: rotational parts: discs or short cylinders $L/D \le 1.5$	long cylinders $L/D > 1.5$	non-rotational parts: flat parts $A/B \le 3$, $A/C > 4$	long parts $A/B > 3$	cubic parts $A/B \le 3$, $A/C \le 4$
	0	**1**	**2**	**3**	**4**	**5**	**6**	**7**	**8**	**9**
8 parts are very small or large but are non-abrasive	2	2	2	2	2	9	9	9	9	9

Fig. 2.9. Difficulty levels for difficult to feed parts

'course' results of a survey have indicated that a surprisingly large number of small parts fall into one of the designated categories. A first digit of 9 indicates a special category of parts for which conveying and orienting are difficult for reasons other than those concerned with basic shape. Difficult parts are those which tangle, nest, are light, are sticky, are delicate, are flexible, are very thin, are very small or large or are abrasive. The various combinations of difficult to feed parts are identified in Fig. 2.9 where it can be seen that they are represented by

CHOICE OF THIRD DIGIT

CHOICE OF SECOND DIGIT		part is ALPHA symmetric (see note 5): part has only one natural resting aspect or end and side surfaces can be readily distinguished by their shapes or dimensions (see note 7)	part is ALPHA symmetric (see note 5): end and side surfaces can be distinguished because of steps, chamfers, holes or recesses	part is ALPHA symmetric (see note 5): end and side surfaces can only be distinguished because of other features, features too small or slight asymmetry (see notes 8 and 12)	part is not ALPHA symmetric (code the main feature or features causing ALPHA asymmetry) (see note 6): steps or chamfers (see note 9): part can be fed in slot and supported by large end or protruding flanges with center of mass below supporting surfaces and the part is not triangular (see note 10)	steps or chamfers (see note 9): part cannot be fed in slot and supported by large end or protruding flanges with center of mass below supporting surfaces or part is triangular	through grooves (see note 9): can be seen in end view (see note 11)	through grooves (see note 9): can be seen in side view (see note 11)	holes or recesses (cannot be seen in outer shape of silhouette)	other geometric features	slight asymmetry, features too small or non-geometric features (such as paint, lettering, etc.) (see notes 8 and 12)
		0	**1**	**2**	**3**	**4**	**5**	**6**	**7**	**8**	**9**
part has 90° or 120° rotational symmetry about the principal axis (see notes 2, 3 and 4)	**0**	0.7 1 0.55 1 0.9 1	NA 0.3 1 NA		0.7 1 0.3 1 0.9 1	0.55 1 0.1 2 0.5 2	0.5 3 0.15 2 0.5 2	0.5 2 0.15 2 0.5 2	0.5 3 0.15 3 0.5 3	0.5 3 0.15 3 0.5 3	

			part is ALPHA symmetric (see note 5): code the same feature or features coded in the second digit: steps, chamfers or grooves can be seen in side view or other features on side surfaces (see note 9)	code the same feature or features coded in the second digit: steps, chamfers or grooves can be seen in end view or other features on end surfaces (see note 9)	end and side surfaces can only be distinguished because of features too small or slight asymmetry (see notes 8 and 12)	part is not ALPHA symmetric (code the main feature or features causing ALPHA asymmetry) (see note 6): steps or chamfers provided by external features (see note 9): features on side surfaces (see note 15)	steps or chamfers provided by external features (see note 9): features on end surfaces (see note 15)	steps or chamfers provided by non-external features (see note 9): features on side surfaces (see note 15)	steps or chamfers provided by non-external features (see note 9): features on end surfaces (see note 15)	holes or recesses (cannot be seen in outer shape of silhouette): on side surfaces (see note 15)	holes or recesses (cannot be seen in outer shape of silhouette): on end surfaces (see note 15)	other features, slight asymmetry or features too small (see notes 8 and 12)
			0	**1**	**2**	**3**	**4**	**5**	**6**	**7**	**8**	**9**
part has 180° rotational symmetry about the principal axis (code the main feature or features causing 180° rather than 90° rotational symmetry about the principal axis) (see notes 4 and 13)	steps or chamfers can be seen in side or end views (see note 9)	**1**	0.4 1 0.15 1 0.5 1.5	0.4 1.5 0.15 1.5 0.55 1		0.25 1 0.1 1.5 0.45 1	0.15 1 0.1 1 0.25 1	0.2 2 0.1 2 0.25 3	0.25 2 0.1 2 0.25 3	0.25 2 0.1 2 0.25 2	0.25 2 0.1 1 0.25 3	
	through grooves can be seen in side or end views (see note 9)	**2**	0.2 1 0.3 1 0.5 1	0.2 1 0.3 1 0.5 1		0.25 1 0.1 1 0.45 1	0.25 1 0.1 1 0.25 1	0.25 2 0.1 2 0.25 3	0.15 1 0.05 1 0.25 1.5	0.25 2 0.1 2 0.25 2	0.25 1 0.1 1 0.25 3	
	holes or recesses (cannot be seen in outer shape of silhouette)	**3**	0.2 2 0.3 1 0.5 1	0.4 1 0.15 1 0.5 2		0.25 2 0.1 2 0.45 2	0.25 2 0.1 2 0.25 2	0.25 2 0.1 2 0.25 2	0.25 2 0.1 2 0.25 2	0.25 2 0.1 2 0.25 2	0.25 3 0.1 3 0.25 3	
	other features, features too small or slight asymmetry (see notes 8 and 12)	**4**										
part does not have 180° rotational symmetry about the principal axis (code the main feature or features causing rotational asymmetry) (see notes 4 and 14)	steps or chamfers can be seen in side or end views (see note 9): external to the degenerated envelope	**5**	0.25 1 0.1 1 0.25 1	0.25 1 0.1 1 0.25 1		0.1 1 0.05 1 0.1 1.5	0.1 1 0.05 1 0.1 1.5	0.1 1 0.05 1 0.1 1.5	0.1 1 0.05 1 0.1 1.5	0.1 1.5 0.05 1.5 0.1 2	0.1 1.5 0.05 1.5 0.1 2	
	steps or chamfers can be seen in side or end views (see note 9): non-external	**6**	0.25 1 0.1 1 0.25 1	0.25 1 0.1 1 0.25 1		0.1 1 0.05 1 0.1 1.5	0.1 1.5 0.05 1.5 0.1 1.5	0.1 1.5 0.05 1.5 0.1 1	0.1 1.5 0.05 1.5 0.1 1.5	0.1 2 0.05 2 0.1 2	0.1 1.5 0.05 1.5 0.1 2	
	through grooves can be seen in side or end views (see note 9)	**7**	0.25 1 0.1 1 0.25 1	0.2 1 0.25 1 0.25 1		0.1 1 0.05 1 0.1 1.5	0.1 1 0.05 1 0.1 1.5	0.1 1 0.05 1 0.1 1.5	0.1 1 0.05 1 0.1 1.5	0.1 2 0.05 2 0.1 2	0.1 1.5 0.05 1.5 0.1 2	
	holes or recesses (cannot be seen in outer shape of silhouette)	**8**	0.25 2 0.1 2 0.25 1	0.25 1 0.1 1 0.25 2		0.1 2 0.05 2 0.1 2	0.1 2 0.05 2 0.1 2	0.1 2 0.05 2 0.1 2	0.1 2 0.05 2 0.1 2	0.1 3 0.05 3 0.1 2	0.1 1.5 0.05 1.5 0.1 2	

Fig. 2.10. Difficult levels for square or triangular prismatic parts

a 10×10 matrix for which each element represents a single or multiple combination of difficulties. The figures in the array indicate what are considered to be typical values of additional cost of the feeding device ('standard' feeder = 1).

Figure 2.10 shows the coding sheet for square or triangular prismatic parts. It can be seen from the figure that for each box in the matrix the figures on the left indicate the modified efficiency of the orienting system for an average shaped part of the type indicated by the classification whilst the figures on the right give an estimate of the cost of the required feeder relative to the cost of a 'standard' feeder (cost value 1).

The modified efficiency of the orienting system is based on three factors, the proportion of parts in each orientation entering the orienting system, the proportion of parts in the required orientation leaving the orienting system and the 'length' of the part in each orientation. Of these, the first can be obtained either by experiment or by analysis. The second can best be determined by experiment and the third can be determined from observation. The output from the orienting system in the required orientation F is thus given by

$$F = \frac{VE}{L}$$

where V is the conveying velocity of the parts
E is the modified efficiency of the orienting system
and L is the maximum length of the part.

It is important to recognise that when determining the useful feed rate of a feeding device, the figure which is appropriate is the actual rate or the required rate *whichever is the smaller* i.e. the fact that a feeder may be capable of feeding 100 oriented parts per minute is not relevant if the requirement is for only 10 parts per minute since the actual cost of feeding the part is governed by the latter.

The difficulty level for feeding and orienting represents the cost of feeding and orienting the required part at the required rate relative to the cost of feeding and orienting a simple part at standard rate (60 per minute). It can be shown that, based on current equipment costs and labour rates and using typical figures for manual handling, if the difficulty level is greater than 10 then manual handling should be considered.

2.5.2 Classification and Coding for Automatic Insertion

If mechanised or automatic assembly is being considered, it is essential that an accurate estimate can be made of how much the assembly equipment will cost in terms of the amount attributable to each product produced by the system. It is only by having this information that a quantitative rather than a qualitative estimate of assembly costs can be made and the effect of re-design of the product for assembly on costs can be determined. Conversations with those responsible for submitting proposals to justify automatic assembly have indicated that, in general, those with ultimate responsibility for decision making on capital expenditure projects are unimpressed by vague assurances of the viability of a project but require a strong indication of the financial implications of various courses of action. Before estimates of assembly costs can be determined, it is necessary to compile a list of the various types of assembly operations which is sufficiently detailed to be representative of the majority of assembly tasks whilst at the same time being sufficiently comprehensive to be usable in a classification system. Accordingly, work at the Unitversity of Massachusetts and the University of Salford has resulted in the development of a two digit coding system for automatic assembly which categorises assembly operations into one hundred types using a 10×10 array.

After evaluating questionnaires sent to builders of assembly equipment with regard to the cost of workheads to carry out these assembly tasks, each element of the array is assigned with a number which relates the cost of a workhead required to carry out a specific task (Wr) to the cost of carrying out a basic assembly task (element 00 in the coding system). Clearly, differences of opinion existed from one manufacturer to another but the general relationships of cost from one task to another were consistent, and it is to be expected that the data presented should be capable of allowing the potential user of automated assembly equipment to obtain a realistic indication of the costs involved. It was particularly noticeable that all the assembly equipment manufacturers questioned were emphatic that the degree of difficulty of the assembly task does not significantly affect the rate at which assembly can be carried out but does cause the cost of the equipment to increase.

As was the case for the classification system for parts handling, it is hoped that the designer will be able to obtain a reasonable estimate of the cost of assembly, but more importantly the coding system should enable the designer to see what type of design features are necessary to reduce the cost of assembly.

The classification system is illustrated in Fig. 2.11 where it can be seen that the assembly operations are broadly categorised into three primary activities: the addition of a part which does not secure itself or any other parts, the addition of a part which secures another part(s) and/or itself immediately, and the case where all mechanical parts necessary for assembly are present but where some other form of activity takes place. The former two categories are further subdivided depending on the direction of insertion.

The type of assembly activity which determines the second digit of the coding system depends on the group of the first digit. For parts which do not secure themselves or others the relevant parameters are considered to be:

1) Does the part need holding prior to subsequent fastening?
2) Is the part easy or difficult to align and position during assembly?
3) Is there resistance to part insertion or can this be achieved with minimum force?

Clearly, a part which needs intermediate holding during machine indexing will require additional equipment and hence extra costs, and should be avoided where possible. Similarly, mating parts which, because of tight tolerances and/or lack of guidance features such as chamfers and tapers, will require more accurate (costly) placing equipment. Parts which have resistance to insertion will also generally require more expensive equipment.

Parts which secure others and/or themselves are basically categorised according to the type of fastening method used, and are further subdivided into those which are difficult to align or position during assembly, and those which offer little resistance to insertion or may be fastened with minimal effort.

The final category of assembly where all parts are initially in position and/or non-solids are added are assigned a second digit which is appropriate to a range of non-standard assembly operations. Obviously many different types of operation could be categorised as non-standard but it is hoped that those chosen are representative of the majority of these types of operation.

CHOICE OF SECOND DIGIT

CHOICE OF FIRST DIGIT

First digit	after assembly no holding down required to maintain orientation and location (4)						holding down required during subsequent process(es) to maintain orientation and location (4)			
	not self-locking				self-locking					
	easy to align or position during assembly (5)		not easy to align or position during assembly (no features facilitating alignment or positioning)		only part inserted is located (4)	part inserted and other part(s) located (4)	easy to align or position during assembly (5)		not easy to align or position during assembly (no features facilitating alignment or positioning)	
	no resistance to insertion (6)	resistance to insertion (6)	no resistance to insertion (6)	resistance to insertion (6)			no resistance to insertion (6)	resistance to insertion (6)	no resistance to insertion (6)	resistance to insertion (6)
	0	1	2	3	4	5	6	7	8	9
0 — addition of any part (1) where neither the part itself nor any other part is being finally secured immediately; straight line initial insertion (part is inserted using a simple single-axis straight line motion) (2); from vertically above	1	1.5	1.5	2.3	1.5	1.5	1.3	2	2	3
1 — straight line initial insertion; not from vertically above	1.2	1.6	1.6	2.5	1.7	1.7	1.6	2.1	2.1	3.3
2 — initial insertion not straight line motion (3)	2	3	3	4.6	3	3	2.7	4	4	6.1

First digit	no screwing operation or plastic deformation immediately after insertion (snap/press fits, circlips, spire nuts, etc)		plastic deformation immediately after insertion						screwing immediately after initial insertion	
			plastic bending			rivetting or similar plastic deformation				
			easy to align and position during assembly (5)	not easy to align or position during assembly (no features facilitating alignment or positioning)		easy to align and position during assembly (5)	not easy to align or position during assembly (no features facilitating alignment or positioning)			
	easy to align and position with no resistance to insertion (5)	not easy to align or position (no features facilitating alignment or positioning) and/or resistance to insertion		no resistance to insertion (6)	resistance to insertion (6)		no resistance to insertion (6)	resistance to insertion (6)	easy to align and position with no resistance to screwing (5)	not easy to align or position (no features facilitating alignment or positioning) with resistance to screwing
	0	1	2	3	4	5	6	7	8	9
3 — addition of any part (1) where the part itself and/or any other parts are being finally secured (7) immediately (part is or acts as a fastener or has a built-in fastening element); part secures itself only; straight line initial insertion (part is inserted using a simple single-axis straight line motion) (2); from vertically above	1.2	1.9	1.6	2.4	3.6	0.9	1.4	2.1	0.8	1.8
4 — part secures itself only; straight line initial insertion; not from vertically above	1.3	2.1	2.1	3.2	4.8	1	1.5	2.3	1.3	2
5 — part secures itself only; initial insertion not straight line motion (3)	2.4	3.8	3.2	4.8	7.2	1.8	2.8	4.2	1.6	3.6
6 — part is securing other parts immediately; straight line initial insertion (part is inserted using a simple single-axis straight line motion) (2); from vertically above	1.2	1.9	1.6	2.4	3.6	0.9	1.4	2.1	0.8	1.8
7 — part is securing other parts immediately; straight line initial insertion; not from vertically above	1.3	2.1	2.1	3.2	4.8	1	1.5	2.3	1.3	2
8 — part is securing other parts immediately; initial insertion not straight line motion (3)	2.4	3.8	3.2	4.8	7.2	1.8	2.8	4.2	1.6	3.6

First digit	mechanical fastening processes (part(s) already in place but not secured immediately after insertion)				non-mechanical fastening processes (part(s) already in place but not secured immediately after insertion)				non-fastening processes	
	none or localized plastic deformation				metallurgical processes					
						additional material required				
	bending or similar processes	rivetting or similar processes	screwing or other processes	bulk plastic deformation (more than 75% of the part material is plastically deformed during fastening)	no additional material required (e.g. resistance welding, etc.)	soldering processes	weld/braze processes	chemical processes (e.g. adhesive bonding, etc.)	manipulation of part(s) or sub-assembly (e.g. orienting, fitting or adjustment of part(s) or sub-assemblies, etc.)	other processes (e.g. liquid insertion, etc.)
	0	1	2	3	4	5	6	7	8	9
9 — assembly process where: all solid parts are in place or non-solids are added or part(s) or sub-assemblies are manipulated	1.6	0.9	0.8		1.2	1.1	1.1	0.8	1.5	

Fig. 2.11. Difficulty levels for automatic assembly processes

Once the relative workstation cost (Wr) has been identified for a particular assembly operation it is a simple matter to convert this to a "difficulty level" for assembly by dividing this by the required production rate (ass/s). The standard by which difficulty level is judged is that of an operation being carried out at the rate of one assembly per second using a device having a relative workstation cost of 1 and this is defined as having a difficulty level of 1. Thus, if an operation is carried out more slowly, the difficulty level for assembly will increase since, in theory, the equipment is being underutilitised and the cost of assembling that particular part in the assembly is greater than it needs to be. Conversely, if an operation is carried out more quickly than standard, the difficulty level will reduce. The maximum assembly rate considered is two assembly operations per second since although the equipment manufacturers were quite insistent that speed of operation had virtually no effect on the cost of the equipment. They accepted that at assembly rates of greater than two per second, mechanical problems would often start to cause equipment malfunctions. In practice, with very few exceptions, assembly at the rate of more than one per second is not usually attempted and the general concept of difficulty level outlined above can be taken to be valid.

Figure 2.12 shows a simple product which can be analysed as shown in the worksheet of Fig. 2.13. It can be seen that for this particular product only one part is actually necessary as determined by the criteria for a separate part. In this

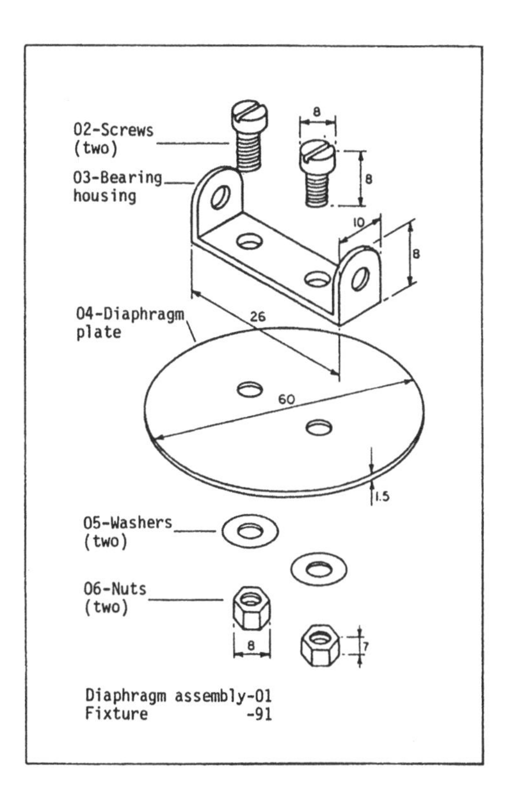

Fig. 2.12. Original design of diaphragm plate assembly

1	2	3	4	5	6	7	8	9	10	11	12	13	14	15	16	17	Required Rate of Assembly F_r (per second)
Part I.D. No.	Number of Times the Operation is Carried Out Simultaneously	Part Feeding and Orienting Code	Modified Feeder Efficiency, E	Relative Feeder Cost, C_r	Feed Rate of Parts min. F_m or F_r	Automatic Feeding and Orienting Difficulty Level	Basic Automatic Handling Cost per Part, C_h	Automatic Assembly Process Code	Relative Workhead Cost, W_r	Automatic Assembly Difficulty Level	Basic Automatic Assembly Cost per Part, C_a	I.D.'s of Parts Added to	I.D.'s of Other Parts Located or secured	I.D. of New Sub-assembly or Group of Parts	Operation Cost	Figures for Estimation of Theoretical Minimum Number of Parts	Remarks
06	2	001	0.7	1	0.5	2	015	00	1	2	03	91		90	0.18	1	Nuts into fixture
05	2	001	0.7	1	0.5	2	015	00	1	2	03	06		90	0.18	0	Washers
04	1	008		M	A	N	U	A	L			05		90	1.6	0	Diaphragm plate
03	1	834	0.15	1	0.13	8	015	08	2	4	03	04		90	0.24	0	Bearing housing
02	2	111	0.3	1	0.5	2	015	69	1.8	3.6	03	06	04,05 04,03	01	0.28	0	Screws
															2.48	1	
															C_{at}	n_{min}	

I.D. No's:
01-89 Parts of the Final Assembly
90 Unsecured Group of Parts
91-99 Tools, Fixtures, etc.

DESIGN EFFICIENCY

$$E_{aa} = \frac{n_{min}(C_h + C_a)}{C_{at}}$$

$$\frac{1(0.045)}{2.48} = 0.018$$

Fig. 2.13. Design for automatic assembly worksheet

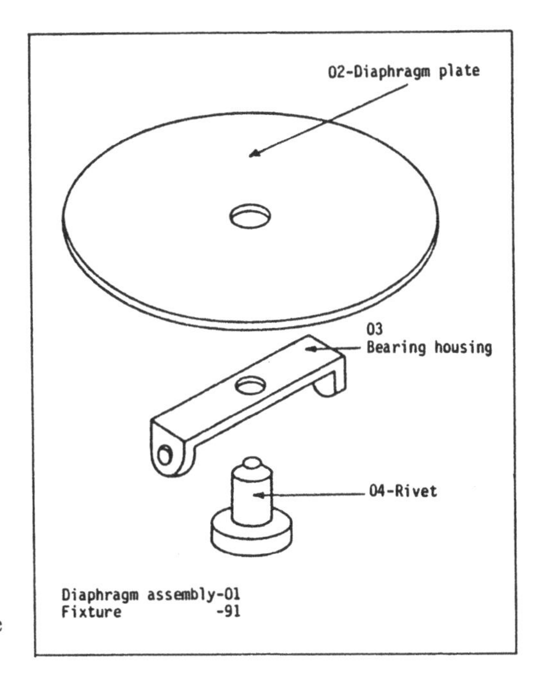

Fig. 2.14. Re-designed diaphragm plate assembly

example, when consideration was given to manufacturing the diaphragm plate and the bearing housing as an integral part it was thought that this would be uneconomic. As an alternative to this, either the diaphragm plate or the bearing housing could be manufactured with an integral rivet but, again, this was considered to be impractical. Yet another alternative would be to glue the diaphragm plate to the bearing housing but it was thought that for functional reasons this would not be satisfactory. It was finally decided that the diaphragm plate would be joined to be bearing housing by a single central separate rivet and Fig. 2.14 shows the proposed re-design. Figure 2.15 shows the worksheet for the new design; it can be seen that the feeding and orienting code for the diaphragm plate has changed from 008 to 001 whilst the insertion code has changed from 08 to 00. Further although the feeding and orienting code of the bearing housing has remained unchanged, its insertion code has changed from 08 to 00. The total assembly cost per part for the assembly of the redesigned product is 0.65 ¢ with an assembly efficiency of 9 per cent compared with an assembly cost of 2.89 ¢ and an efficiency of 1.8 per cent for the original design. This neglects the reduced cost of the transfer system which would be required for the re-designed product, the reduced cost of parts and also the cost of administering the larger number of parts of the original product. With an annual production volume of 2.5×10^6 this would give a minimum saving in *assembly* costs of $ 55,000.

It is important to note that the savings in *manufacturing* costs are significant. Without exception, in re-design exercises carried out to date, the savings in manufacturing costs have always been greater than the savings in assembly costs.

1	2	3	4	5	6	7	8	9	10	11	12	13	14	15	16	17	Required Rate of Assembly F_r (per second)
Part I.D. No.	Number of Times the Operation is Carried Out Simultaneously	Part Feeding and Orienting Code	Modified Feeder Efficiency, E	Relative Feeder Cost, C_r	Feed Rate of Parts min. F_m or F_r	Automatic Feeding and Orienting Difficulty Level	Basic Automatic Handling Cost per Part, C_h	Automatic Assembly Process Code	Relative Workhead Cost, W_r	Automatic Assembly Difficulty Level	Basic Automatic Assembly Cost per Part, C_a	I.D.'s of Parts Added to	I.D.'s of Other Parts Located or secured	I.D. of New Sub-assembly or Group of Parts	Operation Cost	Figures for Estimation of Theoretical Minimum Number of Parts	Remarks
04	1	111	0.3	1	0.5	2	015	00	1	2	03	91		90	0.09	1	Rivet into fixture
03	1	834	0.15	1	0.13	8	015	00	1	2	03	04		90	0.18	0	Bearing Housing
02	1	001	0.7	1	0.13	8	015	00	1	2	03	03		90	0.18	0	Diaphragm plate
								91	0.9	1.8	03	02	04, 03 02	01	0.05	0	Rivetting operation

0.50	1
C_{at}	n_{min}

DESIGN EFFICIENCY

$$E_{aa} = \frac{n_{min}(C_h + C_a)}{C_{at}}$$

$$\frac{1(0.045)}{0.5} = 0.09$$

I.D. No's:
01-89 Parts of the Final Assembly
90 Unsecured Group of Parts
91-99 Tools, Fixtures, etc.

Fig. 2.15. Design for automatic assembly worksheet

2.6 Robots in Manufacturing

Of the four main robot tasks, there is no doubt that for spraying, welding and bulk handling there are many applications where robots can be used effectively, and, more particularly, economically. Indeed, it could be said with justifiable conviction that in the majority of instances where the main concern is with the manufacture of products or sub-assemblies in small to medium batch sizes, these activities will be cost effective. In the fourth important robotic task, however, that of assembly, the situation is different and it is currently far more difficult to find applications for robotic devices in assembly which can be shown to be superior to other more traditional methods. There are several reasons for this; the repeatability required for most assembly work is probably an order of magnitude different than for other robotic activities, the speed of operation is higher than that required for other activities, the complexity of the tasks involved and the associated logic and sensing requirement to perform these tasks effectively put far greater demands on the robot than other activities and, perhaps most importantly, a wide variety of different sizes and shapes have to be picked, transported and inserted.

In the next few years, however, there will certainly be an increase in the use of assembly robots and whilst these activities will often be initially uneconomic, projections into the future would suggest that this will not always be the case and that 'hands on' experience now is essential if full advantage is to be made of the potential of assembly robots.

As has already been stated flexible assembly using assembly robots can be categorised broadly into two types; single arm robots working on a line (programmable assembly) and multi-arm robots working in a nominally fixed location, (robotic assembly). It can be expected that systems of the former type will be the first to be economically justifiable and these will be used for the manufacture of limited ranges of products in the same product family manufactured in medium to large batch sizes where the costs associated with changing from one product to another are comparatively small. However, for products with many parts, the capital cost of such systems will be high and the logical competitor to manual assembly of a wide variety of products produced in small batch sizes is the multi-arm assembly robot. An attempt will be made below to highlight the similarities and differences of these two types of assembly robot and to indicate the direction in which it is thought that development work aimed at improving the performance of assembly robots should progress.

2.7 Characteristics of Assembly Robots

If the similarities of assembly robots are considered, these can best be described by comparing them to dedicated assembly equipment. All assembly robots have the capacity to perform a wide variety of tasks for which changes from task to task require relatively little effort to effect. (This is relevant to the *Robot* tasks and not the *Peripherals*). Thus, they can pick and place from and to a variety of locations but in practice this ability, particularly that of picking, is often not utilised to the full. Robots, because they are dependent for their manoeverability on a complex computer control system, have the capacity to accept and more importantly respond to sensory information. Further, unlike dedicated assembly work stations, the additional cost of this facility is small. The significant advantage of this capability is that whereas a dedicated system will generally be severely restricted in its ability to recover from undesirable situations, the robot can be programmed to deal competently with many undesirable situations without the necessity for manual intervention.

There are three basic differences between single arm assembly line robots and multi-arm robots. These are outlined below:

1) The limited task single station robot will generally not need the same degree of manoeverability, sophistication or 'intelligence' as the multi-arm robot even though, in practice, typical equipment available for purchase would appear to contradict this.
2) The single arm robot working on an assembly line will require the *particular* part it has to assemble for a particular product far more frequently than the multi-arm robot will require a *particular* part for its product. The significance of this feature will be discussed at length later.
3) The multi-arm robot will be far more costly per assembly task than the single arm robot; the equipment for the former is more expensive and its cycle time is generally much longer.

2.8 Requirements for Robotic Assembly

The requirements considered necessary for improved, more efficient robotic assembly are listed below:

1) Faster robots
2) Limited capability, cheap robots
3) Versatile, inexpensive grippers
4) Identification of assembly families
5) Improved assembly efficiency
6) Low cost feeders.

2.8.1 Faster Robots

Unfortunately, the assembly robot arm has two conflicting requirements; it needs to be light so that rapid movements can be achieved with limited power whilst at the same time it needs to be rigid to ensure minimal deformation under load, particularly under conditions of extreme extension. Generally, the former requirement is considered to be the more important of the two since this gives the prime requirement of speed at the expense of positional accuracy but not repeatability. i.e. The more accurate robots such as would be used for assembly will repeat position to well within the tolerances required for assembly but they will not initially position accurately to programmed co-ordinates; this is an inconvenience which can be overcome when necessary by teaching the robot its required positions. Whatever is done, however, there is a practical limit on how light the structure can be and the power which can be put into the drive motors. As a consequence, there is a limit to what is possible in terms of improved robot speed and it is unlikely that, say, there will be an order of magnitude of improvement. This leads to the general conclusion that significant improvements in the economics of robotic assembly will not be achieved if increasing operating speeds were the only method of producing improvements.

2.8.2 Limited Capability, Cheap Robots

Undoubtedly, in real terms, assembly robots will become cheaper as the rate of their production increases. However, assembly robots tend to be more sophisticated than other types of robot and there appears to be a tendency to develop assembly robots with the type of capability of movement associated with operator assembly. This has led to assembly robots having, typically, at least five and usually six degrees of freedom. Whilst it is clear that under many circumstances this type of flexibility is necessary, this is usually because the design of the product being assembled has not been constrained by limitations imposed on it by the capabilities of the assembly equipment. Thought should be given to the design of a cheap yet positionally accurate, limited capability robot which should be capable of carrying out the majority of assembly tasks *provided* the product designer is aware of the limitations of the robot.

If this approach is to have any chance of success, it is essential that a set of product design rules be formulated which can realistically be applied in practice to a restricted mobility robot. Clearly, the design and development of the robot is best left to the robot manufacturers since they can call on a wealth of experience in this activity. However, experience gained in the past on design for manual and automatic assembly in academic establishments in collaboration with industry generally should prove useful when developing both product design rules for robotic assembly and a specification for a limited capability robot; work in this area has already been started.

2.8.3 Versatile, Inexpensive Grippers

For multi-arm series assembly robots it is common for all but small assemblies to have to make provision for gripper changes to meet different picking requirements and different geometry parts. (For the various methods of gripping: mechanical, electro-magnetic, pneumatic etc., each is generally only suitable for a small range of part sizes). Clearly, since gripper changing is a non-productive operation gripper changes should be minimised and, in principe, this can be achieved in one of four ways.

(a) Mount more than one gripper on the robot arm and make use of the wrist roll feature to select the appropriate gripper. Whilst this, when suitable, is an excellent solution it is felt that its application will be restricted to only a very small proportion of robotic assembly problems.
(b) Develop a 'universal programmable' gripper which can cope with a much wider range of part sizes and geometries. Much effort is currently being put into this activity and undoubtedly grippers will become available which are

much more versatile than those being used at present, but this will occur at the expense of programming simplicity and cost. However, since analysis has shown that, within reason, the cost of a robotic assembly installation is not critical to the economics of assembly, it is likely that a significant contribution to robotic assembly will be achieved by these developments.

(c) Design the parts of the product so that the number of gripper changes required to assemble the product is very much more restricted. This, like producing a specification for a limited capability robot, is very much dependent on the development of a classification and coding system for robotic assembly which in turn will lead to the formulation of design rules for robotic assembly.

(d) Assemble many products simultaneously so that the gripper change time to production time is significantly reduced. This is perhaps the simplest way to reduce the significance of gripper changes but its main disadvantages are the extra cost of duplicating all the fixturing required, the possible problem of space and the time taken for the robot arms to move the large distances which would be necessary.

The expense of a gripper is a result of two distinct costs. Firstly, the capital cost of the gripper, which is almost certain to increase with increased versatility of the gripper, and secondly the running cost of the wasted time required to carry out gripper changes. Of the two, the latter is potentially the most significant and thus, reducing gripper changes is probably more important than reducing gripper costs. However, if by good product design the gripper required need not be very versatile, or only a limited range of special purpose gripper need be used, then clearly this will be reflected in a reduction in the capital cost of the grippers and this, whilst not of major importance, is useful.

2.8.4 Identification of Assembly Families

When attempting to identify the appropriate assembly process, two of the important parameters are how many styles does the product have and how many product design changes are likely to take place per year. Traditionally, low volume, small batch size, short market life production has had to rely on manual assembly as the only economically viable method of assembly. Clearly, if potentially advantageous special purpose machines are to be used both the volume and the batch size have to be increased. This can sometimes be achieved by rationalising the design of products produced in a variety of styles such that there is a high degree of commonality in assembly methods. One of the objectives of the re-design for automatic assembly exercise, therefore, is to identify product families so that what were originally many styles of small to medium volume production can be reduced to fewer styles of a much larger volume production.

It can be argued that one of the major advantages of robotic assembly is that it is versatile and capable of dealing with a variety of assembly tasks and certainly changing from one style to another should not be difficult or time consuming for the robot. However, even though the robot can easily cope with different assembly problems, automatic feeding devices cannot, and much time and cost could result from too frequent changes from one product to another if these are significantly different. Thus, although identification of product families for *assembly* is not particularly important in flexible assembly, the identification of part families for *handling* merits consideration.

Requirements (2), (3) and (4) are all dependent on the effectiveness of the design for assembly, and it is significant to note that in general this approach is not being adopted, i.e. manufacturers of robots are tending to develop more sophisticated robots with more sophisticated grippers that presumably will lead to a universal assembly system capable of assembling virtually anything of reasonable size. This philosophy is diametrically opposite to that adopted for piece-part manufacture, where it is usual for the product designer to design parts which can be manufactured by existing, often very dedicated equipment. Whilst piece-part manufacture is a simpler process than assembly and a direct analogy is not appropriate, it would seem sensible to try to deskill assembly robots by good product design since this should result in a less expensive more efficient robot which should be easier to program.

2.8.5 Improved Assembly Efficiency

It is well recognised that the efficiency of special purpose assembly machines is mainly dependent on the quality of the parts used. Further, since the quality of parts used in assembly is usually improved by either automatic or manual inspection it is implicit that improved quality inevitably increases costs and that, in the limit, perfect quality can only be obtained at infinite cost. It can be concluded from this that for any automatic assembly process, an optimum quality will exist which will be the result of a compromise between the rate of increase in the cost of the part to improve machine efficiency, and the potential benefits.

From the above it can be inferred that, realistically, less than perfect parts will be presented to assembly machines and that the way to improve assembly efficiency is not by uneconomic improvements in parts quality but by designing the machine and planning the order of assembly such that the effect of poor parts quality is minimised. This argument is equally true for flexible assembly: it will be necessary to ensure that the parts being used are of appropriate quality, that the robot's activities are planned such that it is assembling the product in the most appropriate sequence, and that when a poor quality part is presented to the assembly, the robot responds "intelligently".

It was pointed out previously that the robot has a significant advantage over the special purpose assembly machine in this respect and this facility should be utilised to the full.

2.8.6 Low Cost Feeding

This aspect of robotic assembly is particularly relevant to the multi-arm robot although, of course, any reduction in parts feeding capital and change-over costs would be beneficial to any assembly system. It can be shown that, typically, the cost of feeding parts to a multi-arm assembly robot is between five and ten times the cost of feeding parts to a special purpose assembly machine or a single arm assembly line robot, because of the relatively low rate of utilisation of parts by the multi-arm robot. Thus, if parts feeding costs for multi-arm robotic assembly are to be minimised, it is important that a thorough study is made of alternative feeding systems for a wide variety of products, styles, batch sizes, production volumes etc.

2.9 Classification and Coding for Automatic Parts Handling for Flexible Assembly

As has already been mentioned, the problems of automatic parts handling are completely different for programmable assembly as compared with robotic assembly because the rates of utilisation of parts are very much greater for the former than for the latter. Thus, for programmable assembly using parts at a rate comparable with special purpose assembly equipment there is no good reason why the classification and coding system for automatic parts handling for automatic assembly should not be valid. The only significant difference between the handling requirements is that for flexible assembly, because products may need to be assembled in relatively small batches, thought needs to be given to methods of assisting the speed of changeover of feeding devices, feed tracks etc., when a different part is needed; and, more importantly, the scheduling of work needs to be such that changeovers are minimised.

For automatic parts handling for robotic assembly it has to be accepted that because the rate of utilisation of parts is low, the cost of feeding an individual part using an automatic parts feeder becomes disproportionately high. In many circumstances, it becomes less expensive to manually feed the parts to the robot, which in turn suggests that manual part insertion might also be appropriate. Work done recently at the University of Salford on analysing the economics of a wide variety of feeding systems indicates that dedicated vibratory bowl feeders, by far the most common feeding device for automatic assembly, are usually unsuitable for robotic assembly and that other methods of parts feeding are more appropriate. Thus, the first requirement for parts feeding for robotic assembly is to identify the *type* of feeding equipment which should be used, and then to design the parts such that they can be dealt with by this type of equipment.

2.10 Classification and Coding for Automatic Insertion for Flexible Assembly

As was the case for handling, the requirements of single arm robots used in programmable assembly are somewhat different to the requirements of a multi-arm robot used in robotic assembly. Almost inevitably, for a product of any reasonable size, a programmable assembly machine will be a mixture of manual, dedicated and robot workstations. At the robot stations, at any time and for some considerable time, only one part will be be inserted and problems of speed, gripper changing etc. are not significant. Most users would agree that in this situation, parts handling is the key to the success or failure of the equipment, and that minor modifications to the classification and coding system for automatic insertion will be appropriate for programmable assembly insertion.

For insertion operations in robotic assembly, however, there are many additional problems which become evident. Gripper changes can be time consuming, some operations such as screw and nut running and rivetting can only be carried out by dedicated pieces of hardware, sequences need to be well organised and the positioning of jigs and fixtures and parts presentation points need to be considered to minimise the robot arm movements.

Work has started on a suitable classification system for robotic insertion on the premise that several 'standard' robots will be defined which have limited capability and that these, depending on the insertion operation to be performed, will be penalised for gripper changes, the use of dedicated hardware and the use of extra capability. Thus, a picture will be built up of which particular parts are causing increased cost, and it is hoped that this system will firstly give an indication of the cost of insertion and secondly give pointers as to how good product design could reduce these costs.

2.11 Conclusions

1) For all forms of assembly: manual, automatic and flexible, the key to effectiveness is very much dependent on good design for parts handling and insertion.
2) Methods of systematic assessment of design for assembly have been developed; for manual and automatic assembly it has been shown that re-design exercises carried out on existing products can result in a reduction in assembly costs of between 20 and 50 per cent.
3) Reliably good design for assembly results in a marked reduction in part manufacturing costs.
4) Design for assembly studies are inexpensive and the return on investment in this activity is probably greater than that for any other manufacturing function.

3 Technological Planning for Manufacture – Methodology of Process Planning

G. Spur and F.-L. Krause
In cooperation with W. Turowski, L. Sidarous, H.-M. Anger and W. Grottke

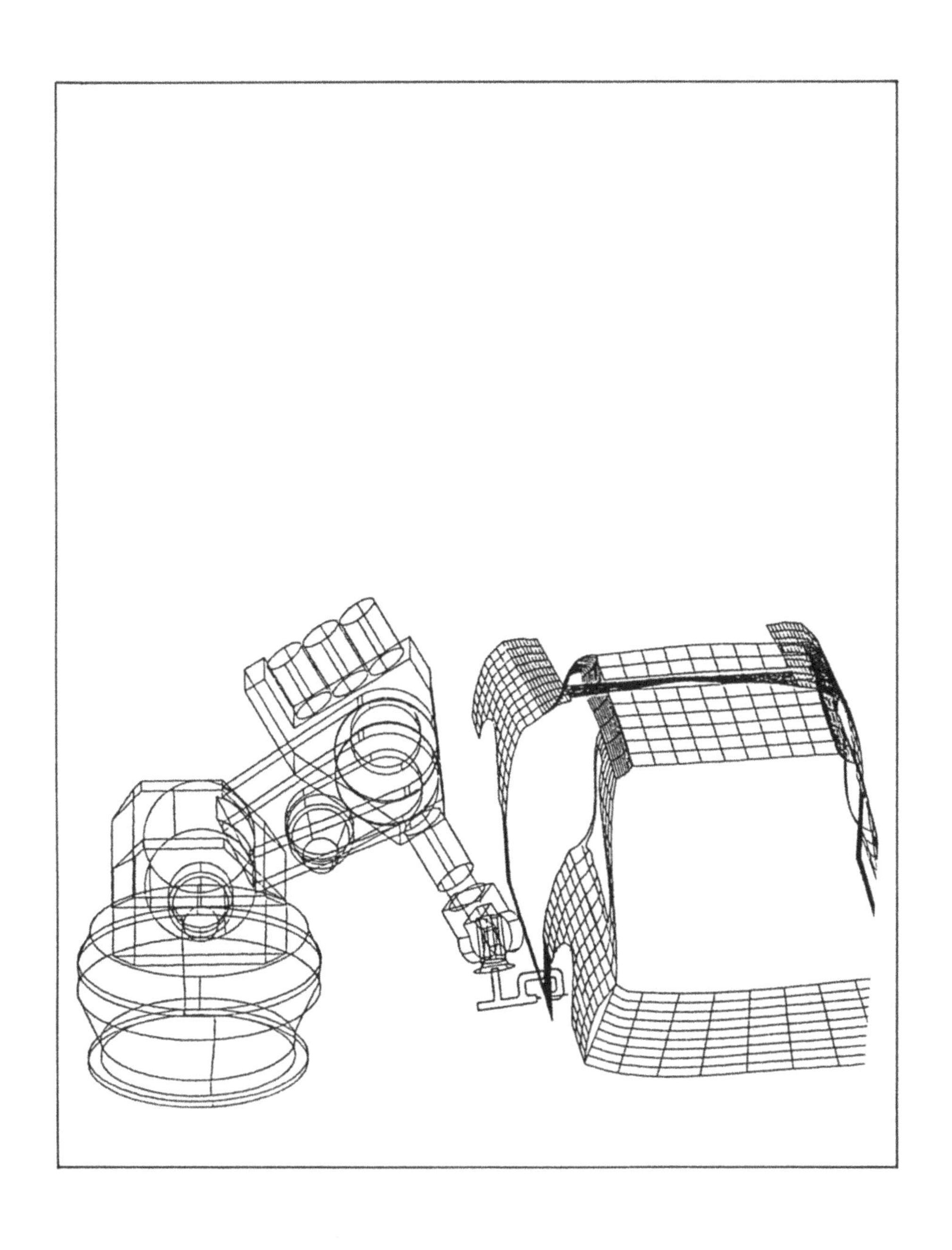

A study of a spot welding application

(Courtesy of Fraunhofer-Institut für Produktionsanlagen und Konstruktionstechnik – IPK, Berlin, FRG)

3.1 Methodology of Process Planning

3.1.1 Introduction

The activity of manufacturing a product is devided into 3 parts [1]. They are design, industrial engineering and manufacturing (Fig. 3.1).

Industrial engineering is the connecting link between design, where the shape and function of the product are defined, and manufacturing, where the design is crafted to a real product [2].

Industrial engineering activity converts the manufacturing papers produced by design (drawings, bill of materials) to working papers (process plans, N/C-control data) for manufacture. The main industrial engineering functions are:

Process planning,
Process scheduling and
Process control

Process planning defines:
What has to be produced
How it should be manufactured and
By what means.

It is the tasks of the process scheduling to determine:
How many items have to be produced
When and where they should be manufactured and
By whom.

Process control compares the actual output against the planned output and corrects deviations.

3.1.2 Tasks of Process Planning

The output of design are manufacturing papers in form of drawings and bills of materials. Process planning generates additional manufacturing papers which are needed for manufacture and assembly. The tasks of process planning can be subdivided into short-range, medium-range and long-range planning [3].
Short range planning comprises processing of the bill of materials, the generation of production plans, N/C programming, and planning of miscellaneous func-

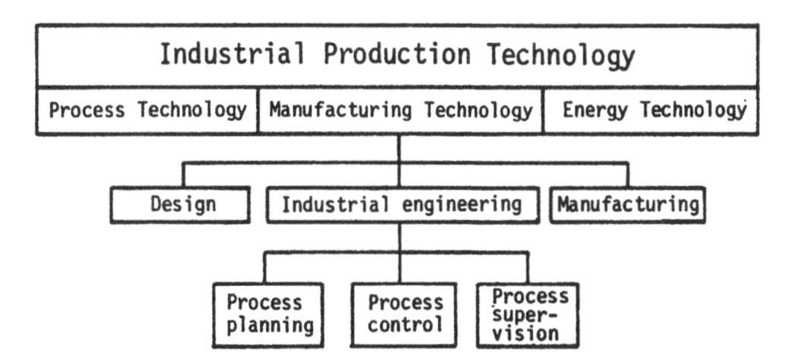

Fig. 3.1. The task of industrial engeneering in manufacturing [1]

tions. Medium range planning includes planning of costs and of quality control. Long range planning comprises material planning, methods planning and investment planning. Depending on their purpose the process planning functions can be basically divided into formal calculations, logical operations and creative decision making. Figure 3.2 shows the connection between planning functions and the planning activity.

With planning of the operation sequences the technical product is decomposed into a series of economical operations which are necessary for manufacture, i.e. a temporal and spatial sequence of related operations is given [4].

The most suitable machining technologies, e.g. forming, shaping, cutting, surface treating etc. are planned and the corresponding production equipment, e.g. machine tools, tools, fixtures are selected. After the most favorable manufacturing process has been found [4] the optimum design of the production equipment is determined by operation methods planning.

By workshop planning is understood planning of the spatial layout of the equipment of the factory. The layout is determined by the operation sequence and is carried out for all processes and subprocesses [4].

It is the main task of production equipment planning to coordinate the production equipment, machine tools, plants, facilities, tools, fixtures, jigs and special production processes (special purpose machine tools). For this reason it is necessary that the production equipment available to the company [4] is known and that this information can be used by operation sequence planning.

Operation time planning determines the times for the planned operating sequence. It comprises:

The times for the individual operation of a processing sequence
The start and finish times for process sequences and if necessary

Tasks of process planning		Planning mainly based on: Formal calculation	Logical operations	Creative decisions
Operation sequence planning	Product structure		•	
	Operation planning		•	
	Method planning			•
Work shop planning	Factory layout			•
	Shop floor and department layout			•
	Work place layout		•	
Production equipment	Machine tool planning		•	
	Tool and device planning		•	
	Planning of floor space for production equipment			•
Operation time planning	Time study	•		
	Lead time calculation	•		
Requirements planning	Material requirement planning		•	
	Prod. equipment requirement planning		•	
	Labour requirement planning		•	
Production time planning	Unit time planning	•		
	Total mgf. time planning		•	
Production cost planning	Determination of material, prod. equipment, and labour costs	•		
		•		
		•		

Fig. 3.2. Process planning functions [1]

The time intervals between successive operations as well as
The total time to perform the tasks (machining time).

Requirement planning has the task of determining the order independent requirements of unit labor and unit production equipment.

For labor the following information must be available:

Number of workers
Qualification
Length of employment
Employment status and
Place of employment.

For material planning it is necessary to know the quantity of parts and the external and internal shape of the workpiece [4].

Production time planning determines the unit times, which are the time slices needed to perform the individual operations of a task. It also does the planning of the total time which is needed to complete an order [4].

From the output of the preceeding activities the unit cost of a product is determined. The value is obtained by multiplying the labor cost for each operation by the values of the time slices and by adding up all cost components for the product.

3.1.3 Generation of the Process Plan

In order to ensure an economic production all information generated by industrial engineering has to be documented on a suitable storage device. It is an essential task of process planning to provide order independent information for processing of an order and for controlling the manufacturing operation. Next to the requirement plan and the production time plan the order independent process plan is the most important output of the process planning activity. The process plan is generated from information of the drawing, bill of materials and the number of workpieces to be produced. It documents the results of operation sequence planning, production equipment planning and operation time planning. The process plan is an order independent document. It shows what has to be produced from which materials and by which machines. It gives information for operating times and the wage group calculations (Fig. 3.3).

3.1.4 Principles of Process Planning

A process plan can be generated by the following three planning principles: They all have their advantages and disadvantages:

1. Generative planning: This is the most expensive planning principle. Here, the process plan is always generated completely a new for each product. No use is made of data from already available process plans.
2. Adaptive planning: In this case a known similar process plan is used as a basis for finding the initial solution. The deviations from the initial solution are taken into account. Only for these is a new plan generated.

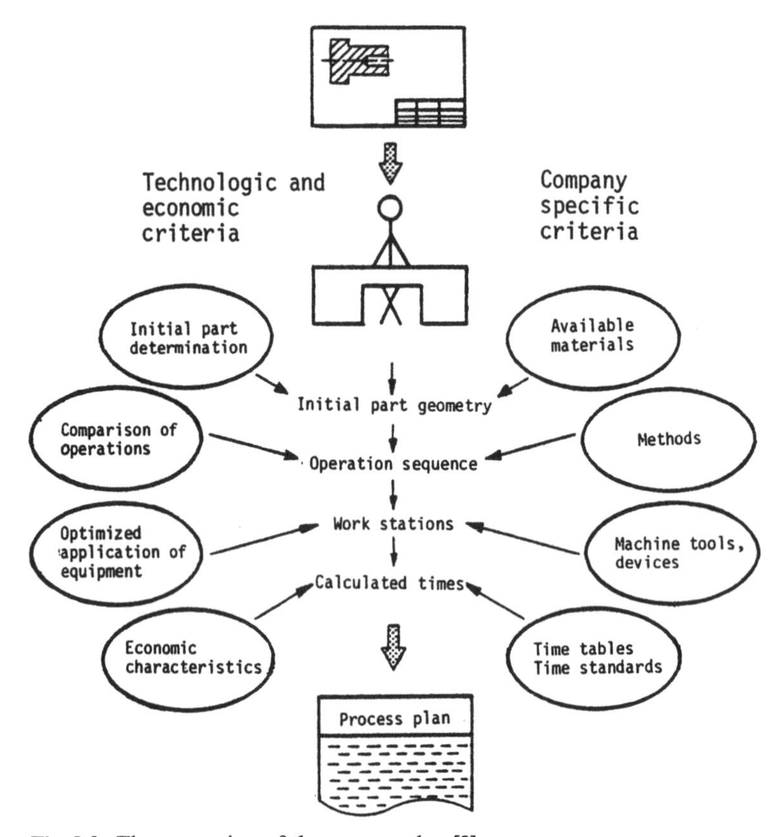

Fig. 3.3. The generation of the process plan [2]

3. Variant planning: By varying parameters within stipulated limits a current process plan can be made from a standard plan. No new requirements with regard to material, tools, production equipment and technolgy are needed.

Frequently a repetition planning method is mentioned. This, however, is not in itself a planning method, but process plan management, in which earlier generated process plans are used.

Valuable features of process plan management are the archive and retrieval functions of the process plans with the help of coding or classification keys.

3.2 Development of APT and EXAPT

3.2.1 APT System

In the mid-fifties the programming system APT (Automatically Programming Tools) has been developed at MIT (Massachusets Institute of Technology USA) and in use since 1959 [5]. Due to its universal applicability to almost all N/C-machine tools and because of its powerful geometrical concept APT has been widely adopted by many industries.

The programming system APT was conceived for complicated three dimensional milling operations required for aircraft manufacturing. Analytical and simple non-analytical surfaces as well as point patterns can be described with the APT-System. However, non-analytical complex surfaces as occur in turbine blades cannot be defined.

The geometric definitions are described by freely selectable symbolic names and are stored under these names. A typical geometric APT statement is:

```
      Symbol=Main  word / Modifier
Example P1=   Point   / 60, 80, 20
```

The modifiers are numbers which describe the parameters of the geometric element on the left side of the slash. It is also possible to define several points on a straight line or on a circle, in the form of a point pattern, and to store these under a symbolic name. After the workpiece geometry has been described the tool paths are defined by position statements, and they are followed by motion statements. Figure 3.4 [6] shows how APT defines the tool path by means of control surfaces. The tool is positioned in relation to drive surfaces and the part surface. The tool is driven tangentially along the drive surfaces, while the tool tip is positioned on the part surface. When a check surface is reached machining is finished along the drive and part surfaces. For every other milling pass a new drive statement is required. Contours and surfaces have to be generated line-by-line. Moving along three-dimensional curves is not possible. Likewise no automatic cut selection is performed.

Technologic data such as the spindle speed and the rotational direction of the spindle, have to be defined by the part programmer. Automatic tool selection is not provided for. The cutting tool is specified in the part program by the word CUTTER, together with the tool parameters.

To simplify programming, loop and jump instructions are provided. The description of repetitive patterns can be done by sub-routines. The APT-part program is generated from geometric information from the design drawing and technologic parameters of the material, and from catalogued tool data. Then the APT-parts program is generated. First the program is punched on cards and entered into the processor in batch mode. Here the first step of interpreting every statement of the part program is carried out. It is analysed, classified and coded. This is followed by an allocation of the geometric elements to memory in computer internal representation. Further processing comprises the determination of the coordinators of the path and of the tool end point. A diagnostic routine detects

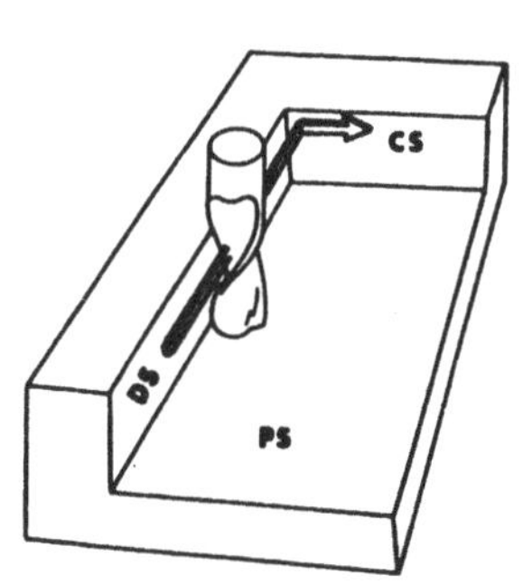

Fig. 3.4. Motion statements. *PS* = Part surface, *DS* = Drive surface, *CS* = Check surface

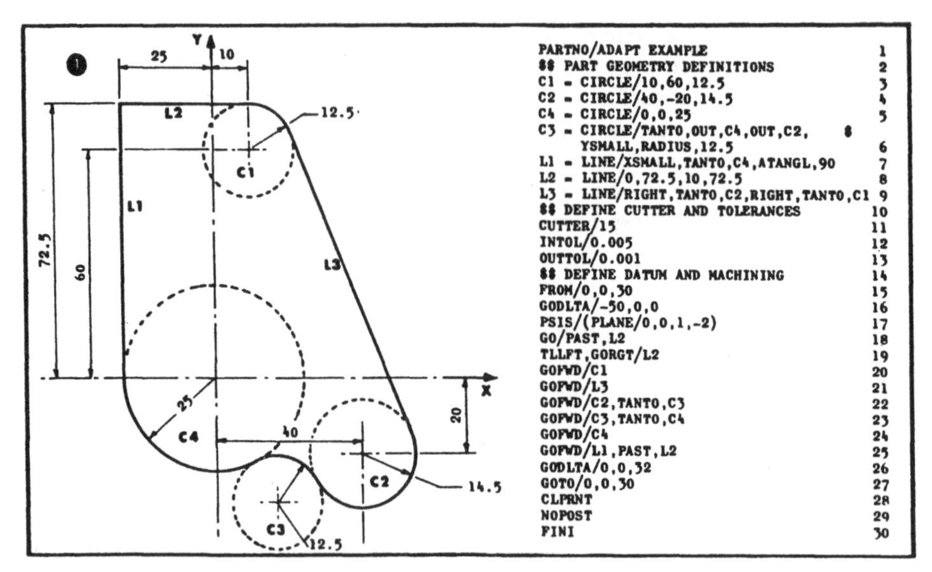


```
PARTNO/ADAPT EXAMPLE                                    1
$$ PART GEOMETRY DEFINITIONS                            2
C1 = CIRCLE/10,60,12.5                                  3
C2 = CIRCLE/40,-20,14.5                                 4
C4 = CIRCLE/0,0,25                                      5
C3 = CIRCLE/TANTO,OUT,C4,OUT,C2,          $
     YSMALL,RADIUS,12.5                                 6
L1 = LINE/XSMALL,TANTO,C4,ATANGL,90                     7
L2 = LINE/0,72.5,10,72.5                                8
L3 = LINE/RIGHT,TANTO,C2,RIGHT,TANTO,C1                 9
$$ DEFINE CUTTER AND TOLERANCES                         10
CUTTER/15                                               11
INTOL/0.005                                             12
OUTTOL/0.001                                            13
$$ DEFINE DATUM AND MACHINING                           14
FROM/0,0,30                                             15
GODLTA/-50,0,0                                          16
PSIS/(PLANE/0,0,1,-2)                                   17
GO/PAST,L2                                              18
TLLFT,GORGT/L2                                          19
GOFWD/C1                                                20
GOFWD/L3                                                21
GOFWD/C2,TANTO,C3                                       22
GOFWD/C3,TANTO,C4                                       23
GOFWD/C4                                                24
GOFWD/L1,PAST,L2                                        25
GODLTA/0,0,32                                           26
GOTO/0,0,30                                             27
CLPRNT                                                  28
NOPOST                                                  29
FINI                                                    30
```

Fig. 3.5. APT-Example

syntax errors, incorrect mathematical statements or erroneous motion statements. An example of an APT part program is given in Fig. 3.5 [6].

The calculated coordinates of the tool path are represented in standardized CLDATA form (Cutter Location DATA) and entered into a post processor, which adapts the program to the controls of the specific machine tool [7]. The APT-system has been repeatedly improved and extended to enhance its geometric and technologic processing capabilities.

3.2.2 EXAPT-System

EXAPT (Extended Subset of APT) is a standardized, universally applicable N/C-programming system. It was developed by the Universities of Aachen, Berlin and Stuttgart in the sixties, and since 1961 it has been marketed by the EXAPT-Society, Aachen. This standardized programming language is suitable for all N/C-machining tasks, with the exception of simultaneous machining in three or more controlled axes. In comparison to APT, EXAPT includes the determination of logic parameters for the most common N/C-machining tasks and automatic technologic planning.

3.2.2.1 Programming of N/C Turning Operations

Figure 3.6 shows a typical example of an EXAPT program which uses standard macros. The time and cost of preparing such macros have been considerably reduced by the introduction of new programming techniques to the EXAPT system, some of which were already available in basic EXAPT. The number of macrostatements used in the sub-routines shown in Fig. 3.6 could be reduced by 60% compared to previously described macro-programming methods. The ex-

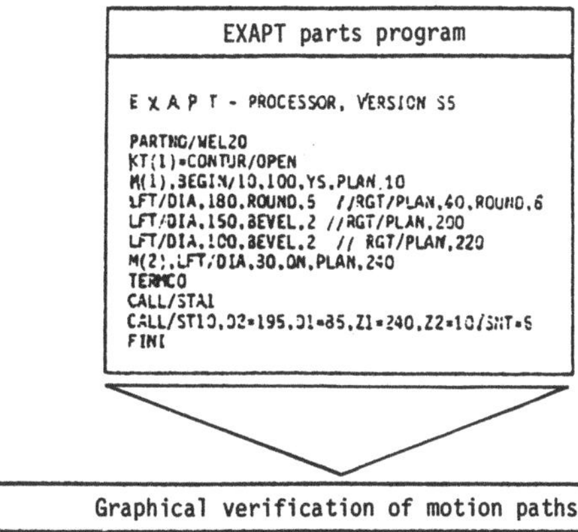

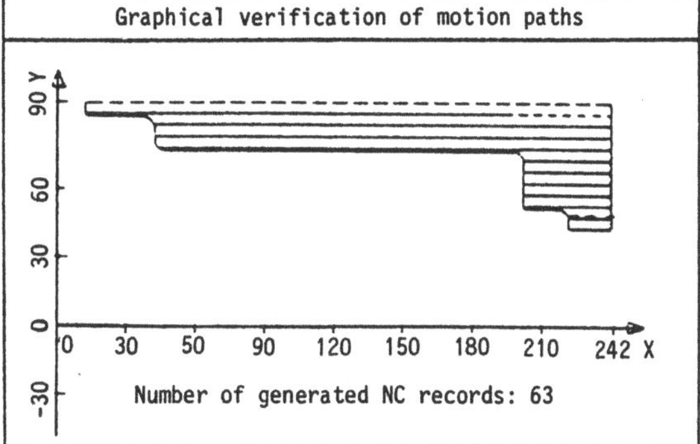

Fig. 3.6. Programming of a simple rotational workpiece with the basic EXAPT system

tented EXAPT-System for turning operations, which includes automatic technologic planning, can be used very economically for programming of rotational parts of complex geometry. An EXAPT program is devided into the following parts:

1. Header
2. Geometric definition
3. Technologic definition
4. END statement.

With the help of this partitioning principle and with the automated planning algorithms included in the EXAPT-system, it is possible to achieve short programming times for the preparation of N/C part programs. Moreover the EXAPT system offers flexibility in programming workpieces for different machine tools.

3.2.2.2 Programming of Drilling and Milling Operations

For these operations the geometric data of the workpiece have to be described accurately. Technologic planning is done automatically. Figure 3.7 shows how the EXAPT-system processes a workpiece with a great number of holes of different diameters. The following steps are done automatically:

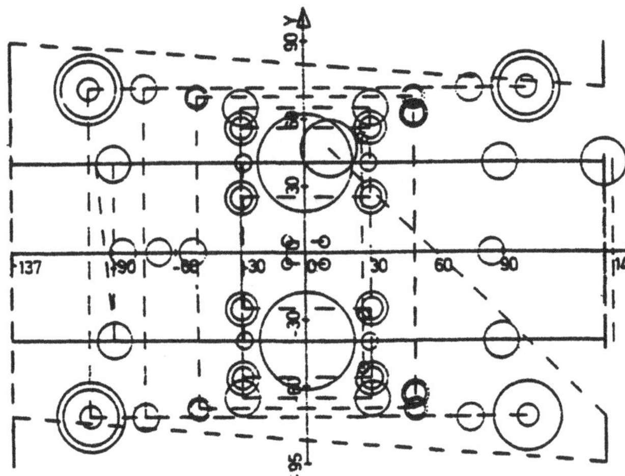

Fig. 3.7. Automatically determined drilling operations by means of the EXAPT system

1. Tool selection
2. Selection of work cycles
3. Determination of the feedrate and spindle speed
4. Determination of the cutter path and collision check.

A special machining operation is N/C-milling of cams. Often the programmer knows only the dynamic movement of the follower, the centre point of a roller and the major geometric dimensions of the cam. With this knowledge the EXAPT system can directly program the cam profile and the corresponding path of the centre point of the milling cutter. The milling cutter radius does not have to be the same as that of the roller. Figure 3.8 shows a path of the centre point of a milling cutter to produce a cam surface. It was automatically generated by the EXAPT system.

3.2.2.3 Programming of Punching, Nibbling, and Flame-Cutting Operations

As in the previously discussed methods the geometric data have to be defined. In addition the programmer has to carry out the following activities:

1. To write sub-routines for parts to be nested,
2. To store these programs and retrieve them until they are needed for manufacture,
3. Description of the processing sequence,
4. Graphical verification of the motion path.

The layout of a flame-cutting operation with the software package NESTEX of the EXAPT-system is shown in Fig. 3.9.

3.2.2.4 Programming of Wire-Eroding Operations

N/C-wire-eroding is being used more and more for the production of intricate workpieces. The programming of a geometrically complicated path is rather simple, because the dimensions of the drawing can be directly used. Standard fillet

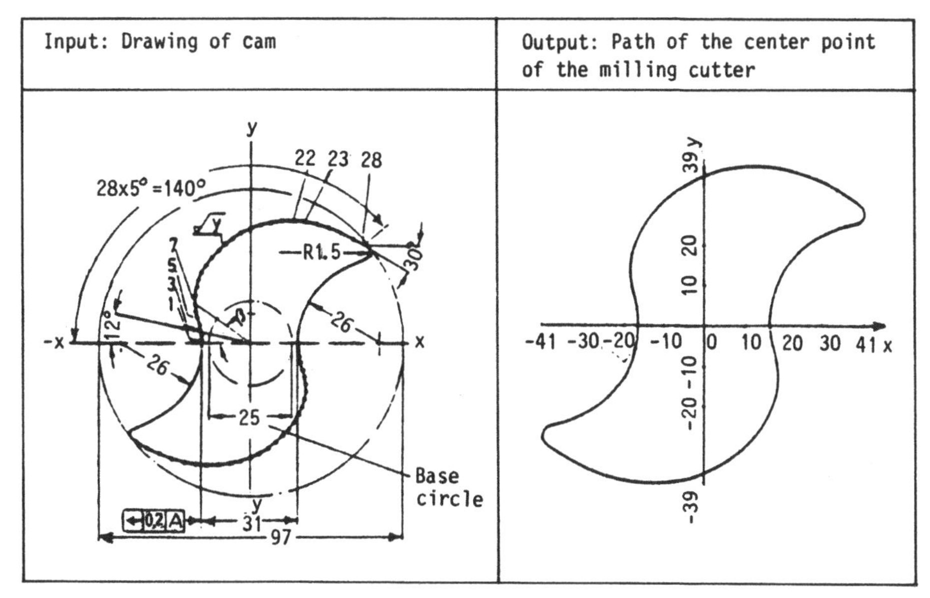

Fig. 3.8. Generation of a cutter path for a cam with the EXAPT system

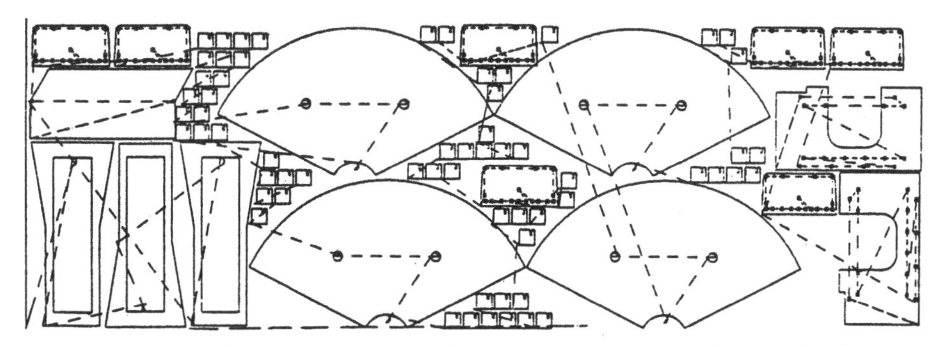

Fig. 3.9. Graphical simulation of a flame-cutting operation programmed with EXAPT

radii connecting geometric element are programmed with one statement only. They are the same for the whole part program.

3.2.2.5 Files for Working Data

The modular structure of the EXAPT programming system enables the user to work with different levels of automation. A new programmer may only use basic elements of the system, he can do that immediately without construction of files. For the preparation of complex part programs it is usually necessary to use common input data that do not change for each part program. To avoid repeated input of these data they are entered into files. The data can be activated for a N/C-program by calling them up under their identification number. The use of these files has the following advantages:

1. There are fewer programs to write
2. A constant manufacturing quality can be obtained
3. Technologic data are improved
4. A high flexibility is obtained when programs are changed [8].

3.3 Techniques of Computer Aided Process Planning

Computer aided techniques are being used to simulate general planning tasks. With the simulation model the effects of new processes, organizational structures and plant modifications can be investigated, without impairing the actual system which is simulated. In general only the statistical behavior of the disturbing variables are known. For this reason the distribution of these variables must be available to represent the behavior of the real system.

Models to handle operation planning problems are deterministic. They are solved analytically. The planning steps for a machining task can be described in detail with them. From practical use of automatic operations planning systems it has been found that dialog operated computer techniques are best suited for this application and hence they will be a determining element of future planning systems [9].

3.3.1 Dialog Aided Planning

Operation planning for expensive and highly automated manufacturing facilities should be done with dialog supported programming systems. A typical example is computer aided operation planning for numerically controlled machine tools. Detailed manufacturing documents are required for setting up the machines, for adjustment of the tools and for the control of the machining process. The generation and correction of these documents at the machines is very costly, because of the high hourly labor rates. Therefore, it has to be ensured by operation planning that the data produced are as accurate as possible. The dialog technique, in connection with the graphical representation of the machining operation allow simplification of the processing of the planning tasks [10].

In Fig. 3.10 different man/machine communication dialogs with data processing equipment are compared for several operation planning tasks. The interactive dialog functions can be used to perform and control automated planning sequences as well as to modify partial results.

The user has to select the planning sequence and possible alternatives out of a range of different solutions, and he has to understand them. The task of a dialog system is to help the planner to arrive at a good decision. A precondition is that the obtained results are represented in an easily understood form.

Computer aided planning systems use different methods to find a solution for the multitude of decisions which have to be made. Figure 3.11 shows different possibilities for solving the planning task. It can be seen how the manual method can be improved in different steps by automating successively the individual processing phases. With increasing use of software and interactive hardware the capability of decision making improves.

Increasing Flexibility →

	Batch mode	Dialogue mode	
Hard-ware		Alphanumerical	Graphical
Applications	Automatic program processing	Alphanumeric output of intermediate results and alternative solutions Control of the planning process	Graphical presentation, visual control of intermediate results Graphical input
	- Part determination - Geometric calculation - Process plan management	- Process planning - Alteration - Information system	- Tool path - Collision check - Presentation of planning steps - Complex geometric problems

Fig. 3.10. System flexibility depends on the processing mode [3]

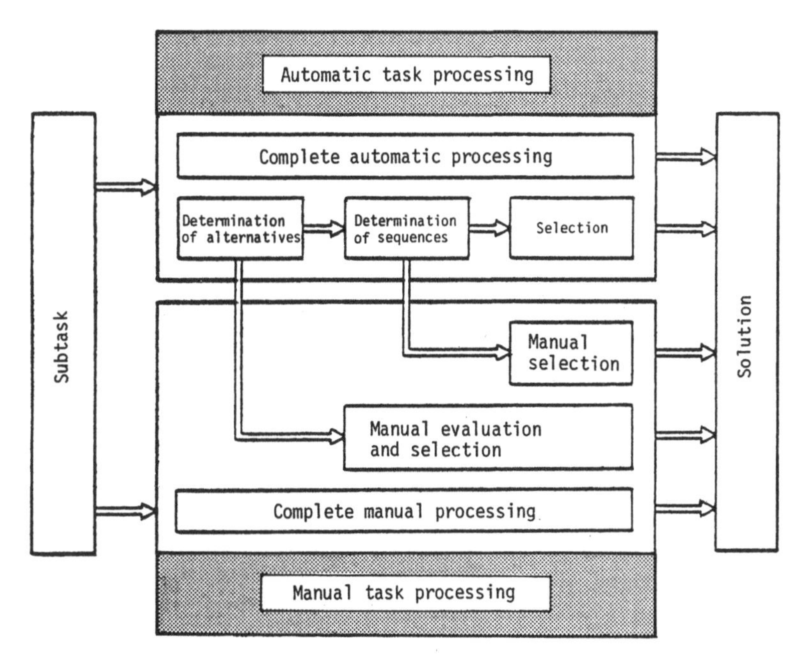

Fig. 3.11. Alternative task processing possibilities using interactive programming systems [3]

For computer aided operation planning the user guided dialog via alphanumeric and graphic means has been successfully used. Figure 3.12 shows different features of dialog systems. An important criterion for the application of an interactive graphical system is that its functions have been optimized for the user.

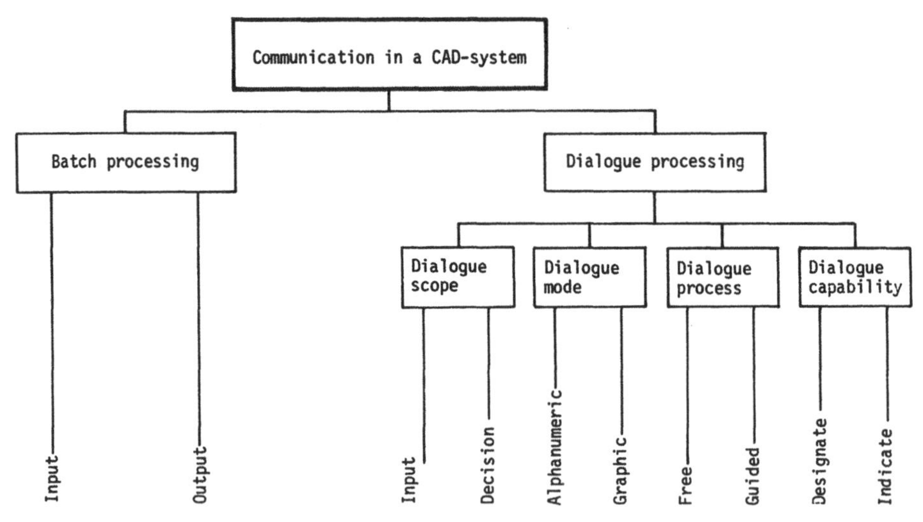

Fig. 3.12. Different features of dialog systems [4]

3.3.2 Algorithms

An algorithm permits the transformation of given input information, with the help of defined rules, to output information. In order to be able to formulate an algorithm the following basic conditions have to be fulfilled [5]:

1. The variables which have to be transformed must be defined
2. The transformation of the variables is done in steps and every step represents the application of a rule
3. The description has to be complete and finite
4. All permissible operations, as well as the language in which the rules are formulated, have to be stated
5. The algorithm must lead to a unambiguous solution and must terminate.

To implement process planning on a computer requires that the planning system is computer oriented. The basis for this are generalized models for the individual planning phases. They must have structures, properties and functions in which the tasks can be described in a formalized way. The formalization is obtained by means of mathematical, descriptive or graphic representations. Some purely mathematical formulations already exist today, however, they do not cover the entire planning range.

In general for computer aided process planning an algorithm can only be applied to specific tasks. Examples for detail planning tasks are:

1. Layout of cutting sequence and pattern
2. Determination of cutting parameters (depth of cut, feed, speed) and
3. Calculation of machining times.

The advantage of general algorithms is their universal applicability and the high reproducibility of the planning results. The formulation of general algorithms for all planning tasks is extremely difficult, since the decision criteria and parameters depend, as a rule, on non-quantifiable variables. Computer aided planning, which usually is a course planning function, uses decision table techniques extensively. Course planning functions are the selection of machine tools, determination of fixturing means, assignment of machining areas and selection of tools.

3.3.3 Decision Tables

Algorithms for planning must be designed so that the system can be expanded easily. For building of an expandable system the decision table technique is often used [13, 14].

The elements of a decision table are:

1. Conditions
2. Actions
3. Rules in form of an allocation matrix
4. Action entry.

The condition field contains in its rows a list of rules in the form of tests and questions, Fig. 3.13. The condition entry quadrant has in its columns for each rule a set of answers, which are assigned to the questions listed in the condition field. All actions to be specified are contained in rows of the action field. In the action entry field the corresponding actions to be performed by a given rule are marked in columns. The conditions and actions are described by a finite set of alphanumerical characters and special operators. The entries for the rules to be applied are abbreviations such as *y* for yes or *n* for no. The actions to be taken are defined in the action entry field by an *x*.

The formulation of the conditions is done in a standardized format, which may have three sub-formats as in Fig. 3.14. Two representations are required for the use of decision tables. One in coded form for computer internal processing and the other one in alphanumerical form for the user. The latter simplifies the documentation and the test of the information for correctness. It also aids the design of an optimized user dialog during the modification of the decision tables.

The generation of the actions can be made in two ways. In the case of very good system knowledge the input can be in coded form by the dialog. This, however, is limited to data and operations for which dialog processing is possible.

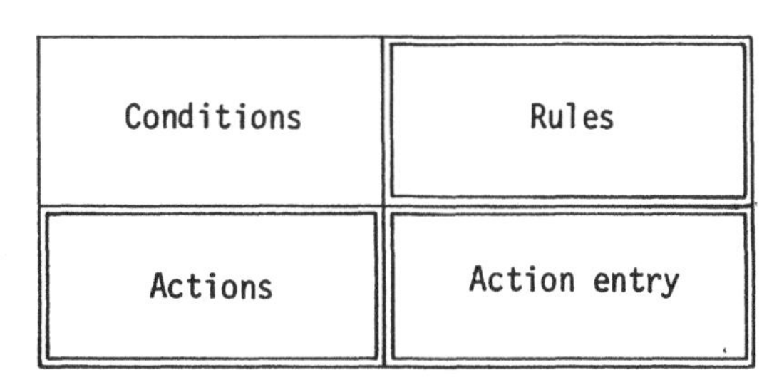

Fig. 3.13. Decision tables with conditions and actions [1]

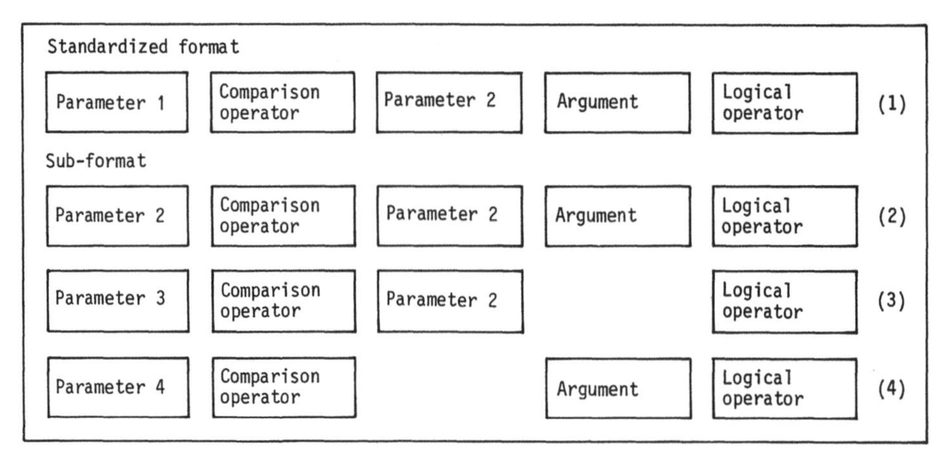

Fig. 3.14. Standardized formats to describe the conditions in a decision table

The actions of a decision table can be described by an algorithm. For computer internal processing the actions and conditions are represented in coded form. The documentation and the interactive representation is made in textual form.

The use of decision tables may serve two aims:

1. To determine all permitted actions
2. To specify all actions to be performed.

If for example a condition is answered with a *y* (yes) in a column of the rules field a corresponding *x* can be found in the action entry field. The *x* specifies the action to be taken in the action field. Thus for every condition the corresponding actions can be found.

3.3.4 Data Files

Computer aided process planning requires that the information for performing the planning activities is readily available and that there are suitable editing and updating possibilities. Practical experience gained with manual systems which use planning data from catalogs and other instructions shows that mere methodic collection and documentation will not meet the current requirements. In order to gather process planning data by computer systematic data collection and presentation has to be carried out.

The planning information collected is suitably classified into function orientated data modules. The computer aided process planning functions are the basis of this classification. How detailed such a classification has to be depends on the degree of detailing of the relevant computer planning operation. The data file can be divided into the following modules:

1. Machine tool file
2. Fixturing file
3. Tool file
4. Material file

5. Measuring instruments file
6. Planning time file

Such a classification permits practice orientated interlinking of functional elements. A precondition is that meaningful interlinking possibilities can be recognized from the parameters of the individual data. Therefore, logical as well as formal interfaces have to be conceived when the data file is designed. Another important point of view is how many planning data should be included. The selection should be as comprehensive as possible, so that sufficient criteria are available for the planning and that the effort in data collection and data processing is as small as necessary. This will keep the frequency of transactions, updating and the need to expand the data file to a minimum.

For every data module written master cards are prepared. They serve as graphic masters for the storage of data. The cards are divided into graphic and data fields, Fig. 3.15.

These file cards are designed such that the description data can be easily recorded. They are maintained in a central card file which is the heart of the data file systems for computer aided planning. Such a planning system has to include the following functions:

1. To store data for fast access
2. To record data
3. To updata data
4. To load data.

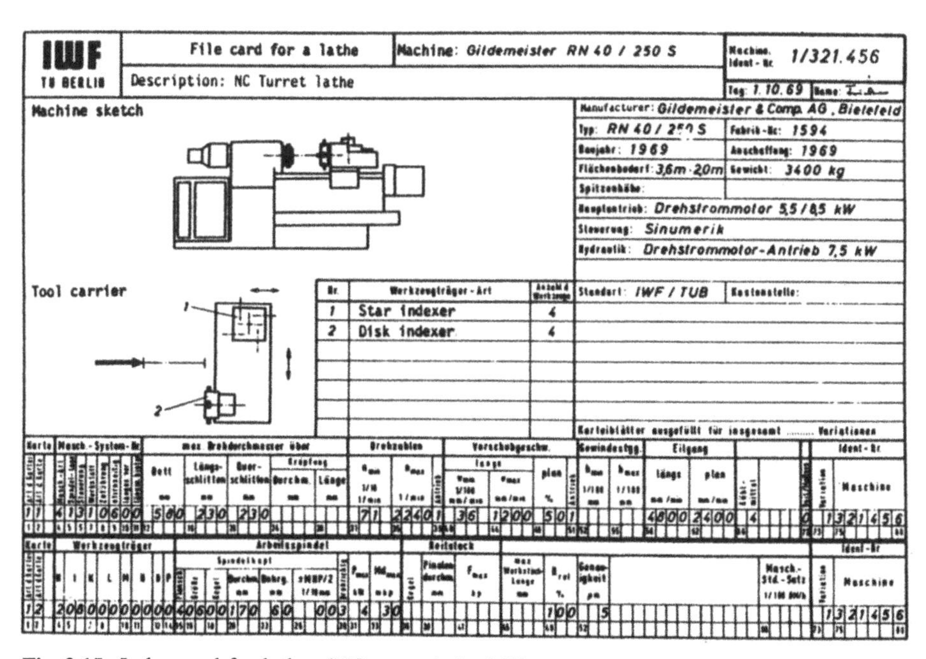
IWF TU BERLIN
File card for a lathe
Machine: Gildemeister RN 40 / 250 S
Machine-Ident-Nr. 1/321.456
Description: NC Turret lathe
Tag: 1.10.69
Machine sketch
Manufacturer: Gildemeister & Comp. AG, Bielefeld
Typ: RN 40 / 250 S
Fabrik-Nr: 1594
Baujahr: 1969
Anschaffung: 1969
Flächenbedarf: 3,6m · 2,0m
Gewicht: 3400 kg
Spitzenhöhe:
Hauptantrieb: Drehstrommotor 5,5 / 8,5 kW
Steuerung: Sinumerik
Hydraulik: Drehstrommotor-Antrieb 7,5 kW
Tool carrier

Nr.	Werkzeugträger-Art	Anzahl d. Werkzeuge
1	Star indexer	4
2	Disk indexer	4

Standort: IWF / TUB
Kostenstelle:
Karteiblätter ausgefüllt für insgesamt .... Variationen
Karte | Masch.-System-Nr. | max. Drehdurchmesser über | Drehzahlen | Vorschubgeschw. | Gewindestgg. | Eilgang | Ident-Nr.
Bett | Längsschlitten | Querschlitten | Greifweg: Durchm., Länge | n_{min} | n_{max} | Antrieb | längs | plan | Antrieb | h_{min} | h_{max} | längs | plan | Kühlmittel | Variation | Maschine
1 1 | 4 1 3 1 0 6 0 0 | 5 0 0 | 2 3 0 | 2 3 0 | 7 1 | 2 2 4 0 | 1 | 3 6 | 1 2 0 0 | 5 0 | 1 | 4 8 0 0 | 2 4 0 0 | 4 | 0 | 1 3 2 1 4 5 6
Karte | Werkzeugträger | Arbeitsspindel | Reitstock | Ident-Nr.
Masch.-Std.-Satz | Maschine
1 2 | 2 0 6 0 0 0 0 0 0 0 0 0 4 0 6 0 0 1 7 0 | 6 0 | 0 0 3 | 4 | 3 0 | 1 0 0 | 5 | 1 3 2 1 4 5 6

Fig. 3.15. Index card for lathes (NC-turret lathes) [7]

The required handling and maintenance programs have to handle the following tasks:

1. Filling of data in the central data file
2. Correction of data within the data file
3. Reading of data from the data file.

A central data management program is necessary for the application of a data file system. This units manages, checks and corrects the data files and organizes the access to them. Thus decentralized tasks processing such as computer aided process planning is possible.

3.4 Graphical Simulation of Manufacturing Processes in Process Planning

The complexity of process planning is essentially characterized by the necessity to process different data for technological and operational activities. For the automation of process planning operations the dialog technique has become an excellent tool.

The available dialog form is an important feature in selecting a dialog orientated system. Systems, which in addition to pure alphanumerical dialog, permit the use of graphically interactive techniques have a particularly high flexibility. Graphically supported process planning systems lead to reliable planning results. They provide suitable output devices and permit transparent communication. For this reason they are also readily accepted by the user.

Graphical computer support for process planning is mainly concerned with the pictorial representation of manufacturing tasks, manufacturing auxiliary equipment and the animation of manufacturing processes and tool paths (Fig. 3.16) [16].

By integrating the representations of objects and the relevant sequence of motions it is possible to simulate the machining progress of a workpiece and to perform complex collision checks. General basic graphical functions of a good system permit a problem-oriented representation, thereby enhancing the communication with the user in a dialog orientated fashion.

Graphic applications				
Production task	Production equipment	Auxiliary devices	Operation program	Auxiliary elements
- Finished part - Blank part	- Machine tool	- Tools - Jigs - Fixtures - Measuring - Devices	Presentation of: - Tool pathes - Intermediate workpiece shape - MFg cell layout - Tool and fixture arrangement	- Scales - Notes - Menue

Fig. 3.16. Information sources for graphical simulation

The objects of process planning can be represented in multiple ways. (Fig. 3.17) [2]. The elements for the graphical representation, such as points, lines and surfaces, may be of different color. Small systems use binary black or white representation for the pixels. With more expensive systems it is possible to assign different colors to the individual pixels, according to their technologic, geometric or functional importance. Color representation of surfaces is also possible. Another method of differentiating information can be done by drawing different types of lines. They can be shown as solid or broken lines. Surfaces can be differentiated by hatching or by halftoning. Also the bright and dark contrast of lines and surfaces may be varied by halftoning.

Depending on the objects and processes to be represented, two-dimensional or three-dimensional image structures are possible.

The simulation may be static, quasidynamic or dynamic. In case of a static simulation only defined discrete intermediate states of the process are represented. The quasidynamic simulation permits a static representation of the objects or process at different time intervals or states of the process. With a dynamic simulation objects and process sequence are changed continuously.

The possibilities for using the above mentioned graphic applications depend to a great extent on the given hardware capabilities; e.g. a multicolor image can only be generated on a display device that has these color functions, or a dynamic simulation is only possible with a peripheral that allows refreshing of the image (partial change of image) [18] or dynamic vector graphics.

Depending on the amount of information to be used and on the available computing equipment the graphical representation and the simulation has to be done in different degrees of abstraction (Fig. 3.18).

The more complete the information available for the graphic representation the better the graphic display can describe the object and the process. The neces-

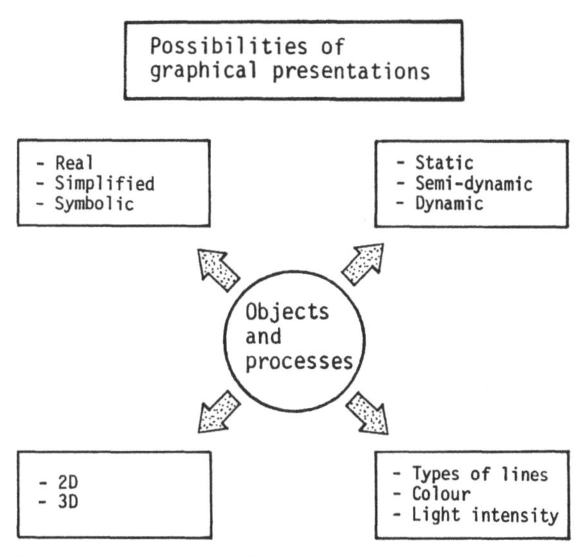

Fig. 3.17. Possible representation of process planning-specific objects and processes

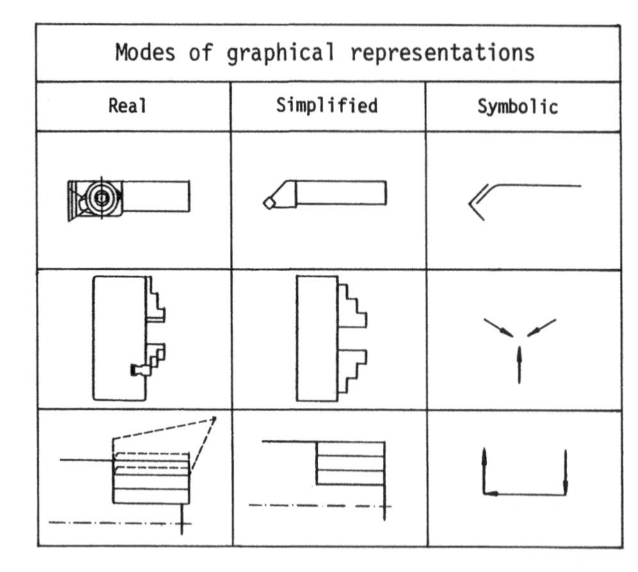

Fig. 3.18. Degree of abstraction for graphical representation

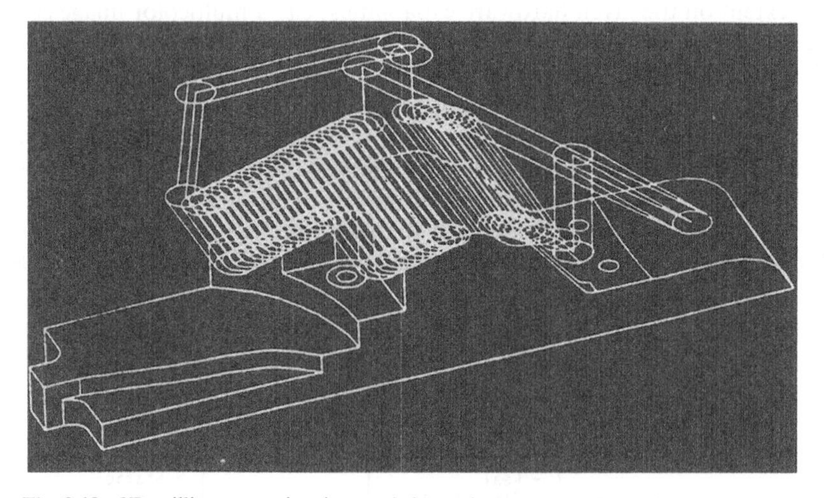

Fig. 3.19. 3D-milling operation in quasi-dynamic representation with GERBER

sary information for the planning process and the graphic simulation can be activated or generated. The information for the display is obtained from existing computer internal planning tasks, data files or manual input data of the operator. The provided data contains mainly the workpiece definition, description of the manufacturing means or of auxiliary equipment, standard values and control data of the operator for determining the process data. The information is generated via formulas, decision tables or look-up tables of various data files.

Next to the description of the workpiece the predominant application of graphic aids is the representation of the tool path for simple visual control [19]. For NC-programming there are a number of solutions available for two and three

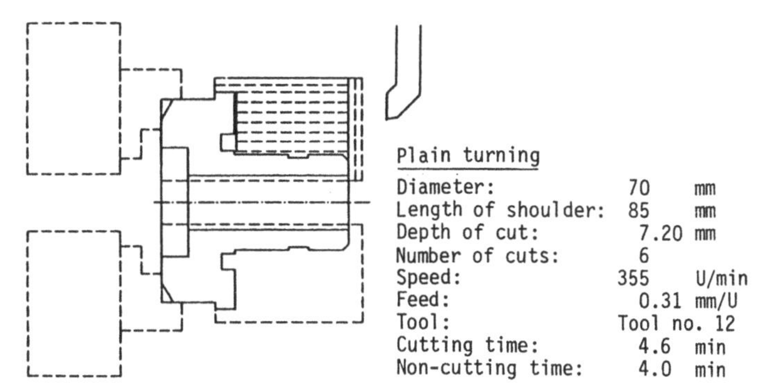

Fig. 3.20. Process planning by simulation

dimensional machining operations. With them the travel path of the tool can be made visible on a graphic display or on a plottter (Fig. 3.19).

In process planning the system CAPSY permits graphical planning and simulation of turning, boring or drilling operations of rotational symmetric workpieces. The graphic display can be programmed by variable input commands to present the relevant machining progress. This permits the user to perform a current check of the planning results and a transparent control of the planning sequence. Figure 3.20 shows the intermediate states of a planning and simulation activity, including the display of process data determined for this machining operation.

Future integration endeavors in process planning and the development of advanced graphical input and output devices will open up further applications for simulating process planning tasks. This will contribute greatly to a high degree of user comfort.

3.5 Systems for Computer Aided Process Planning Including Quality Control

Due to the wide range of applications for computer aided process planning systems and their close connection to company specific requirements, this section can only dicuss a few systems which are relevant to process planning and quality planning.

3.5.1 AUTAP System

The AUTAP-system was developed by the Institute of Production Systematics of the Technical University RWTH Aachen. It is a system for the automatic generation of operation plans [20]. On the basis of non-recurrent input data all necessary manufacturing operations are determined and the sequences of the machining operations are given. Also the necessary planning data for every operation such as machine tools, tools and times are determined. Figure 3.21 shows the principle structure of AUTAP.

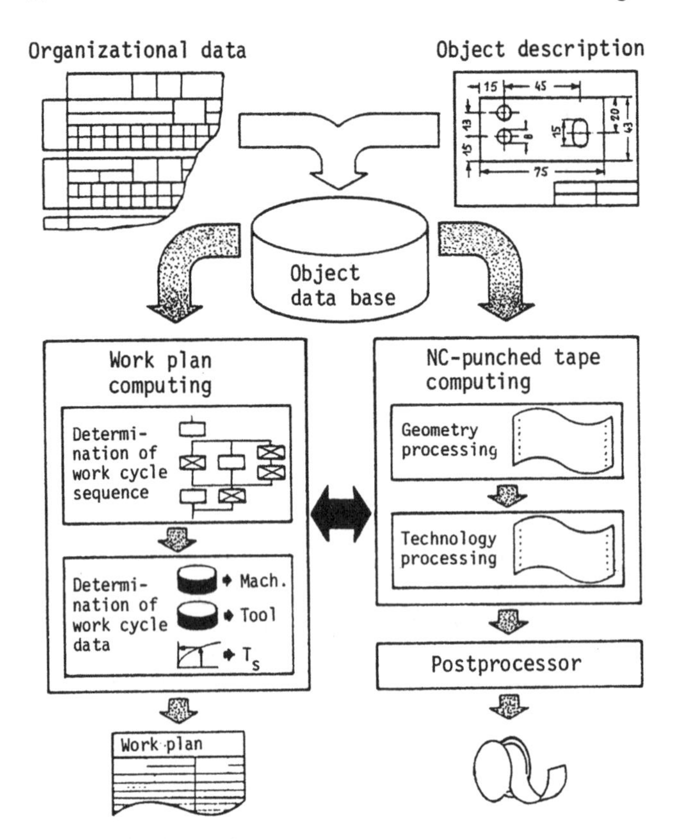

Fig. 3.21. Structure of the AUTAP-system [2]

The workpiece description for the AUTAP-system includes the grouping of parts into similar shaped elements. This manufacturing oriented workpiece description system is suitable for defining any sheet metal and rotational part, including some non-rotational parts, i.e. grooves, spline shafts, and eccentric holes. The only prerequisite is that the descriptions of the parts are stored in the library.

In the first planning step the AUTAP-system adapts the workpiece data to the special requirements needed for the generation of the operation plan.

On the basis of the manufacturing orientated workpiece description method a check is made to determine whether auxiliary elements for the manufacture, e.g., center bores or fixturing positions, have to be defined. Also the total length of workpiece and its maximum outside diameter are generated to determine the dimensions of the blank and the selection of the machine tools.

In the next step the sequence of operations is defined. The machining methods and the manufacturing machines are determined, as well as the auxiliary operations, such as deburring, inspecting and marking.

With the results obtained from the preceding steps the operations data are determined. Thereafter additional data, such as tools and times for machining, set-up and idle states, are generated for the individual machining methods. In the last step the quality control text is described and the operation plan is issued.

3.5.2 ARPL System

The programming system ARPL (Arbeitsplan) is a development of the Institute fo Machine Tools of the Technological University of Stuttgart [22]. The system is designed for companies with single part and small batch production. The main goals are to free the planner from routine work when preparing the operation plan. Thus he can concentrate his activities on other planning tasks. Figure 3.22 shows the planning sequence with ARPL.

The planning is done with the help of an alphanumeric display. Similar to the conventional generation of an operations plan, the organizational data are entered first. The user enters his data in a specified sequence under the guidance of the terminal.

The determination of the operation sequence is done according to the German standard DIN 8580. For every operation a specific module is called up and the operation data are determined and stored. For the machining methods turning, boring, drilling, milling, and grinding a detailed manufacturing description on the basis of the classification system (FBKS) is obtained. This classification system uses a number code with six digits, allowing the description of the workpiece fixtures, kind of machining, kinematic data, and the machining accuracy [24]. The operations are stored by the planner with the help of menu tables. A process specific allocation logic is activated by striking a key. Thus the applicable machine tools are determined and the information is checked for correctness. For every machining operation a special time estimation module is called up to deter-

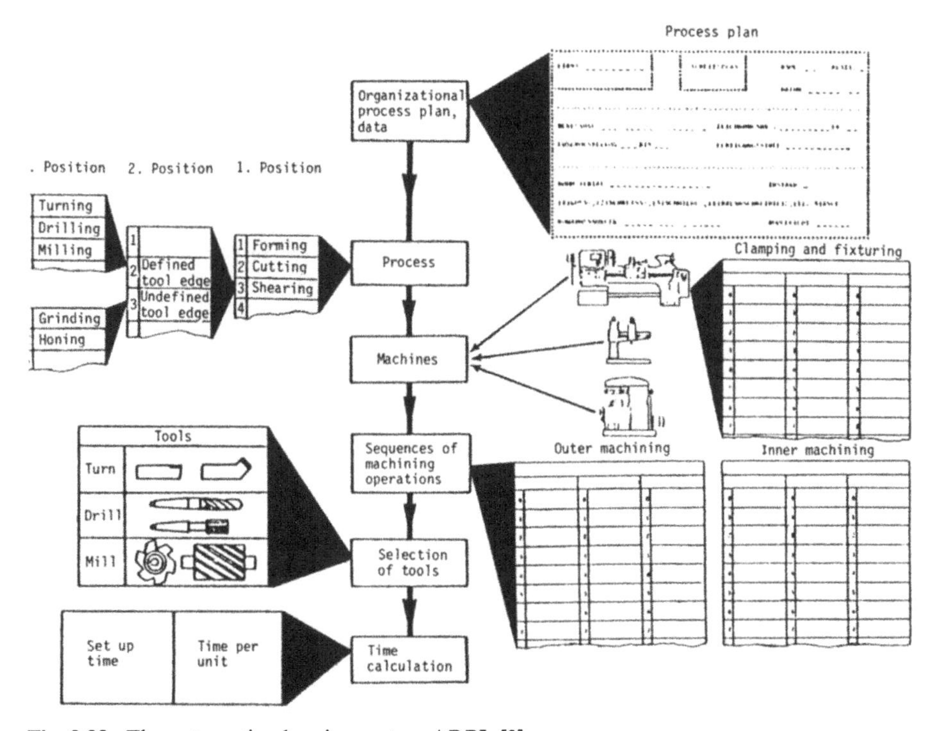

Fig. 3.22. The automatic planning system ARPL [3]

mine the allowable times. A precondition is the availability of machine and cutting data files. Finally the feed rate, required accuracy, and the surface finish are determined.

3.5.3 CAPEX System

The process planning system CAPEX (CAP based on EXAPT) is being developed by the EXAPT-Society. It will extend, the existing EXAPT N/C-programming system to include operation planning [25]. The concept of the system starts with a basic manufacturing schedule which can be supplemented by company specific data. For this purpose the user has a definition language available for which no particular computer knowledge is needed. The CAPEX standard functions include forms to determine processing times, possibilities for handling data, files, table processing techniques, and methods performing mathematical calculations and logical operations.

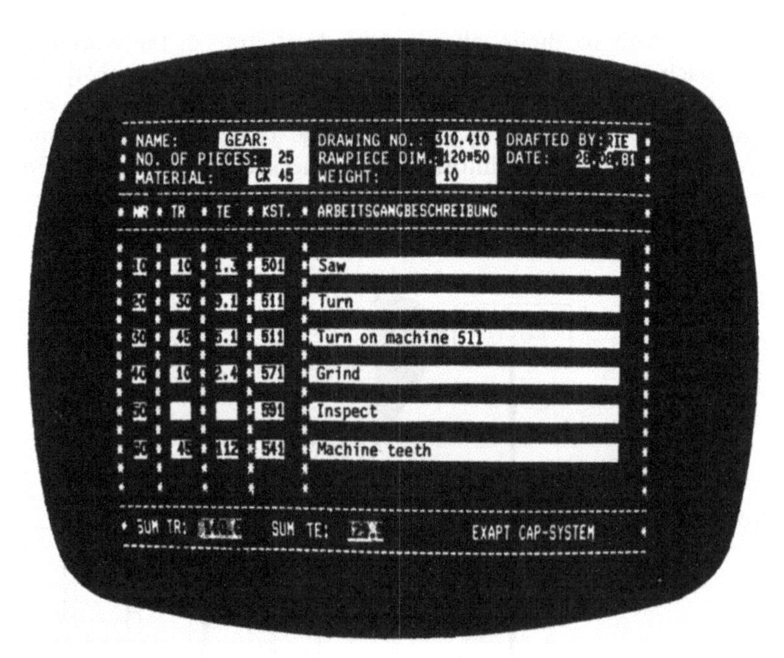

Fig. 3.23. Operations plan of the CAPEX-system [6]

Planning is done with CAPEX by a dialog masking technique. With it the input data are entered into a form on the CRT, this is similar to conventional planning. Figure 3.23 shows such a form displayed by a work station. Company specific data stored in a file can be inserted during the planning operation. Repetitive and variant planning can also be supported by the dialog technique. To increase the degree of automation similar tasks can be defined as macros and retrieved during the planning operation.

3.5.4 CAPP System

CAPP (Computer Aided Process Planning System) is a computer aided operation planning and management system. The work base may be any classification system provided by the user [26, 27]. The main functions of the system are:

1. To help the user to set up the operation plan
2. To perform corrective functions
3. To store and manage the operation plan
4. To retrieve the operation plan

The operation plan is drafted via menu techniques in a user/system dialog. The basis of this planning method is a standardized operation plan which has to be defined by the user. Likewise, the part family data, together with the classification code and all necessary part groups of the standardized operation plan are identified via a part family number.

When an operation plan is prepared the data from the standardized operation plan are used. This is done in three steps. First the general workpiece data are entered in the header of the operation plan. Thereafter the individual manufacturing steps are determined. This is done by using the given sequences of the standardized operation plan. Hereby, every operation can be checked and corrected if necessary. New operations may be added, if they were previously defined in the system.

In the last planning step data for the individual machining phases are entered. These are, e.g. data for machine tools, fixturing, and parameters for machining. After all data have been obtained the generated operation plan is stored. Figure 3.24 shows the principle structure of the CAPP-system.

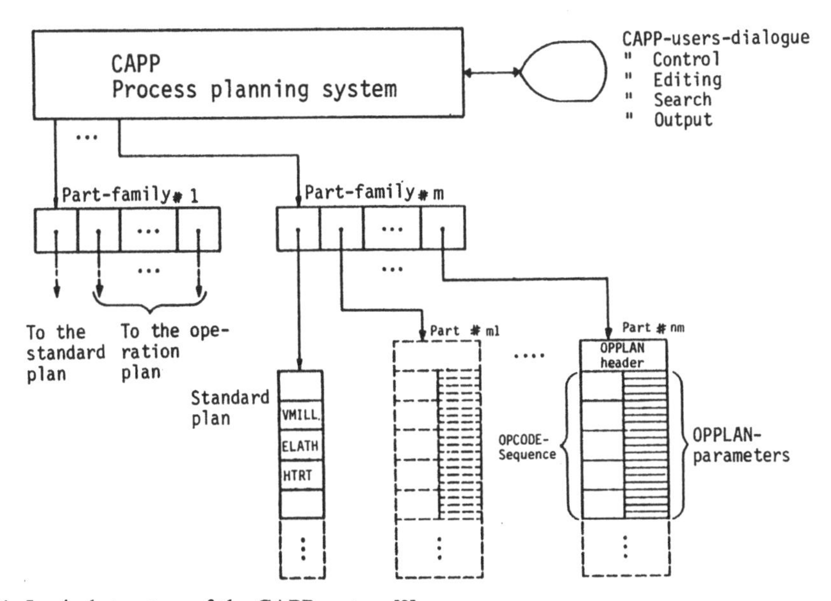

Fig. 3.24. Logical structure of the CAPP-system [8]

3.5.5 DISAP System

The process planning system DISAP is a development of the Institute of Production Systematics of the Technical University RWTH Aachen [28].

This is a very universal system, it can handle any workpiece and any applied manufacturing method. The planning depth can be adapted to company specific conditions. In an introductory stage the system is informed of all basic operations which are required to plan the lowest automation level. The structure of the system and the user orientated organization of the data file helps the planner to realize the next higher level of automation on his own. The system can be expanded in further steps to provide a manufacturing orientated operation plan for the entire range of workpieces of the company. With the help of dialog masks, standardized dialog control commands and well designed check and control functions a user optimized and easy to learn man/machine dialog is possible. Figure 3.25 shows the structure of the DISAP-system.

With the basic version of the system all workpieces can be planned by the same number of interactive operations. An input module permits the description of the workpiece before and during planning. Thereby geometric and technologic workpiece data can be manipulated.

The determination of manufacturing times depends on the planning method of the company. An adaptation of company specific planning sequences and of existing planning aids is possible.

In the last planning stage the data of the operation plan are compiled and printed as the manufacturing plan which can be directly used in production.

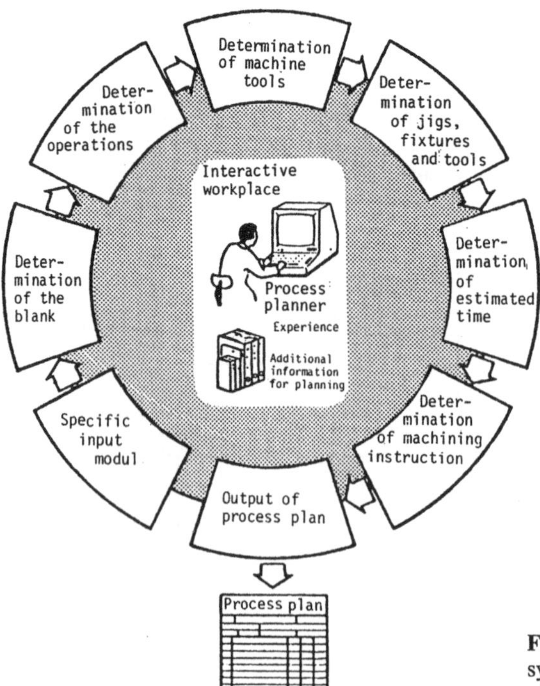

Fig. 3.25. Structure of an interactive system for the generation of a process plan (DISAP) [9]

3.5.6 DREKAL System

The programming system DREKAL was developed at the Institute for Manufacturing Technology and Metal Cutting Machines of the University of Hannover. It is used for planning and selection of the manufacturing tasks of turned parts in single part and small batch production [29]. Figure 3.26 shows the structure of DREKAL.

The system offers a simple conversational operating mode with high user comfort and great flexibility. The data processing is performed by routine tasks leaving to man that decision making, which is difficult to formulate by algorithm. The planning tasks are immediately processed so that the results are available at once. In this way input and decision errors can be corrected quickly.

For the description of the manufacturing tasks so called fabrication elements (FE) have to be defined. The main FE have to be entered in the order in which the workpiece is processed. The lower order FE can be entered at any point in the processing sequence of the workpiece. As a result the input effort is reduced. For the estimation of manufacturing times it is of less importance where for the workpiece the second order elements are to be manufactured. Planning is performed with the aid of a menu technique which selects the appropriate manufacturing method from given manufacturing alternatives. Data regarding machines, clamping locations, operations sequences, and machining areas are obtained via dialog. The calculations for the manufacturing plan are performed automatically on the basis of the workpiece and company specific planning data. Cutting parameters, machine data, machining and down times are calculated and entered in a data file. After the time calculations have been done the results can be checked on a terminal. The output routines are separated from the rest of the program so that they can easily be adapted to the needs of the user. Cost and cost comparison

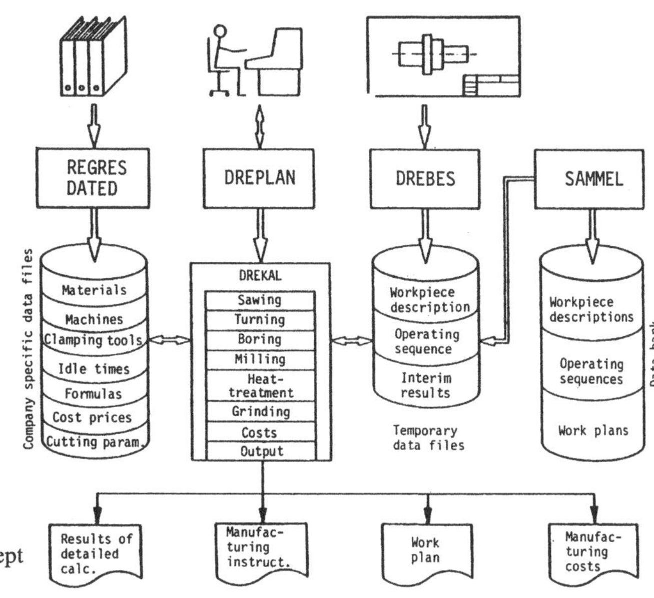

Fig. 3.26. General concept of DREKAL [10]

calculations can be performed after the time calculations. For this purpose data on lot sizes, annual output of parts and planned or estimated total numbers of work pieces are required.

The manufacturing costs per piece, consisting of direct and indirect manufacturing cost, are calculated. A cost comparison can be performed if different machines are available for manufacturing. By repeated cost and time calculations for alternative machines and by comparison of the results the cost optimum and the marginal lot size can be determined.

3.5.7 PREPLA System

For small and large batch production, inspection planning of quality control procedures, capacity, and personnel is very important. Similarly planning of inspection means for standard inspection problems must be carried out. For this purpose the PREPLA-program was developed by the Institute for Industrial Manufacturing and Manufacturing Operations of the University Stuttgart. With it the operations planner can determine for every operation the inspection procedure and the suitable inspection means (Fig. 3.27).

Depending on the configuration of the data processing system and the program module the inspection means can be determined either by interactive mode, via a display terminal, or by data input with punched cards. In the current configuration the inspection means and inspection task master files are stored externally on magnetic tape, whereas the data on the workpiece, the description of the inspection procedure, and additional information on inspection are entered via punched cards [30]. The data file for the inspection includes standard inspection means for dimensional measuring tasks and material testing.

The workpiece identification data, such as part name, number of drawing, operations plan, and the itemized parts list form the header of the inspection plan. With the help of a so called AFO (Arbeitsfortschritt = work progress) number, information on the work progress, the machine group, the performance of the cost center, and the cycle time can be retrieved.

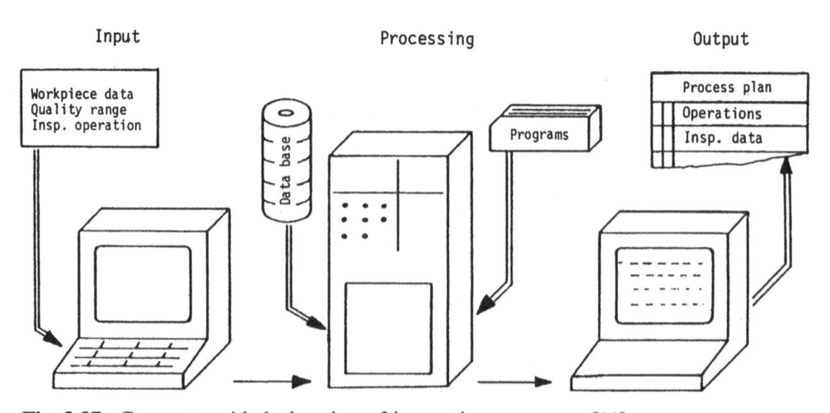

Fig. 3.27. Computer-aided planning of inspection resources [11]

The criteria for the selection of inspection means are described by user input. Typical rules may be to provide the lowest cost solution or to select inspection having the minimum permissible measuring accuracy. The inspection cost is determined by evaluating the hourly system cost. Calculation parameters are the instrument, overhead and labor costs.

The output of the selection procedure shows the five best suitable inspection means on a screen or printer. With this representation the operator can select the inspection means according to his subjective decision.

3.5.8 CAPSY System (Inspection Planning)

This system was developed at the Institute for Machine Tools and Manufacturing Technology of the Technical University of Berlin [31]. In its current configuration the system represents the basic version.

Starting with the machining documents, such as drawings, itemized lists and the operation plan, the inspection planner selects critical quality features or inspection procedures and loads these into the data processing computer for further processing (Fig. 3.28). The inspection operations which are listed in a separate inspection plan are entered in the operation plan at the place where the quality test has to be done and they are also marked on the drawing.

All other steps are performed in interactive mode with the computer. The information needed by the planner to make his decisions is stored in the data file. The data file includes company specific and company independent parameters. This relieves the planner of manual activities and he can concentrate on optimum planning of inspection sequences. Computer aided inspection planning by inter-

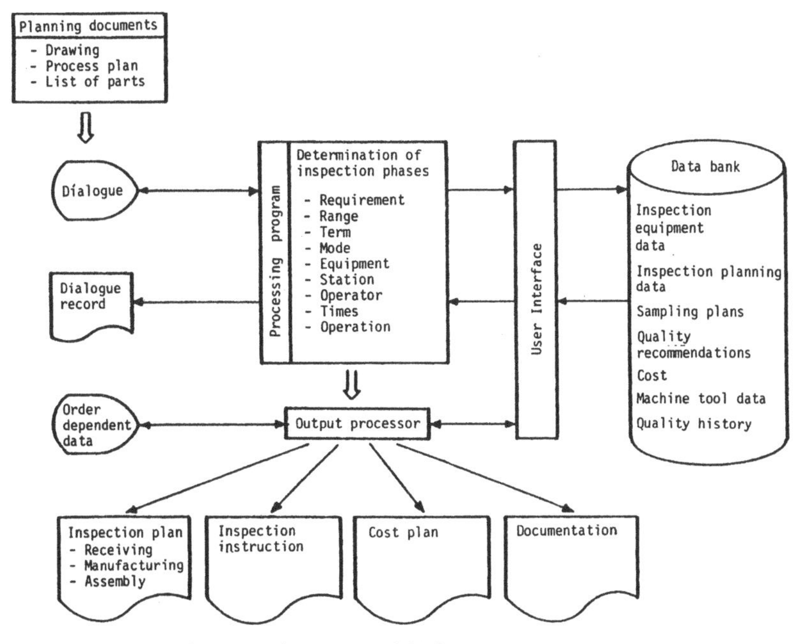

Fig. 3.28. Independent inspection planning system CAPSY

active means starts with a call to the programming system to determine the inspection data. In addition, organizational and technological data have to be entered, such as the name of the planner, workpiece description, number of the drawing, article number, and material to be used.

The inspection planner takes the data describing the workpiece from the planning documents. In the next step the necessary inspection parameters are derived from these data. Parameters may be length, surface finish, shape, or position dimensions. The adherence to these parameters is measured against rated values or upper and lower limits. Inspection determines the deviation of the actual dimensions obtained with the manufacturing process from those given on the drawing. Where a manufacturing process has an error trend due to problems with certain machining parameters this will be explained, and the inspection parameters can be corrected accordingly. However, the final decision to approve the parameter is left to the planner. The computer merely avails him of all information necessary to make a decision. The inspection necessary for the quality control task are selected automatically according to the technical characteristics of the workpiece, such as measuring range, measuring accuracy, rated value, and tolerance.

Where several suitable inspection means are found the most economical solution is proposed.

Having determined the suitable inspection means on the terminal the planner has the possibility of looking at further criteria, e.g. geometric restrictions which may be present due to unaccessibility of measurement points. Finally the planner has to state the kind of inspection, number of pieces to be inspected and time of inspection.

The task of the system is to furnish a basic independent inspection plan. It contains, in addition to general data in its header, detailed data describing the inspection operations. Whereas the basic inspection plan is order independent for a workpiece, all other documents generated by the output processor are order dependent.

3.6 The CAPSY Process Planning System

The concept of the process planning system CAPSY [32, 33, 34] which was developed at the Instituted of Machine Tools and Manufacturing Technology of the Technical University of Berlin can be outlined as follow:

1. To provide a universal process planning system which can be used by different companies.
2. To include a wide range of workpieces and standardized planning methods for conventional and N/C machine tools.
3. To design a planning system based on the generation principle.
4. To use the dialog technique to determine the machining sequence and to support modification of the results.
5. To provide a graphic output of the machining cycles to improve visual control and documentation.
6. To allow graphic simulation of the manufacturing process.

The process planning system CAPSY supports the following five functions, Fig. 3.29:

1. Definition of the planning task
2. Provision of planning data
3. Communication between user and planning system
4. Determination of the manufacturing data
5. Generation of the manufacturing documents

Different approaches are taken to the description of the manufacturing task to present data on blanks and finished parts. The workpiece to be planned can be defined by a description processor in a dialog mode. By interlinking the module GEOMETRY with the 3D-system COMPAC or the 2D-drawing system COMVAR it is possible to provide the workpiece information for the planning operation from the input of design. The description of a task for a part program can also be performed by modules of the EXAPT-programming system.

Company specific data files are required for the determination of the manufacturing data, such as information on adjustments, tool paths and auxiliary devices. Data on machine tools, devices, tools, reference time values, and reference cutting values also have to be made available. The dialog between the system and the user offers a flexible execution of the program and the possibility of using the creative ability of the user. Planning acitivities which cannot be decomposed by defined rules can also be performed by means of the dialog. The manufactur-

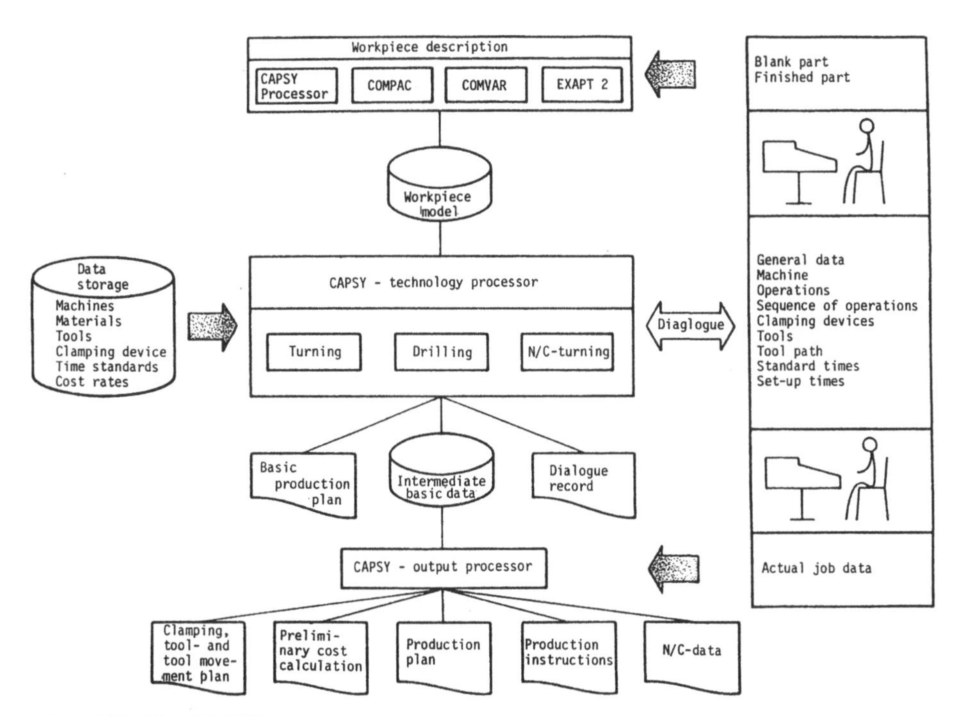

Fig. 3.29. The CAPSY-system

ing data are determined by technology processors for the various manufacturing processes and technologies. Company specific operations can be described with a selectable degree of detailing and data volume.

The overall CAPSY system is hierarchically subdivided into modules for various machining processes which are called up by a supervisory control program. Similar to the overall hierarchical structure of CAPSY, a hierarchical structure was selected for its individual modules.

With the CAPSY system planning is done in a strict dialog. The user determines the next execution phase of the program by entering a short command every time a new planning step is reached. Upon request the system issues a list of possible responses. If the user wants a correction, then the actions performed up to that point can be erased, and the initial state is re-established.

The execution of the planning starts with a call for the workpiece data. After the data of the blanks and finished parts have been read in from the disk storage, the planning-specific part for the manufacturing process starts by a processor request for information on the machining process. The user defines the machining

```
IWT TUB          B A S I C   M A N U F A C T U R I N G   P R O C E S S   P L A N           20-FEB-80

         PART.........: FLANGE                          MATERIAL.....: C 22
         PART NUMBER..: W012                            REFEREE......: PISTORIUS

SEQ  DEP  STA  WAG                                                                   ST    DT    CI

  02    10    3   SAWING                                                            3.0   1.0   8.0

  PRODUCTIVE TIME                                                                   3,0   1.0   8.0
  DELAY ALLOWANCE                                                                   0.3   0.1   0.8

                                                                                    3.3   1.1   8.8

  123    12    7    TURNING                    D   N    S     F

  PLAIN TURNING     A N11 D  170.00          5.0  2  560.0  0.224  TURNING TOOL,                 1.0
     READING PLAN AND DRAWING                                      R2525      DIN4980       1.2
     SPEED CHANGE                                                  -P20            12       0.1
     IN-FEED, TZRN ON CUT, RESET                                                            0.5
     FEED CHANGE                                                                            0.2
     GAUGING                                                                                1.1
  DRILLING              D    8.50 T  65.00       1  710.0  0.071   DRILL           31             1.3
     REVERSE    4 TMS                                                                       0.1
     READING PLAN AND DRAWING                                                               0.8
     SPEED CHANGE                                                                           0.1
     FEED CHANGE                                                                            0.1
     GAUGING                                                                                0.7
     POSITIONING                                                                            0.1
  RECLAMPING

  FACE TURNING      A M11 L   80.00          5.0  2 1400.0  0.090  FACING TOOL,                  0.8
     READING PLAN AND DRAWING                                      R2525      DIN4977       1.2
     SPEED CHANGE                                                  -P20             8       0.1
     IN-FEED, TURN ON CUT, RESET                                                            0.5
     FEED CHANGE                                                                            0.2
     GAUGING                                                                                0.6

  PRODUCTIVE PLAN                                                                    5.0   7.6   1.1
  DELAY ALLOWANCE                                                                    0.5   0.9   0.4

                                                                                     5.6   [illegible].5   3.5

   12     15    9   VERTICAL MILLING                               MILLING CUTTER  31   10.0  30.0  15.0

  PRODUCTIVE TIME                                                                   10.0  30.0  15.0
  DELAY ALLOWANCE                                                                    1.0   3.0   1.5

                                                                                    11.0  33.0  16.5

  991     5    9   MEASURE                                         MICROM. CALIPERS      5.0   3.0  10.0

  PRODUCTIVITY TIME                                                                  5.0   3.0  10.0
  DELAY ALLOWANCE                                                                    1.5   1.3   1.0

                                                                                     5.5   3.3  11.0
```

Fig. 3.30. Basic production plan

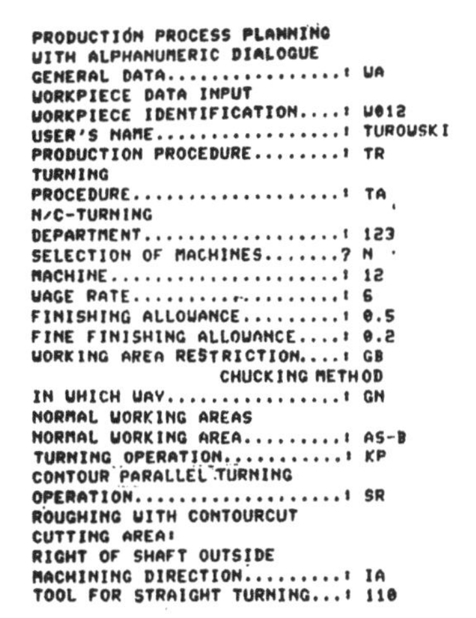

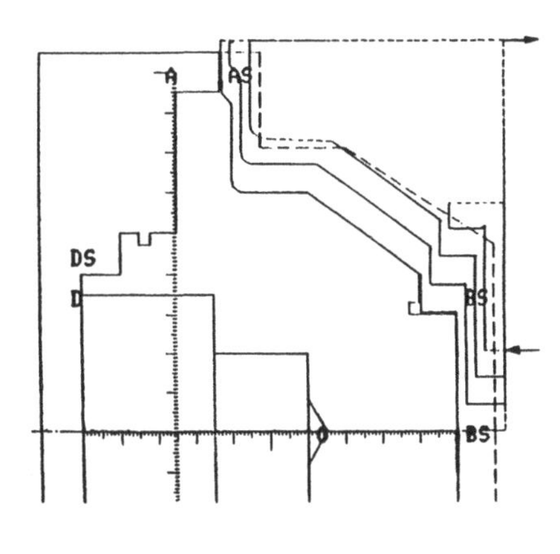

Fig. 3.31. Contour parallel N/C-turning with CAPSY [4]

operations and fixturing requirements for each succeeding operation. For this activity the graphic features are particularly helpful. With all necessary information entered the modules for the main and sub-operations are called up.

An important planning output is the basic production plan (Fig. 3.30) which is generated with the help of general technological information stored in memory. Other manufacturing documents can be drawn up with information from the basic production plan. With the help of external storage of text the CAPSY system allows generation of any plan or form in a company specific language, without interference with the program. The user only has to exchange the text file. A preliminary calculation of the production costs is possible by means of the cost data stored in the data file system using data from the basic production plan.

With N/C-Programming the most important task is the determination of the tool path. The process of generating the plan and the dialog used for planning are the same as for machining operations for conventionally and numerically controlled machine tools. There are, however, deviating requirements which need other solutions. Different strategies are used, e.g. for the distribution of cuts; for conventional machining a shoulder-directed distribution of cuts is *suitable* and a shoulder-oriented distribution of cuts is to be preferred for the N/C-machine tool.

For more complex workpiece geometries the calculation of the tool path or the determination of the contour-parallel tool path (N/C copy turning), (Fig. 3.31), is very time-consuming with manual programming of N/C-machine tools. In these cases the computer can calculate very quickly the cutter data for straight, *axis* parallel, longitudinal, or transversal machining operations.

3.7 Planning of Assembly Sequences

If one considers the entire manufacturing process of a product then it becomes apparent that assembly planning is an essential part of preparing a manufacturing plan.

The planning functions necessary for drawing up an assembly plan are different from parts manufacturing, since the two operations are fundamentally different. In parts manufacturing a workpiece is transformed from the unmachined into the finished state by a stepwise change of its shape by means of suitable machining processes. The assembly, however, can be visualized as a process in which individual components such as parts, flexible material, and sub-assemblies are added to the finished product by means of assembly processes and assembly fixtures.

The planning process for an assembly operation can be subdivided into the following subtasks:

1. Determination of the assembly parts and sub-assemblies
2. Determination of the assembly operation sequences
3. Allocation of the assembly stations
4. Allocation of the assembly fixtures
5. Calculation of the allowable assembly
6. Calculation of costs

For the solution of assembly planning tasks different decision paths may be taken. They may be formal, logic or creative, (Fig. 3.32).

The itemized list which shows the parts and subassemblies of a product can frequently be determined from an itemized file which was specifically structured for assembly (itemized list of modules). It is used in conjunction with the general assembly drawing.

There are different methods of performing assembly planning. They are related to the required planning effort. The methods are analogous to those used to draw-up the operation plan for machining.

1. Repetition planning
2. Variant planning
3. Adaptive planning
4. Generative planning

Assembly planning tasks		Planning mainly based on		
		Formal calculations	Logical determinations	Creative decisions
Assembly system	Assembly structure			●
	Work place layout			●
Assembly sequence	Product structure		●	
	Assembly operation planning		●	
Assembly equipment	Machine tools, tools		●	
	Spec. equipment			●
Assembly times	Times study	●		
	Calculation of allowed times	●		
Assembly requirements	Assembly equipment	●		
	Labour requirement planning	●		
Assembly costs	Determination of cost of assembly equipment	●		
	Determination of labour cost	●		

Fig. 3.32. Decision modes for assembly planning

For the automation of assembly planning a program was developed at the Institute for Production Systems and Automation of the University of Stuttgart. With it assembly plans of predominantly manual assembly operations for small and medium size production runs can be drawn-up with the aid of the computer [35, 36]. The system is called REMP; it is based on an exact description of the assembly tasks, assembly functions and the main parameters which are necessary to perform the assembly. These tasks together with stored production data of the company are the input to assembly planning.

Figure 3.33 shows the sequence in which the computer aids the generation of the assembly plan. Starting with the design and manufacturing data the assembly sequence is determined in the first planning step. The assembly sequence states the kind and order of the individual functions required for the assembly task. In the next planning step every assembly function is described exactly. Here, factors such as size and weight of the parts to be assembled, the precision of the joining process, the dimensions of the work stations and the arrangement of the tools are considered.

To aid the planner's work a coding sheet was developed with which he describes the functions of the assembly sequence with as little data as possible. Corresponding to the actual assembly functions, this form was divided into a coding sheet for "Handling" and "Fastening". The coding sheets are filled – in by the planner by hand and the data are entered into the computer and processed

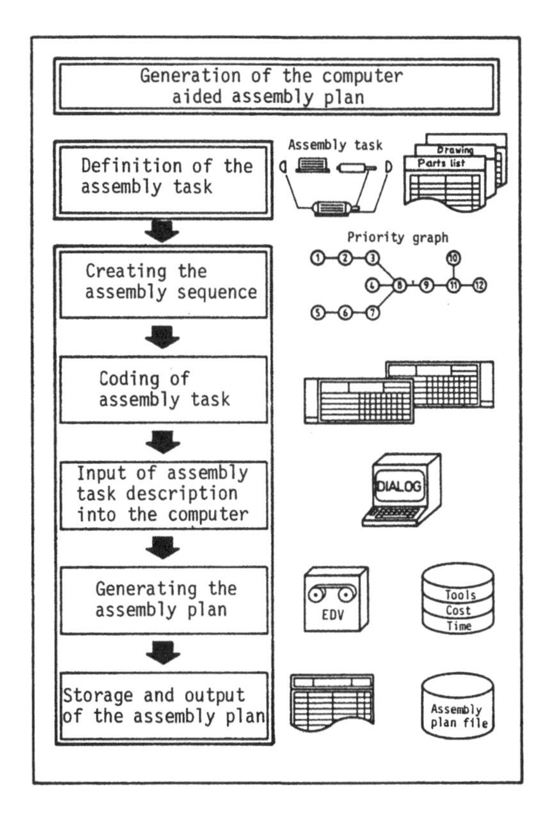

Fig. 3.33. Structure of REMP-system

in a dialog mode. With the aid of the entered code numbers the assembly functions and the assembly sequence are processed sequentially. For the fastening and clamping operations the computer selects from the tool data file the possible fastening tools for the described assembly. For every selected tool the assembly time is determined by the Methods Time Measurement (MTM) method and the assembly costs are calculated.

Planning of the assembly fixtures and special purpose tools for an assembly sequence cannot be performed by the computer. The creative activities necessary for this function have to be performed by the planner as before.

As a result of the computer aided planning activity a complete and reproducible plan is drawn up.

At the Institute for Production Systematics at the University RWTH Aachen another programming system for the automatic generation of assembly plans was developed. Here planning is done on the basis of similarity features with the help of planning documents [37, 38]. A description of the planning task for an assembly object is shown in Fig. 3.34.

The planner defines the shape and dimensions of the assembly object and of the individual parts or sub-assemblies. Additionally, order data, such as number of pieces, time of delivery, and organizational data are given. The individual functions for drawing-up the assembly operation plan are realized in this system by standard company independent programs. They contain all the control func-

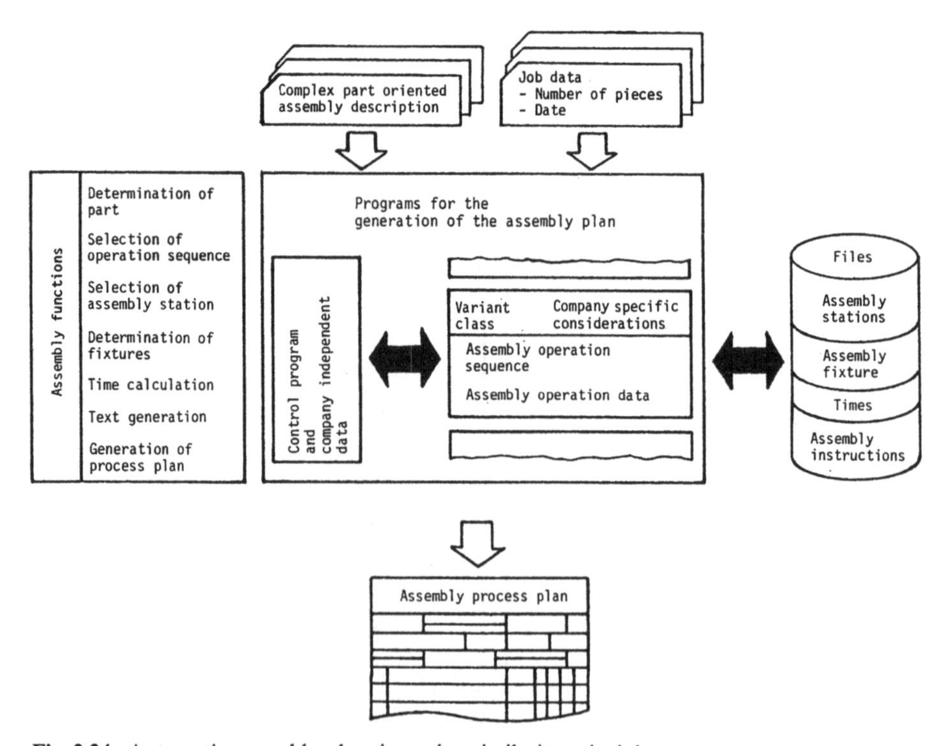

Fig. 3.34. Automatic assembly planning using similarity principle

tions and the process planning logic which can be formulated for general use. For every class of variant, standard company specific programs are given, including planning information for determining the sequence of the assembly operations and the operation data. The programming system uses data files which are partially independent of the variant classes.

The planning operation is subdivided into several functions. First the parts to be assembled are inspected and classified. Thus the variant classes and the planning tasks are defined. Assembly objects with design or manufacturing similarities are grouped as variant classes (Fig. 3.34). Other grouping criteria may be based on assembly tasks or assembly stations. The next step is to determine the operation sequence of the assembly with the help of a structured tool like a network plan. It contains all possible assembly operations for a class of variants, including application criteria. After having determined the sequence of the assembly, the operation data for each task are determined.

3.8 N/C Technology

The economic application of N/C machine tools depends strongly on the technical and organizational capabilities which the programming system offers [39]. For this reason a production system must be considered in close connection with the factory information system (Fig. 3.35).

Information is generated for all operations of a machine tool which contribute to the direct or indirect machining progress to convert a workpiece from a raw into a finished part. The output of planning system is a plan describing the sequence of operations for the manufacture of the product. It comprises all control, tool path, and auxiliary information necessary to operate the manufacturing system. Thus, it is a decisive factor for the performance of the manufacturing process.

The generation of the operating instructions for numerically controlled machine tools can be performed manually or automatically [40, 41]. In case of manual programming the preparation and calculation of the manufacturing program is done with conventional aids, such as tables, monograms and calculators. The data obtained from these aids and from experience as well as from empiric

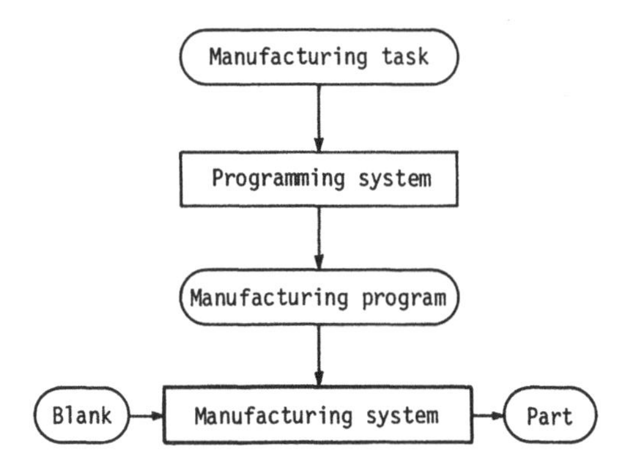

Fig. 3.35. Interlinking of programming system and production system

rules lead to subjective results. It is obvious to try to automate recurring operations and calculations with an electronic data processing (EDP) system (automatic programming). Hereby, for the description of the manufacturing task the part programmer uses a problem oriented language for the solution of his specific task. They are the so-called higher programming languages and can be used easily by an EDP system (Fig. 3.36).

The description of the manufacturing task, the so-called parts program, contains all the necessary geometric and technological information to manufacture a part. The EDP system determines the manufacturing operations with the help of a given algorithm, which is implemented as a program (= processor). The object code is stored either on a punched tape, an operation list, or a magnetic memory device.

If it is the principle task of a program to determine the technological data then one speaks of a technology-orientated programming system [42]. The task ends with the generation of a machine independent manufacturing program (CLDATA, DIN 66 215). A post-processor converts this program to machine tool specific code and generates a punched tape.

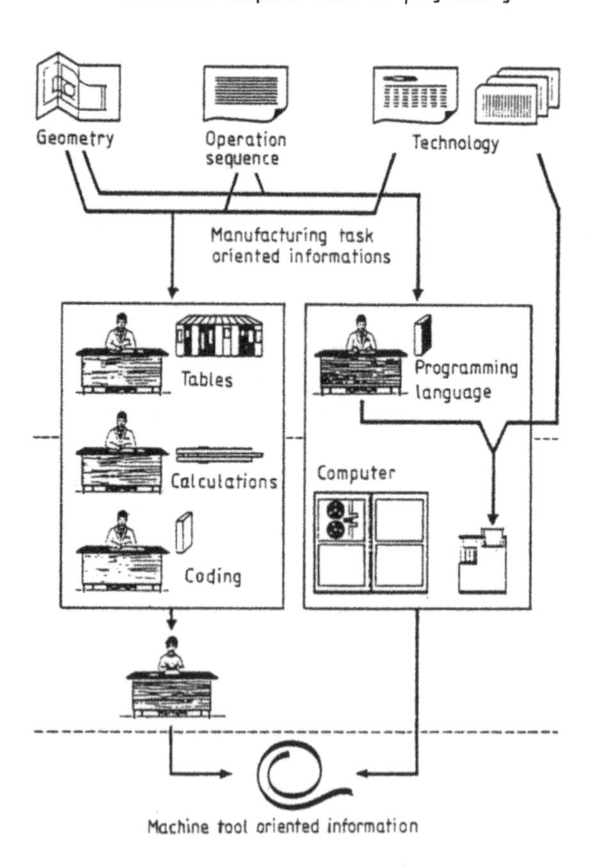

Fig. 3.36. Manual and automatic programming

Efforts to automate N/C-programming started very early when N/C machine tools became available. The first programming systems which were developed only determined the tool path for a workpiece which was difficult to program by the manual method.

Most programming systems are geometry orientated, which means the calculation of the tool path along a given workpiece contour or surface is performed by the computer. The determination of technological data, such as cutting parameters and cut patterns, have to be performed by the programmer. Automating the generation of technologic data is very difficult. For this reason the development of many programming systems was guided by the special requirements of the individual manufacturing process. The turning operations for lathes became a main point of interest, since this machining process occurs most frequently. In addition most N/C-lathes have been equipped with continuous path control. The knowledge gained from the dimensional geometric processing capabilities for the description of rotational workpieces led to the conception of control and programming systems for flame-cutting, nibbling, and wire eroding. To evaluate such

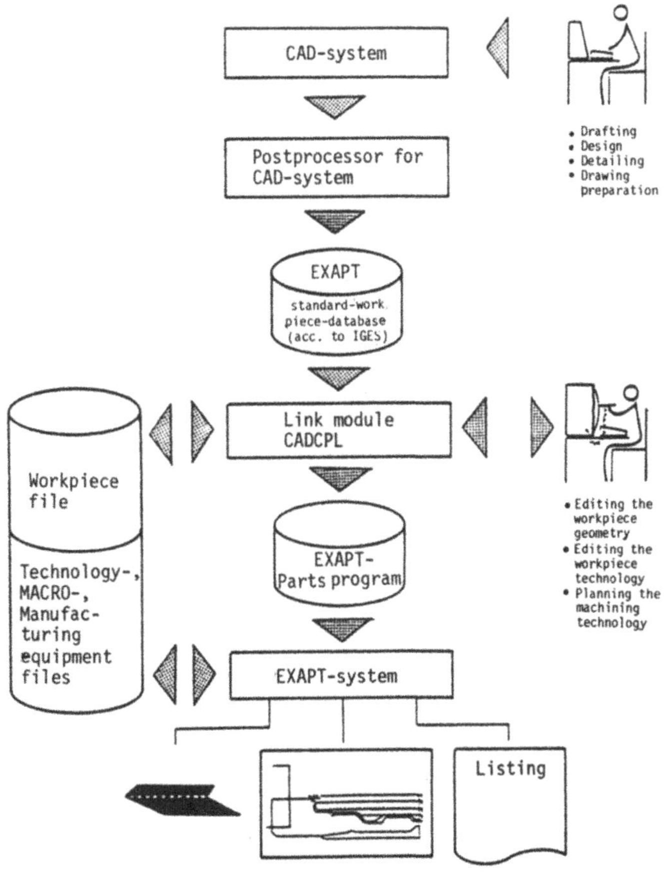

Fig. 3.37. CAD-EXAPT interface

a programming system its range of application and degree of automation is of importance.

In addition to the automatic programming systems the so-called workshop programming systems have gained an important position within N/C-technology. This was spurred by the introduction of CNC and the microprocessor. Because of the advanced development stage of the new control systems, it was possible to generate part programs with the help of the dialog method and to simulate the sequence of operations on a graphic display terminal [43]. The simulation can be done statically or dynamically, and geometric information can be enhanced by different color shadings. By integrating the programming system with the control system the N/C technology can easily be introduced without major organizational changes in smaller companies.

As the CAD and CAM technology was integrated into CAD/CAM systems during recent years it became possible to use the computer internal workpiece representations of a CAD system for N/C programming. As a result the input effort and the error possibilities were substantially reduced. In addition to the turnkey CAD/CAM systems supplied by the CAD equipment manufacturers existing CAD systems were also interfaced with automatic programming systems. Figure 3.37 [44] shows how an interface module enhances the EXAPT system with CAD capabilities.

3.9 N/C Programming on the Shop Floor Using Graphical Simulation Techniques

The increasing efficiency of modern CNC-controls makes it possible to apply the techniques and capabilities of automatic programming directly at the machine tool [45]. The division of the operation task into programming and controlling an integrated system permits programming of new parts during the operation of the machine [46]. The inclusion of graphic control outputs and the dynamic simulation of the machining operation on a display terminal constitute essential improvements in user comfort during program generation.

With a dynamic simulation system developed by the Institute for Production Systems and Design Technology of the Frauenhofer Gesellschaft (FhG) Berlin a N/C-program can be represented on the graphic display in real time [47, 48, 49]. Updating of a new image is done every time the tool has traversed on the screen a complete cutting path, which represents the actual cut [50]. The advantages of this process are:

1. Optimum verification of the N/C-program
2. Quicker implementation of a new N/C-program
3. Reduction of tool breakage and of damage to the machine tool
4. Improved safety at the machine tool [3]

Figure 3.38 shows the simulation run for turning. Various parameters can be inserted during the simulation phases [47].

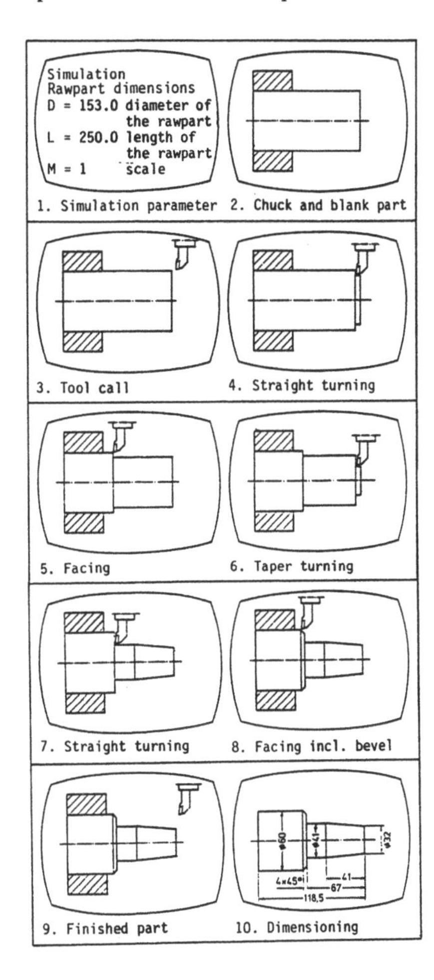

Fig. 3.38. Dynamic simulation of an N/C-program on a graphic display

Before a machining operation can be shown on the graphic display the contour of the blank of the workpiece has to be entered. In the case of a cylindrical blank the information is its diameter and length. After entering the scaling factor the blank is shown with the clamping chuck on the display. The clamping of the workpiece may be performed by various means, e.g. internal and external clamping. The available clamping tools are stored in the simulation system. In addition a cutting data file is maintained. When a call is made for a cutting tool a picture of it appears on the display.

After the start of the simulation program the individual N/C instructions are interpreted and the incremented steps necessary for the tool to produce the workpiece contour are determined [47]. When the tool simulates the metal removal process the workpiece contour is modified accordingly. At critical machining points the user can change the speed of the simulation run with an override switch. With the real-time representation of different cutting speeds the user can observe the machining operation and can perform corrections to the N/C-program, if necessary.

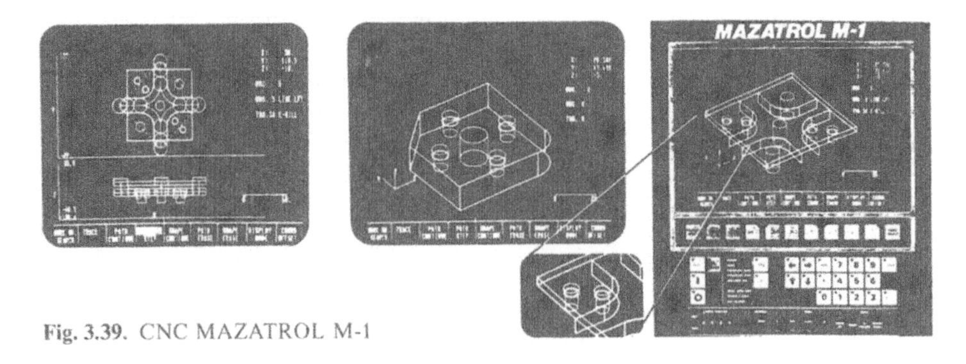

Fig. 3.39. CNC MAZATROL M-1

In addition to geometric data processing the simulation system can also interlink with the technological supervisory program to access or alter technologic data. If the value of important parameters of the machine tool or the workpiece are exceeded the simulation will be interrupted and corrective action can be taken.

Another approach to preparing programs and testing them was demonstrated with the MAZATROL M-1 system of Yamazaki Machinery Works, LTD., Japan.

Programming of a machining centre is done with the aid of a dialog system and a colour display. The geometry of the workpiece to be manufactured and the selected tool motion path can be shown on the display in various colors and from different view points. In addition to general functions, such as windowing and scaling, a number of technological functions are offered. For instance, cutting parameters can be determined automatically for different materials and cutting tools. Figure 3.39 shows some features of the MAZATROL M-1 system.

3.10 Programming of Robots Using Graphical Techniques

Programmable material handling devices, such as industrial robots, open up new modes of automation for small and medium size batch production [51, 52]. An economic utilization of such handling devices requires extensive planning work. In most cases planning is based mainly on the individual experience of the planner, and therefore frequently leads to unsatisfactory results. Important tasks in planning the application of industrial robots are the determination of the number of required handling devices, the layout of the machine tool installation, the decision as to the position of the handling device, the determination of the optimum work paths, and the provision of control data. The application of more than one handling device results in the additional task of coordinating the number and sequence of the partial tasks to be performed by the individual robots. The aim when planning handling tasks for a machine tool installation with robots is to use as few motion axes as possible and to achieve short work paths [53].

The determination of optimum work paths depends on following parameters:

1. The geometric facts
2. The kinematics of the handling device
3. The motion speed of the individual axes

4. The operating sequence
5. The gripper design

The calculation of the optimum work path includes the task of providing the control data for the programmable handling device. Various programming tools are available for programming of handling devices. Figure 3.40 shows how programming of robots can be classified by the different methods presently used [54].

With the manual method the working task is programmed by mechanically interconnecting the control circuit of the robot via a plug board of the handling device. With the play-back method the robot is programmed by guiding its toolholder or gripper through its work path. The actual position values obtained hereby are recorded in defined time intervals and then stored. They produce the user program. Programming by teach-in is done by commanding the robot through its trajectory with the help of a teach pendum which contains switches and push-buttons to move the robot. Off-line programming is done with robot independent peripheral computing devices. Once the program is completed it is loaded into the control computer of the robot. These methods offer the following advantages [54]:

1. Reduction of the idle-time of the robot including its peripherals
2. Comfortable program generation by applying higher programming languages
3. Possible programming via a CAD-system
4. Simple modification possibilities by using editing programs and
5. Simple documentation methods

The existence of uniform interfaces and defined basic functions are the preconditions for the application of universal off-line programming methods. In Germany there is an ongoing standardization effort for programming of robots [55]. Most textual programming methods include on-line and off-line programming. With this hybrid method the frame and motion-specific data are generated off-line and

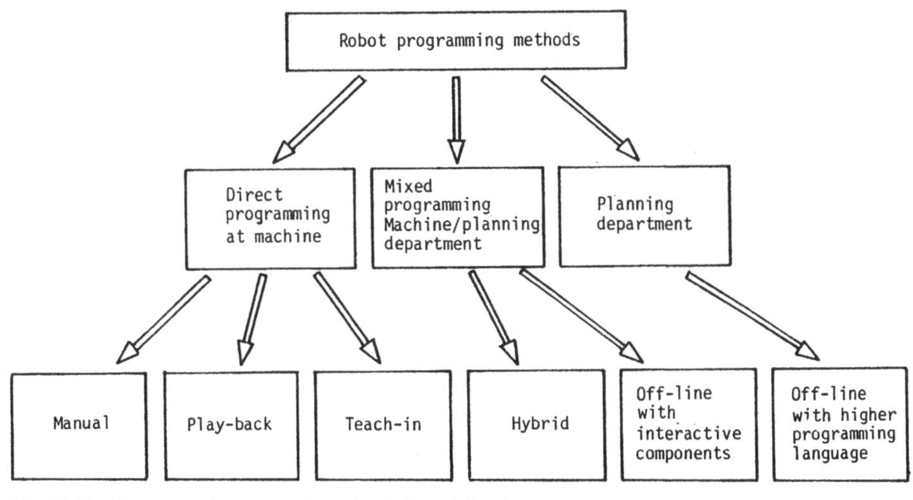

Fig. 3.40. Programming procedure for industrial robots

the position and orientation of the gripper are inserted on-line by means of the teach-in method. With the interactive off-line method external program frames and internal motion data are generated. The program frames and motion data are then interlinked by an interpreter.

With the off-line method two main groups of programming languages are to be distinguished [56]:

1. Explicit or motion-oriented programming languages
2. Implicit world model or task-oriented programming languages.

With the explicit languages every motion of the robot for a task has to be described individually. The exact definition of the positions and orientations in three-dimensional space require the programmer to have an extraordinary imaginative capacity to view objects in this space, which cannot always be presupposed. The same applies to collision avoidance when objects are in the work space.

The advantage of the implicit, world model oriented languages over the explicit languages lies in the fact that the description of complicated motions and logical relations is omitted and programming is application oriented. This is one of the main reasons why these kind of languages are well suited for programming of interlinked manufacturing facilities. In particular with assembly processes it is necessary to define program sequence instructions, error detection and correction routines and monitoring functions by a problem oriented language.

Implicit programming languages should have the following features:

1. Description of the motion trajectory
2. Description of the operation sequence
3. Description of the workpiece and its environment
4. Monitoring and supervision.

Figure 3.41 gives an overview on well known robot programming languages.

The increasing use of off-line programming systems for the application of industrial robots requires that checking and testing of the generated control information must be possible. The graphic simulation of the robot motion is a tool to generate fail-safe control programs by off-line programming. With graphic simulation the motion of the robot is displayed and monitored on a screen. In addition the necessary stations of the manufacturing cell are represented. The graphic system can not only by used to check the effects of the program but can also generate control data for the design and selection of robots as well as for planning of the manufacturing cell. Examples are given in Figs. 3.42 to 3.44.

Advanced developments in the field of N/C-programming for machine tools lead to the integration of the programming language in the control computer (CNC). Thus the term "Workshop Programming" was introduced. It is now the trend to make similar software packages available for on-line as well as off-line programming of robots.

Due to the complexity of the motions of the robot their controls are equipped with program storage devices and some of them even with on-line programming languages.

The potential reduction of the program development time for the application of industrial robots in small-batch production as well as in CAM-systems gave the impetus for the development of off-line programming languages.

Name	Denotation	Developer	Remarks
MAL	Multipurose Ass. Lang.	Uni Milano	(BASIC) structure
VAL	Variable Assembly Lang.	Unimation	On-line
SIGLA		Olivetti	Integrated in the control system (on-line)
SRI-Sprache		SRI	(FORTRAN) structure
ROCOL	RObot-COntroL	Uni-Leningrad	(ALGOL) structure
CINCINNATTI		Cincinnatti	On-line
EMILY			
Robot Control		IPK	On-line
ROBOTLAN	ROBOT LANguage	KAWASAKI	
ALFA	A Language For Automation	General telephone electronics	
ML	Manipular Lang.	IBM	
URI	Language for IR URI-mark II	Uni Rhode Island USA	
TL	Toyota Language	Toyota	On-line
AL	Assembly Language	Uni-Stanford	
RAPT	Robot APT	Uni-Edinburgh	APT
ROBEX	ROBot EXapt	WZL-Aachen	EXAPT
AUTOPASS	AUTOmated Parts ASSembly system	IBM	Subset of PLI
LAMA	Language for Automatic Mechanical Assembly	MIT	Similare to "AUTOPASS"
SAIL	Stanford Artifical Intelligence Language	Stanford	
PLANNER		MIT	
UNI-BUDAPEST	Robot language	Uni-Budapest	On-line
HELP	High Level Robot Language	DEA	Off-line

Fig. 3.41. Robot programming language

Fig. 3.42. Graphical simulation of a loading task [7]

The most plausible way of off-line programming of robots is to put the programming systems into a computer, located in the process planning department. However, this solution makes any further standardization of the programming language difficult because the output of this computer has to be entered into the existing plant control system.

Programming languages which are not dependant on a particular type of control permit standardization. The present development effort with programming languages shows that a great number of systems are already available and, consequently, standardization will become very difficult in the future.

In connection with CAM systems standardization may be possible by using off-line programming languages based on APT or EXAPT [52].

3.11 Integrated Aspects of Technological Planning

For computer aided process planning the link with computer aided design is very important. Thus the workpiece description can be obtained directly from the drawing [59]. It is the central task of the design system to provide workpiece models which can also be used by process planning. For this purpose information such as part geometry, surface finish, tolerances, and production data have to be contained in the workpiece model.

In addition the computer aided process planning system should also be linked to N/C-programming and manufacturing control. The aim of the integration is to reduce the amount of data and the number of files to describe the workpiece. Figure 3.45 shows a summary of the various tasks which can be performed by an integrated CAD/CAM system [60].

Various approaches can be taken to interlink the process plan with the N/C programming system. In the first method all available systems are interfaced. This may not be easy when different information formats and protocols are used in the modules to be combined.

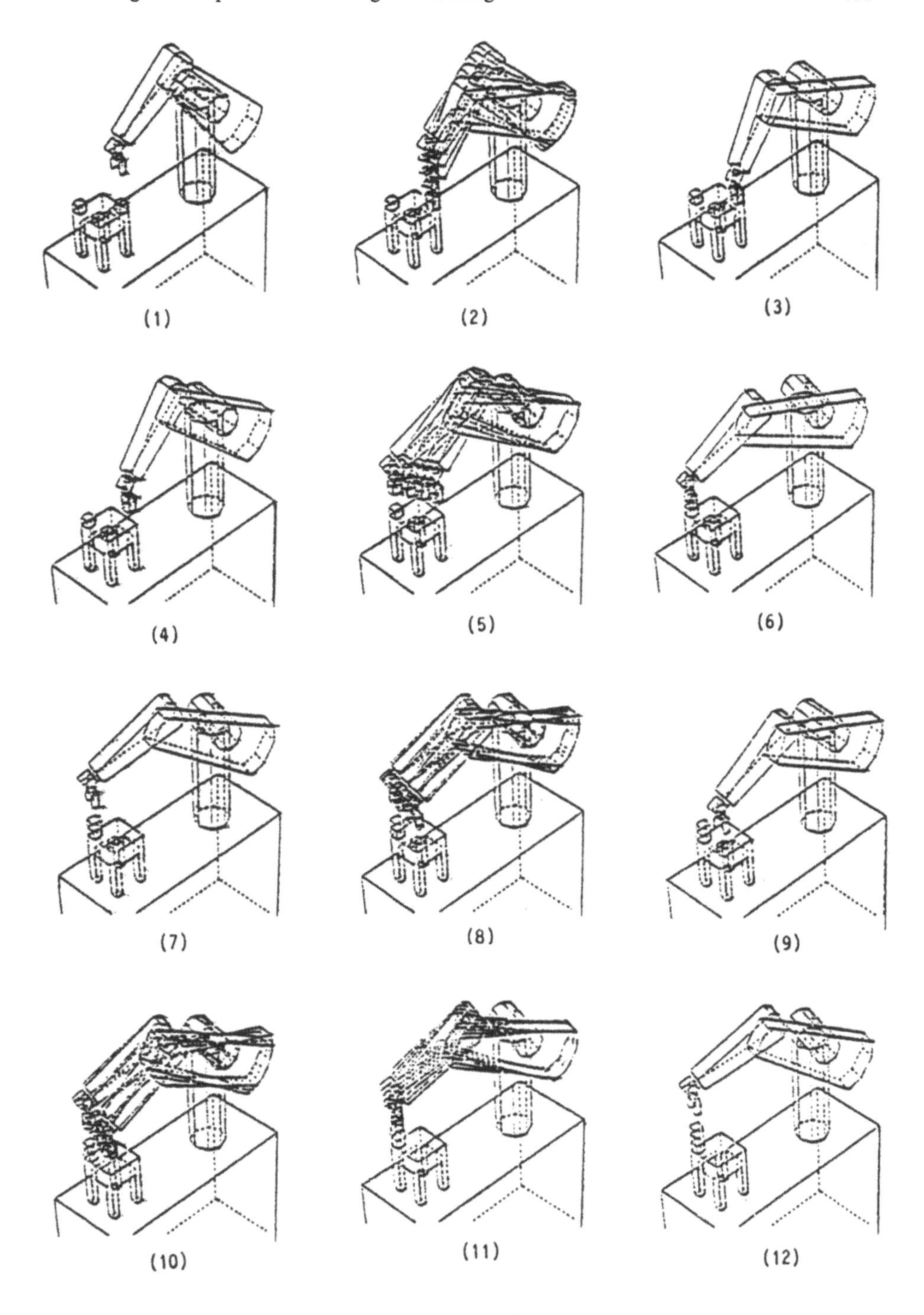

Fig. 3.43. Graphical simulation of a handling task [8]

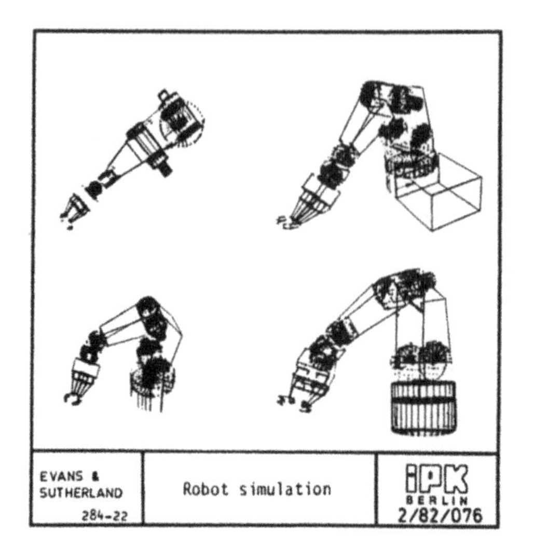

Fig. 3.44. Robot simulation. (Source: Evans & Sutherland)

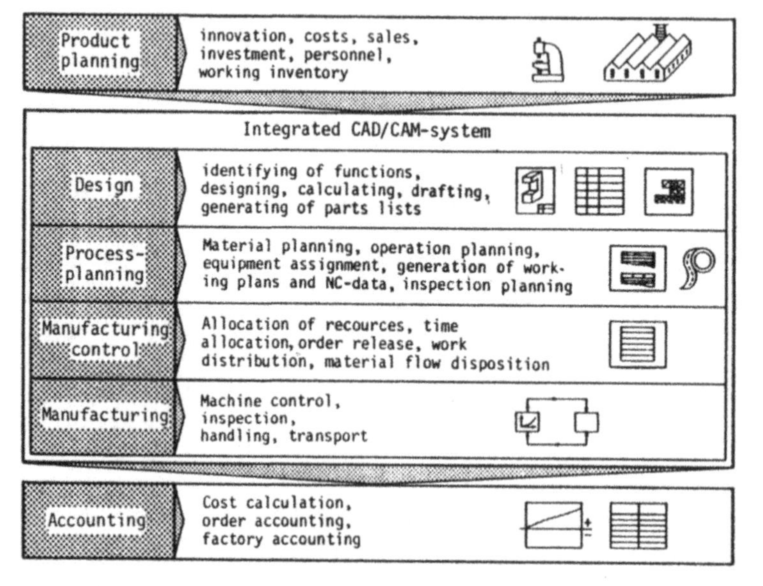

Fig. 3.45. Tasks of an integrated CAD/CAM system

With the second method the transfer of the computer internal representation of the workpiece to the N/C-programming system is determined by the manufacturing process or operation. However, to interconnect the part program requires the availability of a sophisticated processing module. In this case no adaptation is necessary to the N/C-system on which the part is machined. Actually the link with the computer internal representation of the object requires only little programming effort. However, difficult modifications have to be performed within the N/C-system, since a standard interface is not used for the data input.

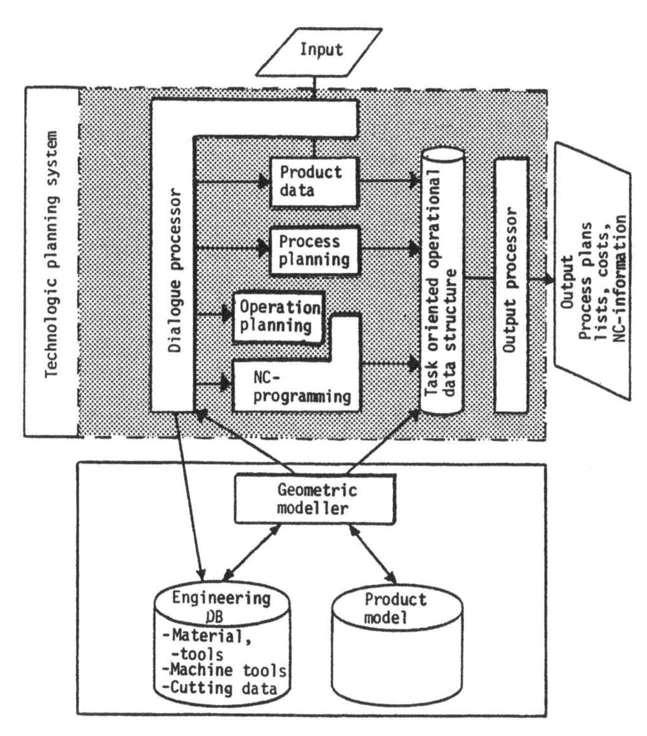

Fig. 3.46. Structure of integrated technological planning system within APS

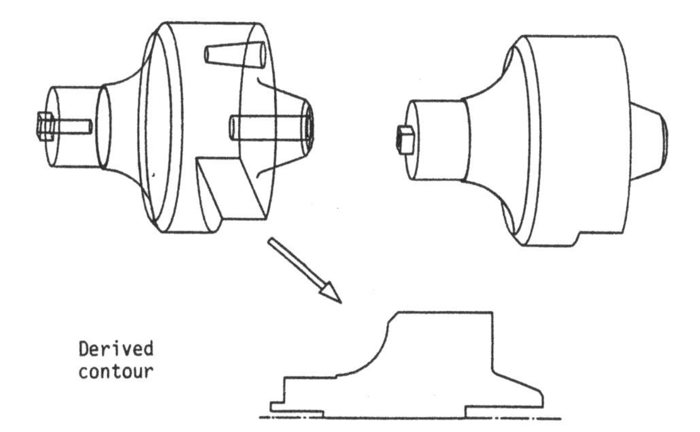

Fig. 3.47. Generation of the task oriented geometry of a part from a 3 D-model

Another method is to utilize common data files for technological data, such as tools, materials and times. This means that all functions of the individual program participating in the integrated technology planning system have to have access to the same data file. This leads to a reduction of memory size and maintenance effort. A concept on which this idea is based is used by the technologic planning system APS (Advanced Production System) (Fig. 3.46) [61].

The entire system is built from basic modules which are interconnected and which have access to a common engineering data base. The integration of the APS-system is realized via a common product model generated by the APS-geometric modeller. This offers the possibility of generating the product model during the design phase and integrating it with the entire system by special functions provided. Figure 3.47 shows an example of the 3D-representation of a part as a wire frame model. The hidden lines are removed and the turned part contour is outlined. The information was derived from the computer internal model representing the part. The model is the basis for planning the turning operation [62].

In order to interface process planning with production control the following aspects are important [63]:

1. Information for the blank material and for the individual parts can be taken from the operation plan or directly from the part list.
2. For time scheduling and capacity planning information on the sequence of operations, allocation to the individual work stations, and manufacturing times for the individual operations are of importance. The timing and capacity requirement data can be derived from the operations plan.
3. Data on auxiliary production equipment, their functions and on subprocesses are part of the operations plan. They can be provided on time, within the frame of the workshop control.

The planning operation may have to be modified by feedback information obtained from production control.

3.12 References

1. Fricke, F (1974) Beitrag zur Automatisierung der Arbeitsplanung unter besonderer Berücksichtigung der Fertigung von Drehwerkstücken. Dissertation, TU Berlin
2. Spur, G (1979) Arbeitsvorbereitung. In: Enzyklopädie Naturwissenschaft und Technik. Verlag Moderne Industrie, München
3. Eversheim, W (1980) Organisation in der Produktionstechnik. Bd 3: Arbeitsvorbereitung. VDI-Verlag, Düsseldorf
4. N.N. (1978) REFA. Methodenlehre des Arbeitsstudiums, Teil 1–6. Hanser, München
5. N.N. (1964) APT-PART PROGRAMMING MANUAL. IIT Research Institute, Chicago
6. Shaw, R (1979) N/C-Guide. VDI-Verlag, Düsseldorf
7. N.N. (19791) APT-Reference Manual, Control Data. Corporation Sunnyrale, California
8. Budde, W, Adamczyk, P (1980) Application techniques of the new EXAPT-Modul SYSTEMS. ZwF 75, H. 6, pp 280–284
9. Arndt, W (1980) Eine Lernmethode für automatisierte Arbeitssysteme. Reihe Produktionstechnik Berlin, Band 7. Hanser, München
10. Spur, G, Stuckmann, G (1976) Rechnerunterstützte Fertigungsplanung – dargestellt am Beispiel der Mehrschnittdrehbearbeitung. ZwF 71, H. 5, pp 179–182.
11. Stuckmann, G (1978) Bildschirmunterstützte Arbeitsplanung für programmgesteuerte Drehmaschinen. Diss. TU Berlin
12. Krause, F-L (1980) Systeme der CAD-Technologie für Konstruktion und Arbeitsplanung. Reihe Produktionstechnik Berlin, Band 14. Hanser, München Wien
13. Strunz, H (1977) Entscheidungstabellentechnik. Grundlagen und Anwendungsmöglichkeiten bei der Gestaltung rechnergestützter Informationssysteme. Reihe Betriebsinformatik 2. München

14. Thurner, R (1972) Entscheidungstabellen. Aufbau, Anwendung, Programmierung. VDI-Taschenbuch T 33. VDI-Verlag, Düsseldorf
15. Tannenberg, F (1970) Automatische Ermittlung des Arbeitsablaufs bei der maschinellen Programmierung numerisch gesteuerter Drehmaschinen. Diss. TU Berlin
16. Pistorius, E (1983) Graphische Simulation von Drehprozessen. CAMP '83 – Proceedings, Berlin
17. Grottke, W (1982) Graphisch unterstützte Arbeitsplanerstellung. In: Rationelle Arbeitsplanung. BW 33-51-02. VDI-Bildungswerk, Düsseldorf
18. Kaebelmann, E-F, Krause, F-L, Müller, G (1987) Aufbau, Funktion und Anwendung graphischer-interaktiver Sichtgeräte. ZwF 73
19. Streckfuß, G (1980) Vorteile graphischer Systeme für die maschinelle NC-Programmierung. tz für Metallbearbeitung, 74 Jg., H. 10, pp 39–46
20. Eversheim, W, Fuchs, H, Zons, K-H (1980) Anwendung des Systems AUTAP zur Arbeitsplanerstellung. Ind Anz 102, H. 55, pp 29–33
21. Wessel, H-J, Steudel, M (1980) Gegenüberstellung von Systemen zur automatischen Arbeitsplanerstellung. VDI-Z 122, H. 8, pp 302–310
22. Tuffentsammer, K, Ruoff, F, Wolf, M, Lueg, H (1977) Vorgabezeitermittlung und Arbeitsplanerstellung im Dialog mit dem Rechner. tz für Metallbearbeitung 71. H. 7, pp 49–52
23. Ruoff, F (1980) Arbeitsplanerstellung und Vorgabezeit am Bildschirm für konventionelle spanende Fertigung. Ind Anz 102, H. 34, pp 23–24
24. Lueg, H (1975) Systematische Fertigungsplanung. Vogel, Würzburg
25. CAPEX (1982) Firmenschrift der EXAPT-NC-Systemtechnik GmbH, Aachen
26. Link, CH (1977) CAM-I, Automated Process Planning System (CAPP). Technical Paper, Dearborn, Michigan
27. Automated Process Planning System (CAPP) (1976) Systems Manual. Version 2, Release 1. Arlington, Texas
28. Eversheim, W, Loersch, U, Esch, H (1977) Arbeitsplanerstellung im Dialog mit dem System DISAP. Ind Anz 103 H. 97, pp 29–32
29. Tönshoff, H-K, Ehrlich, H, Meyer, K-D, Prack, K-W (1979) Arbeitsplanung im Dialog mit dem Rechner für die Anforderungen mittlerer Unternehmen. Ind Anz 101, H. 73, pp 40–42
30. Babic, H-G, Bläsing, J-P, Lang, H (1976) PREPLA – Ein Baustein zur rechnerunterstützten Prüfplanung. wt-Z ind Fertig 66, H. 3, pp 155–158
31. Spur, G, Hein, E (1981) Ergebnisse zur rechnerunterstützten Prüfplanung. Endbericht P 6.4/28; B-PRi/2. KfK-BMFT
32. Spur, G, Stuckmann, G (1976) Automatisierte Arbeitsplanung für die Einzelteil- und Kleinserienfertigung. Proc. CIRP Seminars on Manufacturing Systems 5 (4)
33. Spur, G, Anger, H-M, Kunzendorf, W, Stuckmann, G (1978) CAPSY – A dialogue system for Computer Manufacturing Process Planning. 10. MTDR-Konferenz, Manchester (England) Sept
34. Pistorius, E (1983) Grafische Simulation von Drehprozessen. Tagungsunterlagen CAMP '83, Berlin, März
35. Turowski, W (1982) Einführung in das System CAPSY; Vortrag RKW-Seminar, rechnerunterstützte Konstruktion und Arbeitsplanung für kleinere und mittlere Industriebetriebe, 29./30. September
36. Hirschbach, D, Hoheisel, W (1978) Rationalisierung der Arbeitsplanung durch Einsatz der EDV, Teil 2 AV 16, H. 3, pp 71–74
37. Hirschbach, D (1978) Rechnerunterstützte Montageplanerstellung. Reihe Forschung und Praxis, Krauskopf, Mainz
38. Steudel, M (1978) Automatische Arbeitsplanerstellung. Ind Anz 100, H. 84, pp 42–43.
39. Eversheim, W, Steudel, M (1977) Automatische Montagearbeitsplanerstellung für Unternehmen der Einzel- und Kleinserienfertigung mit Hilfe der EDV. Forschungsbericht des Landes Nordrhein-Westfalen, Nr. 2693. Westdeutscher Verlag, Opladen
40. Spur, G (1972) Optimierung des Fertigungssystems Werkzeugmaschine. Hanser, München
41. Kief, B (1974) NC-Handbuch. NC-Handbuch-Verlag, Michelstadt
42. Shaw, R (1972) NC-Guide. VDI-Verlag, Düsseldorf
43. Schultz, R (1982) Rechnerunterstützte NC-Programmierung. VDI-Bildungswerk, Berlin

44. Spur, G, Potthast, A (1981) NC-Programmkontrolle mit dynamischer Simulation bei der Drehbearbeitung. ZwF 78, H. 4, pp 153–155
45. CADCPL: CAD-EXAPT-INTERFACE. Firmenschrift EXAPT, Aachen
46. Meier, H (1981) Werkstattprogrammierung mit CNC-Steuerungen am Beispiel der Drehbearbeitung. Produktionstechnik Berlin. Hanser München Wien
47. von Zeppelin, W (1981) Parallelprogrammierung und graphische Simulation erschließen weitere Möglichkeiten der Werkstattprogrammierung. ZwF 76, H. 8, pp 355.
48. Spur, G, Potthast, A (1981) NC-Programmkontrolle mit dynamischer Simulation bei der Drehbearbeitung. ZwF 76, H. 4, pp 153–155
49. Spur, G, Potthast, A (1981) Grafisches Simulationssystem für die NC-Drehbearbeitung. ZwF 76, H. 8, pp 387–390
50. Spur, G, Potthast, A (1982) Dynamic Simulation System for NC-Turning Programs. Annals of CIRP, vol 31/1
51. Rahmacher, K, Heßelmann, J (1983) Erfahrungen bei der Entwicklung und Anwendung eines grafischen Prozeß-Simulationssystems. ZwF 78, H. 6, pp 276–279
52. Spur, G, Tannenberg, E, Dicke, KG, Weisser, W (1976) Industrieroboter in der spanenden Fertigung. ZwF 71, H. 1, pp 4–7
53. Spur, G, Duelen, G, Adam, W (1980) Entwicklungsstand von Industrierobotern. ZwF 75, H. 11, pp 522–526
54. Schmidt-Streier, U (1982) Methode zur rechnerunterstützten Einsatzplanung von programmierbaren Handhabungsgeräten. Springer, Berlin Heidelberg New York
55. Prager, KP (1983) Kopplung externer und interner Programmiersysteme für Industrieroboter. Reihe Produktionstechnik Berlin (in Vorbereitung) Hanser, München Wien
56. VDI 2864, Entwurf (1983) Programmieren numerisch gesteuerter Handhabungssysteme – Adressierung von Koordinaten und Funktionen. VDI-Verlag, Düsseldorf
57. Gini, G, Gini, M, Gini, R, Guise, D (1979) Introducing Software Systems in Industrial Robots, Proc. 9th International Symp., Industrial Robots, Washington, DC. 13–15 March 1979, pp 309–321
58. Simon, RL (1983) The Marriage between CAD/CAM Systems and Robotics. Tagungsunterlagen CAMP '83 Berlin, März 1983
59. Sata, T, Kimura, F, Amano, A (1981) Robot Simulation System as a Task Programming Tool. Proc. 11th International Symp. Industrial Robots, Tokio 7–9 Oct 1981, pp 595–602
60. Spur, G (1982) Technischer Informationsfluß beim Einsatz von EDV. Tagungsunterlagen FTK '82. Stuttgart, 7–8 October 1982, pp 45–55
61. Spur, G (1980) Neue Forschungsschwerpunkte für die Automatisierung der Produktionstechnik. SFB 57, Kolloquium 1980, ZwF-Sonderdruck, pp 2–8, Hanser, München
62. N.N. (1982) APS-Advanced Production System, Progress Report 2. Kernforschungszentrum Karlsruhe
63. Spur, G, Germer, H-J (1982) Three dimensional solid modelling capabilities of the COMPAC system and some applications. ACM-'CAE 82', Computer Engineering, Milano 15–17 February 1982
64. Spur, G, Hirn, W, Seliger, G, Viehweger, B (1980) Simulation zur Auslegungsplanung und Optimierung von Produktionssystemen. ZwF 77, H. 9, pp 446–452

4 Evolutionary Trends in Generative Process Planning

R. Srinivasan and C. R. Liu

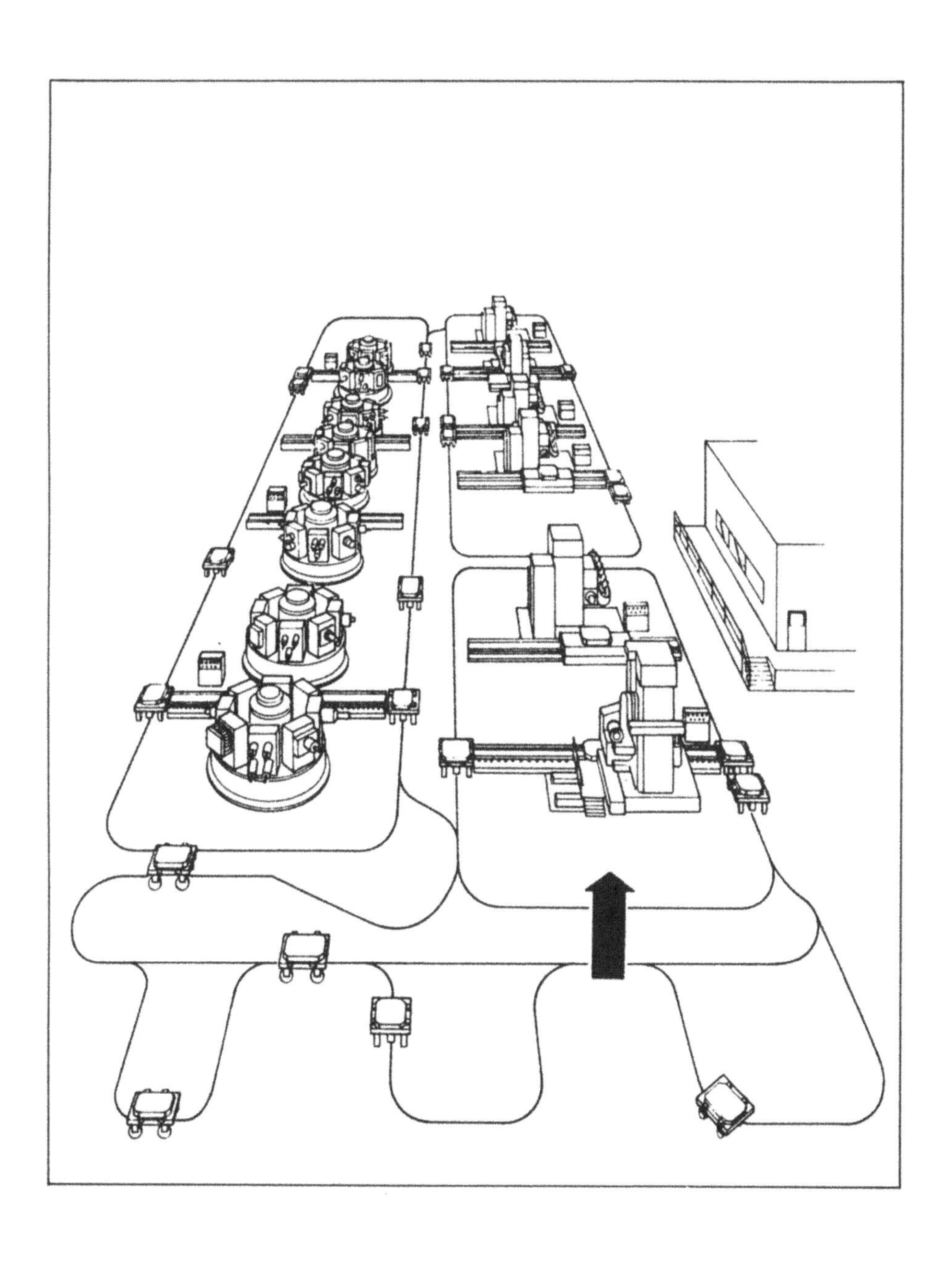

A flexible manufacturing system

(Courtesy of Kearney & Trecker Corporation, Milwaukee, Wisconsin, USA)

4.1 Introduction

Since 1952 when MIT first demonstrated a servopositioning system driven by hole designs punched on paper tapes, numerous developments have occurred resulting in the revolutionized machine shop scene of today; currently almost every other metal forming as well as metal cutting process is numerically controllable. Soon after the beginning of the automation of the machine shop functions followed the automation of the design functions. The workpiece drawing and design operations are increasingly being automated with the aid of cathode-ray tubes. Yet, complete production automation has been elusive until recently when efforts were directed towards the integration of the basic production planning functions, viz.

1. Material and equipment capacity planning
2. Production drawing and design
3. Manufacturing

The force behind the drive towards total integration of the said production planning functions stems from the realization that these functions are not entirely independent of each other. This necessitated the development of efficient interfaces among the production subsystems (Fig. 4.1).

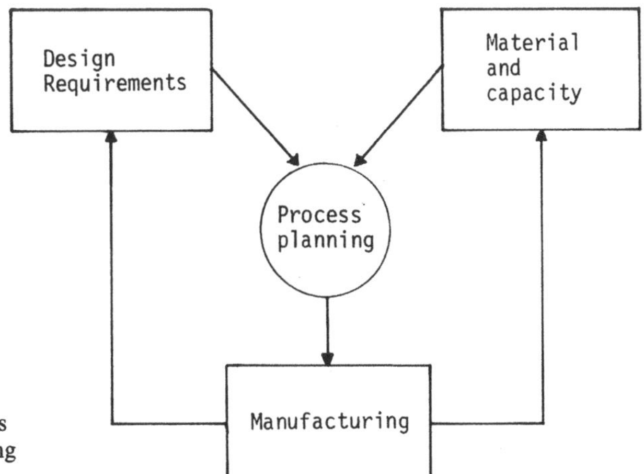

Fig. 4.1. The place of process planning in the manufacturing environment

Ideally, the process planning function should serve to translate the design requirements into manufacturing process details in the context of the current machine shop load situation [1]. The main objective of this paper is to trace the evolution of a class of process planning systems designed to meet the above requirement, with the intention of calibrating a representative few of them against the ideal requirement as posed above. We will also briefly present a novel system representation which, we feel, will help overcome the deficiencies of the system representations employed in the systems to be discussed later in this paper.

4.2 The Principal CAPP Methodologies

The necessity for the automation of process planning resulted from the need for systematically producing accurate and consistent process plans. Two principal methodologies have evolved over the past two decades, namely the variant and generative methods. They are briefly defined as follows [2]:

"A variant system is one based upon the retrieval and extension of a standard manufacturing plan, with the identification of such plan resulting from an established decision rule. A standard plan in this case is a permanently established ordered sequence of fabrication steps for a specific category of mono-detail parts".

"In contrast to the variant system, the generative system is able to construct an optimum fabrication sequence of its own accord through a series of more refined and sophisticated decision algorithms which operate on the basis of input obtained from a 3-dimensional CAD part model as well as a comprehensive technology data base".

The variant method is based on the group technology principle, whereby parts are classified into part families on the basis of their geometric and/or manufacturing process similarities. A standard plan is established for each family. The major disadvantages of this method are [1]:

1. The difficulty in accomodating various combinations of geometry, size, precision, material, quality and shop loading
2. Enormous on-line data base requirements to accommodate stored plans
3. The difficulty of maintaining consistency in editing practices

The interested reader is referred to [1] and [2] for more details.

4.3 Generative Process Planning

Logical decision making procedures are the basis for generative process planning systems. The origin of generative process planning (GPP) can be traced back to the late 1950's when the use of decision tables in manufacturing decision making was formalized. Since then, planning algorithms have progressed from simple decision procedures to today's sophisticated pattern recognition techniques. Two factors have predominantly influenced the growth and performance of GPP systems:

1. The degree of informational completeness of the system representation employed
2. The amount of human intervention (for interactive systems) required in the decision making process

The trend in the evolution of GPP systems is to have more and more of factor 1 and less of factor 2. Also, it is worth mentioning here that the growth in the GPP systems has more or less paralleled that of the machine shop automation from stand-alone NC machine tools to today's distributed control Flexible Manufacturing Systems (FMS). In the following sections, an attempt has been made to trace this growth of GPP systems, placing strong emphasis on the most recent developments. Before describing the developments in GPP systems, a brief mention is made of extended part programming systems which come into the picture at the output end of the CAPP systems in a computer integrated manufacturing system.

4.3.1 Extended Part Programming Systems

Since 1957 when APT (the first part programming language) was developed in the United States, several extensions to APT have been made in order to shift part of the geometrical computational load off the shoulders of the part programmer. A need to develop a standard part programming language and an associated computer processor, which was independent of particular machine tools or control systems or computers, arose in the mid-1960's [3]. In 1965 work began at NEL (National Engineering Laboratory, Scotland) on the NELNC processor (2 C, L) for two dimensional contour milling operations, which was later extended to operations on lathes, drilling machines, punching machines and burning machines. The major improvement made in NELNC over APT was that the former allowed the part programmer to describe the path to be taken by a cutter using English-like statements describing the part geometry taken from the drawing, leaving the processor to generate vast amount of tool and other information required. Also NELNC has limited area clearancing and tool selection capabilities.

Along similar lines, but with much advanced procedural or production technique orientation, the EXAPT series of systems were developed at the Institute for Machine Tools in Aachen, W. Germany and the COMPACT system at MDSI (Manufacturing Data Systems Inc., USA). The EXAPT series of systems is composed of:

1. EXAPT 1 for the preparation of control tapes for drilling and simple milling operations
2. EXAPT 2 for turning
3. EXAPT 3 for contour milling.

The system architecture consists of data files for tools and materials, modules for cutting parameters calculation and macro modules to determine operation sequences. The machining operation types and surface finish requirements are explicitly provided as input, i.e., the input is in procedural form (statements like DRILL, TAP, BORE, FACMIL etc.) [4]. Further, the EXAPT system has the capability to select tools automatically, determine machining parameters, cut divisions and cutter paths [5]. The philosophy of EXAPT 1 was later extended to develop a preprocessor called MAPROS (Macro Preprocessing System) [6] which translates the workpiece description (parametric) into EXAPT 1 commands. This served to integrate NC drawing and manufacture of mechanical workpieces.

The process of CAD/CAM integration started with extended part programming systems; further growth of this process is described in the following sections.

4.3.2 GPP Using Decision Tables and Tree Structures

Manufacturing decision tables were the starting point (late 1950's) for nonpart-family type of process planning systems. Decision tables are made up of conditions, data and actions which are the principal elements of all computer programs [7]. The decision tables are useful in reducing problems to their simplest form. The "If-Then" relationship of the decision rules is the most significant feature of decision table logic. The interested reader is referred to [7] for a more detailed description.

The decision tables can be converted to decision trees, which offer the following advantages: Trees are easier to update and maintain [8], the required branches alone can be extended to considerable depth. The most common types of tree structures used in manufacturing planning are the E-tree (exclusive path selection), N-tree (nonmutually exclusive path selection), and D-tree (decision tree). The DCCLASS system [7] uses the D-tree structure (Fig. 4.2). The planning is done interactively and the output is a list of ordered processes. Apparently, since the data base of the CAD system is not directly interfaced with the process planning system (DCLASS) in the generation of cutter location data (CL data), the DCLASS type of systems contribute very little towards CAD/CAM integration.

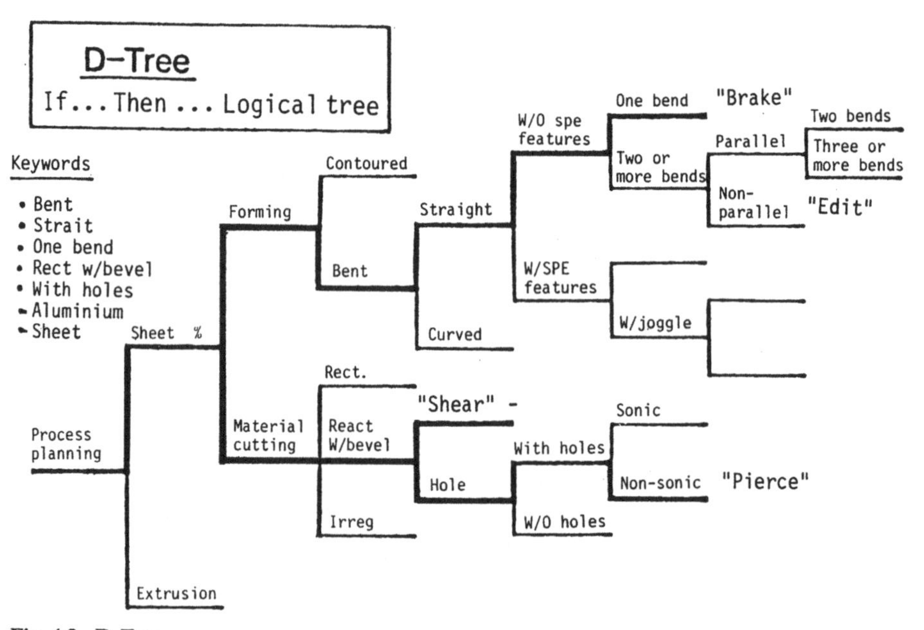

Fig. 4.2. D-Tree

4.3.3 Iterative Algorithms

For most industrial parts, the entire process planning cannot be done in just one stage [8], because various decisions have to be made starting from the design of the part up to the generation of CL data, and each of these decisions has some

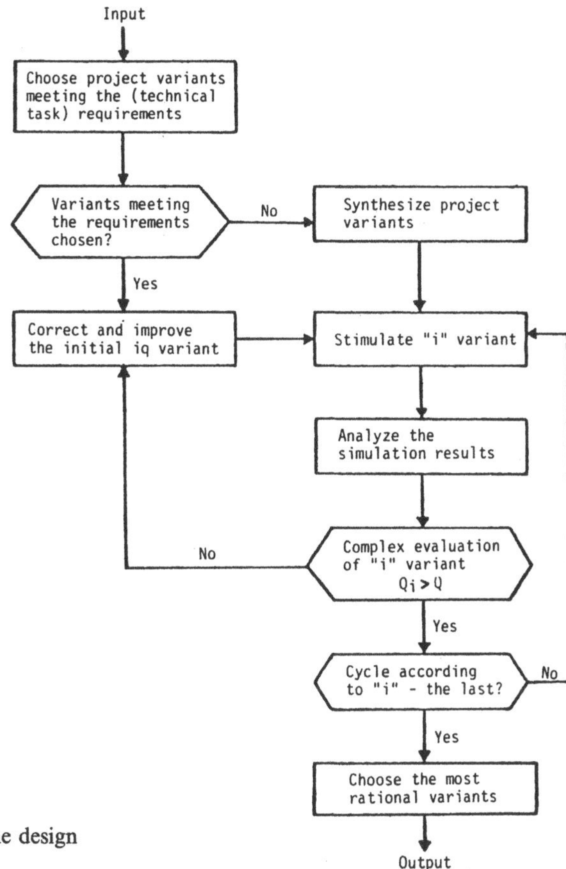

Fig. 4.3. Iterative algorithm of the design process at each level

interaction with the other earlier ones. Therefore, the manufacturing decision process is a multilevel recursive or iterative [8, 9 and 10] decision process as shown in Fig. 4.3. This becomes clear when one considers the example where there could be more than one process (route) to transform the workpiece from on state to another, and that each particular process (route) chosen may influence the subsequent process decisions differently. To give a specific example [10], consider the following: for the machining of a plane various processes are available, such as planing, milling, stretching and grinding. This multivalued transformation can be mathematically defined as follows:

$$\text{transformation } \phi: \quad c_{i-1} \xrightarrow{k} c_i \qquad k = 1, \ldots, n \text{ routes}$$

The algorithm shown in the figure includes the "units for searching and synthesis of the variants, the units for simulating the functioning process of each variant of the technology, the units of analysis and evaluation of the simulation results, the units for optimizing the initial variant for selection of the most rational variants" [10]. A recursive algorithm such as the one above has been applied to the process planning of simple holes [8].

4.3.3.1 Recursive Process Planning [8]

Each step in process planning is viewed as a transformation from one state to another state, represented by the following expression:

$$F'(\text{lwh}, \text{opr}) \rightarrow \text{mwh}$$

where lwh is the less machined surface, mwh is the more machined surface, opr is the operation or tool that produces the transformation.

It should be noted that there could be more than one operator opr that can produce the same transformation. Since process planning could be conceived of as the reverse of machining, the various steps in process planning can be derived by starting at the final machined state, and through inversions of the above function arriving at the initial material state (raw material). However, this is seldom possible because the inversion process is not unique. This is because the operator 'opr' is multivalued. This problem is tackled by the authors using the following procedure: narrow down the number of outcomes of the inversion process by restricting the operation choices, for example, by placing tool life constraints on the tool to be used, by considering other attributes of the lwh, etc.

This approach seems to have a high potential in process planning applications provided that suitable geometrical representations are developed to recognize and mathematically handle the 3-D geometry of the material to be removed. In fact it is the major breakthrough that is yet to be made in the integration of CAD and CAM!

4.3.4 The Concept of Unit-Machined Surfaces

The methodologies discussed so far have had only limited success in their objective of integrating CAD and CAM, mainly because the workpiece representation employed in these systems was not amendable for complete extraction of manufacturing details. The representations used were useful only for classification of components by geometry and size. The quest for more complete object representation resulted in the evolution of the concept of viewing the workpiece as a set of unit-machined surfaces [11].

4.3.4.1 COFORM

COFORM (Code for Machining) [11] is a coding system to describe machined surfaces. The information that can be coded by COFORM consists of information both on machined surface geometry and process relevant details. Thus COFORM can serve as a formatted input medium for computer-aided process selection and planning systems. Figure 4.4 shows the example of a coded workpiece.

The types of surface that can be coded using COFORM are holes, simple and complex, slots, and other surfaces. Since the major objective of such a coding system is to serve as an input medium for CAPP systems, the coding is done in great detail. The major criteria which COFORM has been designed to satisfy are:

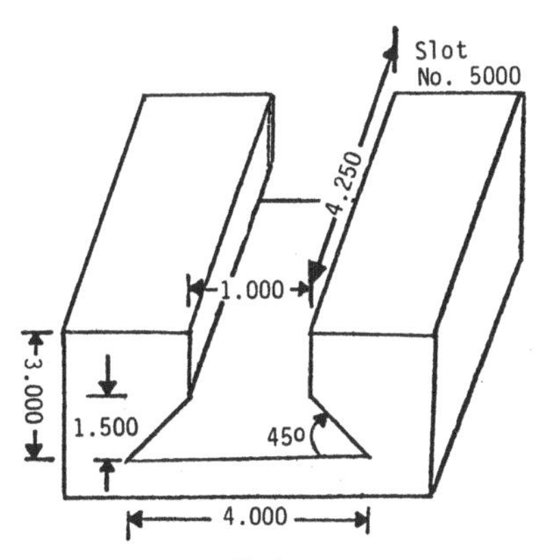

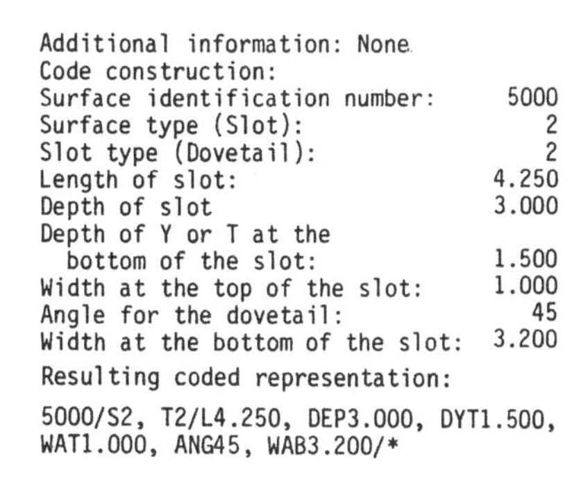
Additional information: None
Code construction:

Surface identification number:	5000
Surface type (Slot):	2
Slot type (Dovetail):	2
Length of slot:	4.250
Depth of slot	3.000
Depth of Y or T at the bottom of the slot:	1.500
Width at the top of the slot:	1.000
Angle for the dovetail:	45
Width at the bottom of the slot:	3.200

Resulting coded representation:

5000/S2, T2/L4.250, DEP3.000, DYT1.500, WAT1.000, ANG45, WAB3.200/*

Fig. 4.4. Dovetail slot

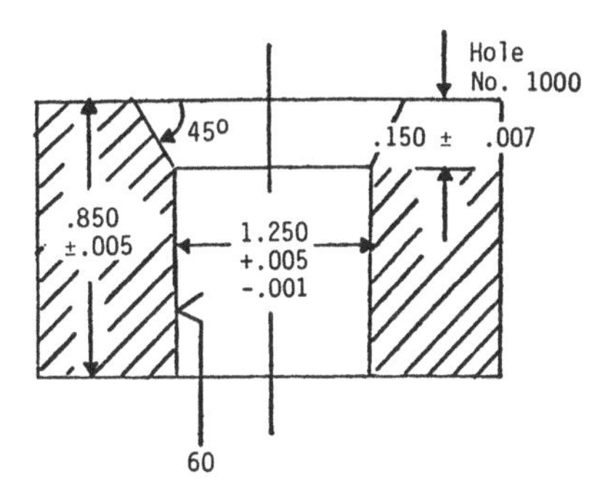

Additional information: None

Code construction:

Surface identification number:	1000
Surface type (Hole):	1
Changes in diameter (may be defaulted):	1
Dimension of diameter:	1.250
Tolerance on diameter - positive:	.005
Tolerance on diameter - negative:	.001
Side surface finish:	60
Diameter length:	.850
Tolerance on length (±X):	.005
Chamfer code:	1
Chamfer angle	45
Chamfer length:	.150
Chamfer tolerance:	.007

Resulting coded representation:

1000/S1, D1/DIA1.250, TDP.005, TDM.001, SSF60,
CA45, CL.150, CT.007/*
DL.850, TL.005, CC1

Fig. 4.5. Single diameter hole

1. The ability to describe completely the part spectrum in general part manufacturing
2. Completeness of description for process decisions
3. Simplicity of the code versus the economy in using the code
4. Reliability of the coding scheme
5. Naturalness of the coding scheme

The underlying philosophy of COFORM is that of viewing every workpiece as a composition of unit machined surfaces, where a unit machined surface is one for which a specific manufacturing process, as well as the process parameters, can be very easily determined. Each of the unit machined surfaces is fully described using geometric as well as process relevant descriptors. The information coded consists of the major type of surface (hole, slot or any other plane), the material type, major machinability data, initial condition of surface (machined, unmachined, etc.), the final condition of surface, tolerance requirements, many other geometric characteristics such as parallelism, symmetry, perpendicularity etc. and several other characteristics specific to a surface, such as reference surface, access (tool) directions, etc. As can be observed from the above listing of the major characteristics that are coded using COFORM, the structure of the coding system is a combination of monocode and polycode structures. A useful feature of the COFORM code structure is the ability to accommodate more than one geometry data block (one for each unit machined surface) in just one statement as in the case of a complex hole (Fig. 4.5). The processors that decode the codes are constructed on the basis of a modular format, i.e., each surface processor controls the decoding of the code pertinent to its specific surface type (hole processor, slot processor and plane processor).

The role of COFORM in computer-aided process selection and planning is discussed in the next section.

4.3.4.2 APPAS

APPAS (Automatic Process Selection and Planning Program) [12] is a generative process planning system developed at Purdue University for the planning of processes done on machining centers and numerically controlled drills. The spectrum of machined surfaces for which planning is done includes holes, simple and complex, slots and other planes, flat and convex. The input of APPAS is a complete geometric and process relevant description of the unit-machined surface (which can also be taken from the output of COFORM). The information on process capabilities of the available machines is stored in several modules. The process selection methodology is based on decision tree type of logic (Fig. 4.6). Based on the surface input information, which is in the codified form as obtained from COFORM, the appropriate surface is called. The set of processes required to preduce a unit-machined surface is chosen by matching the surface accuracy requirements against the accuracy of the available processes, and the specific sequence of the required processes is derived, based on predefined process precedence relationships, as shown in Fig. 4.7. The machining parameters such as speed, feed etc. are determined specific to the surface and tool materials. Also available is an option to determine the optimal machining parameters using modified Taylor's tool-life equations. The system has the additional capability to determine machine requirements, such as horsepower, spindle stroke, axes, and machine time.

A similar approach was used in CUBIC [13], developed at the Twente University of Technology, a software system for programming NC machining centers, with the capability to plan (2 C, L) milling, drilling, boring, reaming and tapping operations. The basic differences between APPAS and CUBIC are:

1. CUBIC uses the relatively more flexible Manufacturing Procedure files, while APPAS uses decision tree type of logic
2. The collision problem has been partially solved in CUBIC:
   (a) The collision due to tool clamping devices is avoided by introducing dummy shape elements to indicate the free space around the shape composition (collection of surfaces)

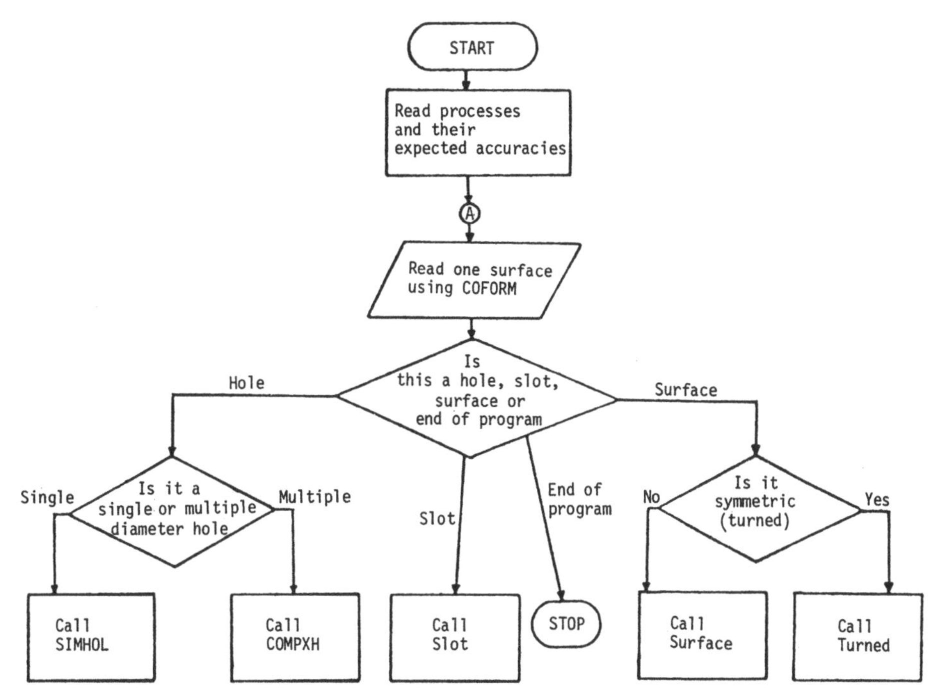

Fig. 4.6. Process planning macro-flow chart

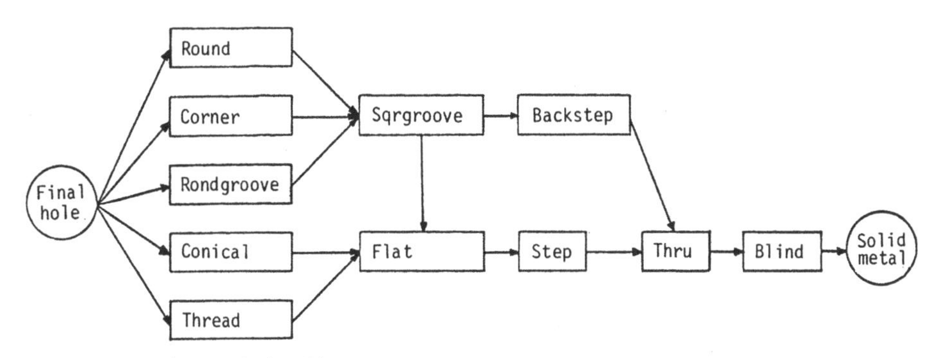

Fig. 4.7. Precedence relationships among EMS's

(b) Collision during positioning operations is avoided by specifying feasible work regions called "pattern groups"

4.3.4.3 AUTAP and AUTAP-NC [14]

The part spectrum planned for by AUTAP consists of rotational and sheet metal parts. The workpiece is described as a string of predefined technological form elements such as cylinder, hole, bevel etc. The process sequence is determined automatically using a set of company specific production rules, very much the same way it is done in CUBIC and APPAS. In addition to this, the chucking and clamping positions are also determined based on the workpiece contour. The manufacturing operational instructions are transformed to CL data by the AUTAP-NC postprocessor.

The type of representation used in AUTAP is sufficient for simple axisymmetric and sheet metal parts. The inadequacy of the object representation used in AUTAP and other systems is discussed in detail in a later section.

4.3.4.4 More Sophisticated GPP Systems

4.3.4.4.1 STOPP

STOPP (Sequential and Tool Oriented Process Planning) [15] is a high level generative process planning system developed at Purdue University for machining centers. The basic philosophy of STOPP is to design the process plans to utilize the available tools, and hence the name 'tool oriented process planning'. Unlike many other existing process planning systems, the process plans produced by STOPP are much more practical because they are produced on the spur of the moment. The concept of 'Elementary Machined Surfaces' (EMS) is employed in determining the machining requirements, but unlike APPAS, the exact shape of the EMS is recognized using pattern recognition techniques and the information is used in the process planning procedures.

The process planning procedure is broken down into three distinct stages.

1. Preprocessing the boundary files representing the workpiece to detect the presence of EMS's (holes, slots or pockets)
2. Determining the machine tool and tooling requirements to produce the EMS's (Simple Processing Cycles – SPC's)
3. Postprocessing the SPC's to optimize the SPC sequence and produce CL data

A brief discussion of each of the stages is presented below.

Preprocessing. In STOPP the workpiece is conceived of as consisting of a set of surfaces, and every surface as consisting of a sequence of EMS's. The surfaces are represented using boundary file representation (also called Constructive Solid Geometry representation) which are obtainable directly from CAD systems. Every surface is represented as a set of edge-lists. Features such as holes on the surfaces are represented as hole edge-lists, each of which is hooked on to the corresponding surface edge-list, where a hole edge-list is defined as the ordered list of edges representing the half cross-section of a hole. Whether the surface itself has to be machined or not is also indicated in the surface data structure. Thus, the workpiece (set of surfaces) geometry information is extracted from the CGS representation of a CAD system, and the geometry of holes is retrieved from each of the surface data structures. Similar procedures are used for slots and pockets.

Construction of SPC Records. Every hole is represented as a linked list of 2-D primitives (Fig. 4.8), where each primitive represents an EMS. The hole specification includes the surface finish, roundness error, positional accuracy, straightness, tolerances on length and diameter, etc. All the relevant specifications are stored in the attribute field of a hole record. From the EMS shape information

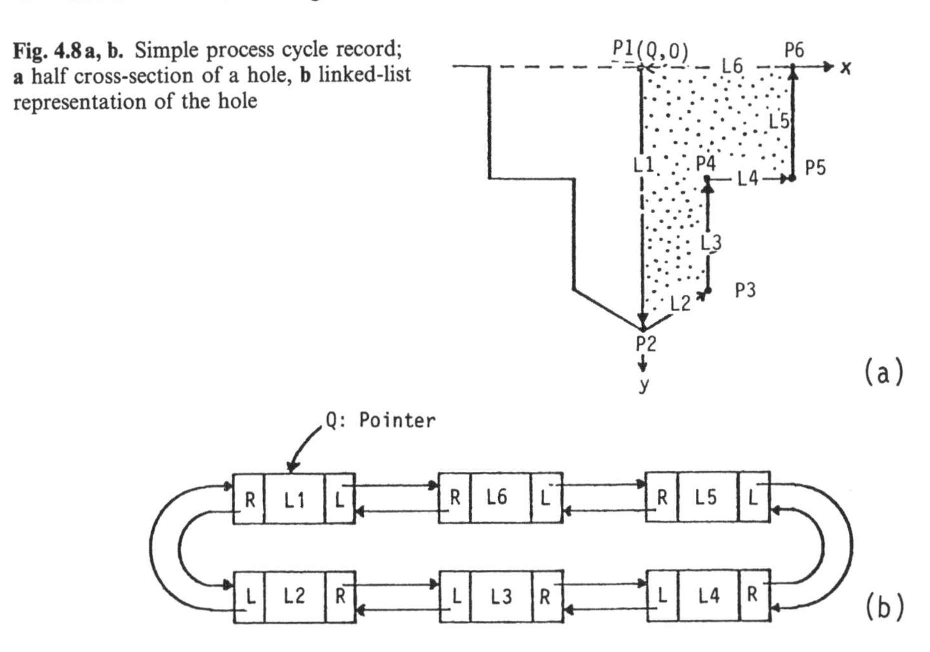

Fig. 4.8 a, b. Simple process cycle record; **a** half cross-section of a hole, **b** linked-list representation of the hole

and the specification listed above, the tool selection is made. The EMS specification and the selected tool number are thus stored in a SPC record. Similar procedures for construction of SPC records are used for slots and pockets also.

Postprocessing. First, the SPC's obtained in the previous stage are sequenced with the objective of minimizing sequence dependent times, such as tool changing time and positioning time. Some simple heuristic rules are used to solve this problem. Second the cutter path for each EMS is determined. The cutter guide points are determined from the hole location data, such as the position, orientation and other data relevant to the EMS. All the data are obtained after allowing for the tool-workpiece clearance. All these data are calculated using procedures defined specifically for each type of EMS. The output of this stage is a series of macro-commands such as RETURN and NEWTOOL for a tool-change, TOOLAXIS for tool axis indexing, GO-RAPID for rapid tool movement, and CUT-LINEAR and CUT-CIRCULAR for actual machining commands. The procedure also considers the number of axes (3, 4 or 5) of the machining center while producing the macro-commands.

Though STOPP has not been implemented on a commercial basis, the basic philosophy of STOPP has high potential for piecewise industrial application if appropriately extended and modified. In terms of the degree of integration of CAD and CAM that is achieved, STOPP ranks high among the various other systems discussed in this chapter.

4.3.4.4.2 CIMS/DEC; CIMS/PRO

CIMS/DEC [16] is a part description system. The 2-D cross-sectional plane of every shape element is represented by a chain of directed line segments (Fig. 4.9)

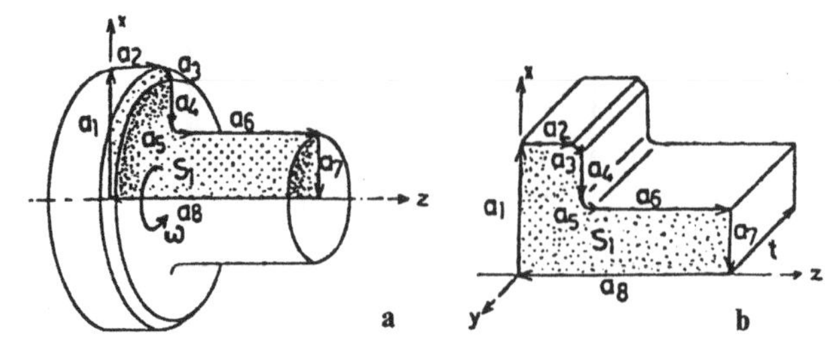

Fig. 4.9 a, b. Solids described by directed line segments $\{a_i\}$, directed surface segments S_1, revolving operator ω, and parallel moving operator t. **a** Rotational solid, **b** box-type cubic solid

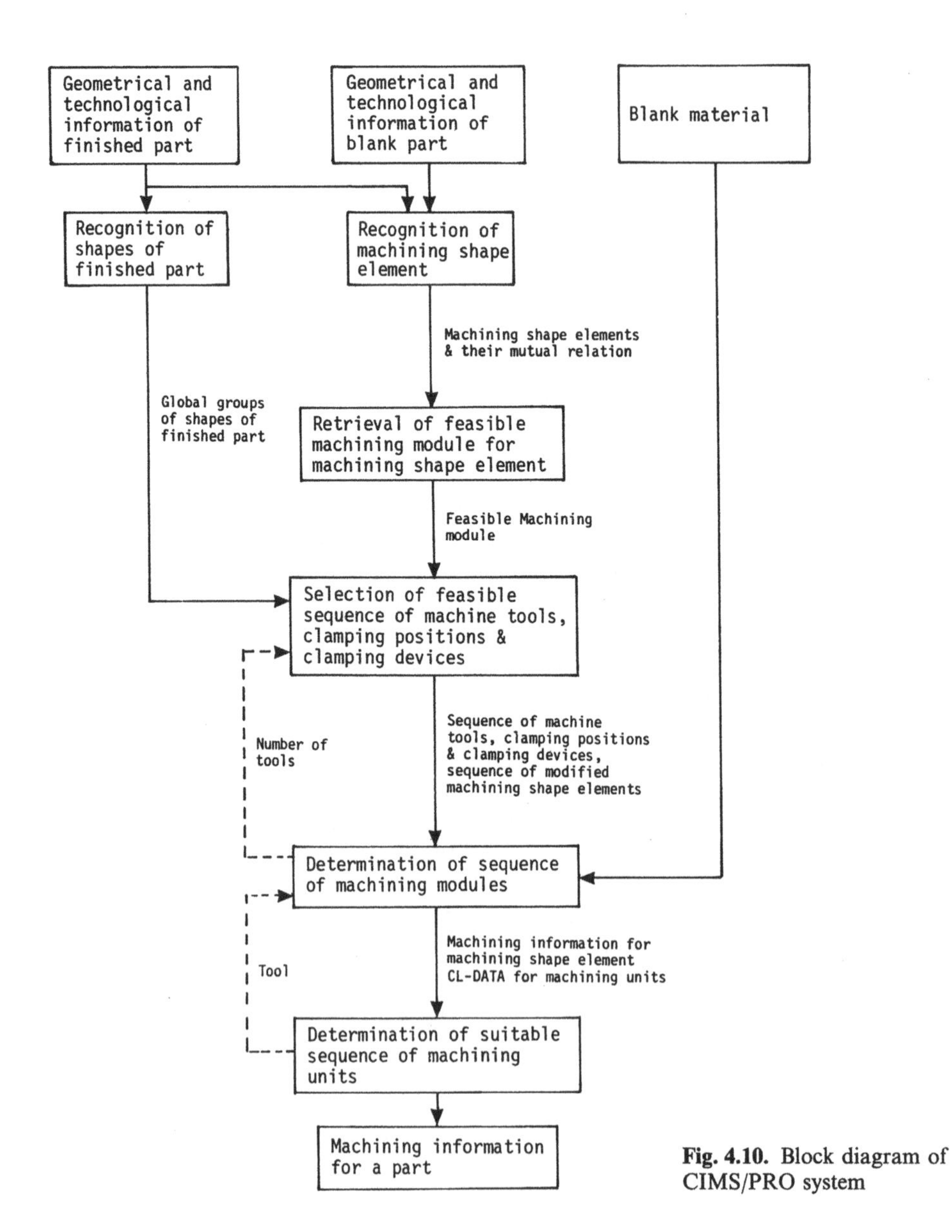

Fig. 4.10. Block diagram of CIMS/PRO system

and then is either rotated about an axis or moved parallel to an axis to form rotational prismatic components. Both geometric and technological information is stored for every plane. The successive changes to the shape through machining are derived mathematically using algebraic contraction rules. The machining sequence is determined using pattern recognition of curves, steps, flanges etc.

CIMS/PRO [17] is a computer-aided production planning system (Fig. 4.10) working with input from CIMS/DEC. Using the recognition of shapes as described above, the system is able to determine the clamping methods and CL data. The operation sequence is optimized to yield minimum tool traverse time using graph theory principles.

The CIMS methodology appears to be one of the most sophisticated process planning methodologies found in the literatur. But the representation used is still lacking in ability to perform the following:

1. Determining the machining sequence imposed by reference dimensioning of surfaces
2. Determining the machining sequence if the raw material used has already been shaped using Nearest Net Shape (NNS) methods
3. Handling of more complex shapes.

4.4 Adequacy of the Existing GPP's in the Wake of New Developments

In the previous sections detailed descriptions and analyses of a representative few of the existing generative process planning systems were presented. In this section we will see how adequate these systems would actually be if they were employed in CMS's in the near future.

4.4.1 Recent Trends in the Design of CMS Control Systems

Lewis [18] has presented an excellent analysis of the characteristics that a CMS Control System should possess. It has been concluded that the control algorithms used in data processing systems are inefficient if applied to CMS control systems, because they tend to ignore the utilization of the peripheral units; actually for a manufacturing system, the peripheral units, which are NC machine tools, have to be well utilized. Also, a centralized control system is unsuitable for a CMS (especially if it is big), because:

1. The control fails if the model is inaccurate, if the processor fails, or if the processor is too slow
2. Large control programs have to be written, making the control algorithm too complex.

These reasons have suggested the use of decentralized or distributed control systems, which would result in faster control responses and higher control reliability.

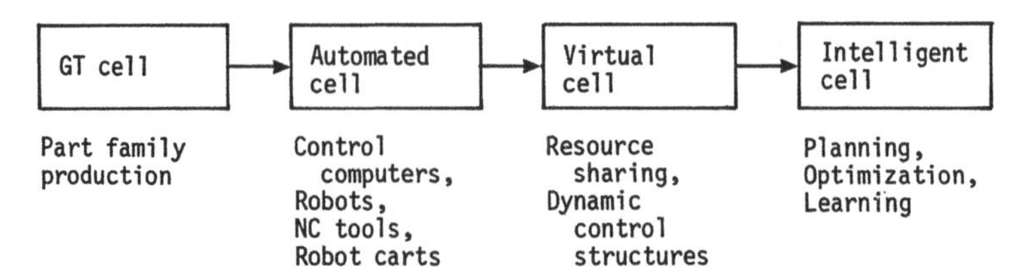

Fig. 4.11. The evolution of the manufacturing-cell

The National Bureau of Standards [19] has come up with similar concepts for CMS systems. The idea of hierarchical control (Fig. 4.11) has been proposed, whereby every control unit is responsible only to its immediate higher level control unit. The shop control system schedules cell activation, and allocates workstations and other resources to virtual or pool cells, which in turn controll the workstations. At every control level, problems are decomposed into smaller tasks and are assigned to the lower level control units using local intelligence.

The ultimate criterion for judging the design of such control architectures is the flexibility in analysis, reporting, routing, scheduling, dispatching and monitoring. These control concepts are designed to perform well under uncertainty, which is so characteristic of manufacturing systems. This design flexibility has to be taken advantage of during actual operation, for example, by postponing the decisions, such as process selection and routing, as much as possible. It would be futile to produce process plans much in advance, because of system uncertainties and because the system is designed to be flexible to adjust to contingencies. Chen [1] has suggested that the process plans be produced in real time (just before machining), if possible. He has suggested that process planning activities be integrated with the material requirements planning (MRP) and capacity planning activities (Fig. 4.12). The suggested integrated planning procedure is as follows:

1. From the forecast of finished units and direct orders for finished units, the master production schedule is set.
2. From the master production schedule the MRP system obtains the demands for the individual components, taking into account the assembly and subassembly requirements.

In this system process planning has been done using the variant principle. The coding system is called APPOCC (poly code, 18 digits) and the process planning module is called COBAPP. Several feasible process plans are produced. The various process plans are ranked according to their steady state throughput times. This is accomplished by modelling the machine shop as a queueing system and using computer analysis of networks of queus (CAN-Q) [20] to analyse the queueing system. The process plans are tested one by one in the ranked order until a process plan which satisfies the capacity requirements (machine tools and man power) is found. The new capacity demands generated by this selection are computed using the MRP module. Thus the two planning modules, namely, MRP and Process Planning, interact with each other to load the machine shop as evenly as possible, minimize the job throughput time, and improve the other production

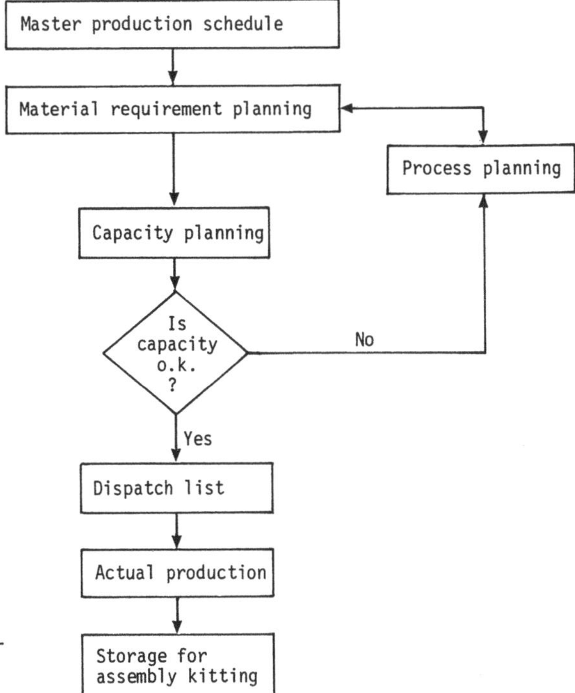

Fig. 4.12. Framework of integration of MRP, process planning and capacity planning

shop measures in general. Chen has also concluded that the system which employs an automated process planning system (multiple process plans) performs significantly better than the one that employs a manual process planning system (usually a single process plan) in terms of the reduction in average tardiness of jobs and the work-in-process inventory.

Other factors that could influence the production economy are the process parameters such as feed and speed and the tool changing scheme [21].

4.5 Dynamic GPP Using Pattern Recognition Techniques: A New Concept

Research is being conducted at Purdue University to explore the possibility of applying syntactic pattern recognition techniques to process planning. Experiments indicate that pattern recognition techniques can contribute in a major way to the solution of the problems of object representation and machine tool representation. They are suitable for the automatic extraction of manufacturing information, such as process sequence, process parameters, machine tool selection, tool bit selection, and chucking and clamping positions. In the following sections, a brief account of the proposed system representation, along with the dynamic process planning methodology that results from this representation, are presented. It is worth mentioning here that the concept used in every subsystem is influenced by that used in almost every other subsystem, and it is also true with process planning systems.

4.5.1 Proposed Representation Schemes

4.5.1.1 Object Representation

The proposed object representation is expected to cover a vast spectrum of both rotational and nonrotational industrial components. Every workpiece is conceived of as a sequence (as opposed to 'a set' as in [15]) of surfaces, and every surface is conceived of as a sequence of EMS's [15]. Thus, every workpiece can be represented as a tree of surfaces, each surface consisting of a series of one or more EMS's. Every surface is represented as one generated by either the translation or revolution about an axis of a two dimensional plane (which is a chain of edges), except where standard technological elements (threads, gears, etc.) are to be used (CIMS/DEC).

4.5.1.2 Machine Tool Representation

Here, the concept of the composite component [22] is used. Every machine tool is represented as a composite component, where a composite component is one which is a collection of all types of surfaces that can be produced by that machine tool. The information about each of the surfaces in the composite component includes the dimensional (nominal) as well as the variational data. Thus every machine tool can be represented by a particular (unique to its type) composite component. Every tool that is used on the machine can be represented as the generator of one or more types of surfaces belonging to the composite component. Thus, a machine capability file can be set up, with some of its records representing the tools that can be used on the machine, and the remaining representing some machine specific data.

4.5.2 Process Planning Steps (Briefly)

4.5.2.1 Flexible Planning Logic

The proposed GPP method is machine tool oriented, i.e., whenever a machine tool becomes available for machining, the workpiece machining requirements are matched against the capabilities of the machine, and the possible operations are selected, sequenced to produce economical machining and performed. This is repeated until all the required surfaces are produced. Since none of the operations are determined in advance, but are done only upon availability of a machine and continued in the same way until all surfaces are generated, this process planning methodology can be applied in Flexible Manufacturing Systems (FMS). The resulting flexibility in process planning will greatly help exploit the flexible planning (routing, etc.) potentials of the FMS.

4.5.2.2 Identification of Surface Precedences

The precedence relationships among the surfaces of a workpiece are identified using two rules: [7] for any two contiguous surfaces, the position of one with respect to the other determines the machining precedence relationship between the two surfaces, [8] for any two noncontiguous surfaces, the reference dimensioning, if any, determines the machining precedence between the two surfaces. Using the above two rules the entire surface precedence structure for the workpiece can be derived in the form of a tree with the surfaces as the nodes.

4.5.2.3 Selection of Machines, Tool Bits and Clamping Positions

As indicated earlier, the entire selection problem reduces to that of testing whether a subset or the whole set of the workpiece requirements (surface generation) is a subset of the capabilities of surface generation of the machine tool. This can be reduced to a set of problems of syntax analysis in formal languages. Readers interested in syntax analysis theory are directed to [23], an excellent text for this and other topics in pattern recognition.

4.5.2.3.1 Theorem 1

The problem of machining process selection can be decomposed into one or more syntax analysis problems.

The above theorem has been verified as holding good for axisymmetric parts.

4.6 References

1. Chen, JS (1981) Integration of Process Planning with MRP and Capacity Planning for Better Shop Production Planning and Control. Unpublished Ph.D Thesis, Purdue University
2. Barnes, RD (1976) Group Technology Concepts Relative to the CAM-1 Automated Process Planning (CAPP) System, presented to the Executive Seminar on Coding, Classification and Group Technology for Automated Planning, 21 Jan; St. Louis, Mo, pp 736
3. McWaters, JF, Welch, DS (1970) *Numerical Control User's Handbook* (Leslie, WHP, ed) McGraw-Hill Book Co., N.Y., pp 236–271
4. Reckziegel, D (1970) EXAPT 1. *Numerical Control User's Handbook* (Leslie, WHP, ed) McGraw-Hill Book Co., N.Y., pp 159–196
5. Stute, G, Eitel, H (1970) The Milling Technology in EXAPT 3. In: *Numerical Control Programming Languages,* pp 88–99
6. Zapomueel, H (1977) MAPREPOS Processing – An Economical Solution for the Integration of NC Drawing and Manufacture of Mechanical Workpieces. In: *Advances in Computer-Aided Manufacture,* pp 329–340
7. Allen, DK, Smith, PR (1980) Computer-Aided Process Planning. Computer Aided Manufacturing Laboratory, Brigham Young University, Provo, Utah, pp 5–25, Oct. 15
8. Barash, MM, Bartlett, E, Finfter, II, Lewis, WC (1980) Process Planning Automation – A Recursive Approach. In: *The Optimal Planning of Computerized Manufacturing Systems,* Report No. 17, Purdue University
9. Griess, V, Priesler, J, Vymer, J (1973) Principles of Computer-Aided NC Programming with a High Automation Level. In: *Computer Languages for Numerical Control,* pp 623–633
10. Tsvetkov, VD (1977) Multilevel Iterative Method of Complex Automation of Designing Technological Processing and Control Programs for Numerically Controlled Machine Tools. In: *Advances in Computer-Aided Manufacture,* pp 329–340
11. Rose, DW, Solberg, JJ, Barash, MM (1977) Unit-Machining Operations – A Code for Machining. In: *The Optimal Planning of Computerized Manufacturing Systems,* Report No. 5, Purdue University

12. Wysk, RA, Barash, MM, Moodie, CL (1977) Unit-Machining Operations, Part 2: Automatic Process Selection and Planning Program. In: *The Optimal Planning of Computerized Manufacturing Systems,* Report No. 6
13. Oudolf, WJ, Enchede, T (1976) Development of a Programming System for NC Machining Centers. In: *Annals of the CIRP,* 25.1, pp 429–433
14. Eversheim, W, Fuchs, H (1980) Integrated Generation of Drawing, Process Plans and NC-tapes. In: Advanced Manufacturing Technology, pp 303–314
15. Choi, BK (1982) CAD/CAM Compatible, Tool Oriented Process Planning for Machining Centres. Unpublished Ph.D Thesis, Purdue University
16. Kakino, Y, Ohba, F, Moriwaki, T, Iwata, K (1977) A New Method of Parts Description for Computer-Aided Production Planning. In: *Advances in Computer-Aided Manufacture,* pp 197–213
17. Iwata, K, Kakino, Y, Ohba, F, Sugimura, N (1980) Development of Non-Part Family Type Computer-Aided Production Planning System CIMS/PRO. In: *Advanced Manufacturing Technology,* pp 171–184
18. Lewis, WC (1981) Data Flow Architectures for Distributed Control of Computer Operated Manufacturing Systems: Structure and Simulated Applications. Ph.D Thesis, Purdue University
19. McLean, CR, Bloom, HM, Hopp, TH (1982) The Virtual Manufacturing Cell, Industrial Systems Division, Center for Manufacturing Engineering, National Bureau of Standards, Washington, D.C.
20. Solberg, JJ (1978) CAN-Q User's Guide (FORTRAN Version). In: *The Optimal Planning of Computerized Manufacturing Systems Report.* School of Industrial Engineering, Purdue University
21. Gupta, SM, Barash, MM (1977) Computer Aided Selection of Machining Cycles and Machines. In: *Optimal Planning of Computerized Manufacturing Systems,* Report No. 8, Purdue University
22. Gallagher, CC, Knight, WA (1973) Group Technology. Butterworths, London
23. Fu, KS (1982) *Syntactic Pattern Recognition and Applications.* Prentice Hall, Inc., New Jersey

5 Design Methodology of Computer Integrated Manufacturing and Control of Manufacturing Units

G. Doumeingts, M. C. Maisonneuve, V. Brand and C. Berard

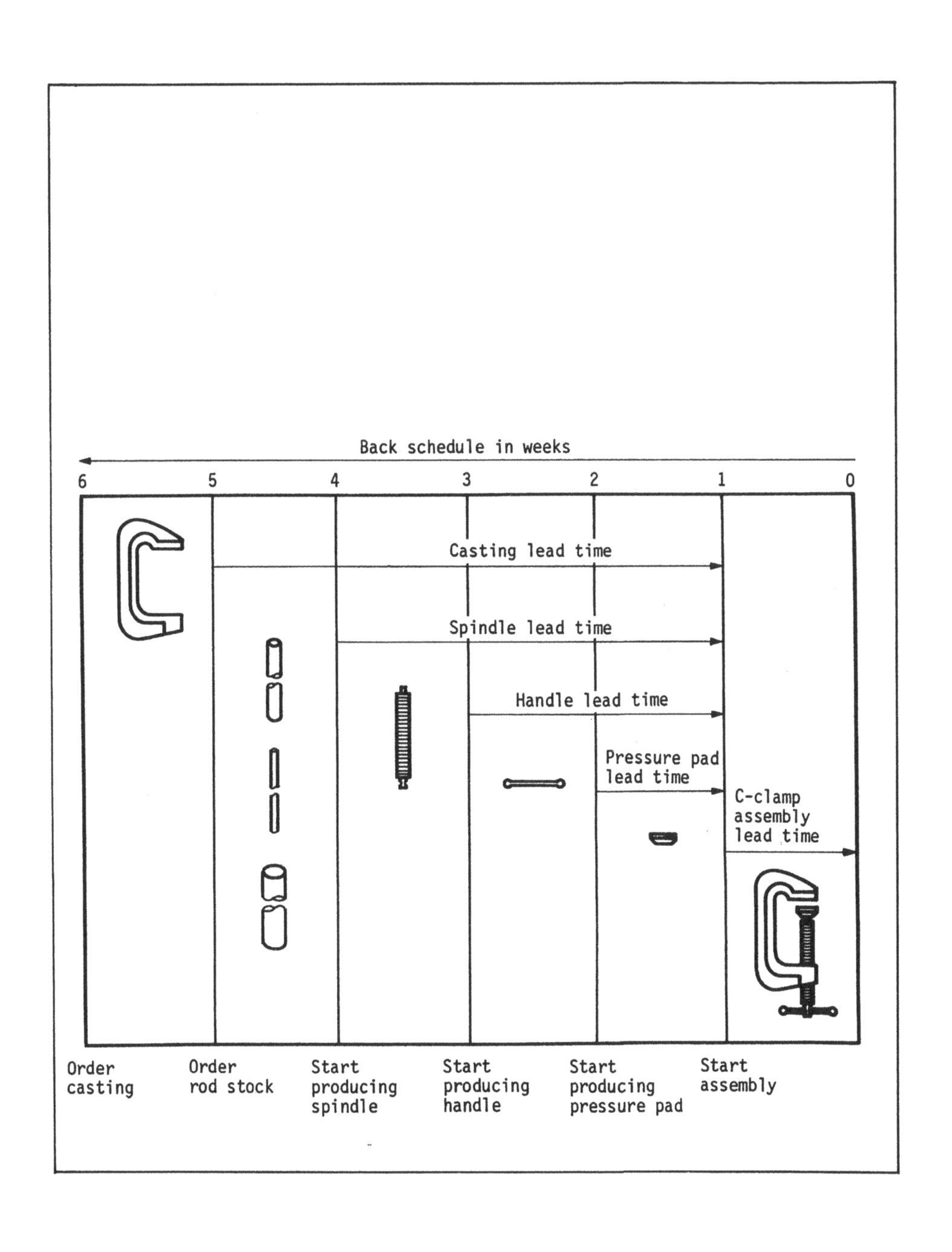

Scheduling of a product

5.1 Introduction

The aim of this chapter is to present design methods and control principles for computer integrated manufacturing systems.

To design a Computer Integrated Manufacturing (CIM) system, it is necessary to have design methods, even if they are needed only for economic or social reasons. The methods must rely on a conceptual model in order to take into account the close interconnection between the product, facility and management. This chapter is divided into the following four parts:

1. A discussion of the need for conceptual models in CIM systems
2. Methods of designing a production control system
3. The design of flexible manufacturing systems using modelling techniques and simulation
4. The principles of CIM systems control

The topics discussed are the main principles of scheduling, the concepts of manufacturing unit control developed by the GRAI Laboratory.

5.2 The Need for a Methodology and a Conceptual Model of a CIM System

In this chapter, we present conceptual models of production and production management systems. These models are the basis of various design methods for CIM. We first show that the need to use such methods results not only from the technical, but also from the economic and social point of view.

5.2.1 The Use of a Design Methodology

For many years the production process was not deemed to require much analytic attention. But the tightening of social and commercial constraints and the improvement in data handling techniques has given rise to an increasing interest in this activity.

A production system is a very complex network of physical activities, decision making and information flows. Hence the complexity of the design of such a system increases with increases in the degree of automation.

For this reason a methodology is needed for guiding the designer toward an efficient and realistic concept of a CIM. It was not until 1970 that designers started to look for methods of improving the design work and automating this activity. Studies made in the United States show the economic impact of the design stage throughout the life cycle of a product (Fig. 5.1). The lower curve shows that the use of a methodology together with the computer improves considerably the cost of the different design phases. Traditional design methods need little investment for the design phase but during succeeding planning and production phases the cost increases considerably. The implementation phase often points out many problems which should have been studied in the previous phase. Thus poor design adversely affects planning, scheduling, production control,

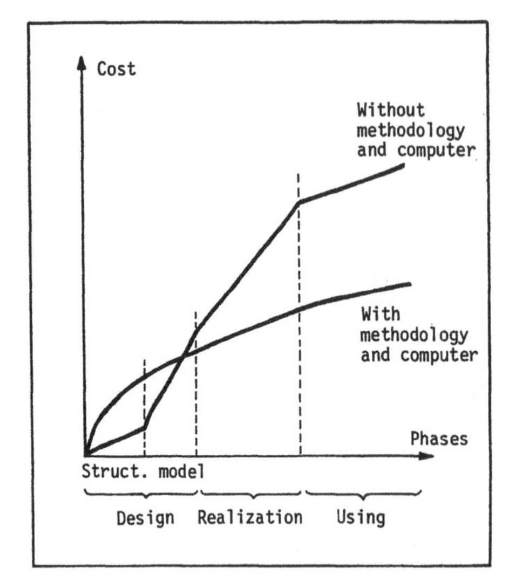

Fig. 5.1. The cost of building a CIM system depends on the automation aids used to realize it

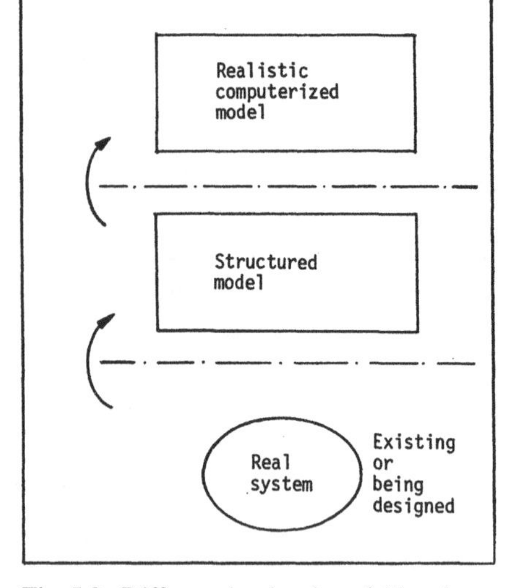

Fig. 5.2. Different levels of modelling for design

quality control etc. The aim of using structured methods is to anticipate and solve during the design stage every problem encountered throughout the life cycle of an integrated production system (Fig. 5.2). This figure shows that the real system should be described by a structured model, which in turn will be mapped into computer memory as a design aid. Every aspect of the present and the contemplated production system must be determined. In addition the data processing methods to be used must be specified.

There are only a few methods of analysis available which can help to find deficiencies. They are not specifically designed for the concept of production systems, but are used for information flow analysis or the design of data bases. Thus, they are very difficult to use for production systems which have different structures and are built from different components.

5.2.2 The Complexity of Computer Integrated Manufacturing

When we look at the individual manufacturing functions we can see strong links between them. For this reason the integration of the individual components within one system is important. There are three major parts of a manufacturing system. They are the product, facility and management (Fig. 5.3). It is very difficult to interconnect these parts for the given constraints of a product. The integration can often be simplified if the product is redesigned for manufacturing. In most cases this will also improve scheduling of production. Figure 5.4 shows a schematic model of manufacturing activities. It contains:

- The information flow
- The physical flow link of the product and its components
- The decision links

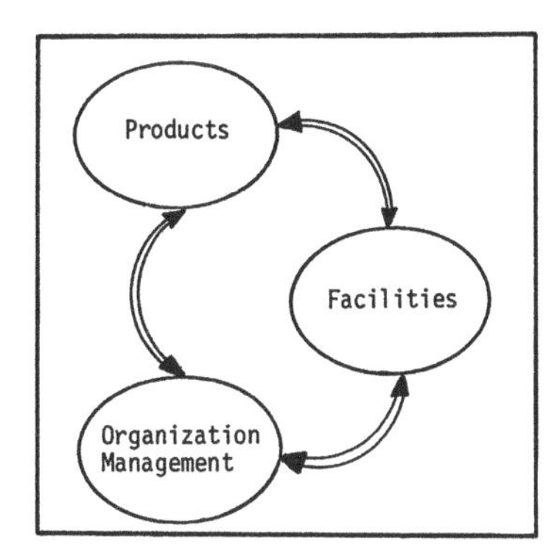

Fig. 5.3. Three major components of a manufacturing system

- The information flow and material flow starts with marketing and ends with the delivery of the finished product. There are several feedback loops which are needed to control the manufacturing activities. The decision links between the functions are the means of integration.

The proper conception and establishment of the links between the functions is a very important factor for the following basic activities:

- Computer aided design helps to specify the product.
- Computer aided manufacturing and computer controlled machine tools produce the product.
- Computer aided production management coordinates and controls the manufacturing system.

With increasing automation two other factors gain significance. They are: Computer aided quality assurance and computer aided maintenance.

Other characteristic features shown in Fig. 5.4 are the feedback loops between the different entities. They are a necessary part of the control system and help to correct any deviation from the desired manufacturing goal.

Figure 5.5 shows the different functions of a production control system. The manufacturing process starts with order forecasting and the customer order. Both functions provide the main input variable to the master production plan. This plan also needs information about parts, production routes, available work centers, and critical parts to be purchased. With the help of scheduling algorithms the production plan is generated. The bill of materials and the production plan furnish the input to gross requirement planning. With this information the available inventory is investigated and the net requirement of parts is calculated. The next activity is a make or buy decision for parts to be purchased and parts to be produced. For the parts to be purchased the orders are issued to the vendors and from there are delivered to material receiving. The parts to be produced are scheduled for production. When all parts are available, orders are issued for assembly.

All these functions can be simulated and studied with the help of a mathematical model. The results may serve as a guide for the design and layout of the computer control system.

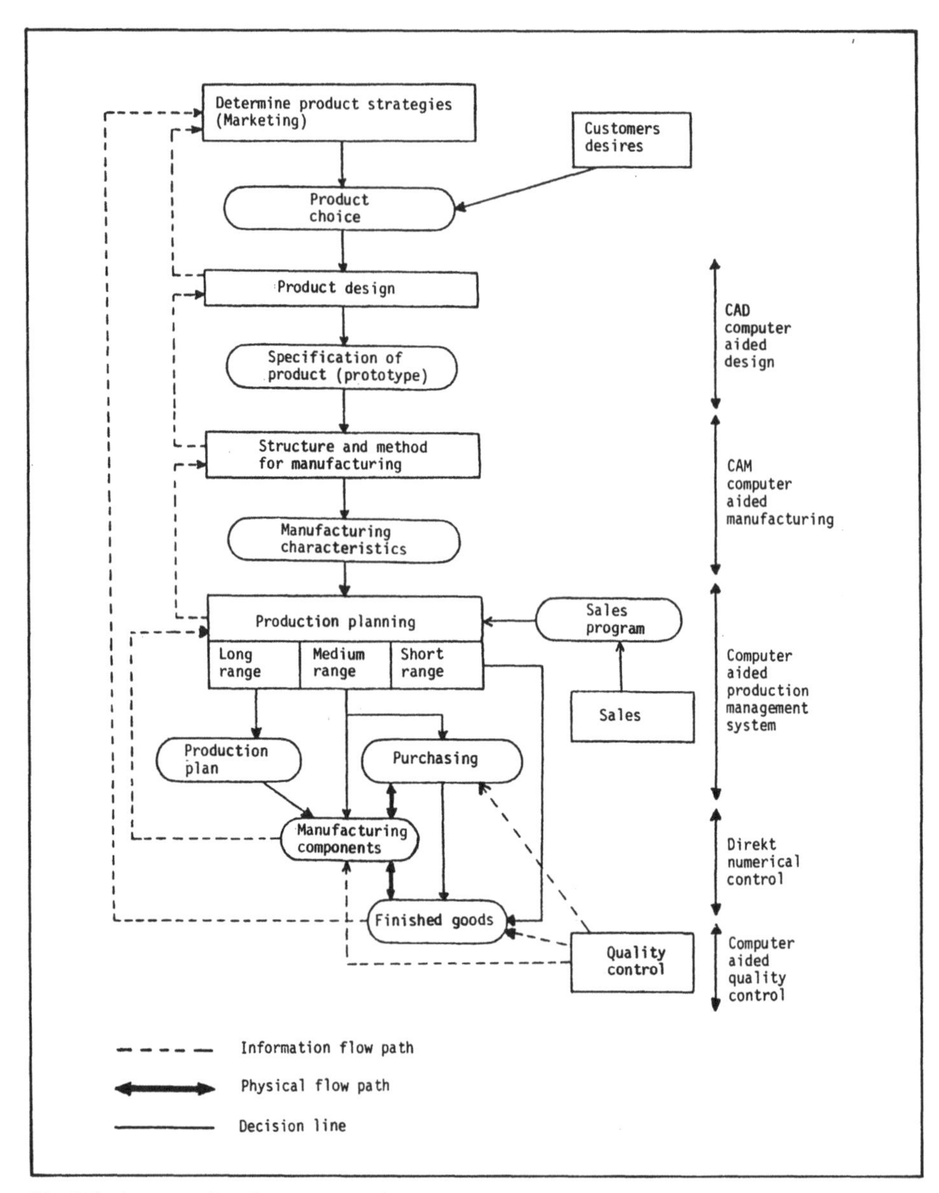

Fig. 5.4. A comprehensive computer integrated production system

A model of computer aided manufacturing activities is based on the concept that any human directed process which leads to a product, tangible or otherwise, must proceed through seven sequential stages, Fig. 5.6. They are described below:

Decide: A decision which actions have to be taken based on requirements and constaints

Design: The development of a conceptual end product; controlled by the decisions

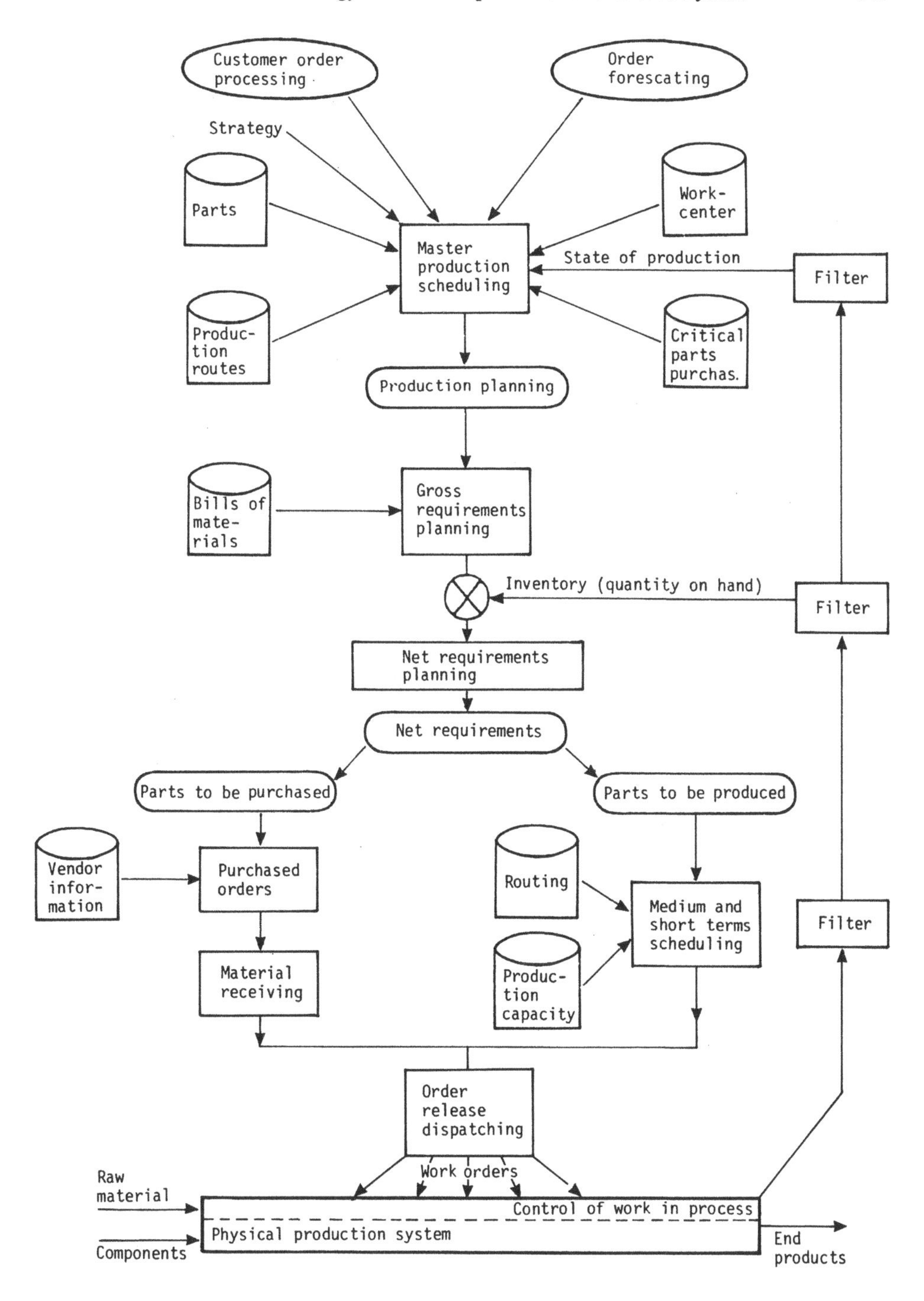

Fig. 5.5. Production control system

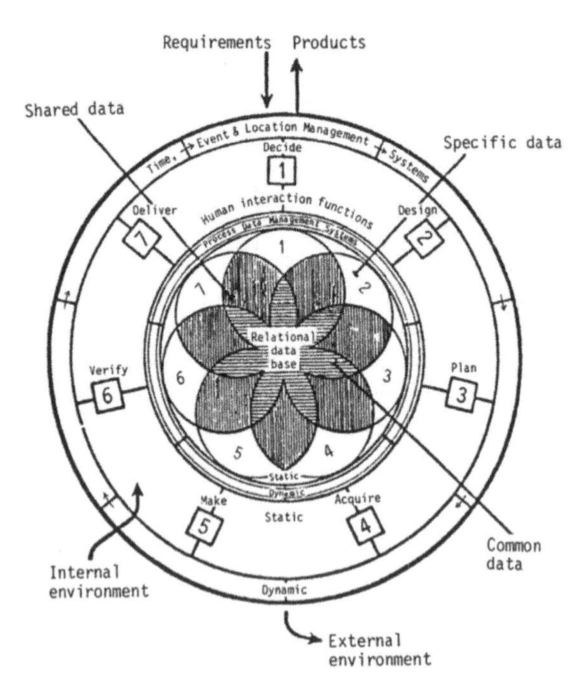

Fig. 5.6. CAM-I: Layer model for computer-aided system planning

Plan: A sequentially ordered set of instructions to achieve the conceptual product

Acquire: Actual acquisition and receipt of objects, both hardware and information used to produce the product according to a sequentially ordered set of instructions

Make: Execution of the instructions; an action that consumes time and energy to produce a product

Verify: Comparison of the actual product(s) manufactured under guidance of a set of ordered instructions with the intended product concept and reporting of the difference

Deliver: Distribution of the completed product(s).

These functions are interrelated and may utilize common or shared information as well as data pertaining to the functions. Four levels of control can be identified in the factory:

1. Factory or plant level at which the production of all products is managed and controlled
2. Job center or manufacturing cell level
3. Cost center or work center level
4. Operator or manufacturing unit level

The hierarchical structure of such a factory management system is illustrated in Fig. 5.7. A shaded disk indicates the fact that a firm may have several factories, a factory may consist of several job shops, a job shop may contain several cost centers and a cost center may have several manufacturing units.

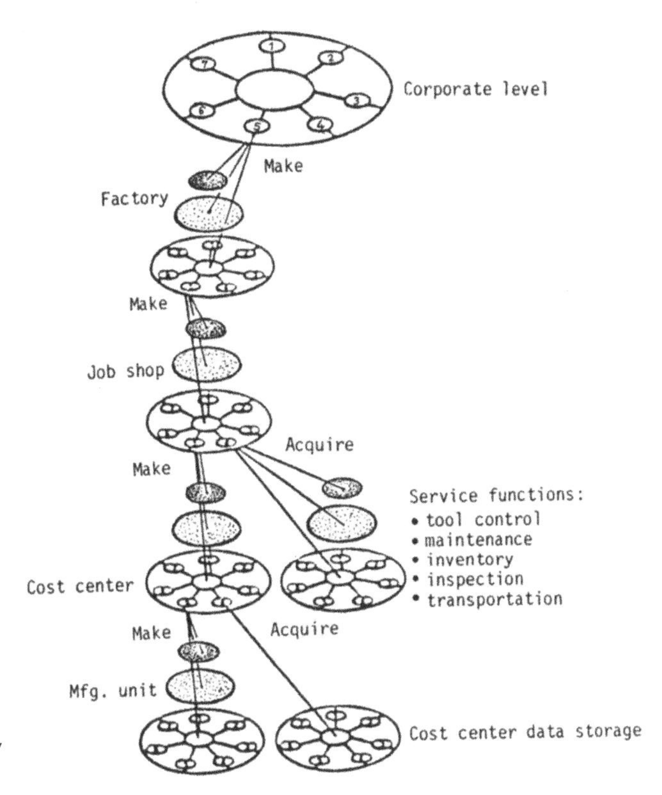

Fig. 5.7. Levels of control of a factory represented by CAM-I

At each level the control functions are decomposed into subfunctions describing available parts and production routes. This is done by a human operator with the help of a computer aided activity model. Operating data are obtained from the data base and allocated to the specific levels.

5.3 Conceptual Model of a CIM System

5.3.1 The Notion of the System

In automatic control theory we are speaking about a "system" which is a collection of different elements for which the interrelations and their functions are defined. These elements are located inside of the system boundary. Within its environment the system has objectives and relations, it may be real or abstract, open ended or closed, static or dynamic. A system enclosed by its boundaries lives and changes with time without losing its particular identity. Thus, the three aspects of a system define: what the system is, what it does and what it will be.

A computer integrated manufacturing activity can be considered as a system. It is an organized system of very great complexity within its technical socio-economic environment. A production system can be represented by a hierarchically structured concept. At each level the control functions are level specific and they support a corresponding level of manufacturing management.

5.3.2 Conceptual Models

In accordance with these comments we must use structured methods to design a computer integrated manufacturing system. The methods are supported by conceptual models which are a collection of concepts representing the schema of the system. Such a schema is the image of the actual system. For example, for a production control system, a conceptual schema is an abstract figure showing the structure of every part of the organization system. We will present here two conceptual models:

1. The ICAM (Integrated Computer Aided Manufacturing) composite view of aerospace manufacturing)
2. The GRAI conceptual model developed by the GRAI Laboratory of the University of Bordeaux

5.3.3 ICAM Model and Architecture

The ICAM program objectives are:

To reduce defense system costs through CAM Technology
To establish means for integrated application of computer technology
To improve the long term competence, efficiency and responsiveness to defence needs
To provide mechanisms for ICAM technology transfer
To validate and demonstrate cost savings and flexibility of the ICAM methodology

The first step of ICAM methodology is to define an architecture for manufacturing which provides a common understanding of a manufacturing system. The objectives of the architecture are:

To present the functions of aerospace batch manufacturing by a CAM model
To show the detailed relationships between these functions
To provide a basis for the design and integration of subsystems.

This structure integrates 3 different views:

The factory view, which is a multi-product and company dependent view of manufacturing.
The composite view, which is based on the essential decisions, actions and activities required to produce a product.
The generic view, which synthesizes the information contained in the composite view. It shows all aspects of the manufacturing function in a single model. The generic view is the focus of the ICAM system and module development phases.

The composite view is an aggregate representation of all activities necessary to design, engineer, manufacture, and maintain a product (Fig. 5.8).

This view may be tailored to various types of products. However, the initial model is primarily intended to represent a manufacturing system producing a complex aerospace product, such as an airplane.

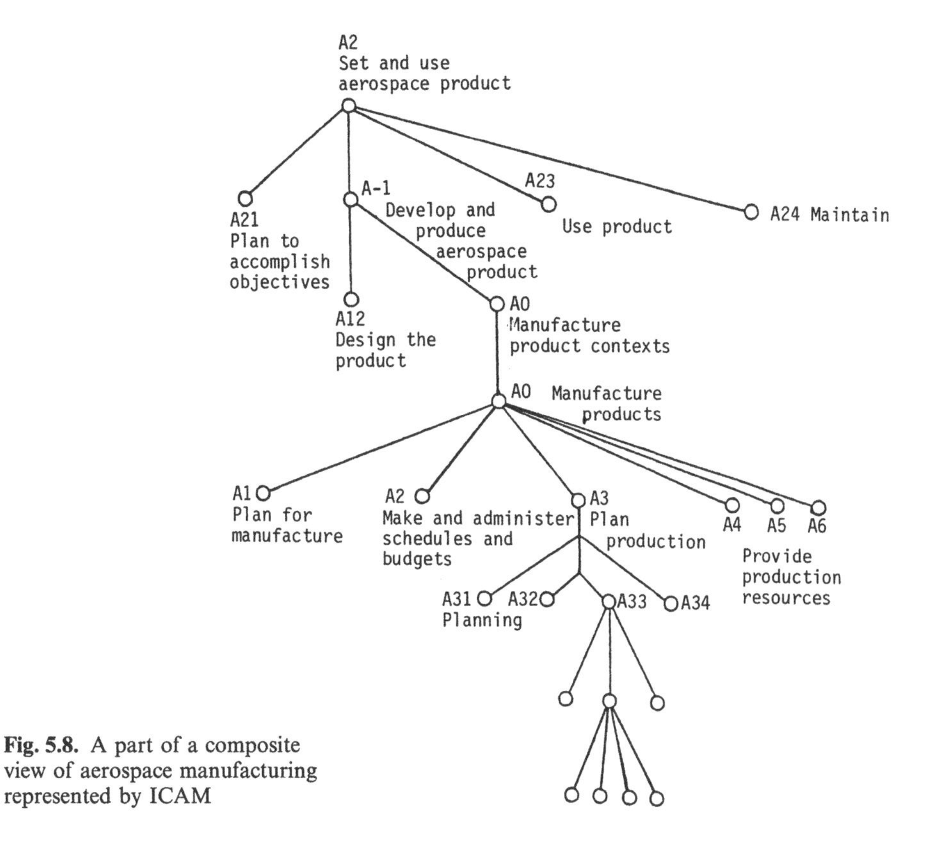

Fig. 5.8. A part of a composite view of aerospace manufacturing represented by ICAM

5.3.4 GRAI Conceptual Model

With this methodology the GRAI Laboratory proposes a concept to structure a production system (Fig. 5.9). It consists of the two parts:

The production control system and
The physical system

The input to the control system is information about orders, recources, energy etc. The physical system has as its input the parts and components, and as its output the product. The production control system is comparable to a machine control system. Like a numerical controller which gives commands to the machine tool, production control gives instructions to the physical system and it in turn passes information to production control. In a control system we can distinguish between:

The decision system and
The information system

Figure 5.10 shows the principle of the production system, consisting of the physical, information and decision subsystems. The information system is the connecting line between the two other systems.

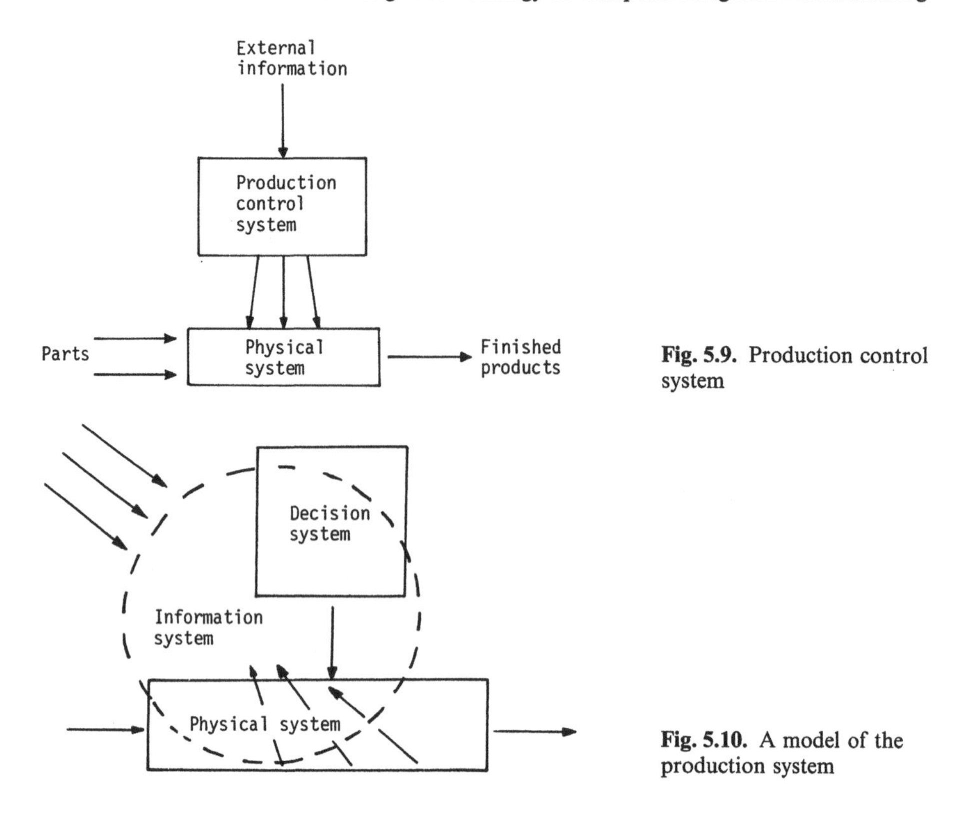

Fig. 5.9. Production control system

Fig. 5.10. A model of the production system

Physical System

The physical system is a set of manufacturing units whose functions are to transform raw materials, parts or semi-finished parts to assemblies, finished parts or a complete product according to a manufacturing process plan and engineering specifications.

The physical system consists of men, machines, tools, and the manufacturing environment (Fig. 5.11). The transformation activities may involve chip removal, forming, joining, assembly, surface finishing, or any combination of these processes. By itself this physical system is not capable of achieving its objectives. For this reason it is necessary to supervise it by a control system.

Fig. 5.11. A model of the physical system

Control System

The efficiency of a control system depends on the consistency and quality of the decisions, and the proper communication of information. The task of this system is to make decisions. Some are obvious and their effects are known. The consequences of other decisions are very difficult to foresee. Unfortunately, they are

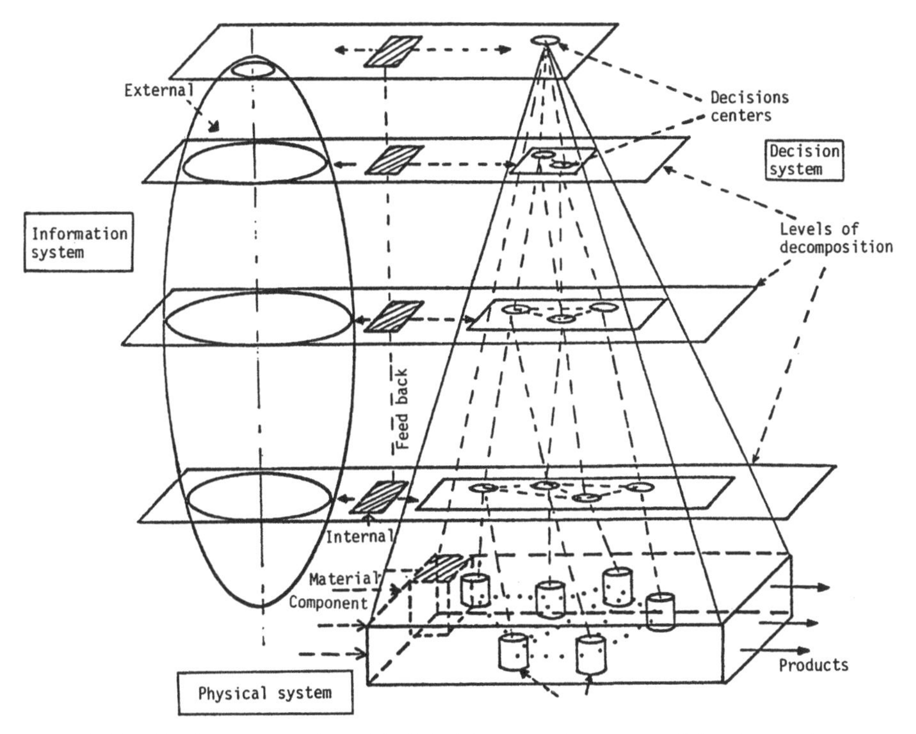

Fig. 5.12. The GRAI conceptual model of the hierarchical structure of decision centers

generally the most important ones. Thus it is necessary to design a control system which makes a minimum of this type of decisions. Moreover, to make decisions, the decision-maker must have information about the system in real time, and about the past or the future. Thus a decision can be of static or dynamic nature and it is necessary that the information system is interconnected with the decision system.

Proposal for a Hierarchical Structure of Decision Centers

In this model we can find 3 sub-systems consisting of (Fig. 5.12):

The decision system to make decisions and perform the control
The physical system executing the manufacturing activities
The information system linking the two others together

The proposed concept has a hierarchical structure. The scope of each level is determined by the task it performs in the system and by the time period for which the decision is made.

The information and decisions are detailed at each level. Thus, the information must be structured to match the decision at each level. The shorter the planning horizon, the closer it approaches the physical system and the preciser the information must be specified. The model allows to structure the system in GRAI connotation. The structure may be represented with the help of a table in which the functions are entered against the planning horizon or periods. The decision centers may be structured according to the following criteria:

The response time to a decision which is made in this decision center
The product cycle
The frequency of decisions

Structure of a Decision Center

After the conceptual model of a decision center has been built it is necessary to describe each activity of a center (Fig. 5.13). The model contains the basic elements, function and action variables or relations of the decision center, and the data on which a decision maker has an influence. The constraints and performance define the frame of the decision. The decision maker defines the allocation of resorces for a desired goal and the responsibilities for the level below. He gives information about states, measured performance and technical data for a given level.

It is also important to coordinate the activities between the decision center levels. A decision center of a given level defines the elements required to coordinate the decision centers of the level below.

Conclusion. The two described models are a very important part of the GRAI method. With their help an overall view of the firm can be given including the hierarchical structure, detailed view of the information, constraints, and the decision making functions of the decision centers.

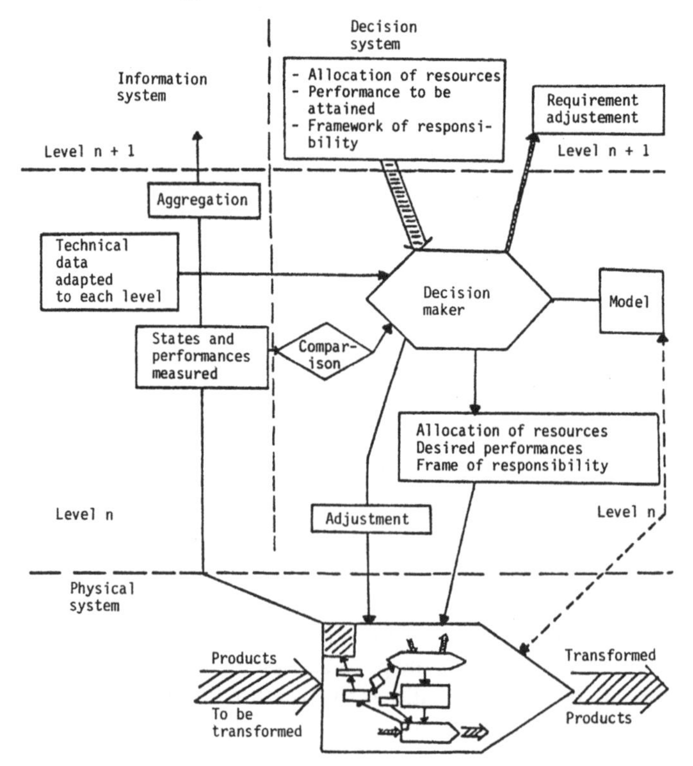

Fig. 5.13. The GRAI conceptual model of a decision center

5.4 Methods of Designing Production Control Systems

In this section three structured methods of designing a production system will be discussed. They are the:

Structured System Analysis and Design method (SSAD)
Integrated Computer Aided Manufacturing method (IDEF)
GRAI Method of Process Analysis

There is another integrated CAD/CAM model of general interest. It was designed by the Computer and Automation Institute of the Hungarian Academy of Science in Budapest. This method is based on the concept of the Structured Analysis and Design Techniques (SADT) and the ISDOS system. It will not be discussed in this book.

5.4.1 The Structured System Analysis and Design Method (SSAD)

This method can be used to modify an existing system or to design a new system.

The study starts with an assessment and analysis of the prevailing situation, and the estimated value of the new system. Thereafter a refined study defines the limitation of the current system to obtain more details and confidence. This leads to an understanding of the functions of the present system and of the structure needed to specify a replacement. Depending on the results, and the facts obtained by the initial investigation, a more detailed study may be authorized. This study concerns itself with the following areas:

A definition of the users of the new system
The building of a logical model of the current system
A refinement of the initial study

The method is supported by a graphical tool, which is a data flow diagram (Fig. 5.14)

It uses the following four symbols:

External entities (square)
Data flows (arrows)
Data storage units (a pair of horizontal parallel lines closed at one end)
Processes (a rectangle with the corners rounded off, optionally divided into three areas)

The steps in drawing up a data flow diagram for an existing or proposed system are:

To identify the external entities
To identify the inputs and the outputs which are expected and needed to schedule the normal course of business
To identify the inquiries and on-demand requests for information

The CAM-I (Computer Aided Manufacturing International) has also used this method for the factory management project.

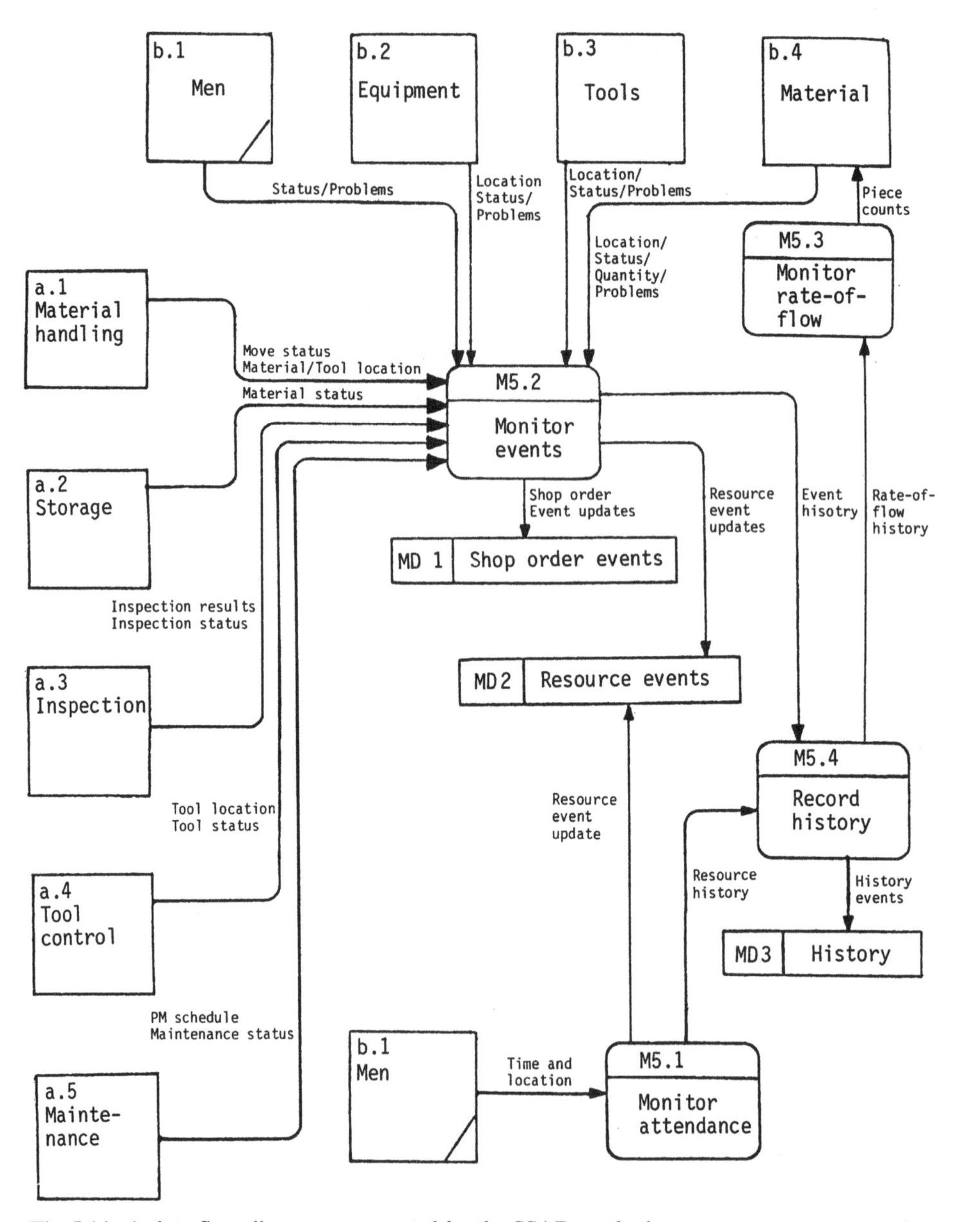

Fig. 5.14. A data flow diagram represented by the SSAD method

5.4.2 ICAM Definition Language (IDEF)

IDEF is a methodology or a collection of tools with which the architecture of a manufacturing system can be described (Fig. 5.15). It was developed for use with the Integrated Computer Aided Manufacturing program (ICAM) and contains the following models:

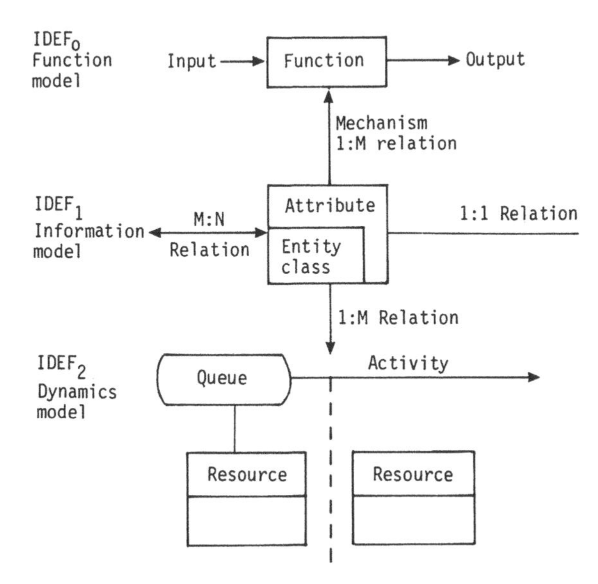

Fig. 5.15. The ICAM definition modules IDEF 0, 1, 2

$IDEF_0$ is used to model the functions of a complete manufacturing facility. It has close similarities with the Structured Analysis and Design Technique (SADT). There are three modelling steps. First the information to operate a facility is identified. Second, the collected information is structured with the help of activity diagrams (Fig. 5.16). Third the results are approved by the user performing the installation.

In an activity diagram boxes represent the system functions or activities and arrows show the information flow. In order to maintain clarity only 6 boxes are

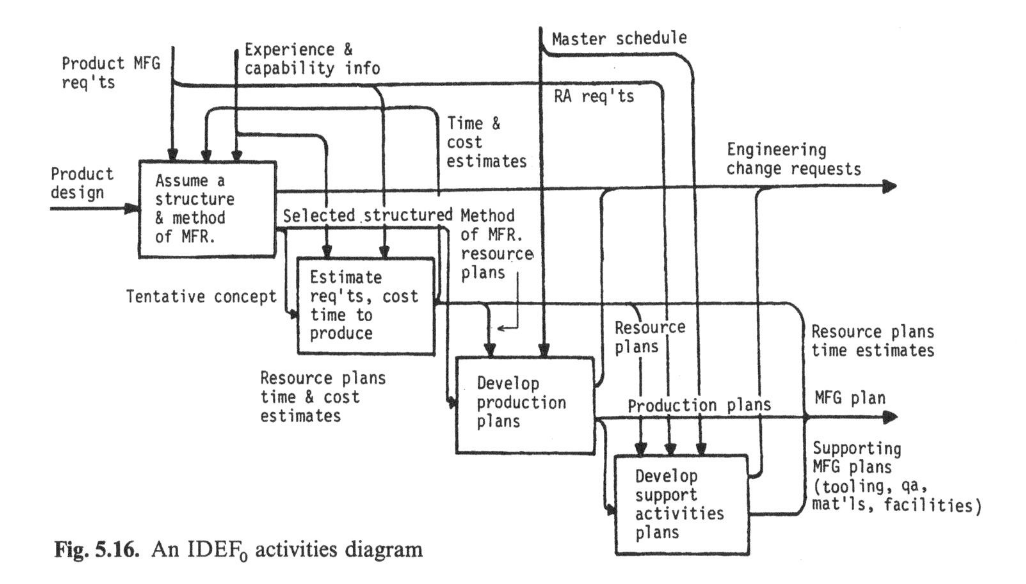

Fig. 5.16. An $IDEF_0$ activities diagram

allowed on one figure. Thus it is necessary to represent different degrees of abstraction. First, the analysis is done with an overview schema to describe the general concept. Successively this schema is broken down into more detailed presentations, until all structural elements which are necessary to describe the system are shown.

$IDEF_1$ describes the modeling of information

$IDEF_2$ serves as a dynamic modeling tool

5.4.3 GRAI Method of Process Analysis

The development started in 1976. It was initiated when industrial firms had considerable difficulties in computerizing their process or systems. This was mainly because of a lack of available methods to analyze a production system and to design a new computer controlled system. Thus the GRAI Laboratory developed an analytic method to fill such a need and conceived a new design method. This method makes use of a structured approach which allows the user to implement his application in a rational way.

The application of this method requires:

1. A group of technicians experienced in the techniques required for the system design. Their job is to advise the system design staff. The GRAI method consists mainly of two phases:
   A phase of analysing the existing system and collecting all data necesssary for the design of the new system.
   A phase of designing the system from data collected during the previous phase. The inconsistencies between the existing system and the ideal system are analyzed
2. One or several analyst-designers whose main job it is to collect all the data needed for the system design
3. A group composed of the main users; this group has to follow the study to check the results of the various steps and to guide the design to ensure technical feasibility and to make suggestions for the new system

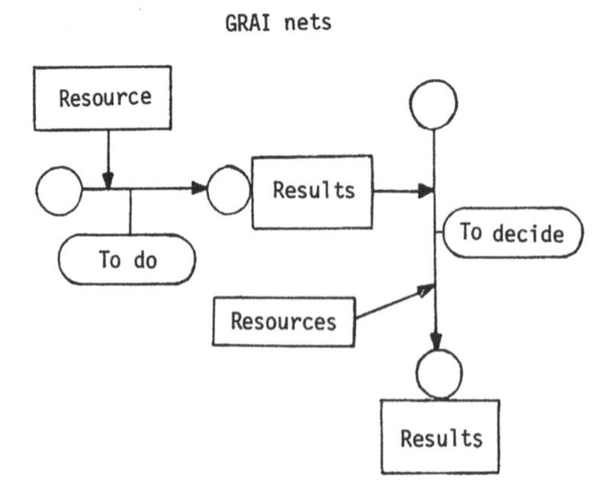

Fig. 5.17 A GRAI net showing activities and results

The high level of detail which can be achieved is one of the main feature of the method. Actually, the method can be used to design the new structure of the system or to specify each element of this structure to obtain a catalog of specifications for the data processing implementation. The degree of sophistication of the method is defined by the design group.

This method tries to point out possible inconsistencies of a structure of the decision centers. It is based on two conceptual models (Fig. 5.12 and 5.13). One is related to the hierarchical structure of the decision centers and the other to the basic elements of decision making. The GRAI method uses the graphic analysis tool GRAPHS. The input is a description of the recources. They are processed by the activities and the results are generated. The planning is visualized with the help of a network (Fig. 5.17).

This process is divided into two phases which are related to the two conceptual models.

Top Down Analysis

The first step of the top down analysis consists in drawing up a frame which shows the structure of the decision centers of the existing system (Fig. 5.18). The different levels represent the decision making horizons specified during the preparation

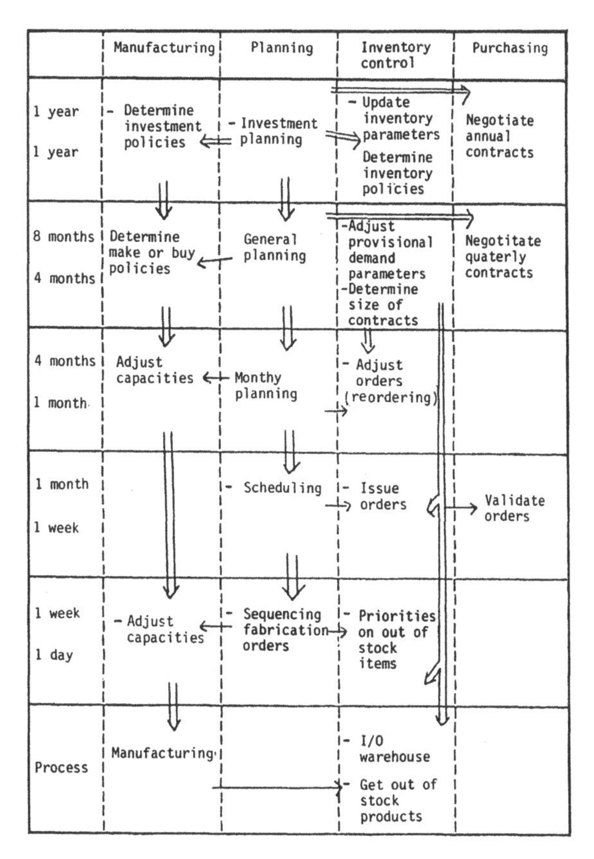

Fig. 5.18. The top-down analysis

of the conceptual model. The columns contain the various functions to production management. These functions are manufacturing, planning, inventory control and purchasing. Some of them may be irrelevant in a particular example. Each decision center has its place in the hierarchy and its relations with others are defined by the two symbols:

Transmission of a decision frame
Transmission of information. (This symbol is only used to represent external or very important information flow)

Structural problems concerning lack of coordination between decision centers can be detected from this table.

Bottom Up Analysis

The second step is the bottom up method which is a detailed analysis of each decision center. It considers:

The decision made in the decision center
The variables used for the decisions
The frame of the decision (constraints on variables, criteria ...)
The constraints imposed by the controlled decision centers
The information used to make a decision
The rules between input and output
The dynamic aspects of production

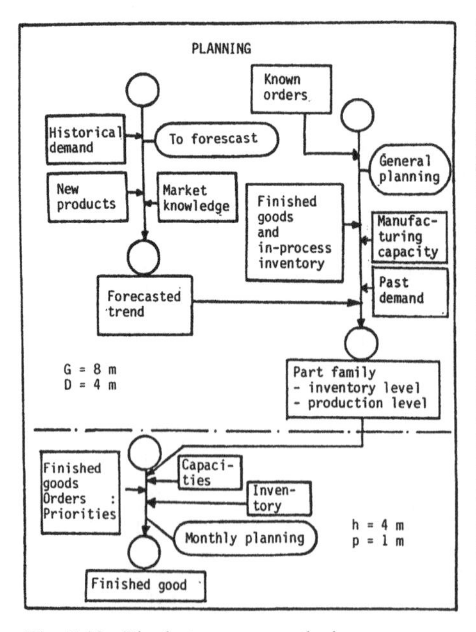

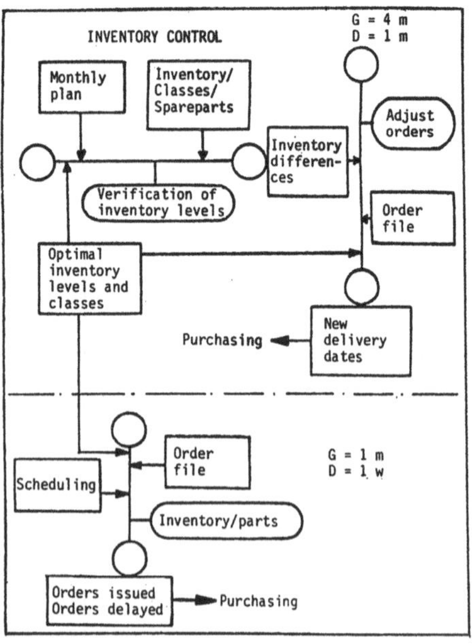

Fig. 5.19. The bottom-up analysis

The system analyst uses the interview technique for this phase.

The analysis is supported by GRAI nets which represent the decision or process activities and the information required by the activities (Fig. 5.19).

Analyzing Results

The aim of the analysis is to find possible faults in the production management system. There are three different types of inconsistencies which may occur.

1. The transmitted information may be faulty:
   There is no updating of lead times
   The knowledge of orders is missing
   There is very slow transmission of information
   The redundance of data
   There are inconsistencies of measurement parameters
2. The problem may be specific to a decision:
   There is no influence on delay times due to purchasing negotiations
   The inventory level is not detailed enough
   There is no follow up on out of stock items
3. The coordination between decision centers may not work properly:
   The frequency of planning and ordering activities may vary
   Manufacturing decisions may cause disturbance to inventory control
   These inconsistencies may be resolved in the design phase when a new structure of the production control system is defined

Designing the Process

The process design phase includes the following two phases:

Drawing up of a frame which represents the new structure of the production control system.

Drawing up of the GRAI net which shows for each decision center the specification of information and decisions.

This design method is based on two conceptual models. First, rules may be defined to guide the design. They have to be adapted to the external constraints of the firm. Second, the internal constraints should not interfere with the design, but the constraints should be related to the manufacture of the products (i.e. cycles).

Conclusion. Various firms have experimented with the GRAI method. It has proved to be a very appropriate tool for analyzing and designing a production control system. It is also a good tool for the design of flexible manufacturing systems.

Further developments will include the implementation of computer aided design and analysis. This point is quite important because at the present time the method relies too much on the experience of the user. The analysis team must include people who are very familiar with the existing production control systems in order to find all possible faults which may be introduced during the analysis.

5.5 Design of Flexible Manufacturing Systems Using Modelling Techniques and Simulation

It is a well known fact that flexible manufacturing systems (FMS) can increase the productivity of a production system. For the design of the various types of FMS several simulation methods have been suggested and developed over the past years. However, most of these simulation methods are not adequate to completely design a FMS. They are only one tool out of many. It will be shown how the GRAI methodology can be used to design a flexible manufacturing system.

5.5.1 What is a Flexible Manufacturing System?

A FMS consists of two main parts: The physical manufacturing equipment and the computer control system.

A FMS may be defined as follows: An autonomous facility which can serve a volatile market with minimum response time from the order input to the saleable product, using the minimum of working capital.

A FMS consists of a group of work stations which are interconnected by an automated material flow system (sometimes tool flow), operating under computer control to produce a part family. The ability to react effectively to changing circumstances is called flexibility. There are several flexibilities which can be defined.

State flexibility: The capability to continue functioning efficiently, despite of a change.
Job flexibility: The ability of the system to react to a change in the work to be processed.
Machine flexibility: The ability of the system to handle changes in and disturbances of the machines and work stations.

The author of this chapter defines a FMS as a facility containing computerized numerically controlled work stations, which are linked together by automated material handling facilities for parts and tools and which are controlled by computers. Such a system has commercial and social objectives.

Recent economic and social developments favor the adoption of FMS. For manufacturing industries it has become necessary to increase the number of manufactured products to serve a competitive market. As a result more products are produced in smaller lot sizes. Also a reduction in innovation and lead times, and inventory and production costs has become necessary. Other factors which have become important are the quality of work life and the technical qualification of the worker.

Modern technology allows an increasing number of intelligent machines to be used in manufacturing operations. This trend is stimulated by the decreasing hardware cost of miniaturized control systems. In addition flexible manufacturing techniques are being conceived. All these developments will lead to the design of numerous FMSs. But there is a conflict between productivity and flexibility as shown in Fig. 5.20. Greatest flexibility is obtained where a man operates a machine, but he is not very productive. The most productive systems are custom

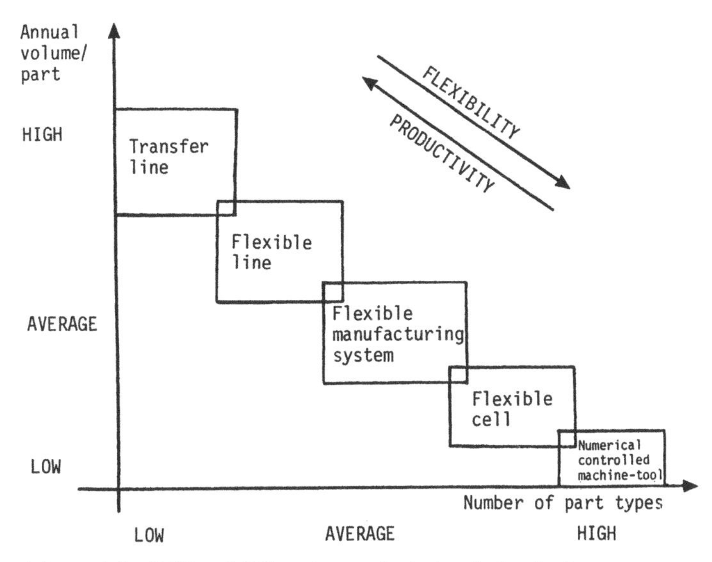

Fig. 5.20. The productivity and flexibility of different manufacturing technologies

designed automatic machines for example as they are used in transfer lines. However, they have no flexibility. A FMS is a compromise between these two alternatives. There are many problems which must be solved in building a good FMS. These problems derive from the different products to be manufactured, the facility, the organization and the complexity of the FMS. Moreover, the dynamic evolution of technological, economic and social contexts must be considered.

5.5.2 Design of Flexible Manufacturing Systems

As in the design of a production control system a methodology is needed for the design of a FMS.

The data required for the development of a FMS are:

The part spectrum
The available machine tools
The objectives of the company, e.g. to minimize inventory, lead time or direct labor, the use of dedicated tooling, indirect labor, maximizing equipment utilization, and flexibility

To solve this problem a detailed description must be made of:

The parts which can be machined on a FMS
The overall characteristics of the machine tools for the FMS
The detailed characteristics of the individual elements of the FMS, including the type of machine tools, the type of tools, the elements of the material and tool handling systems
The characteristics of production management system of the FMS, e.g. its structure and its scheduling and inspection rules
The interaction between the FMS and its environment

The use of the GRAI methodology for the design of a flexible manufacturing system will be discussed in the next section.

5.5.3 GRAI Methodology

Using the GRAI Methodology the design of a FMS can be done in two phases as it was with the design of a production control system, Fig. 5.21. The phases are analysis and design.

In the first stage there are the following activities:

The analysis of the process plan
The analysis of the production control system
The analysis of the quality control procedures

The design phase consists of the definition of the architecture of the FMS and of the determination of the capabilities of the new machines and the control procedures.

The validation of the design hypothesis can be done with simulation techniques and economic evaluation methods.

Analysis Phase

The first step of this study is to analyze and discuss in detail the manufacturing processes under consideration.

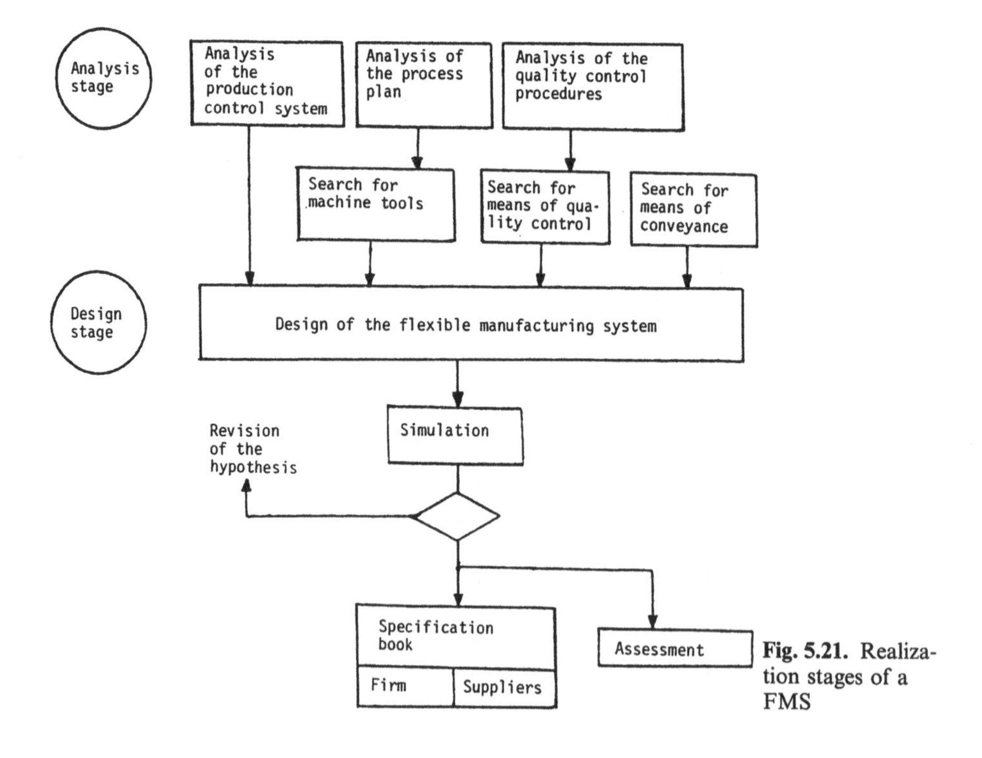

Fig. 5.21. Realization stages of a FMS

Process Analysis

The process analysis starts with a study of the geometry of the part spectrum under investigation, and of the manufacturing processes used in making the parts (Fig. 5.22). Then families of parts are created according to their shape and dimension. This provides a good overview of the diversity of the parts.

Two methods can be applied for the process analysis:

1. If group technology can be used standard processes are identified.
2. If this is not possible the machine tools and machining processes are classified and coded.

In both cases, groups of interchangeable machine tools are formed and identified by a general specification which describes the nature of the process and its performance. A load matrix is drawn up in which rows describe parts and columns machine types. This matrix shows the machining required for each part within a group, as well as the monthly demand for the part. The matrix is used to evaluate the various manufacturing processes. The evaluation is made with the help of a special software package. It allows processing of the data matrix as follows:

Two possible routes are selected for a part and they are compared. The comparison leads to the determination of a macro-route which is a synthesis of the two detailed routes. The macro-route will replace the two routes if they are similar. The routes are compared with other macro-routes.

The results are presented either as a listing or a graphical tree in which each node represents a macro-route.

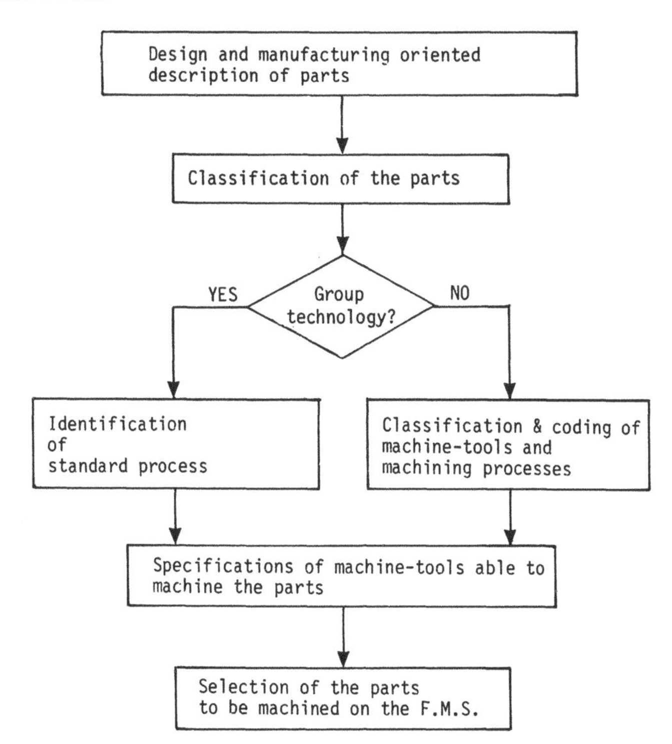

Fig. 5.22. The process analysis sequence

With the help of the macro-routes a study is done to determine which parts can be manufactured on the FMS, which machine-tools are necessary and which cells may have to be designed. This is the beginning of the FMS design. The second step of the analysis is the question of selecting the proper production control.

Production Control Analysis

For this stage, the GRAI method can be used. First an analysis of the existing production control system of the firm is carried out. The results are entered in a table (Fig. 5.23). In this figure all decision centers are identified via the top down analysis. Thereafter a bottom up analysis is conducted with a detailed study of each decision center (Fig. 5.24). The decision centers for the workshop are studied in more detail. The combined results of the top down and bottom-up analyses specify the production control system of the selected FMS. After this stage the design of the FMS can begin.

Design Phase

The design phase is performed in two parts as it was for the analysis. These are the physical design of the installation and the design of the production control system. These two parts are interconnected. Thus when the flexible manufacturing system is designed the selection of the machine tools and auxiliary equipment and the determination of the monitoring and process control systems must be realized at the same time. The design of a flexible manufacturing system is supported by a simulation technique and an economic analysis.

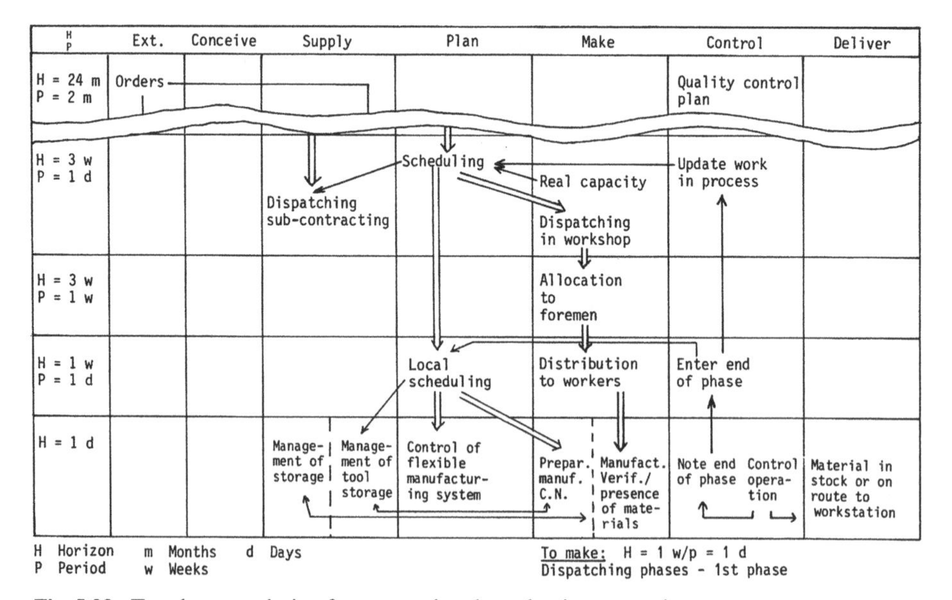

Fig. 5.23. Top down analysis of a conventional production control system

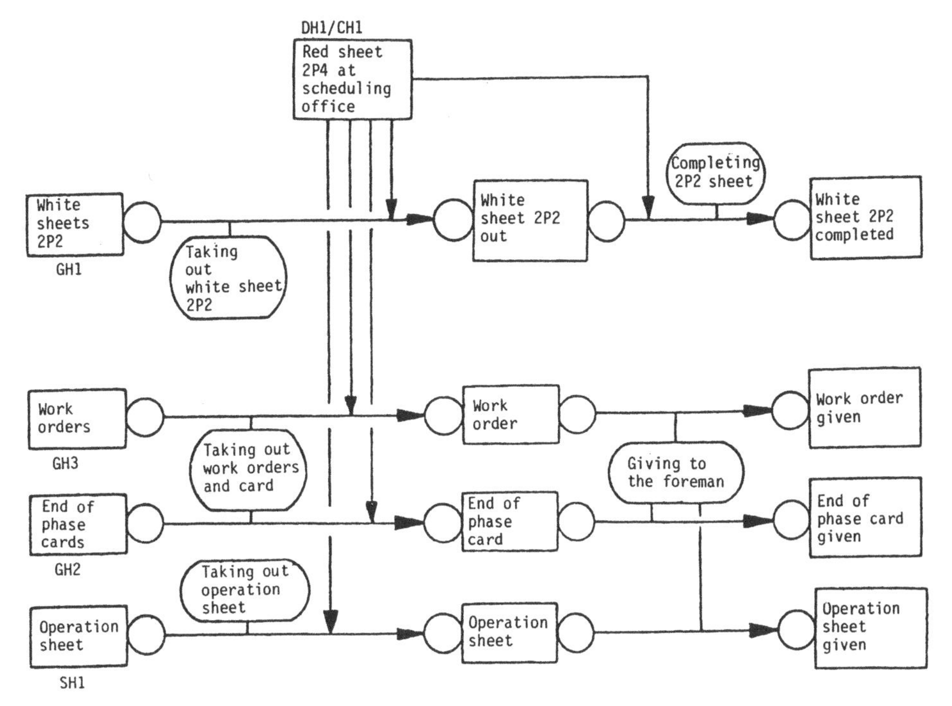

Fig. 5.24. Bottom up analysis of conventional production control system

Determination of the Parts and Machining Process

The results of the process analysis are the first choice of parts to be manufactured, the characteristics of the machine-tools used, the process plan and quality control procedures (Fig. 5.25). With this help the new processes or chosen parts are planned. The new machining times, set-up times, tool-change times and so on are calculated.

A capacity and load balance is done for each machine tool in order to determine the number of machines required. Thereafter that load configurations are created in which the parts are chosen according to their "maximum" or "medium" sizes and machining times. The batch sizes and the number of types of parts may be different for each configuration. The simulation of the automated cell using this configuration will indicate the limits of the FMS and will determine how much it differs from the real system. Depending on the results, the selected facility can be modified in order to improve any unwanted features.

The economic evaluation of the FMS is important in the design phase. This study allows investigation of alternative solutions. The performance of the selected FMS can be compared with that of a manufacturing system which has the same production capacity and the same types of machines, but is structured and managed conventionally. The design of the production control system may proceed alongside this study.

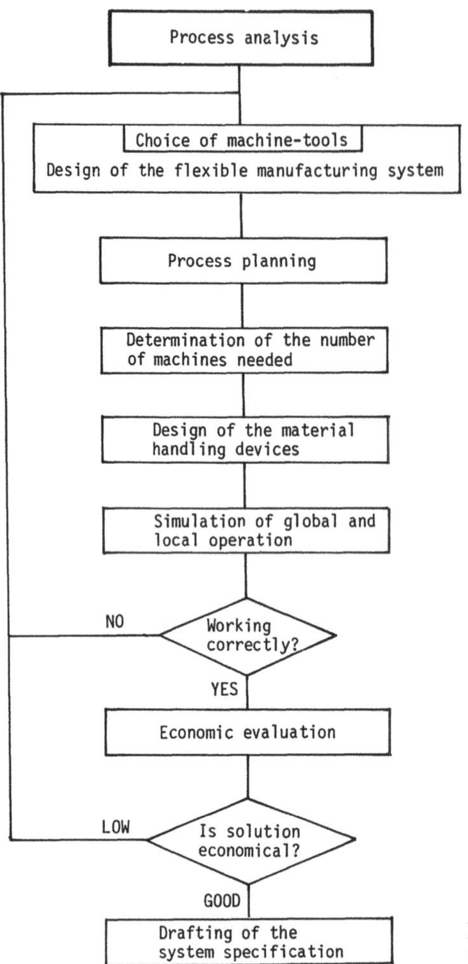

Fig. 5.25. Design of the parts and machining process

Design Stage of Production Control System

There are two phases in this stage. The first phase is a broad definition of the system, which is based upon the analysis that was made and the operating specifications given by the firm (Fig. 5.26). Here the structure of the new system, which is integrated in the existing framework of the firm, is defined (Fig. 5.27). The second phase of this study is the definition of the detailed functions of the system components and the selection of the machine tools. For this purpose the results of the analysis phase and of the general definition phase of the system are used, together with specifications given by the firm, to form a hypothesis. The result of these activities is a preliminary design of the flexible manufacturing system which is represented by GRAI nets. The next step is a confirmation of the results in the simulation stage. The simulation is done for normal and disturbed operations. It allows completion of the study for the new structure and may lead to new hypotheses.

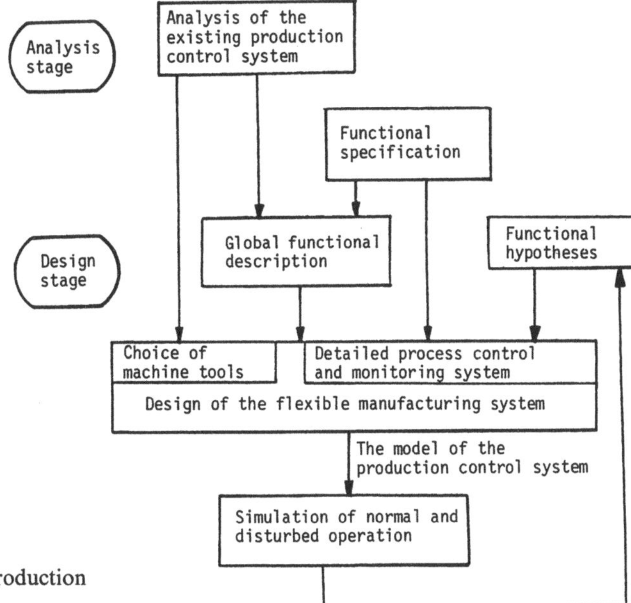

Fig. 5.26. Design of the production control system

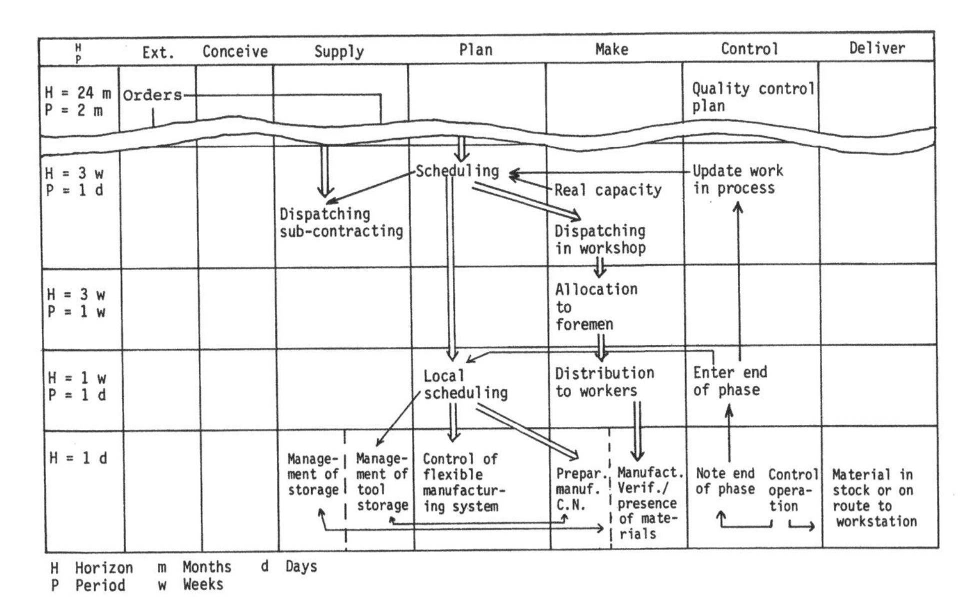

Fig. 5.27. Top-down analysis of the new production control system

The last step is to define the process control and monitoring functions in detail using the GRAI method. This is done by progressing upwards to describe all control and scheduling operations of the workshop (Fig. 5.28).

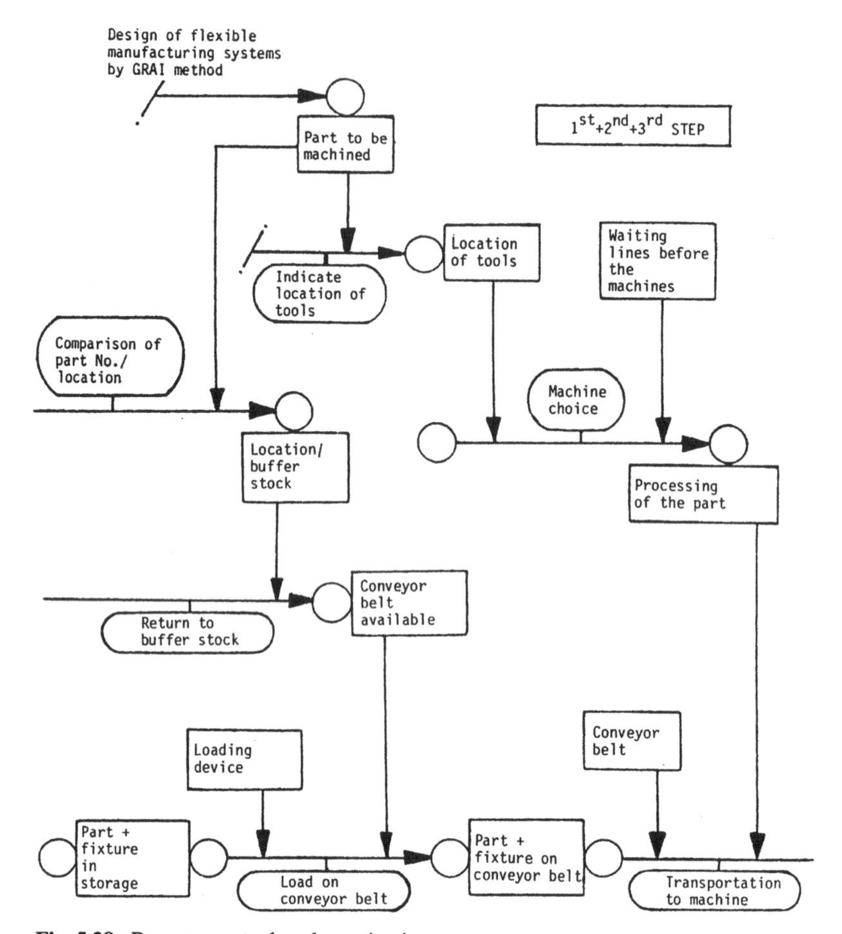

Fig. 5.28. Process control and monitoring

The Use of Simulation Techniques

Simulation techniques are an aid to the design of FMSs. A typical simulation tool which can be used is ECSL-CAPS, developed at the University of Birmingham. ECSL (Extended Control and Simulation Language) is the actual simulation language and CAPS (Computer Aided Programming System) is a programming aid.

The modeling tool used is a diagram of activity cycles. For each important part of the system the process cycle is shown. This consists of active states (activities) and passive states (queues). The various cycles make up the entire diagram (Fig. 5.29).

The system consists of an assembly area where a product is prepared for machining, a palette transportation system which transports parts and tools through the plant and a manufacturing process which processes the parts. Active states are represented by rectangles and passive states by a circle.

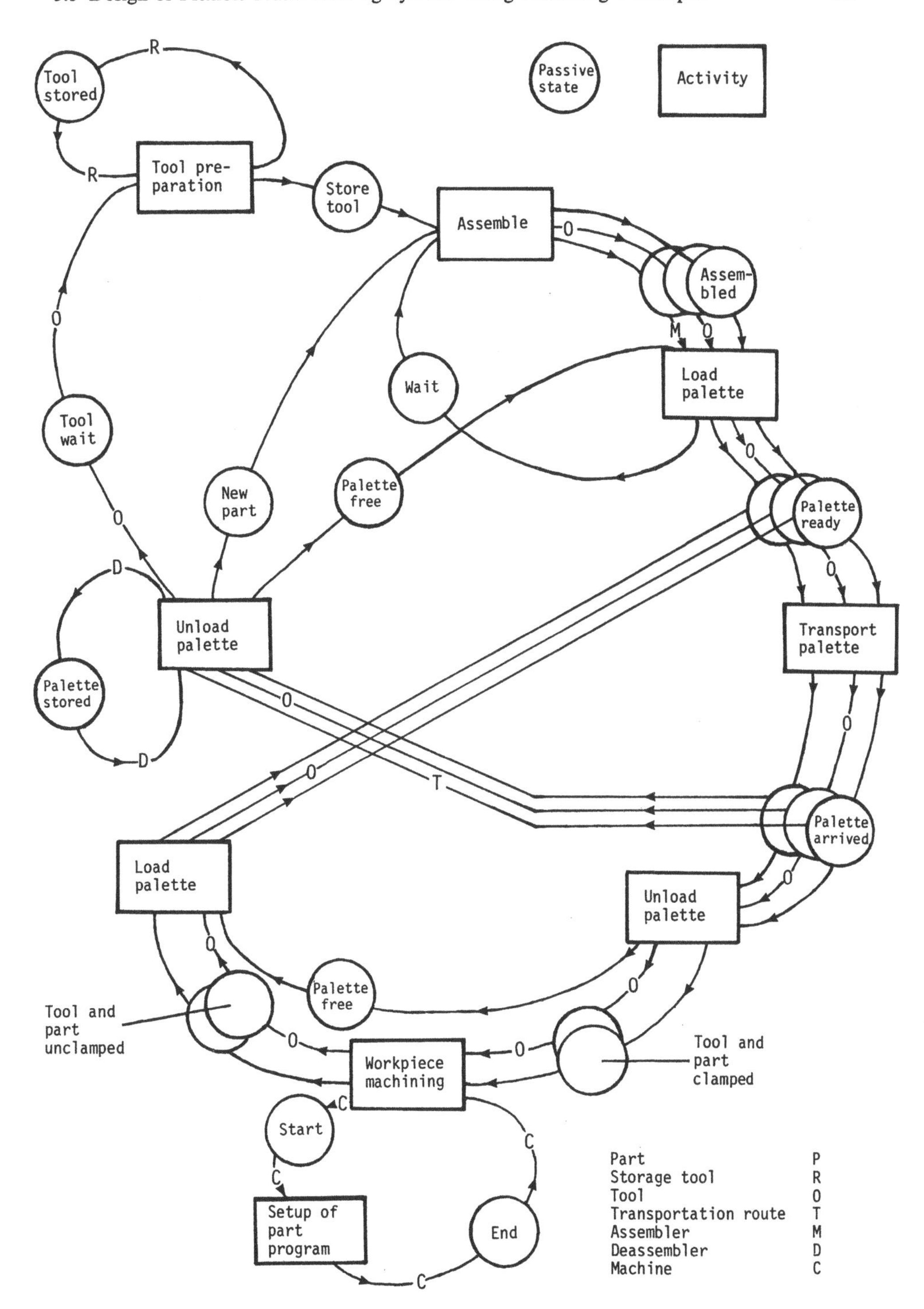

Fig. 5.29. The activity diagram of a simulation

```
      WAS STARTED        36 TIMES
      WAS STARTED        70 TIMES
      WAS STARTED        65 TIMES
      WAS STARTED        40 TIMES
      WAS STARTED        36 TIMES
      WAS STARTED        40 TIMES
      WAS STARTED        65 TIMES
      WAS STARTED       106 TIMES
UTILIZATION OF RETRAN   .1800
UTILIZATION OF DEMONT   .1800

HISTOGRAM OF LENGTH OF QUEUE (NEW PART)
CELL  FREQUENCY
   01141************************************************************
   1  48**
   2   6
   3   5

HISTOGRAM OF LENGTH OF QUEUE (PALETTE ARRIVED)
CELL  FREQUENCY
   01169************************************************************
   1  31*

HISTOGRAM OF LENGTH OF QUEUE (PART AND TOOL CLAMPED)
CELL  FREQUENCY
   9   4
  10   5
  11  41*****
  12 212**************************
  13 459*********************************************************
  14 470**********************************************************
  15   9*

HISTOGRAM OF DELAYS AT (PART AND TOOL CLAMPED)
CELL  FREQUENCY
  15   1*
  21   0
  27   0
  33   0
  39   1*
  45   1*
  51   2**
  57   1*
  63  59***********************************************************

HISTOGRAM OF LENGTH OF QUEUE (PROGRAM SETUP)
CELL  FREQUENCY
   0 190***********
   1 976*************************************************************
   2  24*
   3   5
   4   5

HISTOGRAM OF LENGTH OF QUEUE (PALETTE FREE)
CELL  FREQUENCY
   0 862*************************************************************
   1 338************************

HISTOGRAM OF LENGTH OF QUEUE (PROGRAM SETUP COMPLETED)
CELL  FREQUENCY
   11200************************************************************
```

Fig. 5.30. Results of a simulation

The manufacturing cycle starts with a new part and a tool in a waiting state. The tool is brought to a tool preparation area. From here it can be routed through a holding cycle or placed for pick-up by an assembler. This assembler also selects the part and prepares it together with the tool for transportation. A palette is taken from a pool and the parts and tools are loaded on it. The palette brings both components to the manufacturing process where they are unloaded and clamped for machining and the palette is placed in a waiting state. The program to machine the part is prepared in a special loop. The finished parts and the used tools are loaded onto the palette and brought back, either to the process or to an unload area. Here the palette and the tool are prepared for a new manufacturing cycle.

The results generated by this method are (Fig. 5.30):

The number of times each activity was started
The number of entities found in each queue during a time frame (in form of a histogram)
The value of the entity attributes in the form of a histogram
The work state of the entities in the activities, for example machines and transportation units in operation

The objective of the simulation phase is to verify the operation of the proposed architecture and to quantify the elements of the system. For example to list each type of machine, the conveyors and the minimum number of tools to be used. There are three different variables to be distinguished. They are data resulting from the structure hypothesis, the work hypothesis and variable data on the selected architecture.

With regard to the use of the first type of variables some designers prefer other analytical modeling tools, such as CAN-Q developed by Solberg of the Purdue University. The results depend entirely on the hypothesis applied. Therefore it is essential to explain and comment the hypothesis and the results to the design group. In summary, the GRAI methodology for the design of flexible manufacturing system allows the following:

The determination of the features of new machines
The data acquisition phase for parts, machines and routes
The group technology phase, including the machining process family (size, operations), the parts family (shape, size, raw material) and the matrix of the simplified machining processes.
The selection of the part and machine criteria
The determination of new machining processes and inspection procedures
The design phase, including the architecture (several hypotheses), simulation and economic evaluation.

Conclusion. In the future the FMS will become an accepted tool in producing manufactured goods. However, its design remains a problem. The user has to be part of the design procedures and must be involved with it as early as possible. The GRAI methodology has been successfully applied to the design of three FMS in the mechanical industry.

It will be the task of the system analyst to define the design tools to aid the engineer in analyzing and designing the manufacturing process with advanced design aids. Furthermore a better communication channel must be found to the data processing personnel so that they can learn the structured methodology of designing the production control system.

The goal of the GRAI laboratory is to complete a computer aided design system for FMSs including the following tools:

The GRAI static which structures the decision system
The GRAI dynamic which studies the development of the decision system
An economic analysis to validate the decision structure and
A study of the process and parts to determine the technical specification of the production system

5.6 The Control of the Manufacturing Unit

In this chapter we will describe the state of the art of the techniques used to control a manufacturing unit. First we will recollect the basic scheduling principles and will discuss new control aspects which were conceived for industrial applications with the help of data acquisition or real time monitoring facilities. In the second part we will present an approach to design controls of a manufacturing unit.

5.6.1 Scheduling

During the last ten years scheduling has been a subject of interest to many researchers and manufacturers. They hoped to improve the productivity and competitiveness of a firm. The scheduling software packages were of two different types:

Conventional software and
knowledge based systems

Only few knowledge based scheduling system are being developed. Much additional research has to be done on this subject.

Job shop scheduling is concerned with selecting a sequence of manufacturing operations whose execution leads to the completion of an order. It also assigns times and resources to an operation. The scheduling problem is divided into two separate stages. The selection of process routing is the result of planning while the assigment of times and resources is the output of scheduling.

5.6.2 Classification of Scheduling Problems

The different types of scheduling problems can be divided into three classes which are either a function of time (T), duration (D) or resources (W).

Scheduling can be further divided into 5 subclasses of the types: (T) problem, (W) problem, (T, D) problem, (T, W) problem and (T, D, W) problem.

The (W) problem can be considered as a degenerated problem. The duration is often directly connected with the availability of the resource. For this reason a (T, W, D) problem can often be considered as a (T, W) or (T, D) problem and a

(T, D) problem may sometimes be treated as a (T, W) problem. For each subclass various problems may have to be solved, depending on external, internal, relative or absolute constraints. Internal constraints are often relative and external constraints are absolute.

Another classification method uses the following criteria:

The resource(s)
The number of parts required for a job
The job number

5.6.3 Scheduling Method

There are three different basic approaches used for scheduling of batches:

The first is the chart method. It consists of a calender in which the job starting and finishing dates for each operation are recorded.
The second is a serial method using priority rules. The method uses heuristics to classify the operations according to the priority of an order.
The thrird approach is an analytical one. Here scheduling is regarded as a dynamic programming problem using integers and queues.
The goal of all methods is to minimize the time to process a part through a plant.

The Chart Method

Information and orders are usually centrally processed. Sequencing of a batch is done by assigning the available capacity of the machining centers and machine tools to the work order. In this way the starting dates for each operation are defined.

The best known method for scheduling is the Gantt chart (Fig. 5.31). It describes the status of each machining center, each machine, each worker, and each batch.

Often several manufacturing operations can be executed at the same time. In this way the operation cycle can be shortened. There are numerous possible

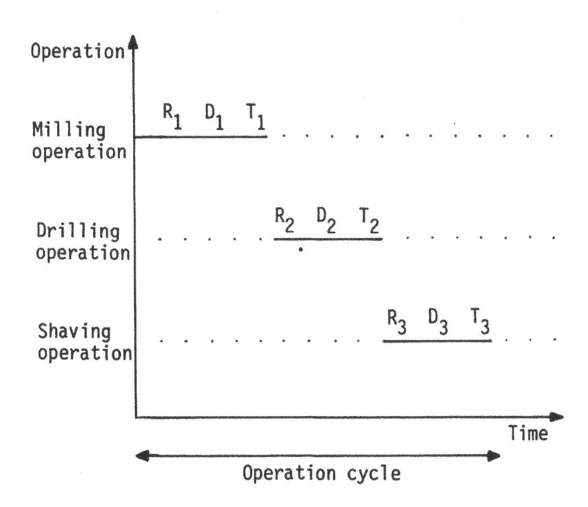

Fig. 5.31. Gantt chart, original schedule

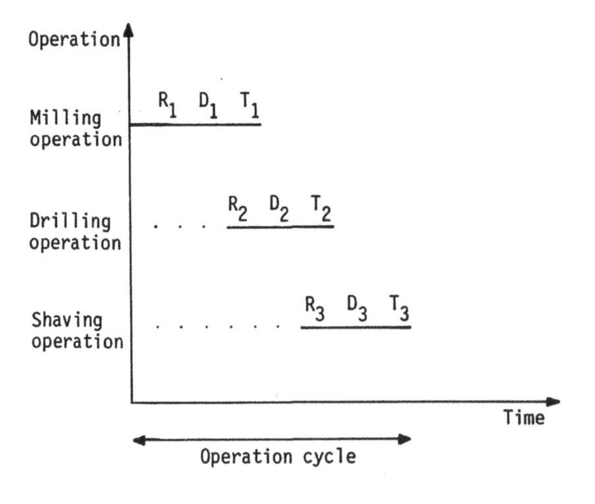

Fig. 5.32. Gantt chart, first revised schedule

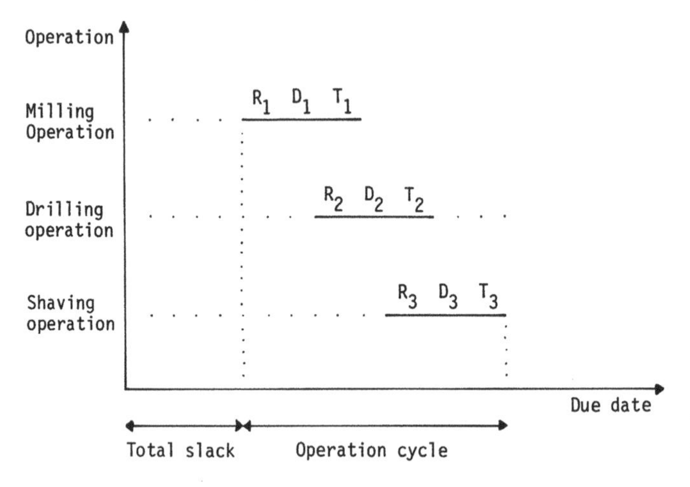

Fig. 5.33. Gantt chart, second revised schedule

operating sequences. Often they must be modified when conflicts exist or an equipment breakdown occurs (Fig. 5.32).

The most frequent scheduling methods are:

Backward scheduling
Forward scheduling, or the
Hybrid method

For backward scheduling the actual due date is the completion date of the last operation for the last batch (Fig. 5.33). Whereas the forward scheduling method starts with the date of the first operation at which the batch enters the workshop. Scheduling slack occurs with the backward method before the operation and with the forward method after the operation. Both scheduling methods have their drawbacks. In the first case due dates frequently overlap. In the second case it ofter happens that an urgent order is added and then the entire planning order must be changed. With the hybrid method slack occurs on both ends (free slack).

For complex scheduling operations the use of Gantt chart is difficult and time consuming.

Heuristic Methods (Serial)

If the scheduling problem is very large or if several problems are considered it becomes very time consuming to obtain an optimal solution, even with the computer. In such cases heuristic scheduling algorithms produce reasonably good suboptimal solutions. The heuristic methods use priority rules for scheduling. As soon as the most important operation is scheduled, the next most important one is chosen and processed. The algorithms stops when all operations have been scheduled. The following different rules are applied:

The processing time rule
The due date rule
Rules using neither processing times nor due dates
Composite rules

Processing Time Rules

Several different processing time rules are used. The most efficient one is the shortest processing time sequencing rule (SPT). If F_j is the flow time, which is the time a job spends in the system, and n is the number of jobs to be scheduled then the mean flow time can be defined as

$$\bar{F} = \frac{1}{n} \sum_{j=1}^{n} F_j.$$

With the SPT method this mean flow time $\bar{F}$ is minimized by scheduling the jobs in nondecreasing order of processing time. In other words the highest priority is given to the operation which has the smallest processing time.

Other rules may be based on the following principles:

1. The highest priority is given to the operation with the longest processing time (LPT)
2. The job with the smallest total work or number of operations is done first (TWORK)
3. The job with the largest number of operations is done first
4. A machine is assigned to the available job with the shortest remaining processing time (SRPT)
5. A job is assigned to a machine on a first come first serve basis (FCFS)
6. This scheduling operation is a combination of the SPT and FCFS rule

The SPT rule is the most effective one to minimize the mean flow time $\bar{F}$.

Due Date Rules

The due date rules determine the point in time at which the processing of a job is due to be completed. For this reason the most significant performance measures are tardiness criteria. To devise a dispatching strategy which employs the due-date rule to attain a good tardiness performance is a much more complex problem than the minimization of the mean flowtime $\bar{F}$. Examples of the due data rules are:

1. The job which has the earliest due date (EDD) is done first
2. Priority is given to the job for which the ratio of slack to processing time is a minimum. Here slack is equal to the difference between the due date and the time the job could be completed if there was no delay. This method is called minimum slack time (CMST)
3. Priority is given to a job whose ratio of slack time to the remaining operation time is a minimum (S/OPN)

The most efficient algorithms use the slack per operation rule.

Composite Rule

These scheduling algorithms contain a parametric family of priority dispatching rules and are generally known as COVERT rules. They are based on two types of information: the project delay cost c_j which is measured as anticipated tardiness for a given job, and the operation processing time t_j. Priority is assigned to the job which has the largest ratio of c_j/t_j. The critical jobs are those with small or no slack.

Analytical Methods

Most analytical methods minimize the mean flow time $\bar{F}$. The most important representatives are the methods of potential constraints. They are based on network analysis. Two widely used representatives are:

The critical path method (CPM)
The program evaluation and review technique (PERT)

Analytical methods have not been able to solve the majority of real and difficult problems. However, they can be of help when batches are processed on a small number of machines and each batch uses each machine. The problem is often simplified as follows: The batches are known and available to each machine and the number of the batches in the queue does not change before all operations are done.

We give an example for MPM method (Fig. 5.34).

In this figure the numbers in the circles represent the number of a job and the number on an arc designates the processing time of a job. The problem consists

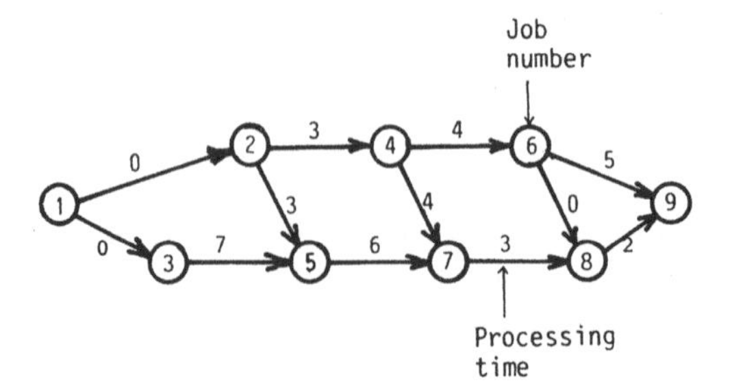

Fig. 5.34. Example of a MPM diagram

of finding the smallest mean flow time. It is resolved by finding the critical route where the sum of the processing times for the operations on this route is minimised. The critical path in this example is 1, 2, 4, 6, 8, 9.

5.7 GRAI's Approach to Manufacturing Control

5.7.1 Introduction

GRAI has been working in the field of production control for ten years. After the various existing control methods were studied the conclusion was to develop a new approach for production control, specifically to solve the scheduling problem.

The scheduling function is very important for production control. It is linked to other functions like order release, data collection and human decision making. The GRAI approach not only concerns itself with the theoretical scheduling problem but with all production control functions which include scheduling, order release, production monitoring, and data collection. It also compares the estimated plan with the actual state of the production and control decisions (Fig. 5.35).

GRAI proposes a new conceptual model based on a hierarchical decomposition method. It gives a general model for production control and thereby introduces the concept of the control module and a structure for a decision aid system (DAS) for production control.

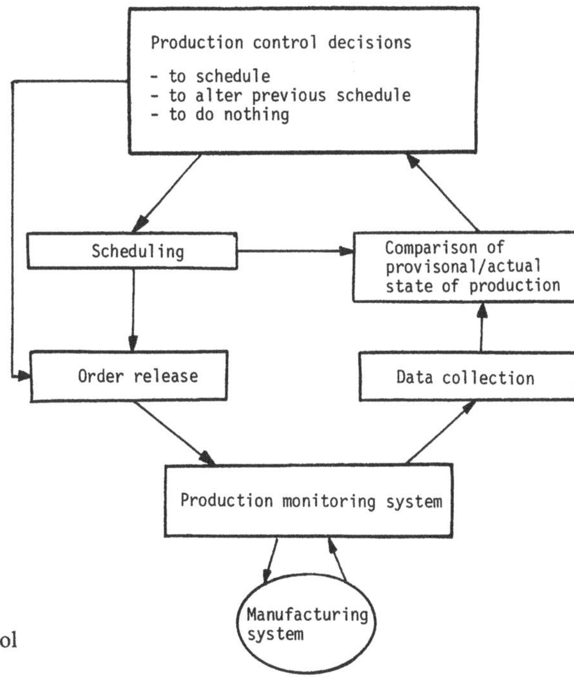

Fig. 5.35. The production control functions

In previous works GRAI developed a DAS for flow line production. A new version of a DAS is for job shop production. This modelling approach will be presented here. It uses a structural, graphical and logical module. The first model is a conception of the general structure of the system, the second is the visualisation of the former model and the third is related to the formalization of the links between each element of the structure and the process inside each element.

5.7.2 GRAI's Approach to Modelling

GRAI has developed a method of analyzing a production control system, in which the production control system is structured as a hierarchy of decision centers. In each decision center the activity concept is the basic element of the structural model. An activity is represented on a screen to easily display a view of the production control structure. With it a decision maker can quickly obtain information on the state of the production system. In addition a logical tool is defined to specify the logical operation of an activity. It is a first order predicate. The structural, graphical and logical tools are related to the structural, graphical and logical models of the production control system. The three concepts will be discussed briefly.

Structural Model

The lowest level of a production system is the shop floor. An activity at this level can strictly be defined by a quadruple whose elements are the components and materials which are required to start the activity, resources necessary for performing the activity and a stock of finished products.

Production rules describe the conditions for performing an activity and explain the relations between the three former elements.

In the shop a job is considered as an activity if its process requires time and resources. The job must be planned by the control system. Planning of a job depends on the product, material, resources and on the production rules. The main types of activities are manufacturing purchasing, material movement and other activities, for instance, timing.

Graphical Model

A graphical tool was developed to represent the different manufacturing activities. There are three different type of symbols used (Fig. 5.36):

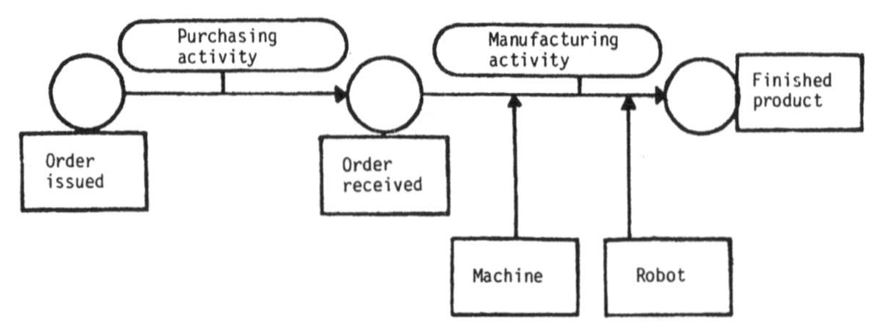

Fig. 5.36. Description of a sequences of activities by the GRAI method

An arrow connecting symbols represents an activity.
An oval shaped box represents the beginning or end of an activity.
A square box with an arrow pointing from it designates a resource needed to operate a manufacturing system.

The graphical symbols can be linked together to show a sequence of operations.

Logical Model

An activity is defined by three types of logical conditions. The initial conditions (for instance the availability of a products), the continuity condition (for instance the availability of resources during processing) and the final conditions (for instance the finished products need sufficient room to be stored). These conditions describe the execution of the activity and are easily formalized by predicates of the type:

PRES (Q, STK (X,T), Y):
There are Q products in process by activity X at time T, they came from activity Y.

DISP (Q_1, X,T, + $Cy(X)$):
There are resources available to produce Q_1 products by the operation following X at time T + cycle of X.

CSTK (Q_2 SUIV (X), $T + Cy(X)$):
There is a simple space to stock the Q_2 products processed by the operation following X at time T + cycle of X.

The Structure of the Computerized Tool

A computerized tool must have the various properties of the three models described above. It must be structured to provide the proper programming techniques to built systematically the shop control system. It must also be logically defined and use predicates so as to have the same operating characteristics as the logical model. If possible it should have graphical features. These prerequisites were implemented in the individual modules of the GRAI system. For each there is an equivalent computerized building block. This set of conditions is important in building a control system which runs close to reality.

Control Module

The computerized tools have similar types of control modules as are used in computer science to solve synchronization of processes. However, in this context it is more difficult to build a model because of the numerous data and constraints involved. The basic idea is the same. In the following we will compare computer control problems with those of manufacturing control. Then we will present the defined tools with their main specifications and show how they can be used in the control system.

Use of Control Modules

In computer science control modules are generally used to monitor tasks to be processed by a computer. A control module is similar to a task. For instance, reading from or writing to a disk or allocating a memory to several users. The module actually controls realtime events, for instance an event which requests that a message be written. It may be triggered by a time sharing programm.

In a shop a program is a routing process. A routing may involve a set of jobs. These jobs are to be carried out by specific activities. Monitoring of these activities is necessary, similar to monitoring of a task in a computer. This is one of the similarities between the approach of computer science and that described here for shop control. The main differences between shop control and the control of a computer are:

1. In the shop the job sequence is influenced by the routing: This is of less significance in the computer.
2. In a shop the tasks are often planned for a long time period. In the computer the tasks are usually queued and executed in realtime.
3. For a shop operation a realtime simulation can be performed which is difficult to do for a computer operation.
4. In the shop disturbances of the planned operation are the only unknown events. In a computer operation there may be several unknown events.

Specification of the Control Module

Figure 5.37 shows the structure of a control module and its input and output lists. In the input list there is a description of the tasks to be performed by the activity. For each task the following five elements may be defined: the name of the product, completion data, routing specification, duration of the operation, and name of the worker. In the output list there are the following three elements of a task: the name of the product, routing specification and the completion date of the operation. This description of these two lists is an example of how a control module can communicate with the next higher level. The execution of a control module depends on the contents of the input list. Whereas the contents of the output list give the state of production at a specific time.

Figure 5.38 shows an example of the use of the control modules by the simulation. There are three main hierarchical levels. All levels of the real production system and that of the structured program correspond to each other. For each type of activity an own type of control module is defined. Each level has prede-

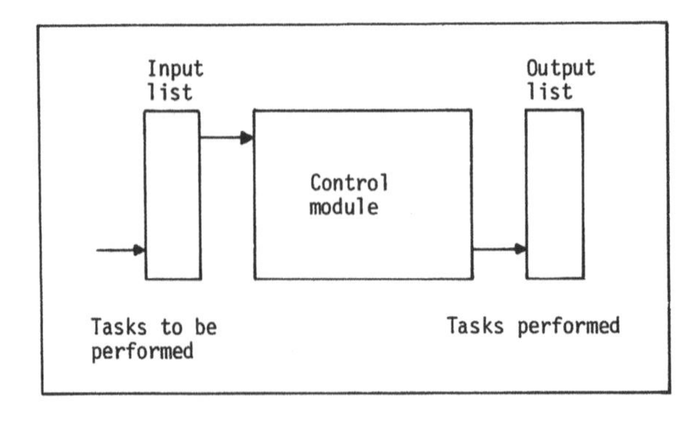

Fig. 5.37. Structure of a GRAI control module

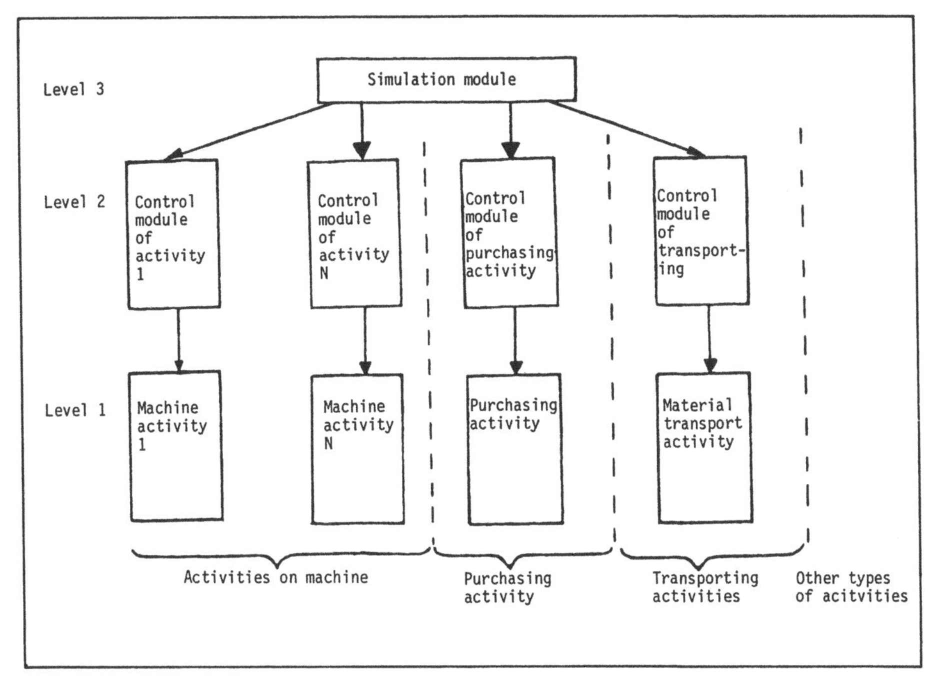

Fig. 5.38. Example of a simulation

fined functions which are part of the algorithms of the simulation. An activity module is a controller for a manufacturing operation. The control module is located on the middle level. From the input list it chooses the product which is to be scheduled first and starts and supervises the activity module. At the highest level the simulation algorithms is located. It manages the input and output lists according to user defined criteria. It contains the optimization method chosen for the analysis of the shop. The three levels will be described in more detail.

LEVEL 1: Activity module

Typical activity modules are for the control of machining, purchasing, material routing etc. An activity module is started by a control module. It receives information from this module, which specifies the operation which has to be performed, the duration of the operation, starting conditions etc.. When the operation is over the activity module transmits to its control module the actual duration time, which was determined from the resources available and the start and the completion times of the operation. If the activity module transmits no information then the operation cannot be done for the time being. When the activity module has finished its job it gives control back to the control module.

The operation algorithms are a set of production rules which are written with the help of first order predicates. A predicate is a construct which may have access to resources that it controls. For example the predicate CSTK shown in the section Logical Modules has access to information on the availability of material in stocks. The parameters of the predicate function determine what has to be tested. For instance when an operation can be started after reading and testing the availability of resources, the activity module calculates the time intervall during which the operation will be performed. From this calculation the availability of different resources is updated. Then the activity module may call its control module to take back the control.

LEVEL 2: Control module

A control module monitors the execution of an activity module. It points to the activity module and determines which job has to be done. In this way it chooses a product according to predefined criteria, e.g. product availability. A typical command would be "take the first product on the list and process it". Such instructions are also written with first order predicates. A predicate defines a function which has access to the input list which is determined by the parameters of this function.

When a control module receives the operation results at the end of an activity, it updates its lists. In this way the completed product is deleted from the input list and the name of the product, the operation number of the routing and the completion date are written on the output list. Depending on the sequence of operations which are needed to make the product, the control module also transfers to the input list of other control modules the name of the produced product, the date when the product should be ready, the next operation number of the routing, the duration of the operation, the name of the worker, etc.. Thereafter the control module gives the command back to level 3.

If an operation cannot be done the activity module transmits no information and the control module examines the input list to find another product which fits the given criteria and tries to schedule it. If it cannot find one it gives the command back to level 3.

LEVEL 3: The simulation module

The simulation is done in real time. The incrementation of the clock control must be given to the highest level. Depending on the user defined criteria and rules the simulation module monitors all control modules. It manages the input lists containing information about which product must be produced on which machine. It performs the simulation by choosing the control modules needed for an operation. The sequence of operations is determined by the simulation algorithm. The third level has knowledge of all activities, available resources and products.

The structure of the simulation module has three important elements:

1. The hierarchy gives the control system a great flexibility. It is possible to add other modules to level 1 and 2 without rewriting the whole simulation.
2. Each level has its own criteria, for instance
   Level 1, availability of resources
   Level 2, availability of products
   Level 3, availability of activities
3. The programming structure may be implemented on a multiprocessor system with the aid of a concurrent programming language.

Application of the Method

The GRAI method represented in the previous sections was used in a small factory (50 persons) of typical job shop fabrication. The shop contained manual and automatic machines (CNC) and manufactures small and medium size lots. The work is dispatched between several work centers. The foreman, his workers and their machines are grouped together in work centers. There are four such centers. Additional work centers exist for CNC machines and quality control. The control system is used to help the production manager decide what to do when a new order is entered or a disturbance (material shortage, machine failure) occurs. It helps the decision maker in his planning work. The data base of the system contains information on the model structure of the shop and the control which has to be performed (Fig. 5.39).

A decision maker has access to the following modules:

1. Interactive communication with the data base. This function allows him to input, alter or display the data related to the shop.
2. Planning. This module plans the activities to be performed in the shop, according to availability of resources and time constraints. The structured programing is based on the control modules concept.
3. Simulation. With this module the decision maker can simulate the shop operation at any time. The control modules are used to structure and program this model.

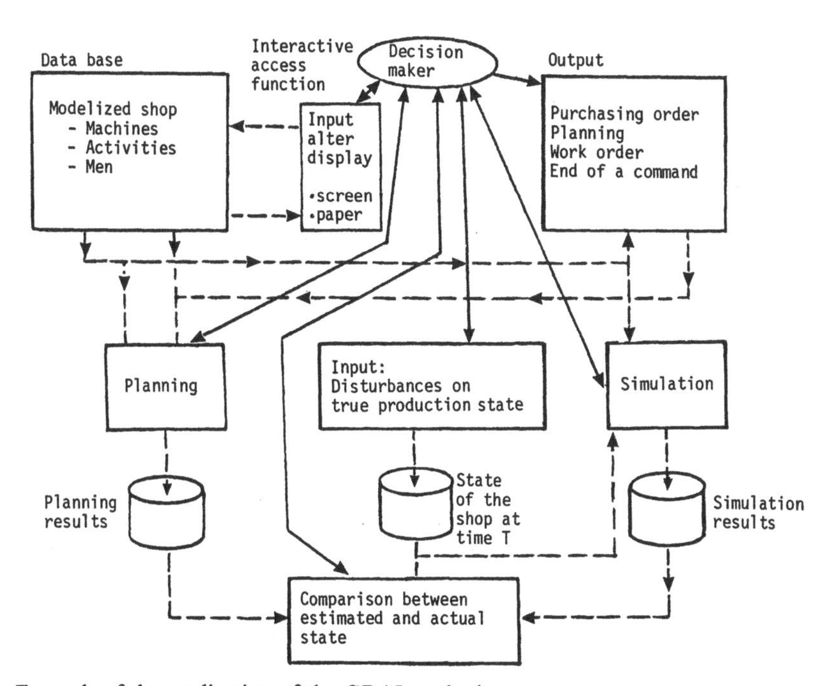

Fig. 5.39. Example of the application of the GRAI method

4. Input of the actual state of production. This module gives the realtime state of production to the control system. It allows comparison and simulation modules to run with actual data. Depending on the results the decision maker can decide to change the data of an order or to do replanning if necessary.
5. Comparison. This module allows the decision maker to compare the estimated and actual states of the production. According to the results the decision maker may decide to replan.
6. Output module. This module gives reports for purchasing, planning, and work scheduling.

Conclusions. The conceptual method shown in this section tries to solve control problems in a job shop-type production environment. The approach uses a new conceptual tool to build the control system. The main part is the control module. Each activity of the shop is related to a control module. In a fully automated shop-shop production system this structured method allows a control system to be designed in which the decision maker is replaced by a computerized module.

6 Computing Aids to Plan and Control Manufacturing

R. Dillmann

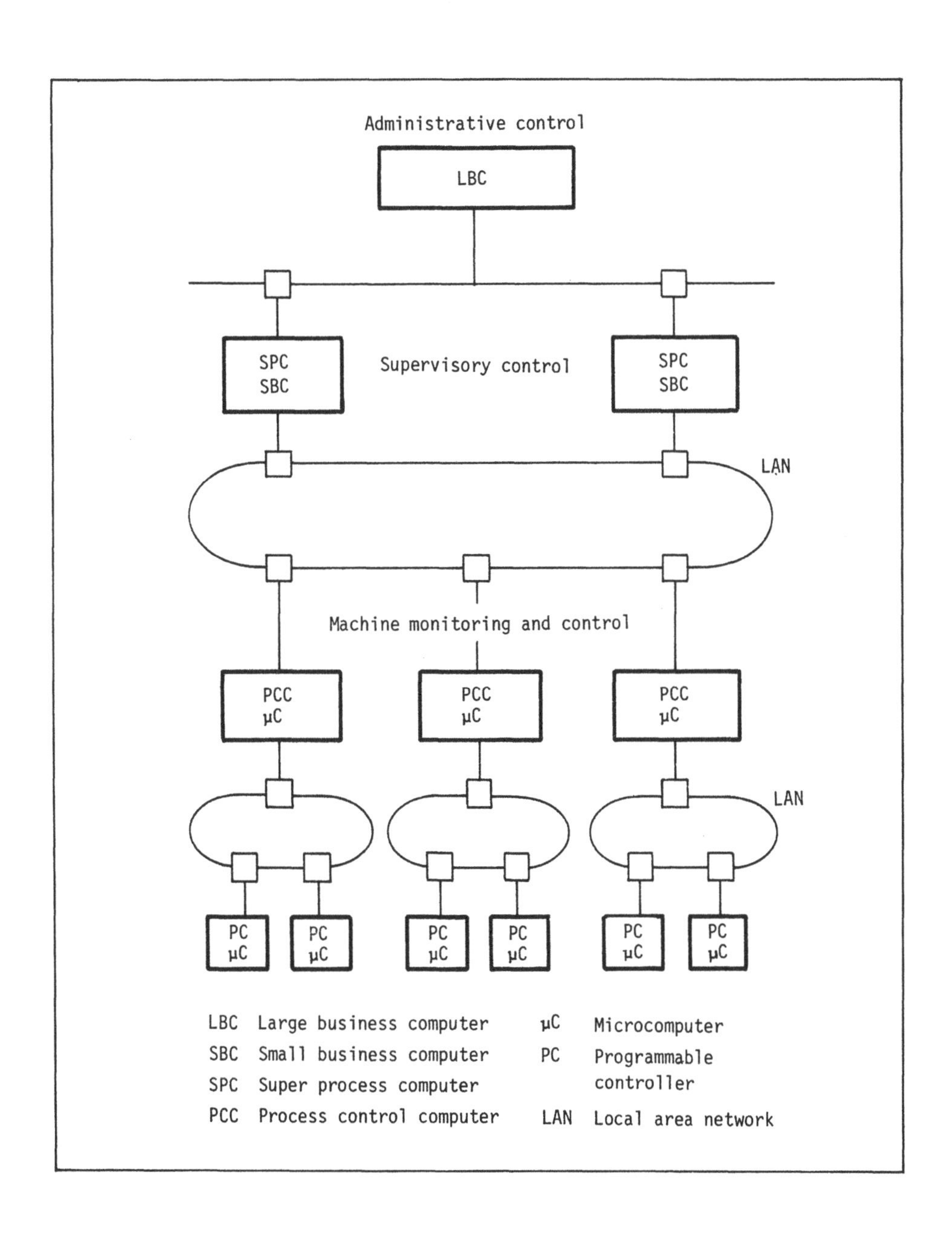

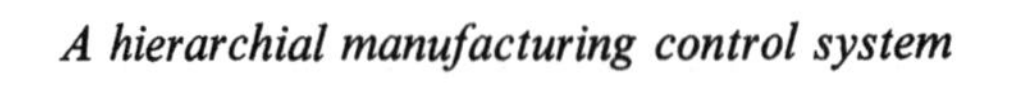

A hierarchial manufacturing control system

6.1 Hierarchical Computer Control Equipment for Manufacturing Systems

6.1.1 Introduction

Current developments in VLSI- and software technology are leading towards the concept of distributed computer based control systems. This development is supported by standard programming languages such as PASCAL, COBOL, PEARL and ADA, high speed communication methods and advanced data base techniques.

Hierarchical control structures support planning and supervisory tasks on different information levels of a plant. Typical functions are production scheduling, equipment control, material flow control, production data acquisition and data processing as well as quality assurance. Figure 6.1 illustrates the wide spectrum of control tasks which have to be performed in manufacturing. To ensure an efficient and versatile computer control architecture, the structured control concept has to be combined with a data base concept on each control level.

Three basic control levels can be distinguished, which by themselves may be structured as a modular hierarchy. The highest level is represented by management information activities where management data is processed. Production scheduling and operational management are a subsystem of the highest control level. From this level horizontal communication to other factories and plants, and vertical communication to local production control levels are also coordinated. The intermediate control level relates to local shop floor supervision and plant coordination, the disposition of material flow, the control of production processes, assembly and quality assurance as well as to the supervision of the lower control levels. Production rate and recourse status information are used on this level to aid production decisions. Horizontal communication with adjacent control areas and vertical communication to lower and higher control levels in the hierarchy are performed via high speed serial data links. The lowest level consists of modules which control technologic and geometric operations. It also contains the machine specific control algorithms (e.g. DDC-algorithms). Future supervision systems will be characterized by increasing intelligence located in the different hierarchical control levels [1–3]. Adaptive controls (ACO, ACC) for machine tools, and self optimizing- and learning controls for assembly robots will be part of the hierarchical control architecture.

Reliability, extendability and flexibility can be achieved if there is a clear separation between all control levels (micro- and macro structure). This modularity will simplify the system implementation. A large amount of effort will be necessary to establish standards in the area of communication networks, programming languages, data bases and VLSI elements. Examples of national and international standardization efforts are work shops organized by VDI, EWICS, Purdue University or the ICAM project.

6.1.2 Definition of Hierarchical Control Systems

The hierarchical manufacturing control system in Fig. 6.2 is separated into different control levels [4]. It has to serve many different manufacturing functions.

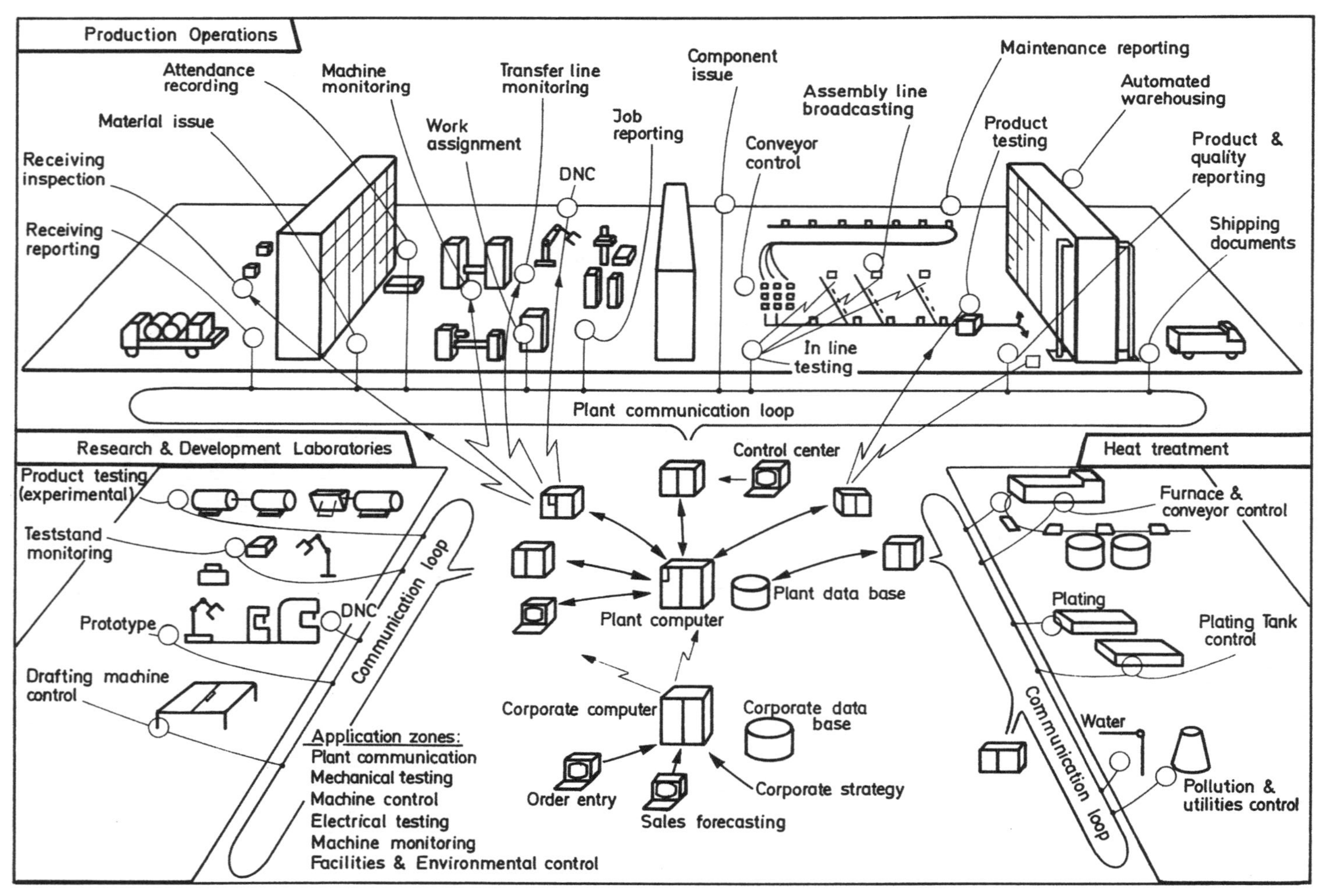

Fig. 6.1. Structure of a computer integrated manufacturing system (according to IBM graphic)

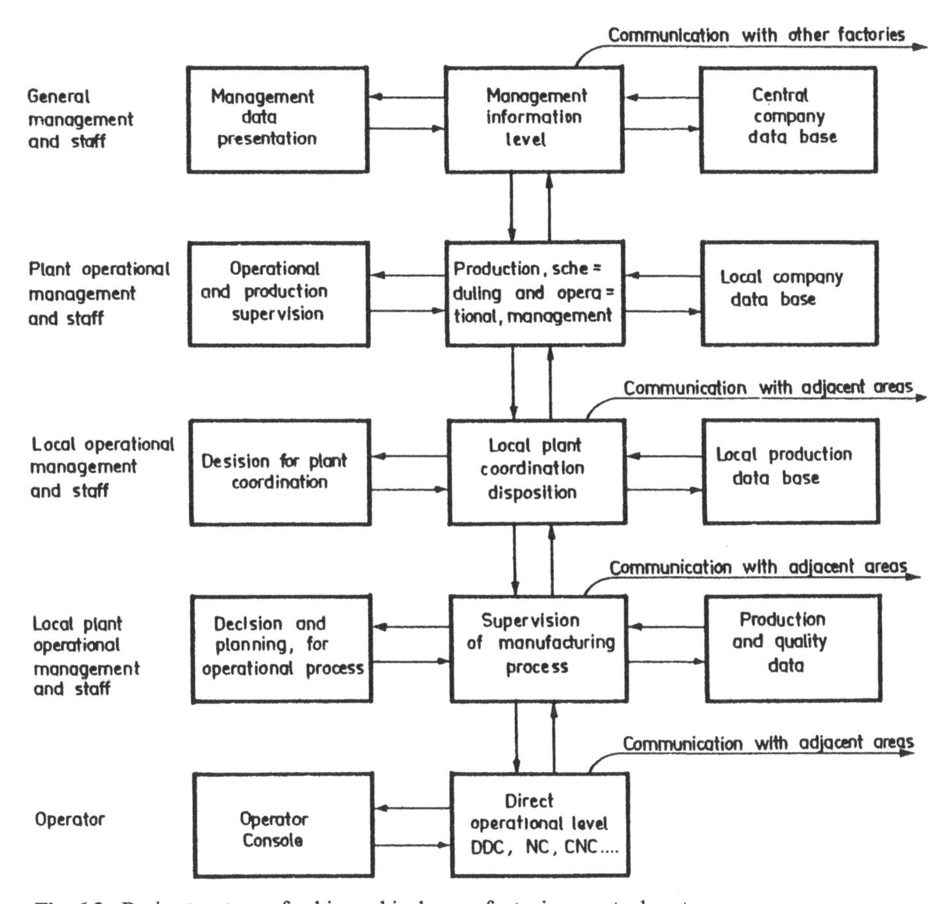

Fig. 6.2. Basic structure of a hierarchical manufacturing control system

There are several hierarchies which have to be considered when a computer control system is conceived:

- Structural hierarchy of the factory: A factory can be composed of multiple subsystems which are located in different areas.
- Organisational hierarchy: Defined parts of a distributed subsystem comprise an organisational unit.
- Management hierarchy: The management of a factory operates on different control levels.
- Hierarchy of information: Decision making is performed using information and data from different control levels.
- Structural hierarchy of manufacturing equipment: The material transport system, machine tools, assembly equipment and quality assurance systems are structured into subsystems which can be decomposed according to the task they perform in manufacturing.
- Hierarchy of operational control: The operational control system has a modular structure and consists of different control levels.

The integration and implementation of control strategies for each hierarchy requires unique programming languages, standardized communication data highways, data base systems and a hierarchy of compatible computer families. The control installation has to ensure the precise operation of each manufacturing unit of the plant at the highest possible efficiency. Scheduling and supervision functions have to be applied to provide optimal synchronisation between all manufacturing units. Any malfunction on the operational unit level must be corrected by the supervisory control units. Disposition and coordination operations have to be performed to assign to each subsystem its control task so that all units work together as one entity in the plant. Proper and optimal performance of each unit has to be ensured. The system has to be able to react systematically to emergencies, e.g., to a decreasing quality. The overall manufacturing system must be able to carry out its scheduling and control functions for the plant to produce the product at minimum cost by efficiently using time, material, labour and energy. The required reliability, availability, flexibility and extensibility of the control installation make a modular, fault tolerant, hierarchical computer system architecture mandatory. It must be able to detect any faults, to locate the problem and to correct it. For this reason redundant computer components may be necessary.

The tasks of the lower operational level are performed by dedicated algorithms which operate the PC, NC, CNC and DDC-controllers and can be done by microcomputers with numeric and arithmetic capabilities. Disposition and coordination tasks require processing of larger data volumes and are usually done with the aid of powerful micro- and minicomputers. The control tasks at the supervisory level are usually very numerous and complex; they need a large business computer.

6.1.3 Control Tasks at Each Level in the Hierarchy

The tasks at each control level are related to those of the manufacturing engineering hierarchy. Common tasks which are carried out in nearly all manufacturing systems and which differ only on the lower operational levels are of the following types:

- Direct control of the manufacturing process
- Product quality assurance
- Production data acquisition
- Plant coordination
- Production scheduling.

Direct Operational Level. On the operational level of every manufacturing process, a wide variety of equipment such as machine tools, material handling systems and assembly lines are directly controlled by dedicated microcomputer based controllers (see Fig. 6.3). Any problems which may exist in these units have to be detected immediately and corrected within the shortest possible time. Self diagnostics and constant updating of stand-by controllers have to be performed on-line. Direct control is done by PCs, NCs, CNCs and robot controllers with the aid of DDC algorithms. The use of standard 16- and 32 bit microcomputers, as

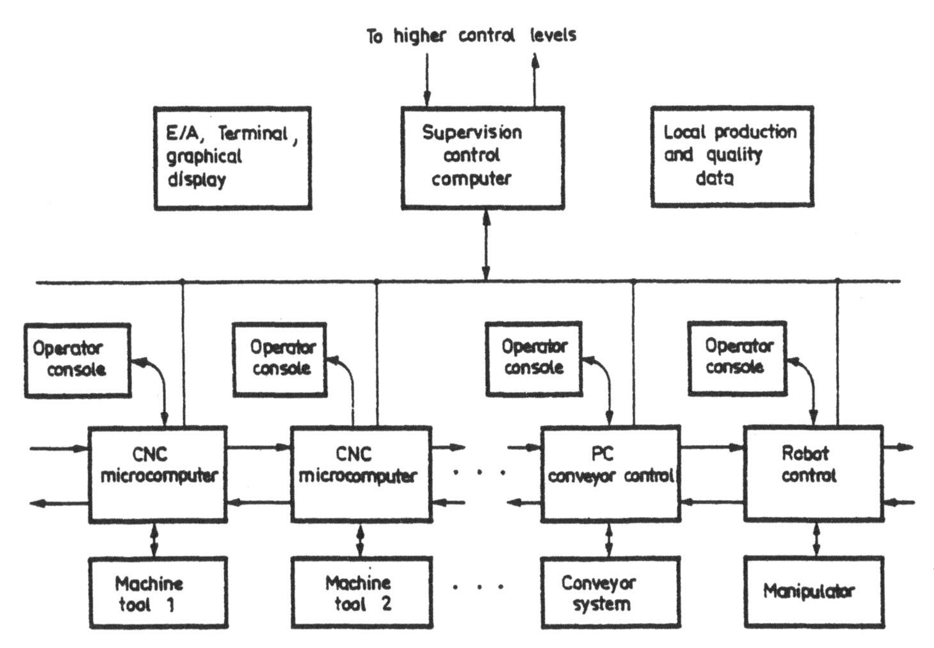

Fig. 6.3. Distributed components on the direct operational control level

well as one chip computers with analog signal processing capabilities, allow structured modular implementation of direct control circuits. Vertical communication with higher levels is necessary to transmit collected information on production, quality, raw material, process parameters, detected errors and nonresolvable problems caused by conflicts.

Programming is performed by low and medium level languages such as Assembler and interpretative languages for machine tools and robots. Control programs generated on higher control levels are written in high level languages such as ADA, PASCAL, FORTRAN, PEARL, APT- and BASIC or their dialects. Compilation, code generation and program tests are done on a higher level which transmits executable programs to each manufacturing unit.

To allow shop floor programming and service facilities for the operator, each unit has a RS 232 interface as part of the programmer console. In the near future, efficient low cost color displays with simple graphic capabilities will be available as standard consoles.

Diposition and Supervision Level of Machine Tools. Manufacturing cells with different machine tools, conveyors and material handling systems, or a group of NC tool machines are controlled by a local plant computer which also collects production data from the manufacturing cell under its supervision. The distribution of control data (i.e. DNC control) and the optimization of all unit tasks are performed on this level. Any emergencies from a lower level control or from this level will initiate correction actions to maintain the production schedule. Quality data acquisition and quality assurance operations have to be performed on-line.

An additional assignment will be to diagnose and update stand-by systems. Because of the number of data to be processed, a micro- and minicomputer configuration with virtual addressing capabilities for disks will perform the tasks. Most of the VLSI manufacturers are designing microcomputer families (32 bit, 16 bit and 8 bit units) which have powerful interfaces and firmware to support communication between the microcomputers. If the higher and lower level control computers are part of a microcomputer family, no code and format transformation is necessary. Otherwise a unique protocol generator with physical – and software layers must be included to do the vertical and horizontal communication.

The supervisory control necessitates a real-time control system which can operate under real-time multitasking mode, and batch operational units which communicate with the supervisory level. To assure economic software development, proper implementation languages should be used. They may be FORTRAN, BASIC and ADA.

Local Plant Coordination Level. The most important task of this control level is the generation of actual production schedules for the local operation. It also has to interact with material flow control and has to process lower level alarms (see Fig. 6.4). The production schedule generated by the management strategy computer on the higher control level has to be adapted to the manufacturing status of the local plant. The efficient allocation of machine resources and the possibility of dynamic response to production problems on lower levels allow cost optimization.

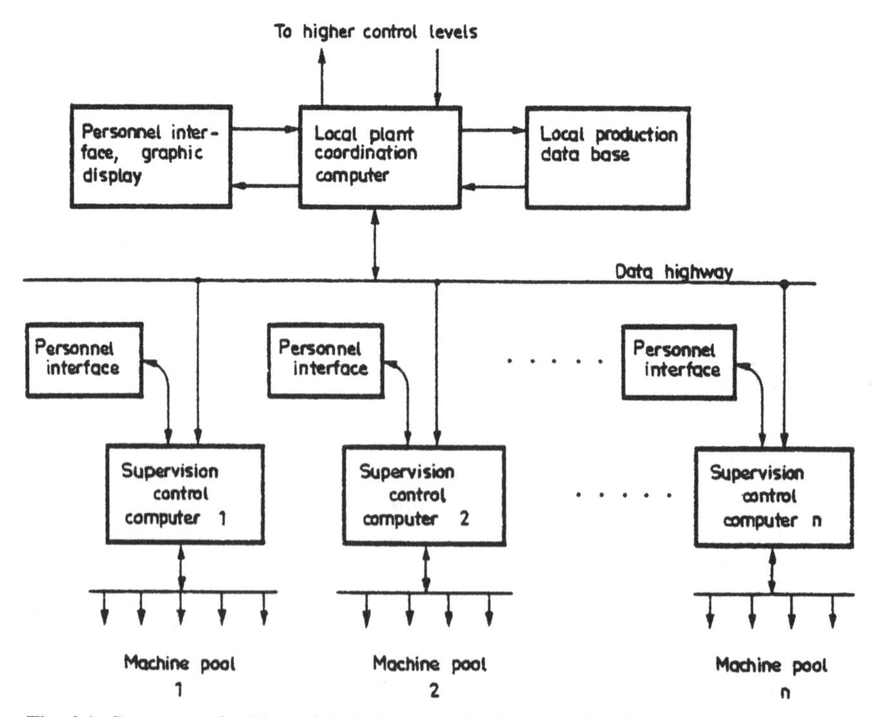

Fig. 6.4. Structure of a hierarchical plant production coordination computer system

A data base containing current information on production, machine tools, the material handling system and assembly lines yields the necessary information for possible rescheduling. Future strategy computers will be able to solve their own production problems with the aid of decision rules and knowledge about the production process. Such expert systems will have a high level interactive man/ machine interface to support planning operations (see Chap. 6.4). The local plant coordination computer has to maintain communication with higher and lower levels as well as with adjacent production areas.

The computing equipment will have a 16 or 32 bit processor and a 100 Mbyte or greater memory. ADA will be the standard language.

Production Scheduling and Operational Management Level. Basic production scheduling and coordination of all manufacturing activities are the tasks of the plant management. The production schedule is drawn up at this level. If properly done, it will efficiently use the manufacturing resources, minimize production cost and speed up the production of the product. In the case of lower level disturbances, or a change of production, a dynamic modification of the schedule has to be made. The warehousing and storage of material has to be directed to avoid unnecessary time delays. Further assignments include the diagnostics of equipment at lower control levels and the updating of stand-by systems. An additional task is the communication with adjacent computers, e.g. for purchasing of material and equipment, sales service and higher management levels (see Fig. 6.5).

Production scheduling has access to an extended data base which contains information on the current production status of the product and its quality. An interactive graphic terminal allows the display of production streams, the location of products and the inspection of statistical production data and cost functions.

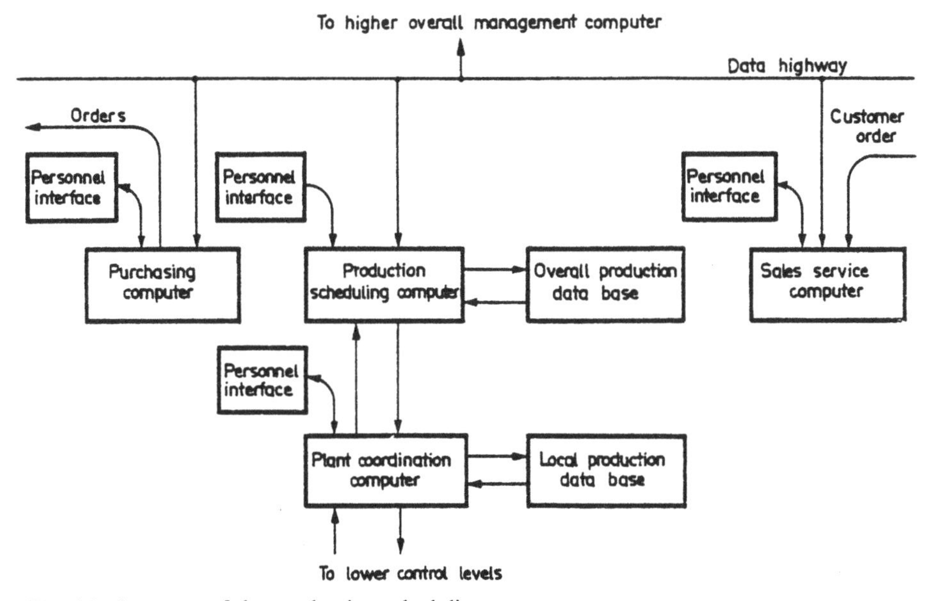

Fig. 6.5. Structure of the production scheduling computer system

Fast access to the production data base is a significant parameter of the scheduling computer. Advanced information management systems with intelligent query capabilities will support the operational management.

The Strategic Management Control Level. From this level, management has access to any control level to obtain data for decision making. Especially the communication with production scheduling, accounting, purchasing, sales and service aid the generation of strategies and operating data for all control levels. The strategic management uses a corporate data base which contains the entire status of the production of the company (see Fig. 6.6). Access to all lower data bases is performed via the hierarchy of control computers. Strategy computers are usually big commercial number crunching machines with a stand-by system to ensure data integrity. Self-check and diagnostic routines assure equipment reliability. Only higher level languages are used for programming. Expert systems with problem solving capabilities are under development. Extended graphic features simplify viewing and inspection of production data.

6.1.4 The Communication Network

The efficiency and flexibility of the hierarchical manufacturing control system depends on the availability of a unique communication network which connects local distributed processing units. Efforts have been made to standardize process

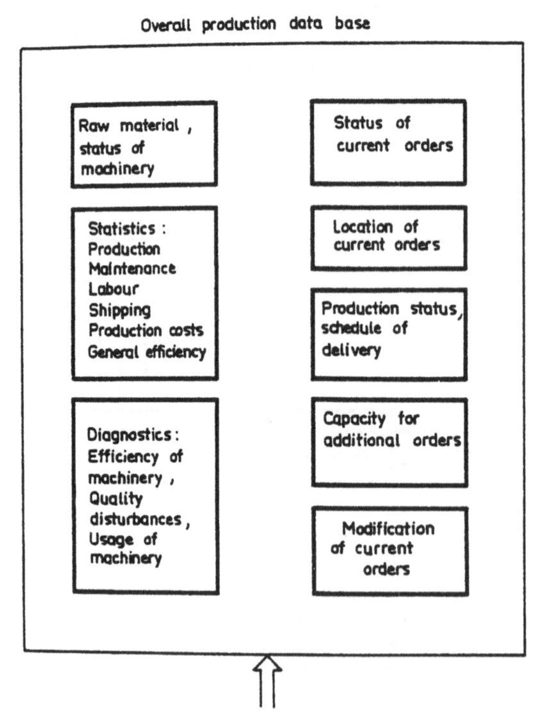

Fig. 6.6. General overall production data base

control communication systems. Different protocols and a great number of communication bus systems have been developed and are used by industry. Most of them employ a protocol hierarchy with different communication levels, including:

- The inter process communication protocol
- The bus access protocol
- The transmission protocol
- The logical link level protocol and
- The physical link level protocol.

The implementation of these protocols is done in hard- and software. Early communication networks were implemented with a star configuration. Today's networks are configured as rings or open or parallel or serial bit lines. They can be configured in hierarchies with interface and control elements (see Fig. 6.7). The nodes in the network can be active, passive or both (send, receive data). Most of the actual plant communication networks operate on a single master basis. This means that only one node has control of the signal processing of the network. In some implementations this master function can be shifted from one node to another. Ring configurations with two bit serial transmission lines are becoming popular. Coaxial cables or fiber optics with adequate interface elements which support logical and physical link levels allow transmission rates of 10 and more megabites per second. Most of the VSLI microcomputer manufacturers supply their systems with communication chips which control access to the bus. They also generate protocols for different link levels, encode and decode signals and perform master or slave functions.

The objects of standardization for an interface are the hardware connectors, their number of pins, pin placements etc. Also the hierarchy of protocols has to be defined, including the codes, message and data format, status words and the method of error detection and correction. For further discussions three representative plant communication networks are selected. They are in an advanced state of standardization

- The IEEE 488 Bus
- The PDV-Bus
- The Ethernet-Bus.

They each have unique features and were designed for different applications. The IEEE 488 Bus was developed for short distances (max 20 m) and high data rate. The communication is performed byte serial via a bidirectional data bus. The PDV-bus is a ring-bus for long distances (1–3 km) with standardized interfaces and bit serial data transfer. The Ethernet, one of the most popular systems, is based on a coaxial cable for bit serial data transfer over long distances and has standardized interfaces. The efficiency of a bus system depends on the following parameters:

- Data rate
- Transmission distance (max. dist.)
- Time delay in the case of interrupts and data requests

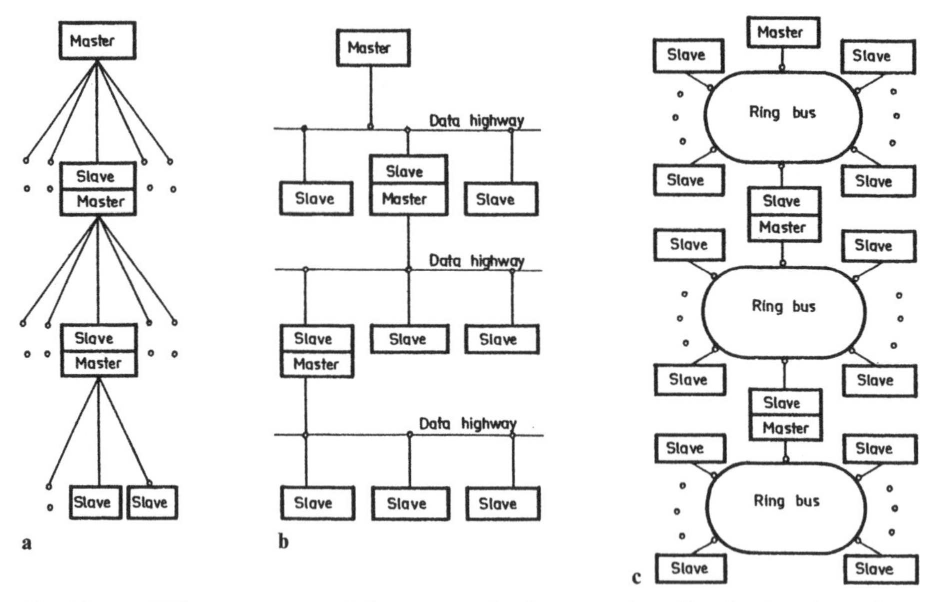

Fig. 6.7 a–c. Different structures of plant communication networks. **a** Star structure, **b** open bus structure, **c** ring structure

- Amount of hard- and software for additional nodes (extendability)
- Reliability, fault tolerance, availability
- Unique logic structure
- Possible geographic distribution of communication process
- Cost of the system components
- Standard plug-in principles.

IEEE 488 Bus

The IEEE 488 bus is a bit parallel-byte serial data transmission system with a transmission data rate of 1 Mbyte/sec. The application area for this bus is local acquisition of process and quality control data as well as general I/O. Most of the instrument manufacturers support their instruments and signal processors with the standard IEEE 488 interface (25 pins, 8 data lines, 5 control lines, 3 lines of acknowledged signals, 8 twisted pair lines). The nodes can have listeners, talkers or both and master functions. The system master controls data transfer between talkers and listeners etc. The data lines use ASCII-code. A unique programming language for the IEEE 488 bus is not defined. BASIC, FORTRAN, PEARL and PASCAL are in use.

If communication over more than 20 m distance is to be performed, a second bus system has to be used. This second bus system may connect multiple distributed clusters of IEEE 488 bus controlled measuring instruments and transfers the data to the next control level. A typical bus system for local distributed processes is the PDV bus.

PDV-Bus

The bit serial PDV bus operates with a transmission rate of 100–500 kbyte/sec over distances of up to 3 km. The bus is capable of hosting 252 stations and is qualified to interconnect hierarchical computer systems. The was developed for an industrial manufacturing environment and was sponsored by the Ministry of Research and Technology (BMFT) of the Federal Republic of Germany. Typical applications are the interconnection of control computers of a hierarchical control system for manufacturing equipment. Real-time and high safety requirements were important criteria for the design of this bus. Currently, the bus is offered for industrial use by several manufacturers as a fast and a slow version.

The fast version is used for direct process communication, and the slower version connects computers at a higher hierarchical level. Process and data peripherals can be connected to the bus by the plug-in principle via an interface control unit. The controller is available as a PDV-bus chip (Valvo MEE 3000). The bus is independent of the communication technology. It can accomodate an electrical or a fiber optic communication cable.

Ethernet-Bus

The Ethernet is presently gaining popularity with local area networks. It is supported by the companies Intel, Digital Equipment and Xerox. Integrated circuits, or so-called transceivers, are available to interface computers and peripherals with the bus. The communication media is a 50 ohm, 10 mm diameter coaxial cable. It allows a transmission rate of 10 Megabit/sec. The maximum length is about 2,5 km and there is the possibility of interconnecting 1024 participants. Figure 6.8 shows the principle of the bus system. A so-called physical layer was standardized. This includes cable diameter and impedance, pin arrangement and plug dimension, the coding method, the voltage level and the noice margin. The next layer is the data link layer. Here, the frames, the protocol, address coding and decoding and error detection are subject to the standard. All following layers

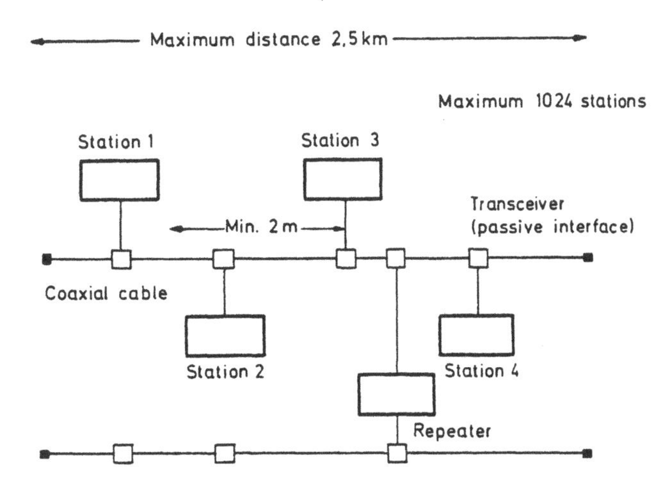

Fig. 6.8. Principle of the Ethernet bus

are to be specified by the user. In this way it is possible to interconnect existing network architectures and communication systems with the Ethernet. With the aid of a transceiver, the user can connect different Ethernet modules. The bus does not employ the master principle. Each participant can independently request access to the bus.

The Ethernet was primarily intended for use in automation, distributed data processing, terminal access and other situations requiring an economical connection to a local communication medium. Originally, the Ethernet was not designed for real-time computer systems; however, this application was not specifically excluded. The lack of an interrupt possibility limits the bus for the control of slow or non-criterial processes. The growth potential of local area bus systems can be depicted as in Fig. 6.9 [6].

A new standard being developed is the Manufacturing Automation Protocol (MAP). This concept was proposed by General Motors for communication within manufacturing facilities.

6.1.5 Influence of VLSI Technology on Hierarchical Control Systems

VLSI technology influences the design of all components of a hierarchical manufacturing control system. It will help to decrease the cost of computational equipment and to increase its speed and efficiency. The main areas of impact will be:

- The minicomputer
- The microcomputer
- Interface chips

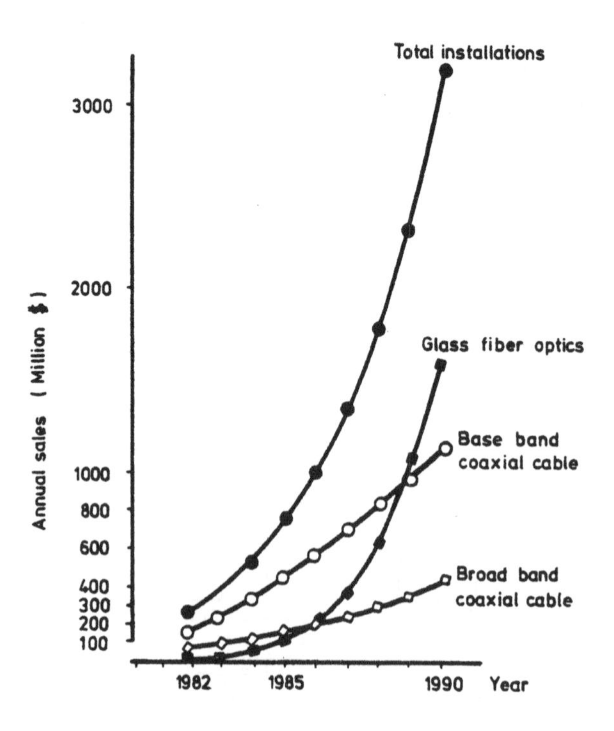

Fig. 6.9. Growth potential of the local area bus systems

- Memory
- Peripheral devices.

There are many different types of computer architectures which can be built from VLSI chips, including new concepts of von Neumann machines, data flow machines, pipeline and array architectures, as well as general purpose multi and poly-processor systems. New developments can be expected for both general purpose architectures for higher level control and special purpose architectures for operational level control.

6.1.5.1 Minicomputers for Higher Control Levels

The higher hierarchical levels of control systems are the domain of minicomputers. However, the distinction between a mini- and a microcomputer is becoming blurred almost to the point where it does not exist anymore. Despite this fact there is still a strong demand for so-called 32-bit superminis. They usually do not contain a microprocessor as a CPU and are designed with a data length, a bus and registers to handle 32-bit words. They allow the user to address a large memory space, to index into long arrays and to specify records in large files. In general, they have a capacity and speed advantage over microcomputers. Typical representatives are the "Eclipse MV" (Data General Corp.), "VAX-II" (Digital Equipment Corp.) and "Prime 50" (Prime Computer, Inc.) computers. Performance data of such mainframes are shown in Table 6.1 [6]. The large address space makes the mainframe very useful for directing the operation of distributed computer systems. In addition it is possible to control manufacturing systems with the help of large process models.

Table 6.1. Characteristic performance data of supermini computers

Number of executed instructions/sec	500,000 to 850,000
Single precision Whetstones (FORTRAN)	600 to 1,200
Main memory (Mbyte)	1 to 28
Mass storage (Mbyte)	up to 4700
Maximum workstations	1 to 128
Typical software	Time sharing, transaction processing, demand paging, database management, word processing, electronic mail.
Networking	SNA, X. 25 and in most cases vendor designed network
16-bit competability	Not provided in all cases
Languages	APL BASIC COBOL FORTRAN PL/1 PASCAL MODULA 2 ADA

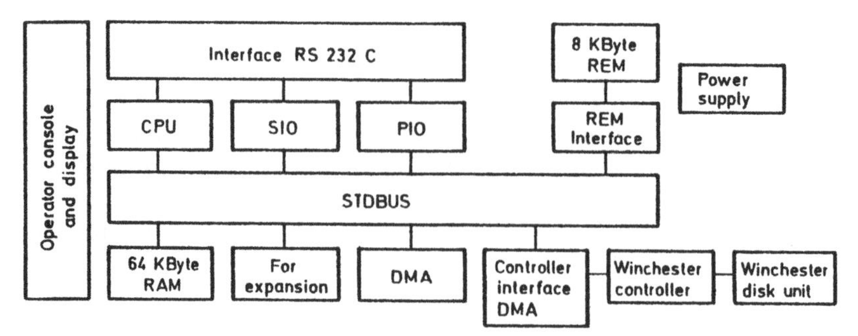

Fig. 6.10. Basic structure of the Synfobase system

The future process computer will have several dedicated processors, e.g. for I/O, to perform mathematical calculations and to operate on databanks. For many applications it is necessary to improve the reliability of a computer network. For this purpose computers with redundant processor configurations will be available. Today only the Tandem computer offers this as a standard feature.

A further innovation will be the associate databanks for minicomputers where the key to locating data is not an identifier but its contents. A special databank computer system which can be connected via a standard interface to another computer is shown in Fig. 6.10 [7]. The associate databank will be an integral part of the future supermini. The operating system of this powerful mini will have capabilities similar to those found in todays large computers. Some of the particular features will be multi-user and multi-programming capabilities and a virtual memory. Unix-like operating systems specifically tailored towards real time application will be the predominant system software.

6.1.5.2 Microcomputers for Operational Control Levels

The 16 bit processor has become a reality and is presently to be extended towards 32 bit capabilities. The Motorola M 68000 has been available on the market for almost 4 years. It has become the basic building block of many new parallel and redundant computer architectures and distributed systems. A competitor to this computer is the 16 bit-processor 80286 of Intel. The chip contains 130000 transistors and has a throughput which is 6 times higher than that of the 8086. The memory controller is integrated together with the CPU on one chip. The physical address space of 16 megabyte can be extended into a virtual space of one gigabyte. Thus it becomes possible to have multi-user and multi-task operations. The pipeline principle was used to increase the throughput of the 80286, Fig. 6.11. The processor is divided into 4 different units, one for bus control, one for addressing, one for instruction decoding and one for instruction execution. All units can operate in parallel mode, thus speeding up memory management and making possible the implementation of virtual memory. The 16-bit, and in the future also the 32 bit-microprocessors will be attacking very strongly the traditional minicomputer market, Fig. 6.12. This diagram shows the expected price trend for micro and minicomputer systems.

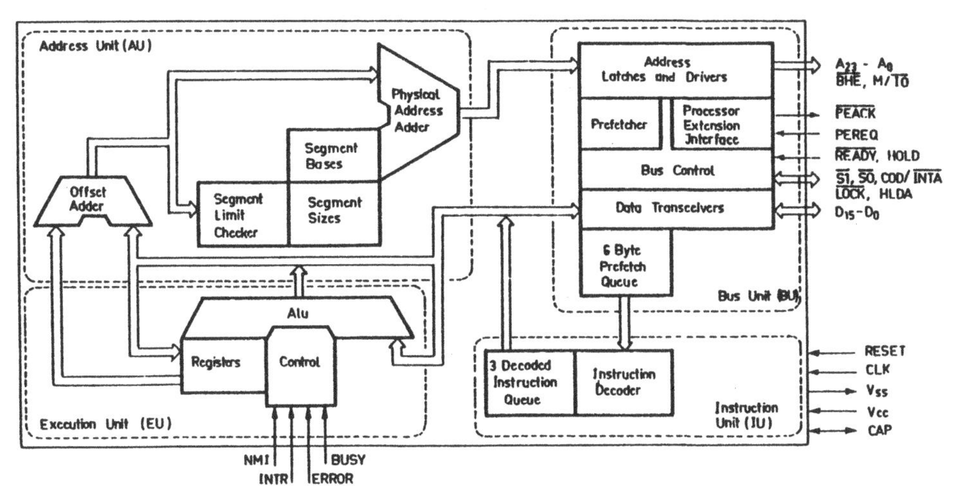

Fig. 6.11. Schematic diagram of the Intel 80286 with integrated memory control

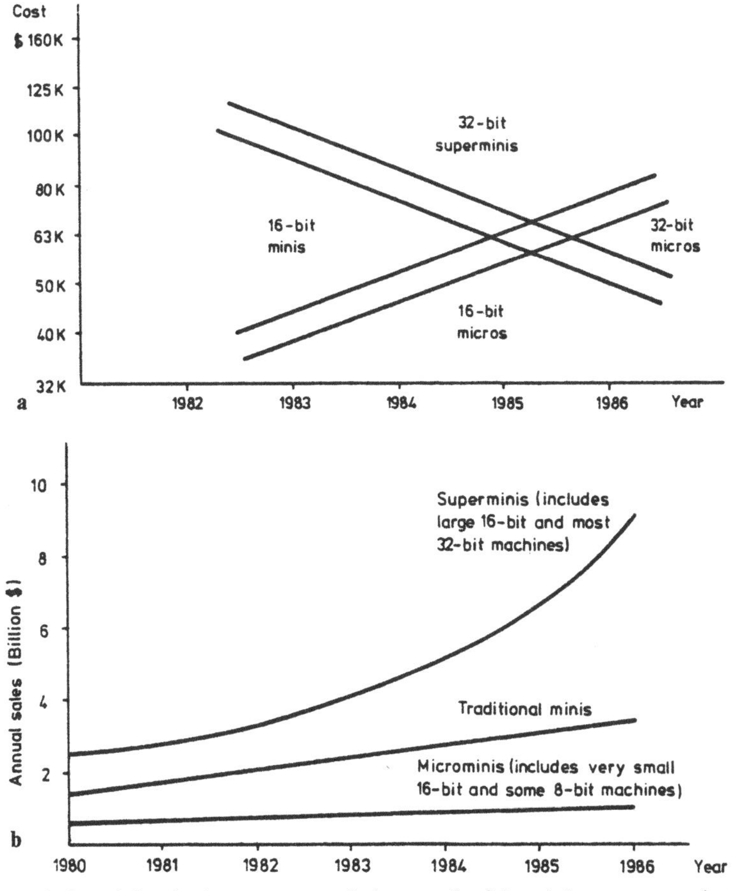

Fig. 6.12. a Cost trends for mini and microcomputers, **b** the growth of the minicomputer market

6.1.5.3 VLSI Interface Modules

With microcomputers there has been a continuous trend to relocate many of the conventional tasks of the processor into the interface components. There are numerous input and output modules available, which are capable of operating parallel to the main processor. Typical representatives are digital input/output channels, control units for floppy disks etc.

Many manufacturers have developed one-chip processors which the user can configure to his specific application. These processors are instruction and bus compatible with the main processor and can easily be integrated into a multi-processor system. Typical important criteria for the user are as follows:

- Own CPU with interrupt circuit
- Up to 4 kbyte ROM and 256 byte RAM
- Some units with EPROM (UV eraseable)
- Own bootstrap loader
- Up to 90 instructions
- Counter and timer
- 2 × 8 bit independently programmable input/output channels
- Communication with the main processor via handshake lines
- Integrated USART circuit
- Semaphor circuit to control access of CPU to memory

Figure 6.13 shows the internal structure of such a one-chip processor built by Motorola.

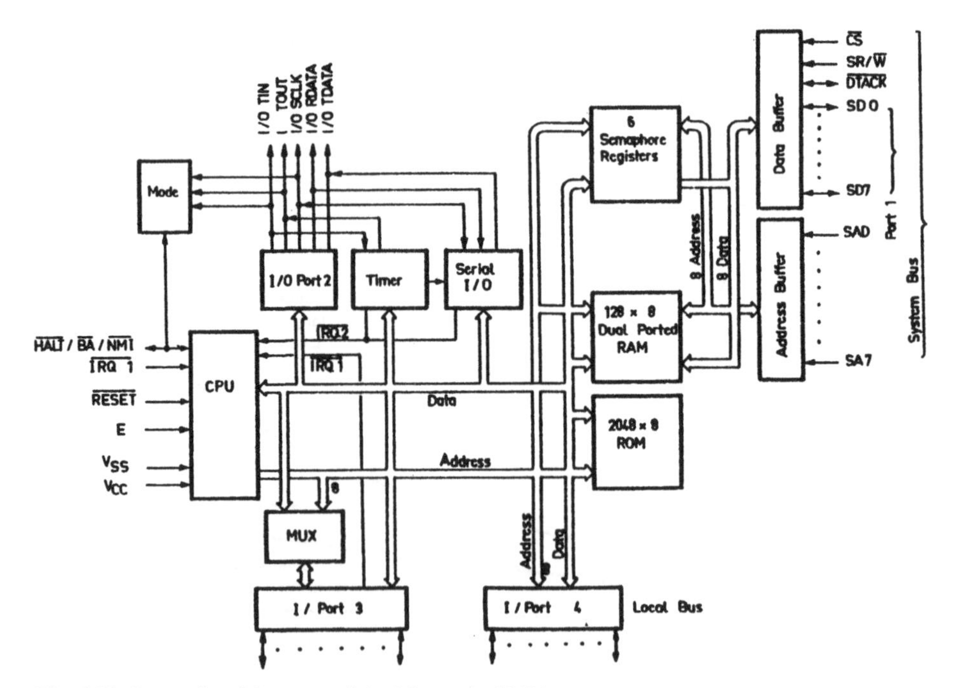

Fig. 6.13. Internal architecture of the Motorola 68120

A typical interface module built for printers, tape punches and floppy disks is the IPT 860 module, which can be connected in different configurations to a Centronics or a RS 232 interface with only little software change in the main processor. The unit is capable of accepting 32 000 characters/sec from the main processor. It then can independently output this information to a slower peripheral.

A more complex module was developed by NEC to control the data transfer between a CPU and a Winchester floppy disk. It can operate 8 disks and has a maximum data transfer rate of 1,5 megabyte/sec. The circuit contains a controller and a microprocessor, 2,5 kbytes ROM and 64 byte RAM as well as a format controller. Error detection and correction is done according to the Fire Code principle. Similar modules were developed for signal processing. A typical example is the 32 bit-processor TMS 320.

For the operation of automation systems, the speech analyzer and speech synthesizer will increasingly be gaining importance as peripherals for the control computer. With these devices, the machine operator can give verbal instructions to the machine, for example he can command a lathe to cut a workpiece to dimensions specified by him. There are numerous speech analyzers already available. However, to date their vocabulary is very limited and their hit rate is unacceptable for most applications. Another problem is the need to have a specific vocabulary for every instructor. In some cases, even the tired voice of the same instructor may lead to recognition problems. The voice patterns must be updated during operation.

The speech synthesizer has been developed to a high degree of sophistication. It is used in dictionaries and other consumer items. When installed in control devices, it can inform the operator about the tasks a machine tool is presently performing, or about difficulties with the control system, and if needed it can give repair instructions. Natural speech processing systems are proposed for the future (see Sect. 6.4).

6.1.5.4 Memory

The development of memory cannot be predicted very accurately since the manufacturing processes are getting increasingly complex. In addition, the required address space for future processors is not exactly known to date. Table 7.2 shows the status of the present memory technology. The 256 kbit RAM has become reality with an access speed of 150 ns. However, it is felt that this speed has to be increased to 100 ns in the late 1980's. There are several firms which will be bringing a 1 Mbit dynamic RAM on the market by the end of 1986. With these devices, redundancy error detection and correction techniques will become a

Table 6.2. Present state of memory technology (bits/chip)

dynamic RAM	1 Mbit
static RAM	64 kbit
ROM	16 kbit
PROM	64 kbit
EPROM	64 kbit

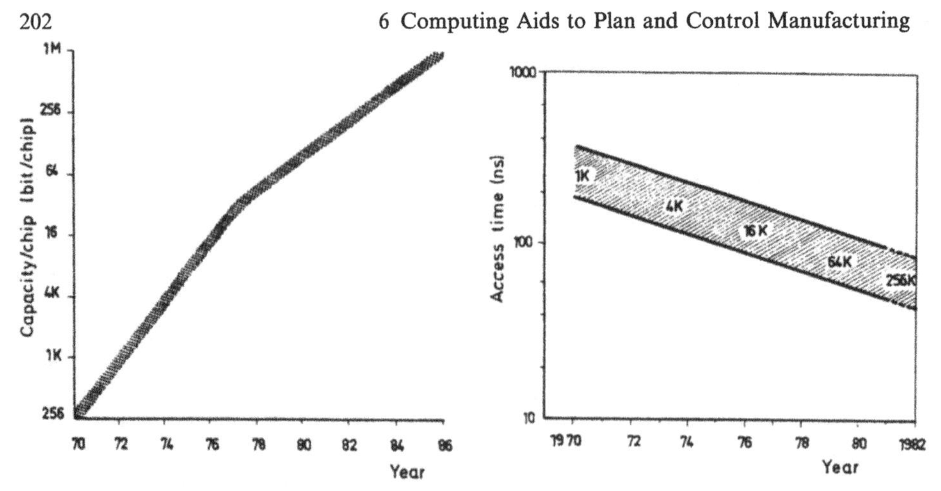

Fig. 6.14. Estimated capacity of a chip

Fig. 6.15. Access time of dynamic RAMs

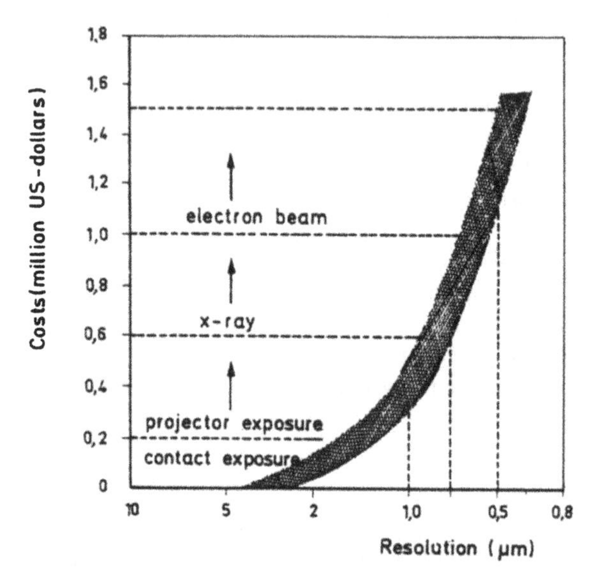

Fig. 6.16. Increasing manufacturing costs for integrated circuits

standard feature. The static RAM memory is also in sight, which is based on CMOS technology. Figure 6.14 shows a projection of the estimated capacity of a chip. The access time which can be expected, is shown in Fig. 6.15. With increasing packaging density, a considerable increase in manufacturing cost can be foreseen, Fig. 6.16. This development, however, will eventually lead to lower memory cost when the new processes are used in mass production. With low cost memory available, higher order languages and better programming aids become more useful.

6.1.5.5 VLSI Data Peripherals

The development of process peripherals will be greatly influenced by microelectronics. The trend leads to autonomous devices which are capable of indepen-

Table 6.3. Performance data of the Burr-Brown CS 450 data acquisition and control system

7	input/output cards
103	normal analog inputs
217	differential analog inputs
22	analog outputs
504	digital input/output channels
504	TTL compatible input/output channels
Floppy disk unit	
Dual cassette drive	
CRT and printer	

dently processing signals in close vicinity to the manufacturing equipment. In many cases, the microprocessor will be incorporated into the sensor. There are a number of manufacturers who market autonomous data acquisition and control devices. A typical product is the CS 450 programmable data acquisition and control system of Burr-Brown. This unit can be programmed with Basic-400. It uses the Intel multi-bus structure and can be expanded from one to seven input/output modules. Important parameters are shown in Table 6.3.

Implementation of complex control algorithms (DDC-algorithms) can be realized with the aid of 8-and 16-bit microprocessors. Provided with fast memories and arithmetic processors (e.g. the Intel 8087 or AmD 9511), efficient I/O interface and multi-channel timers, they are well suited to the solution of many control tasks. They are very accurate, flexible and economical and yield good solutions when CPU and other control circuits are integrated in one chip.

Economical solutions can also be obtained with the use of one-chip processors which have, in addition to CPU, a RAM, a ROM, I/O channels, a timer and A/D or D/A converters integrated in one component. All these features are provided in the one-chip microcomputer COM-87 (PD 78 111) of NEC Electronics. It hosts 8 analog channels with a resolution of 8 bits, 5 I/O ports with 8 bits each, a complex interrupt priority circuit, 8- and 16-bit timers, serial channels and numerous registers. Effective programming can be done with a standard instruction set. There are additional instructions for 16-bit data transfer and for multiplication and division. The block diagram of the 64-pin chip is shown in Fig. 6.17. Integrated circuits of this type are well suited to the control of analog systems, in particular to those where electromechanic control devices are used. Typical applications are for machine drives, robots, airplanes, automobiles and appliances.

In the future, analog devices will be conceived from predesigned functional blocks. This will be made possible by the efficient use of the CAD/CAM technology. This analog cell concept takes advantage of the possibility of storing design data for converters, operational amplifiers, multiplexers and memory circuits in the CAD computer. For the digital circuit design data can be configured by the engineer to different combinations and verified by a simulation program to check the adherance to the customer's specification. It will also be possible to check the economics of the design and to output yield data obtained from stored information about process geometry, number of masks, diffusion and the size of the wafer and the chip.

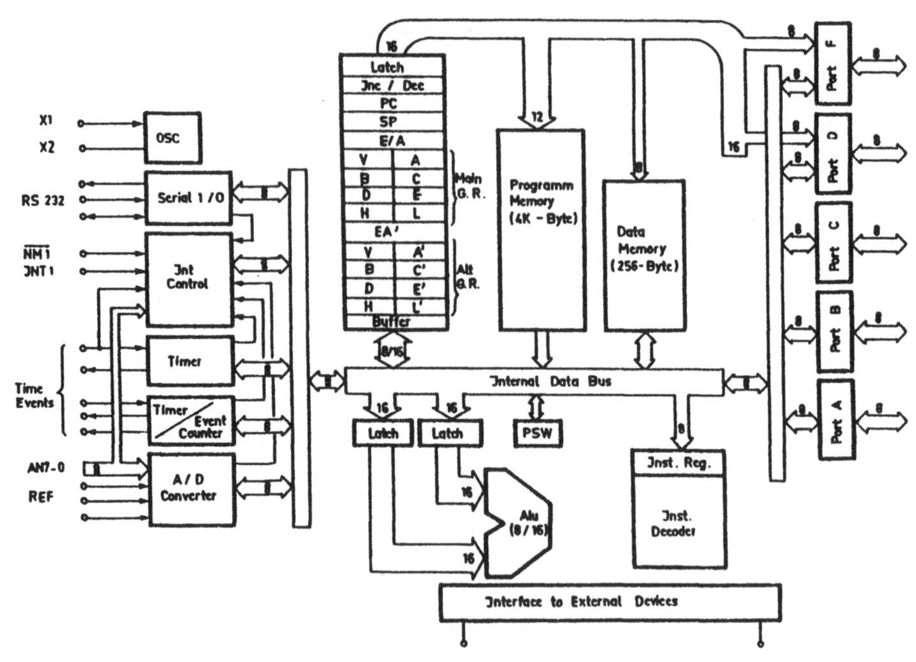

Fig. 6.17. Architecture of the one-chip computer COM-87

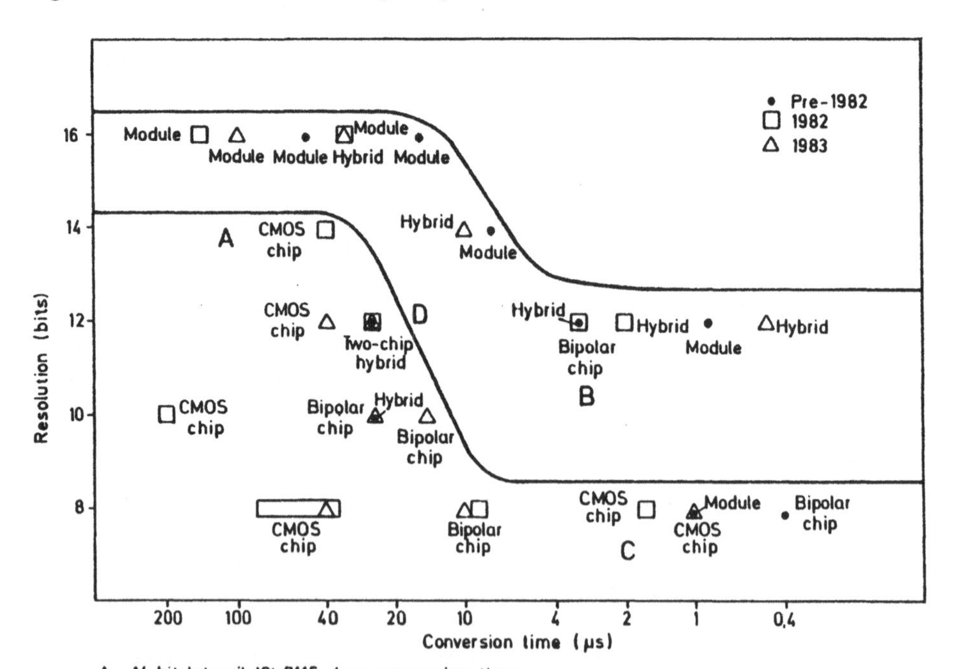

A. 14 bit Intersil ICL 7115, 4 μs conversion time
B. 12 bit AMD Am6112, 3μs conversion time
C. 8 bit National Semiconductor ADC 1020, 1,5 μs conversion time
D. 12 bit prospective industry standard

Fig. 6.18. Modern ADC technology

Further cost increases to improve the resolution, speed and noise immunity of present converters would be of exponential nature. For this reason, suppliers will direct more attention to custom designed circuits which can readily be manufactured with available design tools and manufacturing systems.

The CMOS technology, which previously was mainly used for digital circuits, will have a great impact on analog devices. This will be made possible through the use of the dual polysilicon gate process in conjunction with dual-layer metallization. This manufacturing method allows high yield production of operational amplifiers, voltage references, latches and data converters. Figure 6.18 shows present capabilities of the general purpose analog technology which previously could only be obtained by hybrid and modular circuits [10].

6.1.5.6 Data Peripherals

In the near future there will be quite an improvement in the capabilities of data peripherals. This will occur in particular with external storage devices, Fig. 6.19 [11]. Compared with 1980 performance data, all technologies will show a considerable improvement with regard to access time and storage cost. The cache memory will help to decrease considerably the access time to disk storage devices, even for small computer systems.

The optical disk, originally developed for the entertainment market, will enter the computer field as a low cost storage device, Fig. 6.20. To date, this disk can record information once and it can replay it indefinitely. The information density is 1500 bit/mm compared with 150 bit/mm for magnetic disks. In order to be able to handle the disks more easily, present developments are trying to design disk changers similar to those used in music boxes.

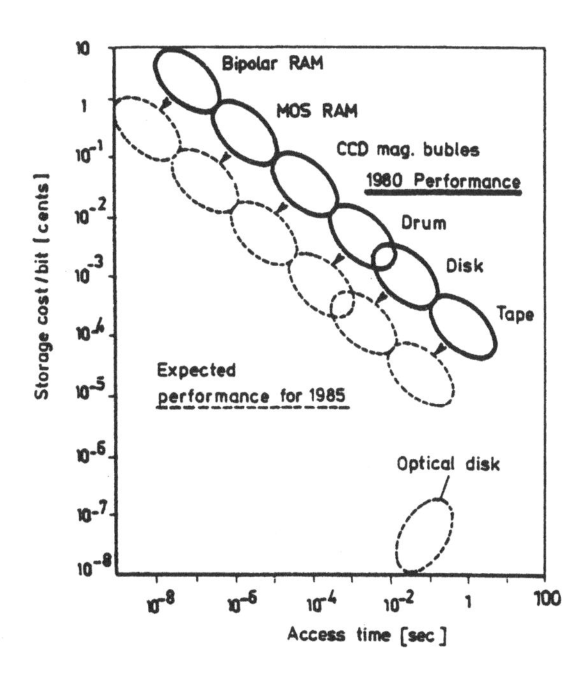

Fig. 6.19. Worldwide sales of external mass storage devices (1980–1990)

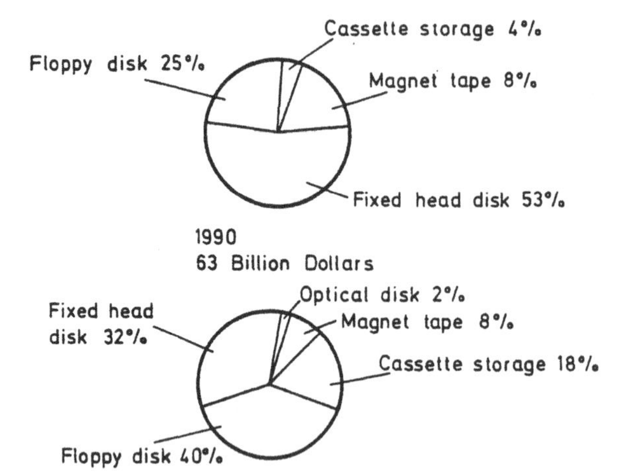

Fig. 6.20. Cost trend for data storage devices

The floppy disk will remain the most popular storage device, because it is very handy and easy to use. Likewise the cassette will gain in popularity.

The color printer will be an important new data output device. Recently, several such printers have been announced. They operate on the laser, matrix, ink jet and thermo principles. To date, the highest quality pictures can be obtained with the laser and ink jet printers.

6.1.6 Software and System Development Aids

With the majority of real-time computer installations, software is the most expensive component. Despite the availability of higher programming languages, on-line control systems are predominantly implemented with the aid of assembly languages. In Germany, PEARL is becoming a popular language for control and microcomputer systems. In the near future, ADA will emerge as a stiff competitor. ADA is a highly standardized language. For this reason, it is registered as a trade name. Translators for ADA must undergo a stringent validation procedure in oder to guarantee the adherance to the language specification. Thus it can be expected that ADA will be highly machine independent, which is necessary for good portability. With this feature software houses will be able to offer a diversity of application programs, and the customer will have the opportunity to purchase reasonably priced programs. The first ADA compilers are being offered to computer users.

Another trend which can be observed, is the implementation of software functions in firmware or even in hardware. In particular this will be true for the kernel of operating systems. It will also be possible to configure firmware with the aid of operating systems, Fig. 6.21 [12].

Currently hundreds of operating systems for microcomputers are available. They range from stand-alone to real-time and multi-user systems. The stand-alone systems are mainly derivatives of the Digital Research CP/M and Microsoft

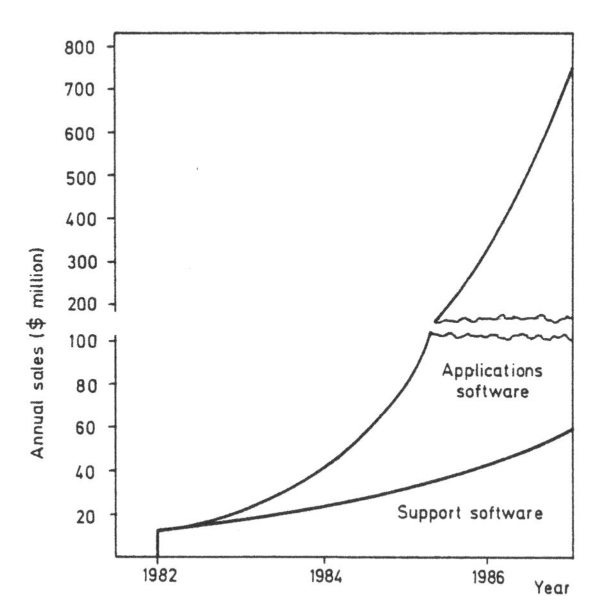

Fig. 6.21. The growth potential of ADA

MSDOS developments. With the increasing use of microcomputers for real-time applications, the new operating systems will become more and more similar to the Unix system of Bell Laboratories. With the help of this operating systems many of the future microcomputer installations will become more powerful than present VAX 780 systems.

New software systems will be available to aid the entire software development cycle. This trend will be stimulated by research results obtained from computer science and artificial intelligence. There are many ongoing research activities trying to produce software development tools for many different applications. Some of them follow the concept of automated software generation, others are only trying to support specific phases of the software development cycle. The first interactive workstations, which will allow computer aided software-engineering, are being offered on the market. They usually provide graphic capabilities to construct hierarchical system diagrams. The EPOS system, developed by the University of Stuttgart in Germany, provides development aids for the design of real-time software. The SARS system of the University of Karlsruhe in Germany pays particular attention to the specification phase of software and hardware for automatic systems.

6.2 Hierarchical Control Architecture for Manufacturing Cells

6.2.1 Introduction

There are various control concepts for manufacturing cells. The most often used concept uses a hierarchically organized control architecture. In this section the control of a robot based manufacturing cell will be discussed. The basic concept can be used for other manufacturing units.

There is an increasing demand in industry for manufacturing cells consisting of multiple robots, machine tools and peripherals which are integrated into a functional unit. Different cell types, like machining cells, welding cells or assembly cells, are to be configured depending on the application. A machining cell, for example, can consist of a milling work station, an inspection work station and a material handling work station. The workstations are functional combinations of robots, machine tools, part buffers, peripherals etc. Vision systems and their combination with other sensors are basic elements in the workstations, which aid high performance finishing, complex assemblies and piece part inspection for quality control. In the context of CIM, manufacturing cells are planned and programmed in an environment which is characterized by an unique data model of the factory with strong data integrity and a defined data transaction management. For this purpose, the integration of CAD/CAM and robotics is expected in this decade. Worldwide national and international projects are being pursued in which robots are to be integrated as a part of a complex production system. One approach is to treat the robot like an NC machine and to use APT like languages for programming. The use of post processors allows interfacing of different robot types. For sensor guided robot applications with intelligent control capabilities, as are needed for assembly operation, a more flexible control concept is necessary. Albus [13] proposes a system in which robots are integrated via a tree type control structure into a manufacturing cell. The CAM-I software project [14] tries to conceive a manufacturing system model which defines a linkage between CAD/CAM, programming and simulation systems and the robot control. Within the framework of the European ESPRIT project, the integration of industrial robots into CIM systems are under study.

The data representation of a robot in a manufacturing cell, the data processing and the data flow to control robots within a cell is called robot architecture. From this architecture a computer soft- and hardware concept for robot controls, and an operational control principle can be derived. This computer architecture fulfills the requirements and integrational aspects for robots in a manufacturing environment. A robot architecture emphasizes intelligent multiarm control, the integration of multisensor systems, peripherals and tool machines. A system architecture is discussed which is hierarchically structured using a distributed polyprocessor system [15]. Robot dependant and robot independant control levels can be identified according to the control task. The use of standard interfaces [16] between the hierarchical control levels allows the integration of each robot type if appropriate interpreters are available. Interfaces with higher off-line control levels, like textual programming system, simulation systems and data bases, increase the use of the robot manufacturing potential. In the following the robot architecture and the control computer concept is presented and discussed.

6.2.2 Robot Architecture

The internal presentation of a robot in a manufacturing cell is assumed to be hierarchical. The control system of the cell is composed of a multitude of individual functions which assure the performance of the manufacturing task, the safety

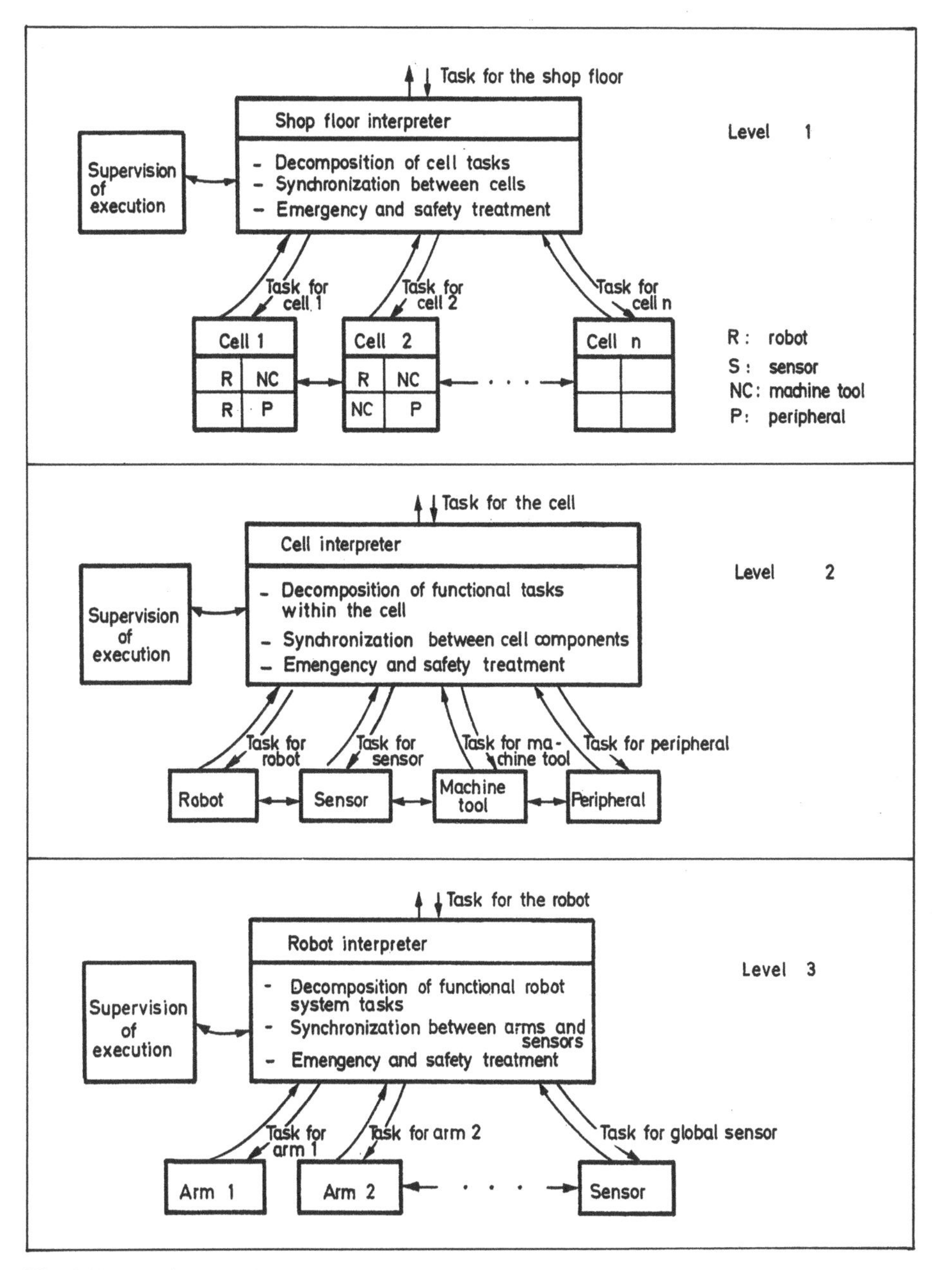

Fig. 6.22a. Basic control structure for a robot oriented manufacturing task (robot independant)

and emergency aspects, the linkages to other manufacturing cells and the interconnection with higher manufacturing control levels. The problem of organizing and coordinating these particular control functions, as well as the problem of defining a logic control structure, implies the definition of five hierarchically

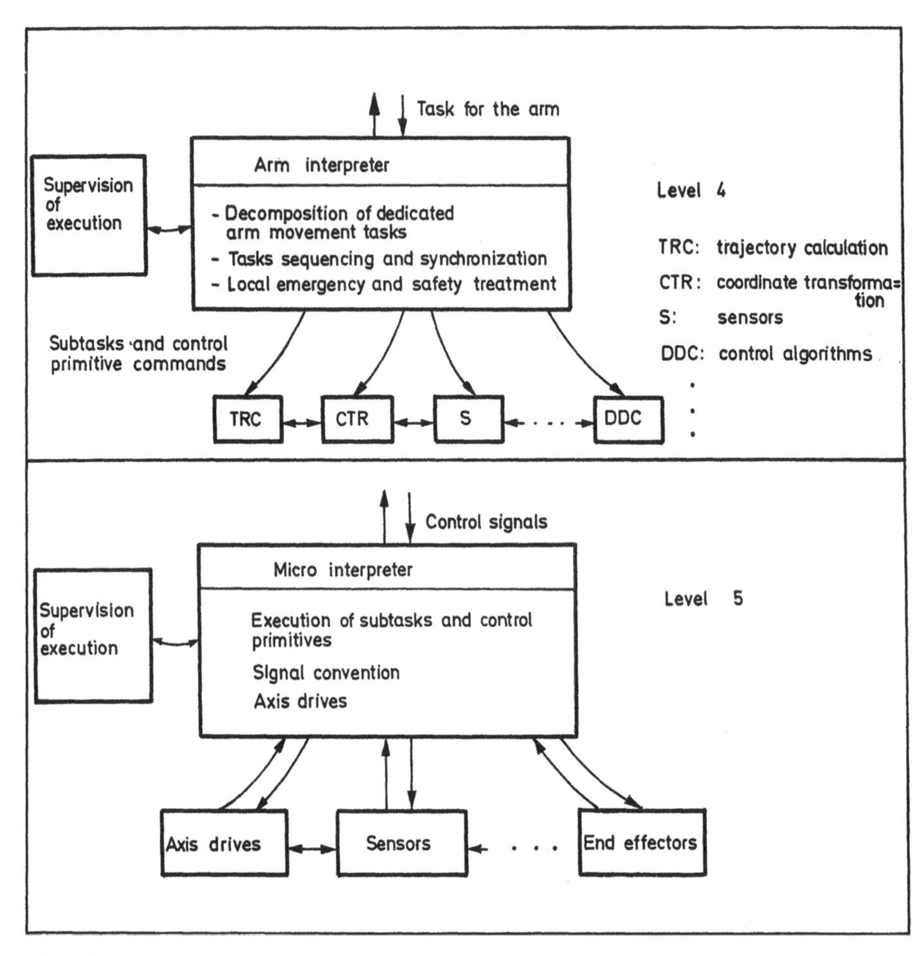

Fig. 6.22b. Basic control structure for a robot manufacturing task (robot dependant)

ordered system control levels. The five distinguished tiers are related to (see Fig. 6.22a, b):

- axis, end-effector and primitive sensor controls
- arm- and local peripheral control
- robot, global sensor and peripheral control
- cell control
- multiple cell control (shop-floor).

The hierarchical tiers are composed of functional modules (soft- and hardware) which perform specific control functions and tasks. The logical combination of these functional modules on each specific control level yields a horizontal control structure. This structure contains manipulator and effector and sensor controls as well as emergency and human safety control procedures. The vertical data flow between the specific control levels is represented by control commands, sensor data and emergence detecting messages. A basic description of a hierarchically organised robot control structure is discussed in [17, 18]. The idea is to use a series of hierarchically ordered task decomposition and task sequencing operators, which have as input on the highest level a manufacturing task description command. This command is to be decomposed via the hierarchically ordered operators into a sequence of basic control functions and primitives, which have to be executed in real time. The system is characterized by the distribution of intelligence. Thus, a sophisticated intelligent control structure can be organized in a hierarchically decreasing order of abstraction and intelligence, and in an increasing order of precision and sampling rate. The use of intelligent modular subsystems for building hierarchically organized manufacturing cell controls has the following benefits.

- Reduction of organisational effort
- Extendability of the system
- Flexible control structure
- Vertical and horizontal data flow between the subsystems is reduced
- Reliability through modularisation
- Reduced complexity of software
- Reduced costs for control software development
- Realization of subsystems in software or hardware
- The structure enables the CAD/CAM/robot linkage
- Alternative design of simple, intelligent sensor driven or highly sophisticated control is possible
- Independent development of robot, NC and peripheral subsystems
- Use of standardized linkages between different control levels.

The architecture of the robot control system is defined by:

- The internal robot data representation
- The task decomposition operators
- The horizontal and vertical data flow and
- The operating principle of the modular subsystems.

The internal robot data representation is related to data objects which are to be processed by control level specific operators. Such operators can be task decomposition operators, sensor and sensor monitoring operators, world model operators, conflict-analysis and decision operators etc. The mechanism of the task decomposition defines the relation between two control levels and the relation between sensor information, action and expected results. The horizontal and vertical data flow defines the software and hardware linkages between subsystems and between different control levels. The operating principle of the subsystems defines how to start, to interrupt and to finish a task.

6.2.3 Internal Robot Data Representation

Within the five tiers defined for the control of a manufacturing cell the internal view of industrial robots is quite different. Seen from higher levels the robot representation is closely related to the manufacturing application. The robot appears on the shop floor level as a component of the different manufacturing cells. The manufacturing cells are designed for a specific task and defined by their functionality. The shop floor is controlled by an interpreter which decomposes the cell tasks according to the shop-floor manufacturing requirements. This includes the synchronisation between cells, the material flow and the emergency treatment. The robots do not appear explicitly on this level. Each cell is defined by its functionality, its task capacity, its elementary operations, performance, etc. To solve the problem of cell task decomposition, the interpreter has to operate on data related to a priori and actual process knowledge and virtual cell descriptions.

The output of the shop floor interpreter is a sequence of instructions for each cell, Fig. 6.23. The manufacturing cells are configured by different combinations of robots, machine tools, peripherals and sensors. The cell interpreter controls the operation of all functional devices within the cell. The input instruction is decom-

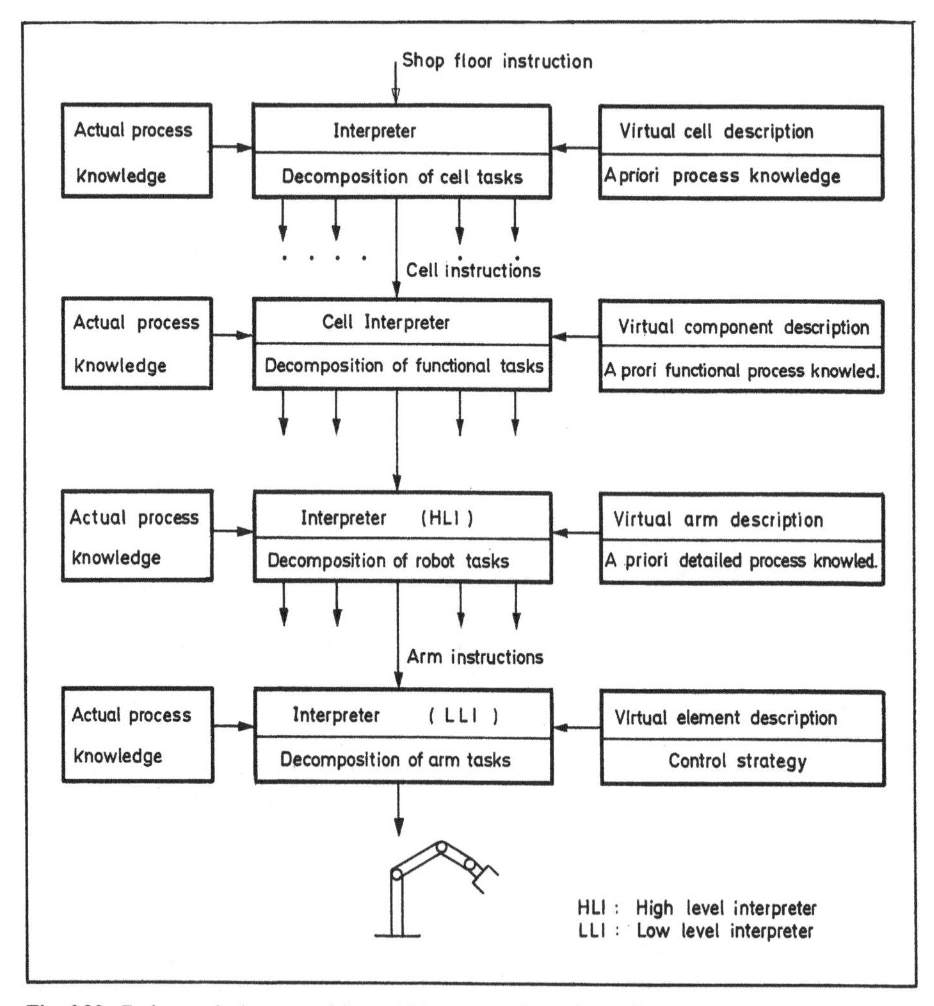

Fig. 6.23. Robot task decomposition within a manufacturing cell

posed into device specific tasks (robot, sensor, NC, peripheral, etc.). Emergency treatment and safety tasks influence the task sequences using actual and a priori process knowledge. Robots are described on this level by their functionality, elementary operations, process capabilities and performance in terms of virtual description. The output of the cell interpreter is a task sequence to be executed by the robot, the machine tool, the peripherals, etc. These tasks are to be executed in parallel and/or synchronized. The hierarchical task decomposition is further refined via the robot and arm interpreters. The principle is to break down a large abstract problem into a set of smaller, more detailed subproblems. At each level of decomposition, details and facts are filled in to make the next lower subtask series more concrete. At the lowest level of the hierarchy an ordered number of independantly soluble problems and functions exists. The details and facts which are needed for task-decomposition have to be represented internally in the control architecture. There is a large amount of required data and descriptions which are

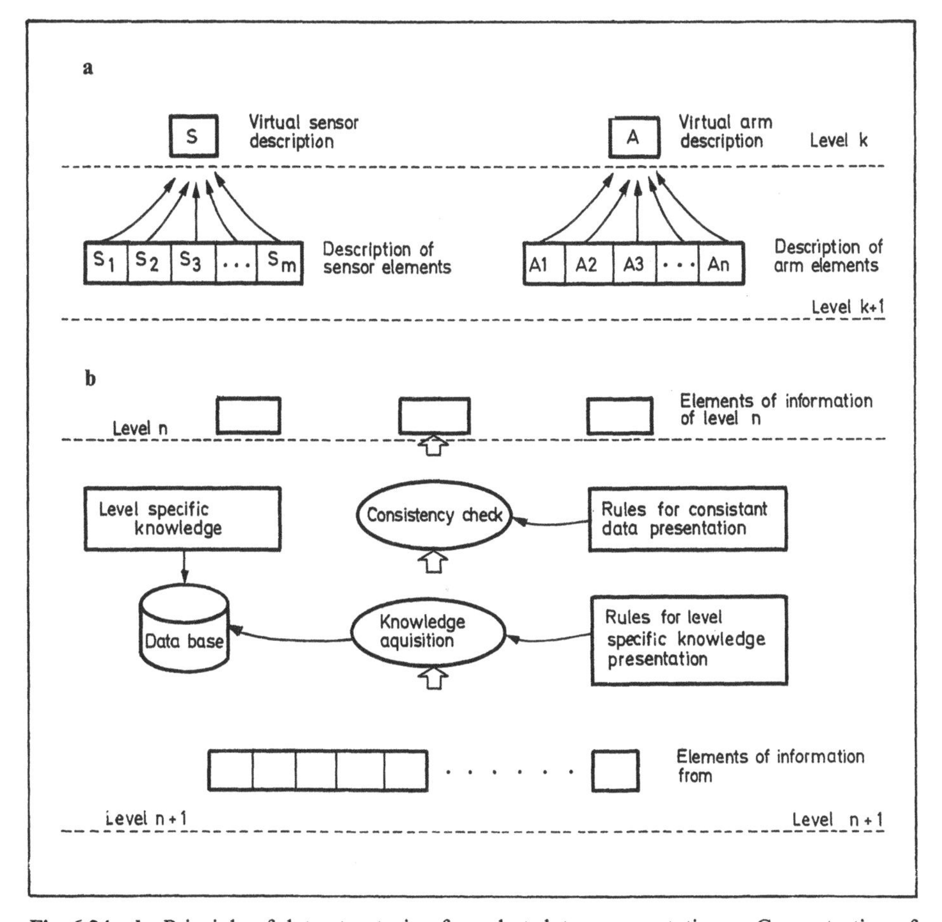

Fig. 6.24 a, b. Principle of data structuring for robot data representation. **a** Concentration of data in terms of virtual description, **b** diagram for consistant knowledge acquisition to virtualise detailed information from a lower to a higher control tier

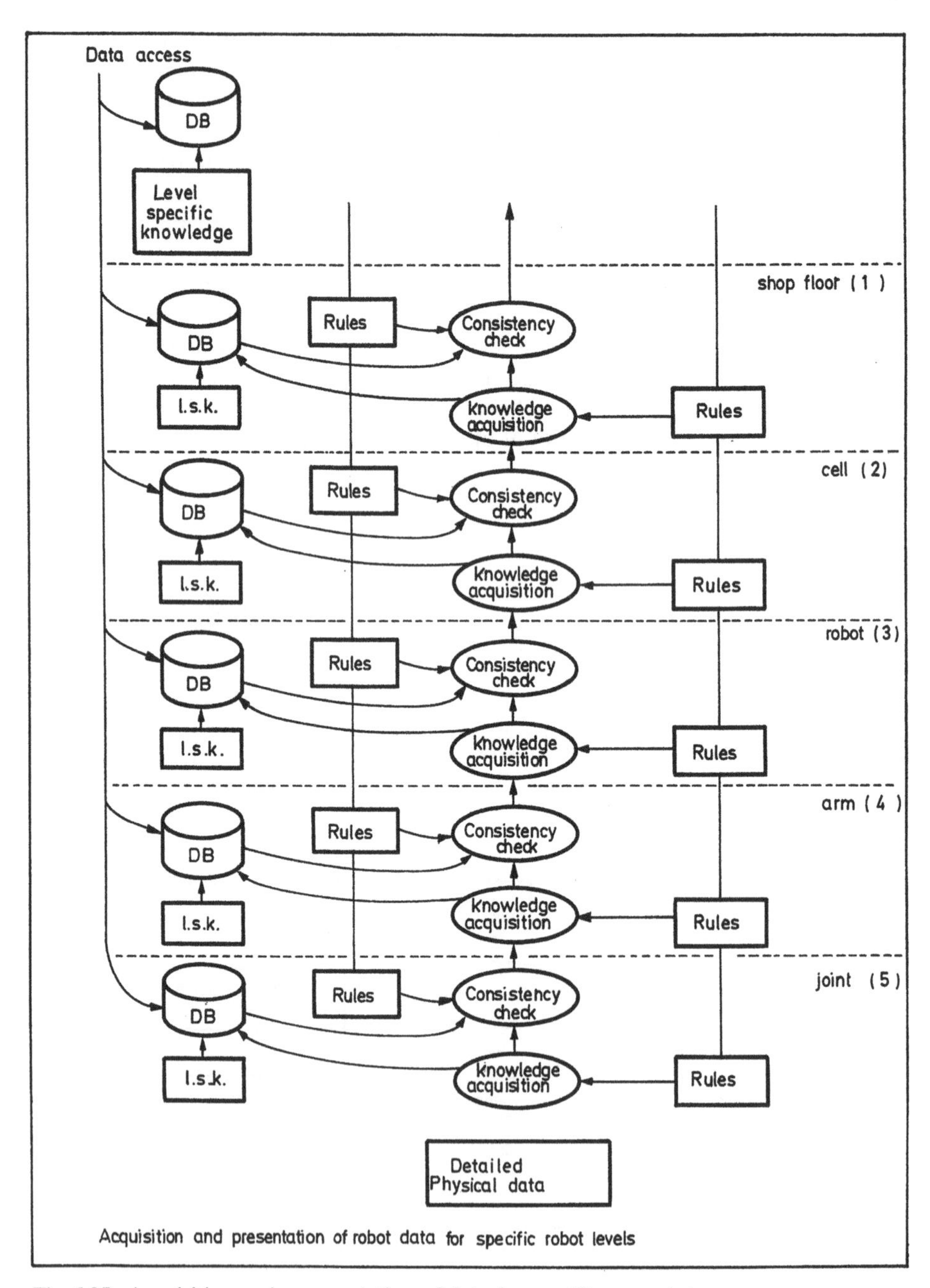

Fig. 6.25. Acquisition and representation of data for specific control tiers

related to the manufacturing cell, to the tool machines, robots and sensors. This data is represented in the form of functional, kinematic and dynamic models. Information on grippers, tools, conveyors and links of the robot, but also on trajectories, positioning tables and uncertainty are some examples of the required data. In addition geometric and kinematic information about the geometry, the application, the resolution of sensor and image processing is needed. The timing of processing and the type of branching according to the sensor information has to be defined. Sensor models for the sensor hypotheses, which allow the decomposition of a sensor plan, must be internally represented. A priori information available from sensors (proximity, touch, force wrist, 2D and 3D visual, compliance etc.), constraints, restrictions and safety requirements must also to be taken into consideration for task decomposition. Depending on the application, the presentation of process knowledge, which might be a catalog of possible actions, reactions and their relationship to preconditions (f.e. sensor pattern) and their effect on the overall system, is needed. All this information is taken together to decompose the tasks. To structure this information in a modular sense, specific processors which operate on the required data types can be defined. Specific processors are:

- A sensor monitor
- A conflict analysis processor
- A decision processor
- A geometric world processor
- A trajectory management processor
- Processors for axis, end-effector and sensor control.

These processors operate on task specific knowledge and produce the inferences necessary for task decomposition and for branching in the decision tree. To limit the amount of data to be processed, on higher levels only virtual information about the lower levels is represented in concentrated form, Fig. 6.24. In other words, instead of an extensive list of details and facts a concentrated abstract description (e.g. functional) is used. This data can be stored in a data base or, for the on-line process, in the memory of the computer architecture. Figure 6.25 shows the principle of data acquisition and representation within the control hierarchy.

6.2.4 Task Decomposition and Execution

The task decomposition in a hierarchically organized control structure was outlined by Nilsson [19]. A hierarchical robot control concept consisting of task decomposition, sensor and predictive world data operators was studies by Albus et al. [13, 17]. Starting from the task description of a manufacturing, a hierarchy of task decomposers operate on problems with decreasing degree of abstraction. They produce a sequence of states or actions which lead from the initial to the goal state in an optimal sense. Constraints and uncertainties are taken into consideration in reaching this goal. The generation of this state sequences requires the simultaneous qualitative and quantitative representation of actual world and robot data. Initial system states and possible actions and reactions supported from the catalog of available control functions are evaluated by a decision operator which has to find the optimal solution in the search space. Decision making is assumed to be a graph searching problem on decision trees whose nodes are possible task states and whose arcs are subsequent actions or subtask sequences.

After the task decomposition the reaching of the goal state within a defined time interval is supervised. If constraints or emergencies are detected a new decomposition cycle will be initiated.

6.2.5 Data Flow and Computational Concept

The horizontal and vertical data flow within the robot architecture is illustrated in Fig. 6.26. Control instructions are transmitted from higher to lower tiers for decomposing task sequences. From lower level sensor data, errors, emergencies, states etc. are transmitted to support the task decomposition at higher levels. The horizontal data flow takes place between the task-decomposition operators, sensory processing operators and world-model operators. Each task decomposition operator is supported with extracted sensor data from the sensor processing modules. The degree of abstraction of the extracted sensor data increases with higher control levels. The sensor processing operators are provided with a stream of task related state expectations and predictions, including uncertainty information from the world-model processor at that level. Each module receives inputs describing the task sequences, action plans and hypotheses generated as output of the decomposition module. Emergency detection and analysis, Fig. 6.27, is

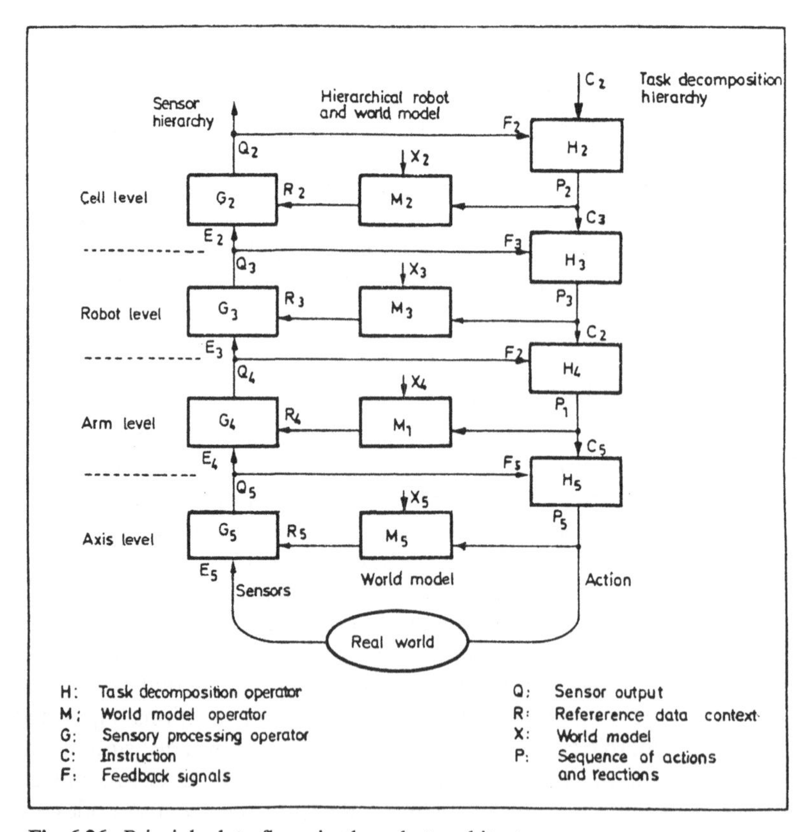

Fig. 6.26. Principle data flows in the robot architecture

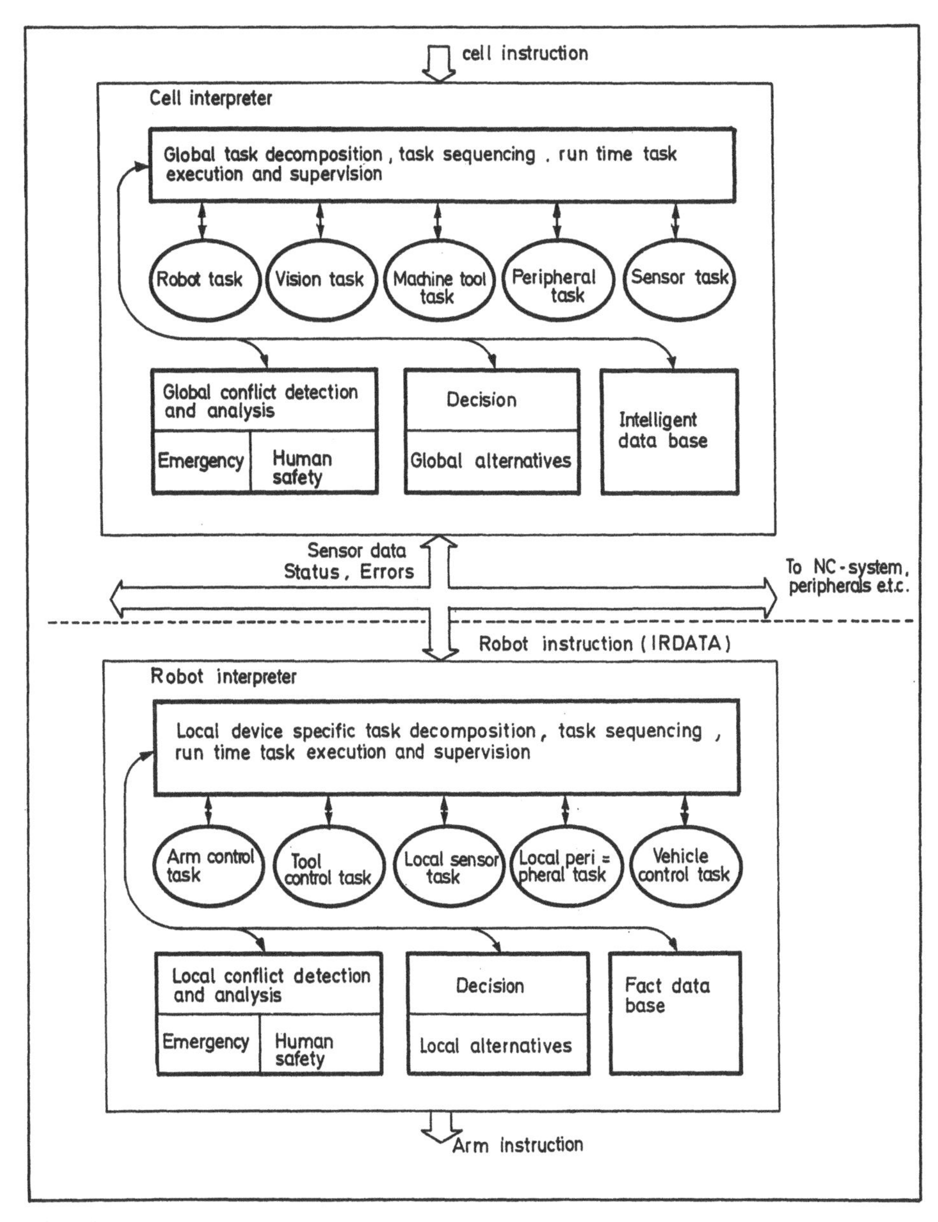

Fig. 6.27. Control structure of the cell and the robot for task execution

performed by the sensor processing module together with the world model module. Decision making and supervision of task execution is carried out by the decomposition module.

Seen from the programming perspective, the hierarchical structure of the cell control is a structured top-down program applied to robots, machine tools, sensors and peripherals. Each block represents a sequence of instructions and a set of structured branches.

In terms of system architecture the computing modules can be viewed as processes or finite state automates. With this terminology each module (operator) is an autonomous process which operates on an input data stream to transform it into an output data stream. All operators can be implemented on a microcomputer which transforms and exchanges date and messages. This structured approach allows modularity in hard- and software. Suitable computer architectures for such a modular system are distributed systems and extendable symmetric polyprocessors [15]. Linkages to CAD/CAM systems are formed by the internal robot and world models, which are presented by frames or a virtual robot and process description.

6.2.6 Conclusion

A hierarchical, modular robot architecture was outlined as an example of a manufacturing cell. The architecture includes the functions of the manufacturing cell, the robot, the arm and the axis in a logical functional internal presentation. Operators for task decomposition, sensor processing and world model manipulation support a data stream, which is necessary for the execution of intelligent manipulation tasks. Such a functional architecture forms a linkage to the formulated application problem. Integration of robots into manufacturing cells means configuring the robot architecture with appropriate machine tool and peripheral control tiers. Efforts are being made by the Karlsruhe Robot Research Group to develop a polyprocessor architecture with the aim of implementing the outlined robot architecture. Linkages to high level robot programming systems like AL and SRL, to a robot data base (RODABAS), to a simulator and the IRDATA interface are basic system features.

6.3 Graphical Simulation Techniques for Planning and Programming of Robot Based Manufacturing Cells

6.3.1 Introduction

In the area of simulation there also exist a wide spectrum of simulation techniques which are applied for planning and operational control of manufacturing processes. Here a method for planning of robot based manufacturing cells will be discussed.

Interactive planning and programming of manufacturing processes are increasingly performed using CAD-modelling software. In addition to the CAD data, technical, procedural and logical data have to be processed in this context, see Fig. 6.28. Interactive planning and programming techniques with graphic simulators support the layout of a manufacturing cell and the definition of programs which determine the manufacturing task. Robot motions and their interaction with the environment, material flow, sensor operation etc. can be displayed on a graphic screen and evaluated by the user. Thus, the operator is allowed to

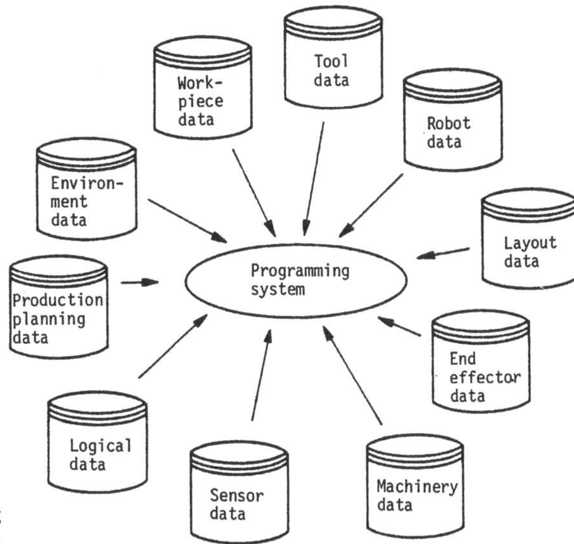

Fig. 6.28. Different types of data necessary for programming robot based manufacturing cells

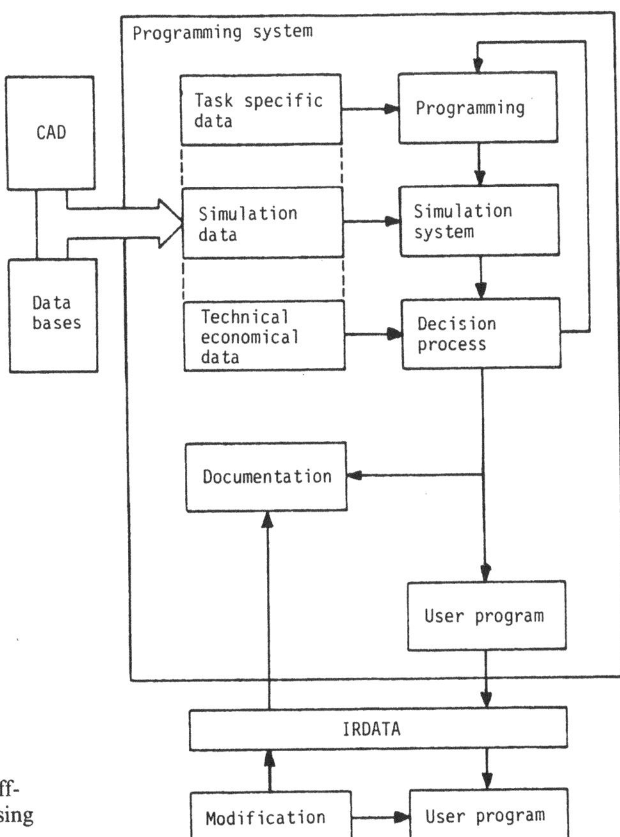

Fig. 6.29. Structure of an off-line programming system using simulation techniques

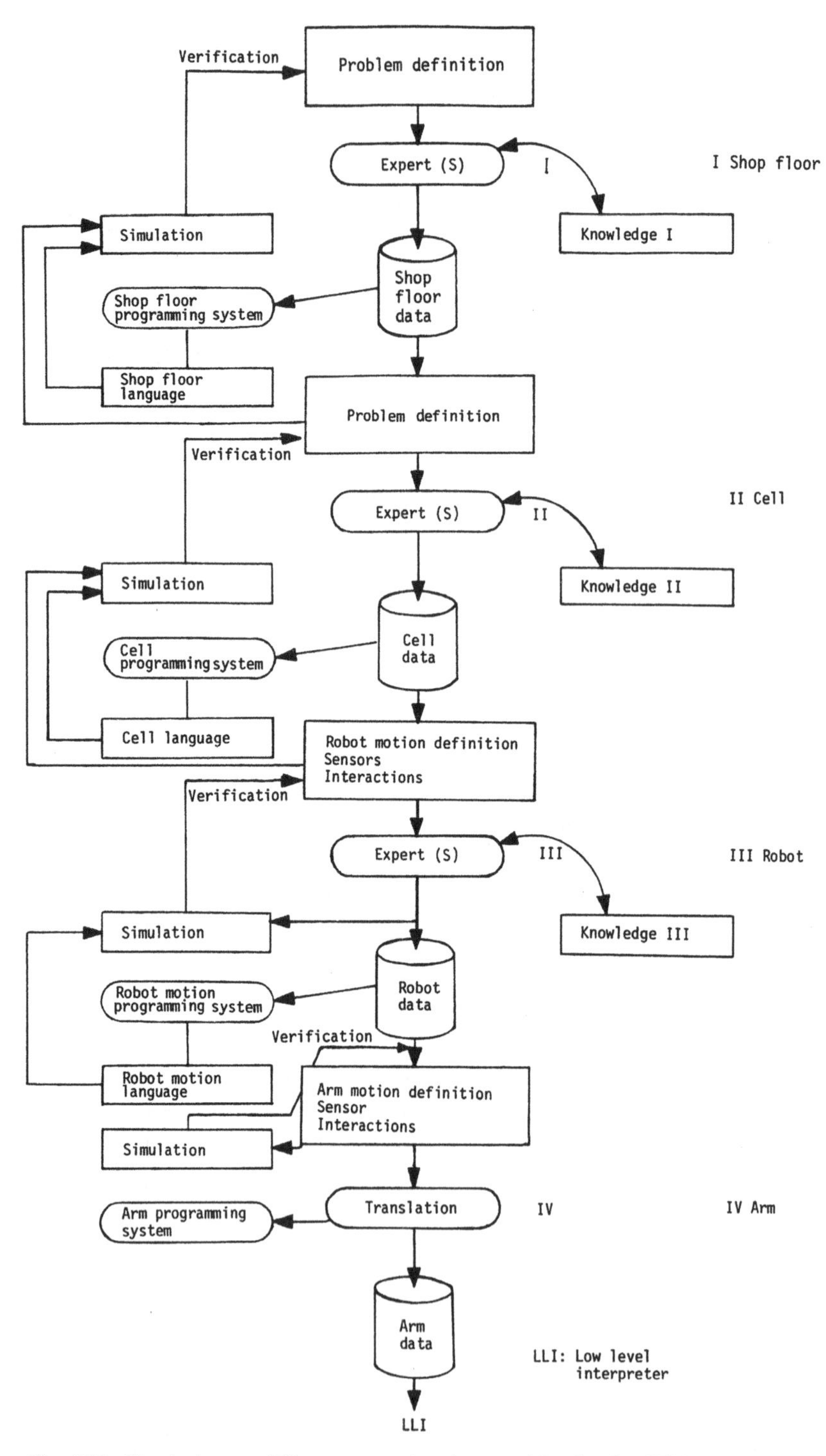

Fig. 6.30. Simulation on different operational control levels of a CIM system

specify input tasks while identifying restrictions, performance of operations and conflicts, like collision of a robot with objects located in its environment.

With an interactive planning and programming system based on a simulator a linkage between CAD/CAM and Robotics is established, Fig. 6.29. The CAD modellers are used to define the geometry of robots, tool machines, transport systems, tools, pieceparts and the overall cell. In addition, the kinematics, dynamics and basic control functions of the units within a cell are emulated. The result is a model of the cell which is called a virtual cell. The user can apply instructions to the virtual cell or to a virtual unit, which are executed by the simulator, and visualized on the graphic screen. Efforts are currently being made to animate the operation as realistically as possible to derive programming standards. Figure 6.30 shows different application levels of simulation procedures for programming a robot within a manufacturing cell. In accordance with the control hierarchy a simulation can be applied to the shop floor level, the cell level, the robot and the arm level. The presentation of the robots is quite different at each level. The degree of detailing increases from the shop floor level to the robot arm level. Simulation of a robot on the arm level necessitates the use of a microscopic virtual robot, consisting of all basic control functions like axis control, end-effector control and basic sensor control. In the case of cell simulation the robot appears as a functional unit which cooperates with tool machines, peripherals and the conveyor system. Here, only a macroscopic virtual robot description is useful.

Emulation of control functions aims to imitate in a virtual sense the control system in software. Emulation is an important tool to support the design of a manufacturing system [20]. In the computer the control system for a manufacturing cell can be emulated as a virtual prototype without any real hardware. With the aid of the emulated control system, the user software can be developed and tested in an early design phase. Concepts for graphical simulation systems, as a basic part of interactive manufacturing cell- and robot programming systems, are under design worldwide. Most of the known systems use CAD based geometry modellers [21–24]. Commercially available CAD robotics systems like Mc Auto-Place (McDonnel-Douglas) [25] or CATIA-Robotic (IBM/Dassault) [26] have a rich library of functions. Collision detection, trajectory optimization, dynamics models and modules for robot movements are examples.

User friendly and problem oriented programming can be realized using methods of artificial intelligence [27].

6.3.2 System Structure for Interactive Planning with a Graphic Simulator

Figure 6.31 shows a conceptual diagram of an interactive planning and programming system with a graphic simulator which is being developed at the University of Karlsruhe. The interaction between the user and the system is controlled via a dialog system. The user can communicate in interactive textual and graphic mode with the system. Via a command selection procedure the user can enter

- the modeller
- the emulator
- the programming module

- the simulator
- the graphical animation system.

According to the system components the dialog functions are separated into functional groups and define specific dialog modes. The modules to be accessed via the dialog have equivalent architectures. The kernel of each module is a library

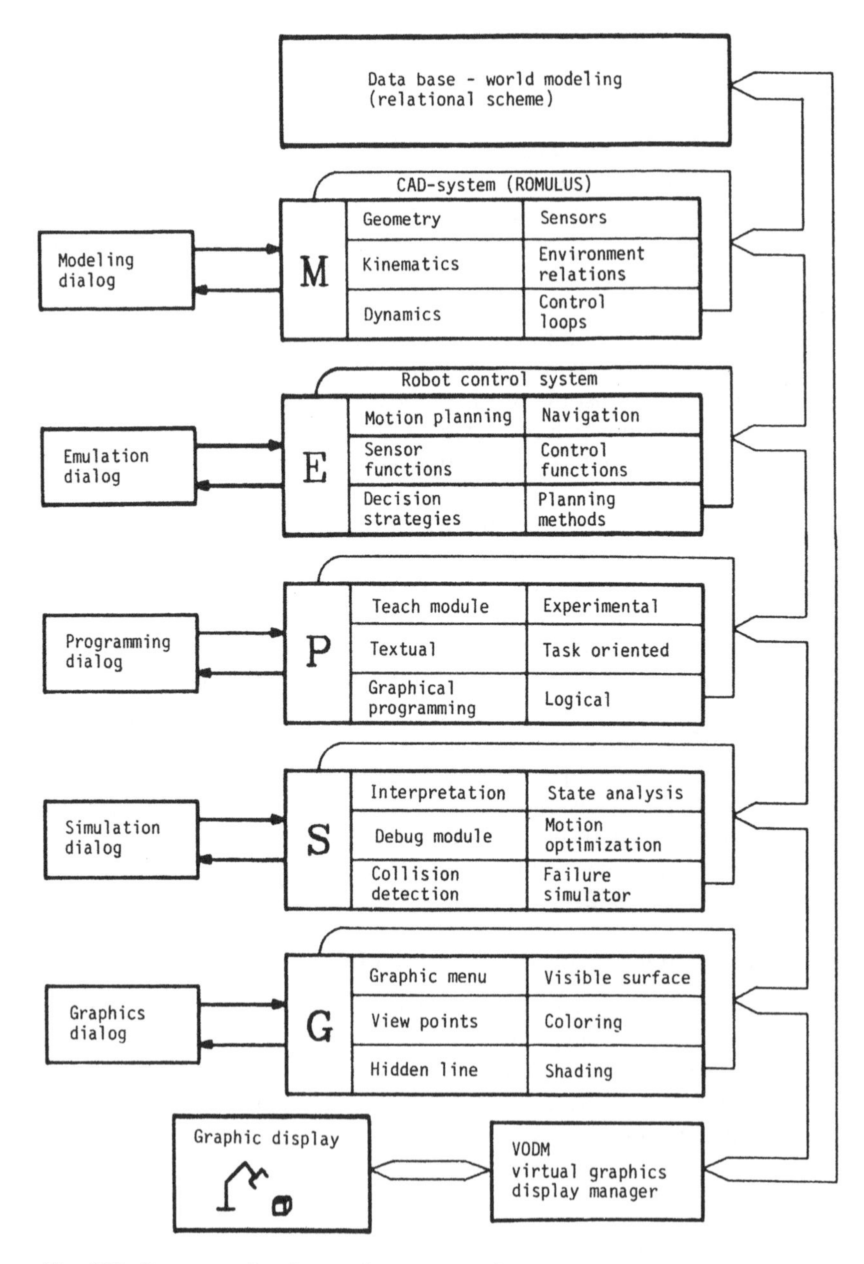

Fig. 6.31. Structure of an interactive programming system for manufacturing cells

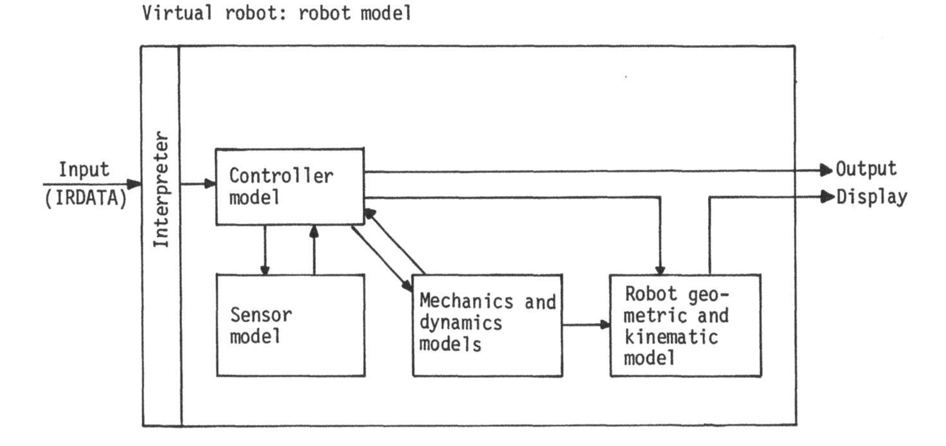

Geometric model:

Wire frame model } : For graphics
Polyedron model }
Solid model : For all applications

Mechanics, dynamics and kinematics models:

Skeleton model consisting of elements with mass, inertia, frictions, center of gravity and kinematic transformation frames

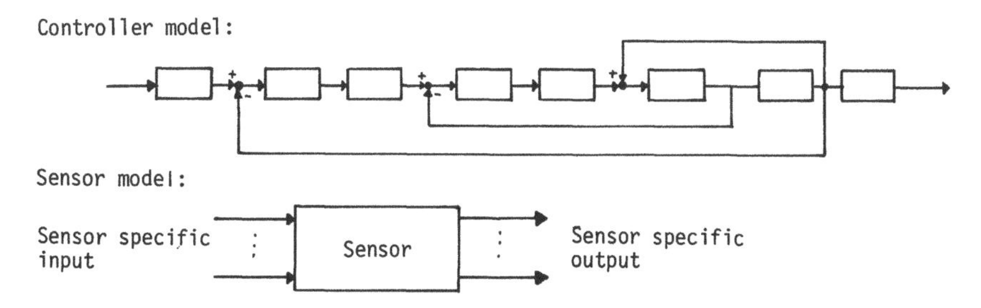

Fig. 6.32. Example of a virtual robot

of programs and program elements, which provide the methods necessary to execute module specific tasks. In addition each module has a method monitor which controls the execution of commands specified via the dialog. The method monitor activates relevant program modules and decomposes them to a method. A central data management system ensures the separation of methods and data. The basic modules are described briefly in the following sections.

The modeller (M) is used to describe the manufacturing cell and the pieceparts with their geometric, physical and functional properties. As a tool for geometric modelling of robots, tools, workpieces and sensors the CAD system ROMULUS is used. The generated CAD-data structures may be extended with additional non-geometric data. Kinematic and dynamic attributes of robots, conveyors and

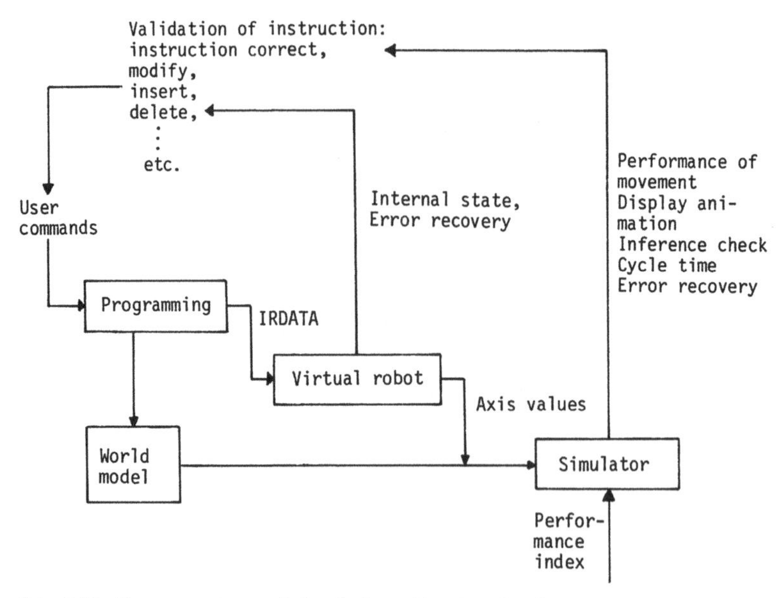

Fig. 6.33. Programming and simulation of a virtual robot

fixtures, functional properties of sensors and axis control loops, and relation between objects within the cell must be defined.

The emulator (E) comprises all methods and functions, which are necessary to control a cell and to define its behaviour. It is also used to build a virtual cell controller which is used for the simulation. Emulated functions are, e.g. trajectory planning, coordinate transformation, sensor functions, control- and decision functions. The virtual cell controller executes the programs defined by the programming module, see Fig. 6.32 and 6.33.

The programming module (P) allows the programmer to develop programs in different modes. The dialog supports interactive textual and/or graphical specification of program statements.

The simulator is used for validation and verification of the generated programs. The execution of the manufacturing program by the virtual cell is visualised on the graphic screen. A performance index (criterion) and a number of analysis functions, e.g. collision check, error recovery procedures, tolerance analysis etc., allows the programmer to analyse and validate the program. If the simulator detects errors, sources of errors or critical states, then they are presented in graphic or in alphanumeric mode on the screen.

The graphic module [6] facilitates the interpretation of the simulated operations. If offers methods which enhance the degree of information for the user with the aid of graphical animation subroutines, like hidden line, visible surface, colouring, shading, zooming, view point transformation etc.

The interface between the simulator and the graphic system is realized via a VGDM (virtual graphic display manager). The VGDM is independent of the type of graphic workstation. It offers, on the simulator side, pseudographic instructions for the construction and manipulation of graphical pictures. With these commands each module can perform changes to the graphical image structure. Dedicated workstation dependant software drivers map the VGDM pseudo-display file onto the workstation specific control structure. Figure 6.31 shows, that each module has access to the central data base. Further, methods and data have to be separated clearly to ensure consistency of the data to be processed. Data management is based on a relational data model. Objects of the same type are allocated to object classes. Each object class is characterized by attributes, of which the values are object names.

6.3.3 Conclusion

Interactive manufacturing, planning and programming of robot based manufacturing cells are tools for integrating robots into CAD/CAM systems. The use of the virtual cell description allows device and computer independant development of software, including machine intelligence. A close relationship between the geometric modelling system and machine tool and robot applications is a major key to effective and economic manufacturing within CIM systems.

6.4 Advanced Computer Architectures (5th Generation)

6.4.1 Introduction

Advanced architectures of future computer systems (5th generation) are currently being conceived by the Americans and Japanese [28]. These systems will be realized in the 1990's and will be knowledge based information processing systems. They will incorporate developments of the following technologies and academical disciplines:

- VLSI technology
- decentralized parallel computing
- very high-level programming languages
- knowledge based expert systems.

The combined soft- and hardware solutions of the 5th computer generation will provide many benefits to the user, such as:

- intelligent interfaces
- knowledge-based management
- problem solving
- inference functions.

Communication with this new computer generation will be possible via natural, graphical or textual languages. These very high level input languages will allow the use of innovative programming methods which implicitly describe the problem to be solved. This is in contrast to traditional programming methods in which each step is explicitly entered into the computer. The knowledge based expert systems use modules of organized knowledge, sophisticated problem solvers and inference functions. They help the user to solve a specialized problem efficiently.

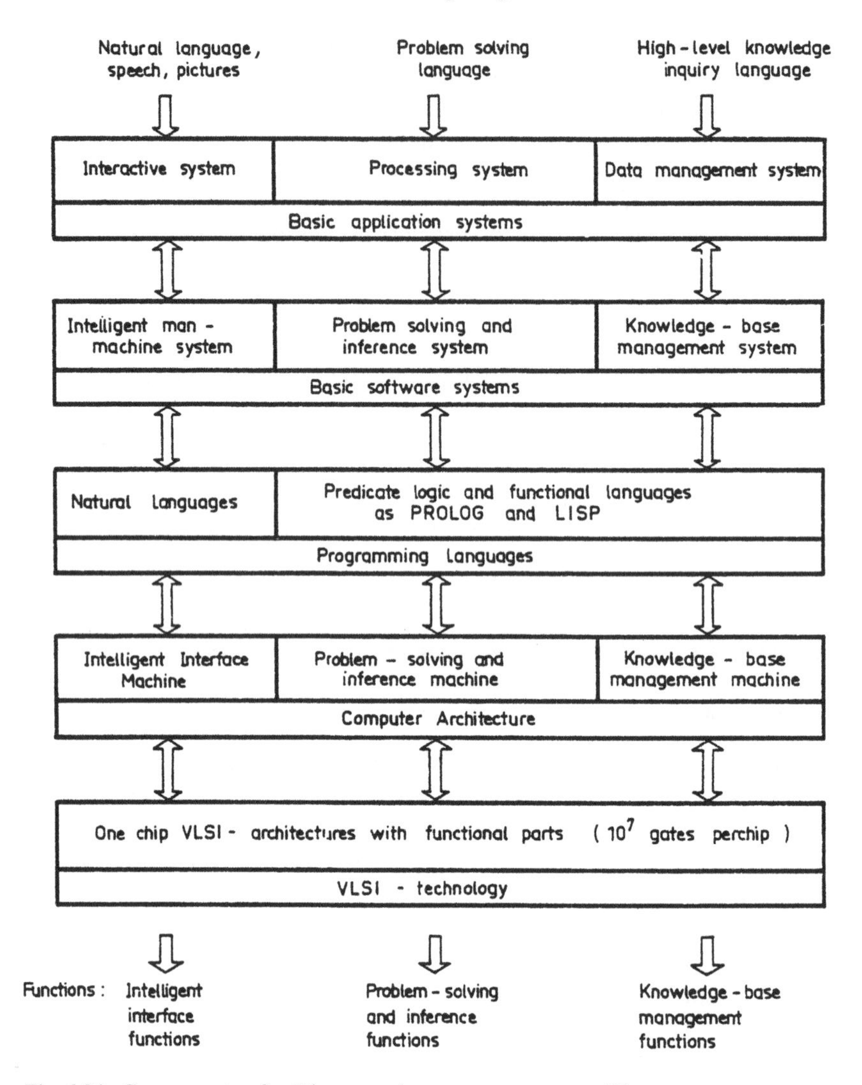

Fig. 6.34. Components of a 5th generation computer system [2]

In the case of CAD/CAM systems, the 5th generation computer will support planning and control of activities at all hierarchical levels to operate a completely automated factory. The components of 5th generation computer are outlined in Fig. 6.34 and will be discussed in the following sections.

6.4.2 Components of 5th Generation Computers

Fifth-generation computers form an information processing network which can be linked together by a unique programming language to provide a powerful architecture for solving problems by inference, with the assistance of a knowledge-based management and intelligent input and output. Each node of the network consists of soft- and hardware modules which perform special functions. The hardware modules are configures from VLSI processors of high functionality and high performance. They can be combined, for a specific application, to perform any desired function.

A basic unit is the intelligent interface which supports communication between the user and the computer by speech, pictorials (graphics) or other natural languages. The interface is built from special-purpose VLSI processors, specially designed for signal and speech processing. They allow communication in a form which is more natural to man.

The problem-solving and inference functions are carried out by a powerful computer system. It is planned to perform 10^7-10^9 locigal inferences per second (LIPS, one LIPS is equivalent to nearly 1000 instructions on a conventional computer). The problem-solving and inference functions make use of the extensive knowledge bases and are connected to the intelligent interface.

The knowledge management functions operate the knowledge data which are required for inference functions. The data base system will have a capacity of up to 10^{12} bytes.

The integration of the three basic functions into a single general purpose computer system allows the performance of wide variety of applications such as computer-aided engineering, manufacturing planning, computer aided design and robotic development and others. This area can be summarized as knowledge engineering.

6.4.3 Applications of 5th Generation Computers

Knowledge processing means utilization of a knowledge base for problem solving and inference. Intelligent man-machine interfaces will help to facilitate communication on the basis of

- natural language processing
- speech processing
- picture and image processing.

The natural language processing system shown in Fig. 6.35 is subdivided into modules which contain knowledge of languages (syntactics, semantics, pragmatics), problem solving domains and the interactive dialog. A basic function of the semantic and pragmatic analysis is to extract structures from the problem description and to build an appropriate tree [30]. This internal tree allows an analysis of the context. From it the problem and its structure are extracted. If the problem description is not complete, a dialog will be initiated to get more input information (detailed questions) until the problem is totally defined. To perform this task, the knowledge about the problem domains its effective use as well as its extension must be available. The answer to the problem definition can then be generated by referencing the application problem solver and the knowledge data base. The result is transformed from its internal structure into a speech signal using a sentence generator. If pictorials or images are used as input, an analog procedure is performed to generate the answer. Examples of the applications of 5th generation computers in the area of CIM systems are knowledge based expert systems to aid scheduling, production planning as well as to perform local and global

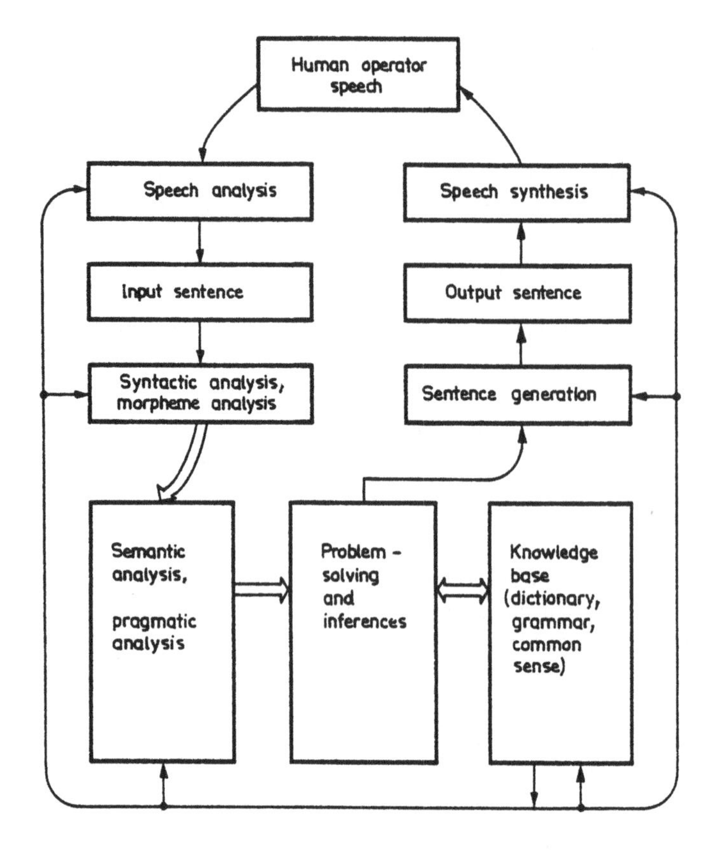

Fig. 6.35. Structure of a natural language processing system

plant control. Also software engineering, design, and material processing will be aided by expert system functions. The knowledge of the system should include control rules for every level of a hierarchical manufacturing system (operational control, coordination level, disposition, scheduling etc.).

6.4.4 The Basic Software System and Programming Languages

The basic software system (see Fig. 6.34) supports the three components of the expert system which are:

- the intelligence interface system
- the problem solving and inference system
- the knowledge base management system.

General functions of the basic software are elementary information-processing functions, interactive processing and management. The intelligent interface software system has to operate on data derived from natural language- and speech processors as well as from picture and image processors [30]. The problem solver includes a semantic analyser, a meta inference system, a dictionary, a grammar, an image data base and a common sense data base. For generating the answer to a problem definition, the question answering system (intelligent answer generator) uses a picture and sentence generator. To solve the problem definition, extensive references to the application knowledge base system have to be made. The application knowledge base will contain a basic program file, a computer architecture and a VLSI-design file. The software includes support systems for handling, editing, debugging, acquisition and consistency checking of knowledge as well as a data base compiler. Besides the application knowledge data base, a system knowledge data base exists, which contains specifications of the system itself (processor operating system and file specifications which facilitate the use of the system).

The programming languages for 5th-generation computers will be of three types [29]:

- natural language, speech and picture type which interact with the intelligent interface
- the high level inquiry language which interacts with the knowledge base management system
- the kernel (logic programming) language which interacts with the problem solving and inference system.

The kernel language is a problem solving language which is based on predicate logic (as Prolog). A program in a predicate logic language is a collection of logic statements. Its execution is equal to a controlled logic deduction of the logic statements defined by the program. The inquiry language for the knowledge base and the knowledge representation language can be implemented by using the kernel language. This language will be the machine language of the 5th generation computer. Other classes of languages such as conventional languages, process control languages, functional languages and object oriented languages will also be supported by this computer generation.

6.4.5 Computer Architecture of the 5th Generation Computer Systems

5th generation computer architecture consists of three basic machines:

- the intelligent interface machine
- the problem solving and inference machine
- the knowledge base machine.

Intelligent interface machines are dedicated special-purpose processors which emphasize operation on speech, pictorials and signals. The first VLSI processors which have the capability to recognize 1000 words are actually available.

The problem solving and inference machine has the role of a CPU in the 5th generation computer. It is based on a logic programming machine with

- abstract data types
- data flow processing and an
- innovative von Neumann architecture.

Sequential and parallel inference machines are under development using data flow machines [31]. The knowledge base machine operates with a relational data base and relational algebra [32]. It holds a large amount of knowledge data which is well structured, so that each knowledge item can be referenced effectively.

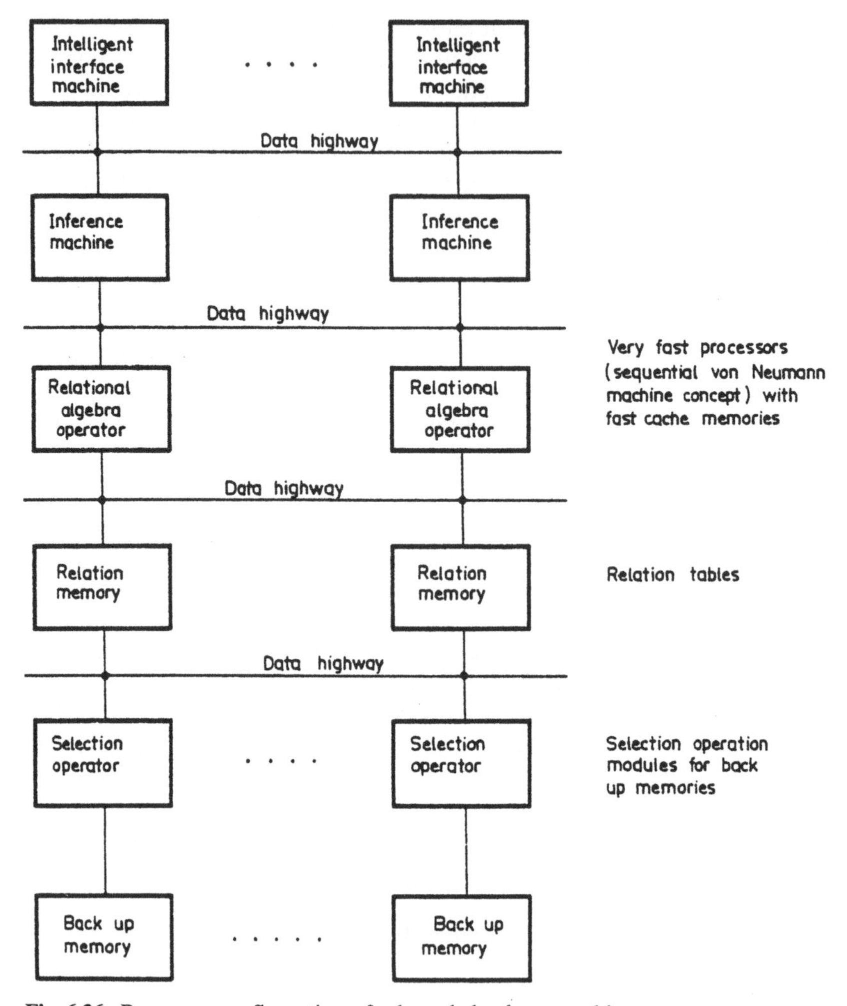

Fig. 6.36. Processor configuration of a knowledge base machine

When the knowledge base machine receives a call from the inference machine, it searches and retrieves knowledge items and hands them to the inference machine. When the knowledge base machine receives knowledge data items from the inference machine, it compiles and integrates them into the knowledge base. The hardware structure of such a machine is shown in Fig. 6.36. New advanced architectures such as that of a

- logic programming machine,
- functional machine,
- relational algebra machine,
- abstract data type support machine,
- data flow machine and the
- innovative von Neumann machine

will enable the 5th generation computer to process the knowledge information at high speed. The VLSI design will support the development of the

- high speed numerical computation machine,
- high level man-machine communication system,
- efficient network architecture,
- distributed function system and the
- data base machine.

This will allow the 5th generation computers to operate with a processing rate of $10^7 - 10^9$ logical inferences per second.

6.4.6 Conclusion

The architecture of the 5th generation computers which will be available in the early 1990's has been outlined. Such computers will work as expert systems in CIM for manufacturing planning and control. Their capabilities can be summarized as follows:

- intelligent man-machine communication capable of understanding speech, images and natural language
- understanding of problem definition and requirement specifications
- synthesizing of information processing procedures
- generation of responses to problem definitions using procedure referencing knowledge.

The knowledge bases contain knowledge about the problem areas, the programming languages and the different machine components of the entire computer.

6.5 References

1. Mesarovic, MD, Macko, D, Takakara, Y (1970) Theory of Hierarchical Multilevel Systems. Academic Press, New York
2. Albus, JS et al. (1983) Hierarchical Control for Robots in an Automated Factory. Proceedings of the 13ath ISIR/ROBOT 7, Chicago, April
3. Rembold, U (1983) The State of the Art and Future Trends of Microcomputer Technology for Industrial Applications. 3rd Symposium on Microcomputer and Microprocessor Application, October, Budapest, Hungary
4. Williams, TJ (1983) Developments in Hierarchical Computer Control Systems. Proceedings of the CAPE '83, Amsterdam, Netherland, April
5. Erste Schritte zur freien Kommunikation. Markt und Technik, Nr. 5, 5. Mai 1982
6. Assoziative Datenbankmaschine. Markt und Technik, Nr. 20, 21. Mai 1982
7. Mallach, E (1983) 32-bit mini press down from above. Mini-Micro System, April 1983
8. Schneller durch Subcomputer. Markt und Technik, Nr. 2, 15. Januar 1982
9. Noch keine Grenzen in Sicht. Markt und Technik, Nr. 1, 8. Januar 1982
10. Godenough, F (1973) Analog LSI, Electronic Design, Jan. 6
11. Klein, R (1982) Optical Memories Increase Disk Capacity. Digital Design, Febr.
12. Softwarekomponenten in Silicon. Markt und Technik, Nr. 14, 8. April 1982
13. Barbera, AJ et al. (1982) Concepts for a real time sensory – interactive control system architecture. Proc. of the 14th Southeastern Symp. on System Theory, April

14. CAM-I Robotics Software Project 1983: RSP Working Group Final Report, Outline of Functional Specification for Prototype Software
15. Rembold, U, Kordecki, C, Dillmann, R (1984) Conceptual Design of Adaptive Multirobot Control. Proceedings of the 1984 ASME International Computer in Engineering Conference, Aug. 12–16., Las Vegas
16. Blume, Ch, D'Souza, Ch, Zühlke, D (1983) Aspects to Achieve Standardized Programming Interfaces for Industrial Robots. Proceedings of the 13th ISIR/Robot 7, April 17–21, Chicago, Illinois
17. Albus, JS et al. (1981) Theory and Practice of Hierarchical Control. 23rd IEEE, Sept. 13–17, Washington, DC
18. Saridis, GN, Stephanon, HE (1975) Hierarchically Intelligent Control of a Bionic Arm. Proceedings of the Conference on Decision and Control, December, Houston, Texas
19. Nilsson, NJ (1973) A Hierarchical Robot Planning and Execution System. Artificial Intelligence Center, TN-76, SRI-International
20. Bloom, HM, Furlani, CM, McLean, CR (1984) Emulation as a Tool in the Design of Factory Automation Systems. Internal Report of the Factory Automation System Devision, Center of Manufacturing Engineering, National Bureau of Standards, Gaithersburg, MD
21. Soroka, BF (1980) Debugging Robot Programs with a Simulator. Proceedings of the CAD/CAM 8 Conference, Anaheim, California, November
22. Wesley, MA et al. (1980) A Geometric Modelling System for Automated Mechanical Assembly. IBM Journal on Research Development, vol 24, January
23. Sata, T, Kimura, F et al. (1981) Robot Simulation System as Taska Programming Tool. Proceedings of the 11th ISIR, October, Tokyo, Japan
24. Computer Graphics for Robot Off-Line Programming (1981) Product description of McDonnell-Douglas Automation Company
25. PLACE (1983) Software Product description Manual. McDonnell-Douglas Company
26. CATIA Robotic. User Manual. Dassault Systems
27. Schmidt-Streier, U, Altenheim, A (1984) Geometriedatenerfassung mit Computer Graphics und Sensoren zur Programmierung von Industrierobotern. Tagungsbericht der CAMP '84, Berlin, pp 155–159
28. Aiso, H (1981) Fifth Generation Computer Architecture. Proceedings of the Int. Conf. on the 5th Generation Computer Systems, Japan, Oct., pp 2–29
29. Treleaven, PC, Lima, JG (1982) Japan's Fifth-Generation Computer Systems. Computer, August, pp 78–88
30. Tanaka, H, Chiba, S, Kidore, M, Tamura, H, Kodera, T (1981) Intelligent Man-Machine Interface. Proceedings of the Int. Conf. on the 5th Generation Comp. Syst., Japan, Oct., pp 3-3-1–3-3-11
31. Uchida, S et al. (1981) New Architectures for Inference Mechanisms. Proceedings on the Int. Conf. on the 5th Generation Comp. Syst., Japan, Oct., pp 4-1-1–4-1-12
32. Amaniya, M et al. (1981) New Architectures for Knowledge Base Mechanisms. Proceedings on the Int. Conf. on the 5th Generation Comp. Syst., Japan, Oct., pp 4-2-1–4-2-10

7 Programming of Robot Systems

G. Gini, M. Gini, M. Cividini and G. Villa

Design study of a mobile assemby robot

7.1 Robot Languages in the Eighties

The scenario of general-purpose programming systems is rapidly changing; what are the consequences for robot programming? Are the programming environments built around ADA, UNIX, and Interlisp useful for robot programming? After introduction the peculiar aspects of robot programming we will discuss some examples of general-purpose languages applied to robots, and languages specifically designed for robotics. Criteria for making a choice between the two approaches should take into account the present state of the art. The need for a strong integration between different components such as robots, vision systems, and other automation equipment could support the first approach. The solution of robot and robot user problems has, until now, supported the second approach, as indicated by the choices of European robot manufacturers. Anyway the expression of actions taken by different intelligent agents, as robots can be defined in the future, will require completely linguistic media. We hope that robot programming in the eighties can provide experience useful for the assessment in the field.

7.1.1 Introduction

Today robot applications are generally carried out in an integrated industrial setting, where robots and other equipment manipulate and sense parts to repeatedly perform a task. We may see a typical setting as consisting of a conveyor belt which transports parts, usually partially oriented and separated, a vision system getting information about the incoming parts, and a robot assembling, inspecting or serving other machines. These robots are sometimes programmed on line, being guided through their task and storing positions and operations by a memory. The drawbacks of these methods have been stressed often enough. Errors during programming require restarting, teaching of many positions is too long and error prone, the programming time is not productive because all the robot related equipment must be stopped during new task programming, synchronization with other equipment is hard, sensors cannot be used to modify actions during program execution. During recent years we have seen a significant change in the attitude of robot manufacturers, and almost every new robot is now sold with an off line programming system. Ten years of experience have passed and manipulator level languages are now accepted. Usually they are used to give the cartesian reference frames indicating where the manipulator hand should be moved in order to accomplish the task. Those frames or their sequence can be modified by run-time events, such as sensor output or external synchronization signals. In some early systems every joint of the manipulator was individually given a value in its own coordinate system. Programming in terms of frames is not yet a good solution because it requires a lot of detailed information be given by the programmer, who is expected to have some skills in mathematics and programming. The writing of robot programs is not as easy as one might think. It is quite difficult to understand and use positions in 3D space, and this is the current basis of robot programming.

An intelligent robot [1] will simply be given a task and the spare parts and materials needed to perform it, instead of programs covering the complete sequence of actions to be taken. An issue to be tackled by more advanced systems is world representation and object modelling. A representation of the world should allow the robot to manipulate parts and to sense the environment. This model should be a copy of the real world. Actually no world model can be complete, but it should at least be detailed and rich enough. Present industrial robots do not have any geometrical world model; their knowledge is encoded as variables and data structures which have only meaning for their human programmer. An integration of modelling used in design and production is highly desirable. Many systems are now commercially available for CAD/CAM applications. Most of them are pure graphics systems, without much interest for robotics. The models used by CAD systems may be useful in robotics, even though robot relevant information, such as the center of gravity, the mass, the grasping point, etc. of the object are not contained in CAD models. The experience gained using the RAPT system is that using CAD systems for world modelling is still long and tedious. Moreover the use of CAD data bases for defining world models to be used also for sensory tasks (for example for vision understanding) is not yet acceptable.

Despite these formidable research issues robots are currently in use. The practical solution taken has been to limit the intelligence and understanding of the robot and to redudce programming of robots to usual programming or, in a few cases, to parallel programming. No use of world models is made in any commercial system.

Many industrial issues in robot programming are still open. How to program robots completely off-line, and how to achieve the integration of different devices, being the most studied. The present status of software and software engineering has its impact here. Having removed the problem of intelligence from robot software has reduced robot software to a sophisticated software engineering problem: software solutions should work here.

7.1.2 Robot Programming

The most obvious ways of implementing robot languages are to adapt a general-purpose programming system, or to develop robot specific programming systems. In the second case new languages may be developed from scratch, or from existing automation languages, such as APT.

The first solution is appealing because education and training of robot programmers can be reduced in time and the development of robot applications will be reduced to the writing of a few routines. The second solution is appealing because robot programmers need not be computer experts, and what they have to learn will be exactly what they need to use. In the case of APT the same NC programmers could be easily converted into robot programmers.

The first solution has a shortcoming in that it heavily relies on what is available for general purpose programming. The computer-user interface is not tailored to the specific needs. The second solution has as a shortcoming that much development effort is often expended only to provide something already available with minor differences. In the case of APT-like languages we may also argue as to whether robot programming is the same as NC machinery programming.

Since a language is a way of implementing or testing different solutions, we may say that all developments made so far have demonstrated what can currently be done at the manipulator level programming. Beyond that, the choice of using existing languages or developing new ones is a problem of market image and acceptance. The need to develop new languages seems obscure. The only important difference is in the programming environment. The development of programs off line requires sophisticated programming environments which are not generally available in standard languages as FORTRAN or even PASCAL. The choice as to whether to develop a complete programming environment is the only inducement we see to the development of new systems. A set of subroutines written in FORTRAN require a long time to edit, write, debug.

If the language demonstrates new solutions or allows task oriented programming then we may still want to maintain manipulator level programming as the target language for planning and sensory activities. The performance of those activities should not necessarily resemble any of the existing programming languages. Graphics, natural language, or sometimes mathematical equations may all be useful in providing ways of performing tasks.

Many papers have reviewed existing robot programming languages among them [2, 3]. We do not intend to review all the existing robot programming sysems. We only want to see what new ideas have emerged from them and whether they have been assessed or are still waiting to be fully explored.

Usually two systems are strictly integrated within a robot programming system. The *user language,* in which application programs are written, and the *run-time system* which executes the code generated by the language translator. This solution is similar, for instance, to the one used in most of the Pascal systems. It may be used as a way of standardizing user languages simply by changing the run-time system, as done in VAL [4], developed for PUMA robots and then implemented on other Unimation robots. Often the run time system is run on micro computers and written in assembly language. This trend could change as ever cheaper computer power makes it reasonable to write all software in high level languages.

An issue requiring attention is that of defining standard software interfaces between the robot and the user level software. What kind of information should be passed to the robot? Joint positions or frames? In which order? How should a point to point execution be requested for a continuous path? There is no reason why a cartesian robot and a polar robot should be programmed in completely different ways. While the run time system is well tailored to the specific hardware in use, the user level language should be problem-oriented more than manipulator-oriented. If this standard software interface were to be provided and accepted by any robot manufacturer we might get any robot language to work for any robot; homogeneity and modularity would also be valuable in industrial settings. We mention here the CAM-I standardization project in robot software [5], which has individuated five main components of robot software: robot language, robot simulator, robot controller, robot modeler, teaching system. They have introduced Artificial Intelligence into the CAD/CAM environment, and this seems more a decision to deal now with problems to be solved by perception and decision making. In our opinion most of these problems should be solved at the robot level, the robot being the flexible and adaptable entity of the FMS.

We do not intend to give here a complete list of reference terms for comparing robot programming languages. Our analysis based on consideration of the following:

1. Expression of movements (joint level, hand level, object level);
2. Expression of trajectories;
3. Use of sensors;
4. Class of the language (Pascal, Basic, functional);
5. Implementation (interpreter, compiler, programming system);
6. I/O: ports, functions provided, integration with other equipment;
7. Multitasking, synchronization, and parallelism;
8. Integration with CAD/CAM for simulation, planning, control.

7.1.3 Languages and Software Environments

We may look at general purpose languages as candidates for robot programming. There is some experience of using Pascal for robot application. See for instance PASRO [6] as a working example of this. Even Pascal, however, has some shortcomings as a language for automation. Among them, Pascal doesn't support cooperation, which can be obtained at the operating system level; moreover, its file management is inadequate and file management would occur very often in integrated manufacturing. Other solutions have been tried, for instance using Concurrent Pascal, a small language developed around Pascal to define and execute concurrent tasks.

Languages for automation did not exist before the introduction of robots. The only exception is APT, the language for NC programming. APT has in fact been chosen as a basis for robot languages in at least two projects ROBEX [7] and RAPT [8]. Nevertheless, it has not yet been demonstrated truly useful in robotics.

Software environments for general purpose programming are now available on most computers. The most complete and advanced software environments are currently UNIX and Interlisp, while the position of ADA is not yet assessed. The main advantage of a software environment over a simple language compiler is that the first provides an unified approach to all the problems encountered in the project, the development, and testing of the program. In the following we will briefly review those three systems, UNIX, INTERLISP, and ADA.

UNIX: The output redirection and the pipeline mechanisms for connecting programs are among the most useful characteristics of Unix for simplifying modular programming. Different programs can be connected in any meaningful way. The Unix operating system is very popular on scientific personal workstations, and some of them are intended for CAD/CAM applications. Unix could become the standard operating system for CAD/CAM applications, but perhaps not for robot programming. The computers used today to operate robots usually run both the run-time system and the user level language. To reduce the hardware costs they are usually stand alone systems, without any standard operating system. Only very sophisticated robots, not currently on the market, could justify the high cost of using a sophisticated computer for their own needs.

INTERLISP: The only experience so far reported of using LISP for robot control and programming has come from the MIT Artificial Intelligence Lab, where Mini [9], an extension to LISP to deal with real time interrupts and I/O, was used to program a robot equipped with a force sensor. LISP has not yet aroused the interest of robot manufacturers. Many reasons for this can be envisaged. Only recently has LISP been made commercially available on mini and microcomputers; it requires a lot of central memory; it is inefficient in mathematical computations and array management. Even though most of those shortcomings are valid, LISP can be considered very interesting for two reasons. The functional style of programming which is at the basis of LISP (even though Mini is not an example of functional programming; it used a lot of SET instructions) makes it interesting because languages embedded in LISP are completely extensible, so that all the manipulation functions can be modified by the user. We will discuss the functional style of programming in the following. A second good reason for using LISP is that robot programming tends to use more and more artificial intelligence techniques, and LISP is still considered the main Artificial Intelligence language. Serious Lisp development requires several software components still not available even in the UNIX environment. For this reason LISP is still an intensive memory user, and this is the reason for developing LISP machines to make the best use of the computer. Unfortunately those machines are much too expensive for current use at the factory level.

ADA: The main motivation for using ADA [10] in robot programming is that ADA is a structured and complete computer language, and offers some advanced tools, such as extensibility, modularity, real time capabilities, and strong type checking to increase the programs reliability. Moreover it has been designed to be the only language of the eighties, and claims have been made that ADA could substitute any language from the assembler to the highest levels. The importance of a complete programming language for robotics applications has been made clear: we want robots to cooperate with other equipment, and this requires task synchronization and coordination of different and sometimes non trivial tasks. Such programming and coordination is not provided by most of the robot programming languages in industrial use today. We do not know of examples of robot run time systems written in ADA, although there are examples of other subsystems developed in ADA. Vision is one of them. A long-standing practice in vision programming has been to choose C or Pascal; ADA is a superset of them and its use should solve more problems than it raises. Experience so far in Ann Arbor [11] has demonstrated that the use of packages (a way of simplifying program encapsulation) and generic packages (a way of implementing abstract data types) can simplify the development of complex software, making it easier to distribute tasks to different people, to integrate them, to modify the manufacturing cell without complex software modifications. On the other hand, the size of ADA can be a problem. Many features of ADA are hardly useful in industrial automation, but their presence makes ADA compilers big and the ADA language difficult to use. In no way is it possible to manually guide a robot, but ADA applications can make it programmable. On the other hand, the definition of the professional requirements and education of robot programmers is difficult to achieve. ADA has a great chance of becoming a reasonable solution for programming robotics cells. The main shortcoming of this philosophy of algorithmic and explicit programming is that it is unsuitable for dealing with a complicated cell in which many events may happen, and time and sequence constraints may be met by different solutions. In this case expert systems, such as for instance GARI [12] seem a more flexible and understandable way of planning and controlling the cell.

7.1.4 Functional Languages and Logic Programming

Since 1980 a lot of literature has stressed the idea that the future of computer languages can be different from the present in some radical way. The evolution of languages, as has been seen up to now from Algol to Pascal to Ada, could be a dead ending for computing. These languages are based strictly on the Von Neumann architecture of computers, and that architecture is unlikely to continue into future computer generations because it has an unnecessary bottleneck in accessing memory. Languages which use assignments access memory one word at a time, and assignments make user languages more suited to the way computers operate than to the way humans think. The next generation of computers should avoid this bottleneck. Many architectural solutions are available: all of them rely on using functional languages. What makes a functional language attractive is its problem-oriented expression, because it performs functions and doesn't care about memory locations, it can be implemented as a very restricted kernel (the function definition and composition operators) and then grow in every way using user defined functions, it is able to run on widely distributed architectures because it does not produce side effects (no global memory is used). We have found an accent on functionality in many robot languages. Mini [9], LAMA-S [13], AML [14], and Lenny [15] have provided some way of achieving functional capabilities, mainly extensibility.

Another language with full functional capabilities is *Prolog*. It is also the most successful language for logic programming. The Japanise Fifth Generation programs mostly rely on it as the basic language for future computers. Its use for robot programming has not yet been tested. In some applications similar to robots, i.e. CAD, it has been demonstrated useful. In a comparison [16] between a 3D graphics program written in Pascal and the same program written in Prolog, the Prolog implementation was more concise, readable and clear than the Pascal version. It also took less storage and ran faster than the Pascal compiled version. Since Prolog has been used for operating systems as well as for plan generation and has various ways of managing arrays its applicability to robotics needs only to be more fully demonstrated. We would expect that a prolog implementation will be accepted both by Artificial Intelligence oriented users and by mathematically oriented people.

7.1.5 European Robot Languages

The European scene in robot programming is very active. The first commercially available language for robot, SIGLA, was European, as are some of the most advanced projects. In the following we make a short presentation of all those we have so far found in the literature.

HELP [17, 18] is the language developed by DEA (Italy) for their Pragma A 3000 (Allegro in USA) assembly robot. It allows concurrent programming and structured programming. The syntax is Pascal-like, all the manipulation functions are provided as subroutines. Signal and wait provide synchronization between different tasks. Any kind of sensors can be connected using a rich set of I/O ports operations. The robot is modular, different arms and different degrees of freedom for each arm can be organized. The coordinate system is cartesian, and two rotary axes can be added to every wrist. The application programs are usually provided with the installation of the robot. Major applications are in the automotive industry, electronic assembly, precision mechanics. It is implemented on DEC LSI 11 computers under the RT-11 operating system.

LAMA-S [13]. The language developed by the Spartacus project, a project aimed at developing robots to help handicapped people in many every day life tasks, such as serving drinks or food. LAMA-S uses APL as implementation language. The user level functions are translated into a low level language, PRIMA, and then executed. Besides move instructions based on the use of frames, LAMA-S provides real time primitives and parallel execution of tasks. The language uses two structures to define the execution order: sequence block, to indicate that all the instructions inside are to be executed sequentially, and parallel block, to indicate that all the instructions inside are initiated in parallel (something as cobegin-coend structure). Other standard control structures are provided. The use of APL demonstrated, according to the authors, that APL is a good implementation tool because it allows functional extensibility of the language. On the other hand they do not recommend it for industrial use because of the following shortcomings: it needs an APL machine to run, the APL syntax is not convenient, the syntax analysis is not perfect, and it is difficult to implement interactive programming using APL.

LENNY [15]. The language under development at the University of Genova (Italy) to be used to describe movements for an emulated anthropomorphic arm, with seven degrees of freedom. It is intended as a language powerful enough to express complex chains of actions, and understandable by humans as a way of representing processes and concurrent computations. One of the key aspects of Lenny is functionality. No reference can be made to any absolute kinematic quantity. References are always to actual mechanical context. In Lenny the robot reference frame is fixed in the shoulder and commands like up, down, right, etc. refer to that coordinate system. Functionality will enable Lenny to use any new procedure as part of the language. Lenny got its name from an Asimov novel in which a robot, named Lenny, accidentally became able to learn.

LM: Language Manipulation [19, 20]. A language development at the University of Grenoble (France). It is implemented on a Robitron robot (4 degrees of freedom) cooperating with a Barras robot (2 degrees of freedom), a TH 8 of Renault, and a Kremlin robot, both with 6 degrees of freedom, and commercially available on the Scemi robot. It is Pascal-like and frame oriented and provides many of the features of AL apart from coordination and parallel execution of tasks. It is integrated with LM-Geo [21], a system used to infer body positions from geometrical relations. LM-Geo produces program declarations and instructions in LM. LM-Geo resembles RAPT but it does not use symbolic algebraic calculus to find the frames which satisfy the equations. It computes the values analytically.

LMAC [22]. A system for flexible manufacturing development at the University of Besancon (France). It was designed to ensure safe control of different mechanical devices in the automated cell; to do this it performs many checks before actually executing code. It offers modularity based on the implementation of abstract data types, it provides generic modules (the types of data belonging to that kind of module can be specified at run time), and object parameterization. External procedures written in any language can be called by LMAC programs. Different tasks representing different real-time processes can be defined and executed. Synchronization is based on Dijkstra guarded commands. Even though its external form resembles Concurrent Pascal it has been completely rewritten in Pascal.

LPR: Langage de Programmation pour Robots [23]. A language developed by Renault and the University of Montpelier (France). It is based on defining state graphs and transition conditions. Transition conditions are also used to synchronize actions. All the graphs at the same level are executed in parallel by the supervisor; every 20 ms an action from each of the graphs at the same level is executed. Up to 24 input/output ports can be used by LPR to provide sensor interface and synchronization with other devices. LPR runs on a VAX 11/780 and produces code for an Intel 8086 microcomputer controlling the robot. It is available on robots produced by Renault and by ACMA Robotique.

MAL: Multipurpose Assembly Language [24, 25]. The language developed at Milan Polytechnic to program a two arm cartesian robot evolved from Olivetti SIGMA. It is a Basic-like system which features synchronization and parallel execution of tasks as well as movement instructions and sensor interfaces. Subroutine calls with argument lists are supported. MAL is composed of two parts, a translator from the input language into intermediate code and an interpreter of the intermediate code. The intermediate code is interfaced with a multimicro hierarchical structure, and all the joints are individually driven by different microcomputers. Force sensing is also controlled by a devoted microcomputer. Photo diodes on the fingers are used as binary sensors. Due to the mechanical architecture of Supersigma collisions between the arms are hardware detected.

PASRO: PAScal for RObots [6] is provided by the German company Biomatik. It is based on the Pascal language to which has been added data types and procedures used to perform robot specific tasks. They are stored in a library and callable by any standard Pascal compiler. It is based on the AL experience. The company may provide assistance in order to modify the coordinate transformations and the control interface for a new kind of robot. Procedures are provided to drive the arm point to point, or along a continuous path. The first implementation of PASRO has been tested on a Microrobot.

Portable AL [26] is an implementation of the AL programming environment created at Karlsruhe University on mini and micro computers. It incorporates AL compiler [26], POINTY [27] and a debugging system. A dedicated operating system has been developed to support I/O and multi-tasking. It runs on a PDP 11/34 and an LSI-11/2 which control the PUMA 500 robot.

RAPT: Robot APT [28, 8]. In its actual implementation RAPT is an APT-like language used to describe assemblies in terms of geometric relations and to transform them into VAL programs. A RAPT program consists of a description of the parts involved, the robot, and the workstation, and an assembly plan. The assembly plan is a list of geometric relations expressing what geometrical relations should hold after a step in the assembly has been done. The program is completely independent of the type of robot used. Sensors are not integrated, movements are not checked against collision avoidance. All the bodies are described as having a position (which is a frame) and some features. Features are plane, cylindrical or spherical faces. A reference system is automatically set in every feature. Against and fits are most used relations. Other relations are used to indicate translation or rotation degrees of freedom left. RAPT builds a graph of those relations and tries to reduce it to the minimum graph using a set of rules. From the reduced graph a VAL program is produced. The Computervision CADD3 system has been used to build the models and to give graphics routines.

ROBEX: ROBoter EXapt [7]. The off-line programming system developed at Aachen (Germany) as a programming tool for FMS. Its main purposes are to develop APT for FMS and for robot off-line programming, and to be independent of the kind of robot used. Applications are in workpieces handling. APT style of programming is used to describe geometry, while the ROBEX extensions are robot movement instructions, interactions with sensor (now only binary ones), and synchronization with peripherals (machine tools, conveyor belts ...). In this APT like system for FMS three languages will be used: EXAPT, for NC part programming, ROBEX, for part handling programs, and NCMES, for measuring programs. The system is portable in two ways: it is implemented in FORTRAN IV and it generates robot independent pseudo-code which is sent to the appropriate robot for further processing and execution. The user inputs coordinates and geometry of the world and programs, either interactively or using a graphic interface.

SIGLA: SIGma LAnguage [29, 30]. The language has been developed for programming Olivetti SIGMA robots. Now quite obsolete and under replacement, it has been available since 1975. SIGLA is a complete software system which includes: a supervisor, which interprets a job control language, a teaching module which allows teaching-by-guiding features, an execution module, editing and saving of program and data. SIGLA has been in use for years at the Olivetti plant in Crema (Italy). Its applications span from assembly to riveting, drilling, milling. The entire system and the application program run in 4K of memory; this compactness was necessary at the time SIGMA was delivered because memory was still expensive.

SRL: Structured Robot Language [31]. The language under development at the University of Karlsruhe. It is a successor of Portable AL and owes something to Pascal too. Data types as in Pascal are added to AI data types. The declaration part also contains a specification of the system components. Instructions can be executed sequentially, in parallel, or in a cyclic or delayed way. Different motions are available, in particular straight and circular motions. The project of SRL is part of a standardization project. The source SRL code is translated into an intermediate code, IRDATA, which is a machine independent code.

VML: Virtual Machine Language [32]. The language developed cooperatively by Milan Polytechnic and the CNR Ladseb of Padova (Italy). Intended as an intermediate language between Artificial Intelligence systems and robot it receives points in the cartesian space and transforms them into joint space. It handles task definition and synchronization as well. It is part of a hierarchical architecture, in which 3 levels are currently implemented.

7.1.6 Conclusions

It is hard to not get lost in the many different languages for robots available. We have tried to review them so as to see what experience they have provided and what problems they have not solved. Many issues have not been addressed and we did not aim to be complete. Most of the new commercially available languages for the North American market have not been included. We wanted to make our overview on the basis of the experience available in Europe now. While most of the attention is now focused on acquiring manipulator level systems we have tried to discover what other trends and experiences are available to expand robot programming toward more ambitious tasks.

7.2 Programming Languages for Manipulation and Vision in Industrial Robots

In this section we discuss issues of design for software systems for industrial robots.

We begin with a short review of the features which have proven to be important in such systems. We examine in particular how various desirable system capabilities, such as the cooperation between manipulation and vision, can be introduced at a reasonable level.

In fact the recent development of industrial robots has produced quite powerful and cost-effective manipulators. The same is generally true of the software available for them. Some crucial operations, such as managing parts with random orientation or realigning pieces, are still done using expensive mechanical devices.

This problem could be solved with the aid of vision systems connected with the manipulator. Various solutions may be adopted to integrate manipulation and vision. If we want to make it easy to program the system as a whole, with the same programming language, we should study solutions in which manipulation and vision use high level languages.

To this end we will discuss how the choice of the programming language is crucial.

7.2.1 Introduction

The simplest programming method developed for robots is the *teaching by guiding*. Only the meaningful positions and a few functions are stored in a memory, and their sequence can be played back to repeat the desired movements any number of times.

Although this method is quite simple it has several drawbacks. An error during the teaching phase requires a restart of the teaching process, unless editing capabilities are available. Teaching of repetitive positions, such as positions on a pallet, is too tedious and error prone. Synchronization of the robot with other systems, such as loaders or moving belts, can be extremely difficult. Interaction with sensors is quite impossible, unless appropriate extensions are made to the basic method [33].

During recent years we have seen a significant change in the attitude of manufacturers of robots with respect to this problem. More and more robots are sold with a sort of programming language, allowing the user to write application programs, or, at least, to integrate the teaching phase with a debugging activity. And more software systems will appear on future robots.

Much time and many resources have been spent in developing different *programming systems,* for each different robot.

Several approaches have appeared. One approach is to take an existent language, FORTRAN for instance, and add to it routines to drive the mechanical devices. This permits the full power of the language to be used, but may require a time expensive process of linking modules.

Another possibility is to write a set of library routines, so that the user program consists of a sequence of calls to these routines in addition to simple control statements.

Yet another approach is to design a language specifically for manipulation.

Most of the efforts have been done using the third approach. One of the reasons is that no available language has the characteristics wanted in a robot programming language. The situation may change now with the introduction of ADA [34].

In the meantime the desirable features of a robot programming language have been identified.

The language should be general purpose, to allow any kind of computation from sensors and vision. It should support cooperation, to express cooperative simultaneous operations of multiple robots and devices. Since robots work in a real-time environment, data must be processed within certain time constraints when they are received; they may be received asynchronously, and the computer should be able to ask for and obtain data at any arbitrary time. The ability to check certain conditions periodically in order to synchronize events which are dependent on those conditions is another important requirement. Specialized data types should be available to ease the expression of manipulator positions in space.

On the other hand, since this great generality may be difficult for the user, a programming environment should be available to support the debugging and testing phases of the application programs.

Many problems require further investigation, such as the proposed independence of the programming system from the physical configuration and the kinematic features of the arm.

The difference between *on line* and *off line* programming deserves further attention. On line systems could provide tools to debug and test programs. However, they require the use of the robot systems, and this fact tends to be rather unpleasant in the case in which the robot should be used in production. Off line programming would allow program development without any interaction with the physical world, at the cost of completely losing any form of interaction with it. It is not clear how to deal with sensors, and sensors are too important to be neglected.

While on line systems may be more directly accessible to non-expert users, the integration of robots and other machines in a flexible manufacturing system will require more and more off line programming to make the whole system work without stopping production.

When a robot is running without any response to sensory data, a simple run time system can be used to control the robot through a fixed sequence of joint positions.

When a sensory response is required [34], computations of arbitrary complexity are required at run time.

Several sorts of response to data obtained from sensors or vision can be envisaged. For instance [8], a discrete choice between one action and another may be made at runtime, sensory values may be used to take an action which depends quantitatively upon them, or sensory data may be used to control continuously the movements of the robot.

The first two kinds of use are generally available, while the third is much more difficult.

Our experience in developing languages at three of those levels will be described, and their use in third generation robots extended.

The most important aspect of the third generation of robots is the integration into the same system of manipulation and vision capabilities.

While the development of the mechanical aspects of the robots has been impressive, most of the robots are severly impaired in their ability to communicate with the external world. Many systems do not have any sensors, others have only simple contact sensors.

One of the most promising sensors is the TV camera able to take pictures and to analyze them according to some criteria. Recent choices for industrial applications indicate an increasing interest in this direction, which in the past has been considered only in research environments.

Commercially available vision systems can solve recognition problems working on binary images of bidimensional objects in a quite short time (one second or less). The constraints are that the parts are separated from each other and that good illumination creates a sufficient contrast between the objects and the background.

We are here interested in discussing how an integrated system for manipulation and vision can be programmed through a high-level programming language for inspection and manipulation tasks. Although every individual module could be programmed through its own programming language, we envisage the availability of a single programming system as an important step in the direction of easing the use of complete automation systems.

We support our presentation with practical examples of integrated systems. In particular we make reference to the solution designed at the Stanford Artificial Intelligence Laboratory [27, 35, 36, 37]; we have considered also the system of Milan Polytechnic [32, 38, 39] and some commercially available systems [40, 41].

7.2.2 How to Classify Robot Programming Languages

Since computer controlled manipulators have been introduced as a general purpose mechanism for industrial automation, the methodology of controlling and of programming them has seen a great deal of development. Some important issues for these systems have gained a wide acceptance [35, 42].

Robots should be programmed in a simple way, without extensive user training. This goal has been partially achieved with many industrial robots, which are programmed by guiding the arm through the motions of the task and storing the sequence of the positions thus obtained for further executions.

Teaching-by-guiding has been successful for tasks where only simple operations or few positions are required. Where complex assemblies are performed, this method does not allow any modification or adjustement of the movements during the execution, and makes it impossible to use force sensors and vision. Even small changes in the assembly station cannot be introduced without repeating all the teaching.

On the other hand, given a task, it is more difficult to write a program to solve it than to guide the arm through the positions it has to reach.

Manipulation and assembly tasks are difficult to program because the expression of movements in terms of manipulator positions requires many details. The intuitive knowledge about physical operations can be hardly expressed by words. For this reason some robot programming language try to allow the system the ability to compute positions and robot control values from a general knowledge about objects and the physical world. The problem is that the user is required to supply this specific knowledge. The use of CAD data bases could offer a good solution.

The increasing interest in the use of sensory feedback, mainly in assembly operations, makes the availability of a robot programming system a real need.

Robot programming language can be conceptually divided into classes, according to the level at which operations are expressed. Different authors [43, 44] have proposed partially different classifications. In our proposal we will consider joint level, manipulator level, object level, and task level languages.

Starting at the lowest level we have *joint level languages*. The description of a task is expressed in terms of the control commands required to drive the individual motors and actuators. Each joint is explicitely controlled. That means that the user should program directly in the joint space instead of in the cartesian space. He should know the geometry of the manipulator, and possibly the law of operations of the motors.

Many languages available for industrial robots fall into this level. This fact shouldn't cause any surprise: it is obviously easier to start with a low level language and then to improve it little by little. Among them let us recall SIGLA [29] of Olivetti Co., and HELP of DEA.

When we talk of "low level" we should not forget that we mean low level with respect to the expression of manipulator actions. So a language which is quite rich from the point of view of the control structures, such as HELP, is considered in this class, only because it requires the expression of movements in terms of movements on single axes. The MAL language that we will present later is in this class too. It is obvious that in the case of cartesian robots the decomposition of movements about different axis is natural, and this may help to explain the large presence of languages for cartesian robots in this class.

Another example is ARMBASIC, the language available on MINIMOVER, which is not a cartesian arm. We should not be surprised to see that in this case the movements are expressed in the form of the number of steps required for each motor for each axis.

At the *manipulator level* we class all the languages in which the user controls the manipulator positions and movements in the cartesian space, independently of the arm configuration. There is no explicit representation of the objects present in the real world. A well known example is the VAL language of Unimation [4, 45]. It allows the expression of manipulator positions in terms of transforms, which correspond to positions and orientations of the end effector in space. Even VML [32], a language designed as intermediate code for robotics, has complex data types such as frames, vectors and rotations.

At the *object level* we have languages which have some knowledge about objects in the world. This knowledge can be partial, because complete object models are not always needed. Only those features which are relevant for the manipulation task are part of the model. Object models can be used to describe the sequence of operations with less details or to compute collision free trajectories [46]. In this class we may consider languages such as AL [26], in which objects are represented through six coordinates as rigid bodies in space, or RAPT [8], AUTOPASS [43], and LAMA [46], in which models are based on geometry.

Most of the unresolved research problems in robot programming languages are at this level.

At the *task level* we are considering those systems which are able to understand and execute descriptions of the task. They are even able to execute tasks by replanning their actions in order to solve error situations [39]. At this level there are no working systems, although some of the systems illustrated in the previous class are oriented towards the task level.

7.2.3 Joint-Level Languages: The Example of MAL

MAL (Multipurpose Assembly Language) is an interactive system, which allows the user to describe the sequence of steps necessary to perform the manipulation task in a BASIC-like language. It allows the independent programming of different tasks and provides semaphores for synchronization. The choice of BASIC is motivated, among other things, by the wide use of this language and the short turn around time after a revision.

MAL has been designed for Supersigma, a cartesian manipulator with two arms. Each arm has three degrees of freedom plus the hand opening. Its mechanical structure is the same as is employed in the Sigma of the C. Olivetti Co., while its electronic control has been completely redesigned and implemented with a set of microcomputers [38].

MAL is implemented in FORTRAN IV, except for a small interface written in assembler. It runs on a LSI 11/02 and requires less than 20 k of memory. Moreover MAL is implemented in such a way that a change in the robot would not require a complete rewriting of the system. For instance, the conversion from a cartesian robot to a polar one should require modifying only a module. Likewise the system should be able to control more than one robot at the same time, possibly working in cooperation.

MAL is composed of two different parts, one devoted to the compilation of the input language into an internal form, and the other to the execution of the intermediate code.

As in BASIC the compilation module provides facilities to create, update and maintain the source programs. The system is line oriented, in the sense that after each line has been typed the compiler checks for syntactic errors and eventually gives the appropriate error message.

The program may be partially executed, by stopping it in correspondance with an instruction or by typing a command. The execution can then be resumed at any point.

MAL uses integer or real variables, and computes general expressions applying arithmetic operators and a few defined functions. The instruction set includes DO cycles, if-then tests, goto's and subroutine calls with argument lists. Different tasks can be defined and synchronized through SET-WAIT instructions. Subroutine calls and wait instructions are handled in the same way, using a stack to push the return address.

Absolute and differential movements are available to move the six axes of the two arms. Forces can be sensed, and actuators, such as the hands, can be controlled. A bell and a light can be activated to alert the user.

The independent programming of different activities to be executed in parallel seems to be an interesting aspect of robot programming. When we want to program robots to execute parallel tasks, first we ought to isolate every logically independent activity to minimize the intersections between them. Every activity is programmed as independent. Then the instructions needed to manage synchronization and information flows between the tasks are inserted. These instructions are tests on common variables, used as semaphores [47].

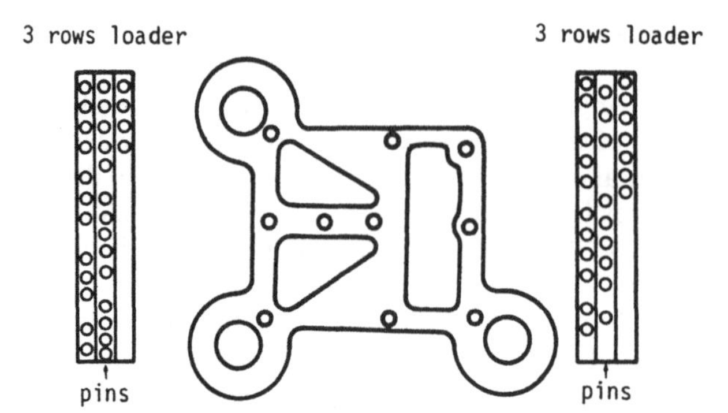

Fig. 7.1. An assembly task

To explain the characteristics of the language, we describe in detail a MAL program. The following example shows an application to an assembly problem, whose component parts are illustrated in Fig. 7.1.

A mechanical part on the working plane has several holes, and a pin should be inserted into each hole. Pins are settled in six rows, three on the left and three on the right. They are randomly positioned within each row.

Each arm of the robot sweeps the rows on its side looking for pins with its optical sensor: when a pin is detected the position of the hole is read.

Before actually inserting the pin into the hole, the program must check whether the area over the hole is free, and must wait in case the other arm is working over it.

This condition is expressed in the semaphore TAKEN, which is set to YES when the central area is occupied and to NO when the area is free. Task synchronization makes it possible to get maximum speed of operations, because most actions can be carried out in parallel by the two arms of the robot. The task for both the arms are the same, the only difference is the position of the three rows.

The MAL program is:

```
 1 . . .                          "standard origin routines
 2 INCR XL=1, ZL=-1, YR=1, ZR=-1
                               "initial positioning of axes
 3 INCR W XL=-1000, YL, ZL, YR, ZR
 4 SET FL=1, XL=0, XR=145, ZL=-195, ZR=-205
                                           "axes are zeroed
 5 SET YES=1, NO=0
 6 MOVE W XR=720
 7 SET CONTA=0, TAKEN=NO
 8 TAPE NUM                                "number of pins
 9 TASK, 2,103
10 "
11 " ________________ LEFT ARM TASK ________________
12 "
13 " +++outer loop:3 rows+++
```

```
14 DO K=0,2
15    MOVE W XL=60*K, YL=2
16    MOVE W ZL=3
17 " +++inner loop+++
18    DO I=1,250
19       MOVE YL=I
20       IF BIT (LOBJ)=0, GO TO 28          "a pin found
21    NEXT I
22 "
23    MOVE W ZL=-40
24 NEXT K
25 "
26 PRINT 'few pins for left arm - I try again
27 GO TO 14
28 INCR W YL=2
29 ACT LH,LH                                "pick up the pin
30 MOVE W ZL=-40
31 TAPE X,Y                            "read a position for it
32 WAIT TAKEN             "wait to access the working area
33 SET TAKEN=YES                        "working area occupied
34 MOVE W XL=X+200, YL=Y                 "put the pin in place
35 MOVE W ZL=3
36 DEACT LH,LH                                 "leave it down
37 MOVE W ZL=-40
38 MOVE XL=60 * K, YL=I
39 SET TAKEN=NO                          "working area is free
40 WAIT XL,XL
41 MOVE W ZL=3
42 SET CONTA=CONTA+1
43 IF CONTA=NUM, GO TO 77                  "all pins in place
44 GO TO 21
45 "
46 " ________________ RIGHT ARM TASK ________________
47 "
48 WAIT XR
49 DO KKK=0,2
50    MOVE W XR=720-60+K, YR=2
51    MOVE W ZR=3
52    DO J=1,250
53       MOVE JR=J
54       IF BIT(ROBJ)=0, GO TO 61             "a pin found
55    NEXT J
56    MOVE W ZR=-40
57 NEXT KKK
58 print 'few pins for right arm. I try again
59 Go To 49
60 "
```

```
61 INCR W YR=2
62 ACT RH,RH                                    "pick up the pin
63 MOVE W ZR=-40
64 TAPE X1,Y1                            "read a position for it
65 WAIT TAKEN               "wait to access the working area
66 SET TAKEN=YES                          "working area occupied
67 MOVE W XR=X1+200, YR=Y1                 "put the pin in place
68 MOVE W ZR=0
69 DEACT RH,RH                                    "leave it down
70 MOVE W ZR=-40
71 MOVE XR=720-60 * KKK, YR=J
72 SET TAKEN=NO                            "working area is free
73 WAIT XR,XR
74 MOVE W ZR=0
75 SET CONTA=CONTA+1
76 IF CONTA=NUM, GO TO 55                      "all pins in place
77 STOP                                            "end of the job
```

The first part of the program moves each arm to its mechanical origin. The movements are obtained giving the final values for each joint: XL, YL, and ZL for the left arm and XR, YR, and ZR for the right arm. LOBJ and ROBJ stand for the left hand and right hand optical sensors.

Movement instructions are expressed in terms of the absolute position for each joint. The W which may appear in the MOVE instruction indicates that we want the arm to complete the movement before executing the next instruction. ACT and DEACT operate the hands.

We have two parallel tasks, one to control the left arm and the other to control the right arm. Each of the two tasks is made of two nested loops; the outer one scans three rows, the inner one sweeps each row looking for pins. When a pin is detected, it is grasped and its destination is read from paper tape. Before moving to the destination the program waits for the working area to be free; the pin is thereafter inserted into the hole and the search is started again.

When all the required pins have been inserted, a normal exit is taken. If the pins in the three rows of an arm end before the job is completed, the search is started again.

The two tasks are executed at the same time, because all the actions not referring to the common area are executed by each task as if they were the only task present in the system.

In a program without synchronization primitives, only one arm could be moved at a time and many of the advantages of a two arm manipulator would be lost.

On the other hand the limitations of the system are in its expression of movements: MAL knows only about joint positions. For a cartesian robot joint positions are in a simple relation with cartesian positions. This is not true for articulated robots.

7.2.4 Manipulator-Level Languages: Mathematical Foundations

The assembly process requires that objects be manipulated; and objects, in elementary terms, are rigid bodies with a position in space. The description of those concepts can be easily made through the language of mathematics. We present here the foundations of object representation.

We consider a right-handed cartesian reference system with axes X, Y, and Z. We call this system STATION. We define the positions of the objects and of the manipulator with respect to STATION, which is the main reference system.

This reference system can be chosen in different ways; typical choices are the basis of the arm, or the intersection of the first two axes, or a corner of the working plane, with the Z axis pointing upwards.

The position of a rigid body in space can be described using six coordinates, three of them representing the position of the body and three the orientation, for instance the Euler angles or roll-pitch-yaw. This system is called frame. It is important to notice that a frame is a reference system, exactly as the main reference system STATION.

A frame F is described by two components, a rotation R and a vector V:

$$F = (R, V) \tag{0}$$

The rotation R can be expressed as a rotation by an angle theta about an axis N

$$R = \text{Rot} \quad (N, \text{theta}) \tag{1}$$

and V is a tridimensional vector

$$V = (X_F, Y_F, Z_F) \tag{2}$$

whose components express the position of the frame F. An example of frame is illustrated in Fig. 7.2.

Rotations are assumed to be positive in a right-hand sense as one looks from the origin along the axis of rotation.

We may note that

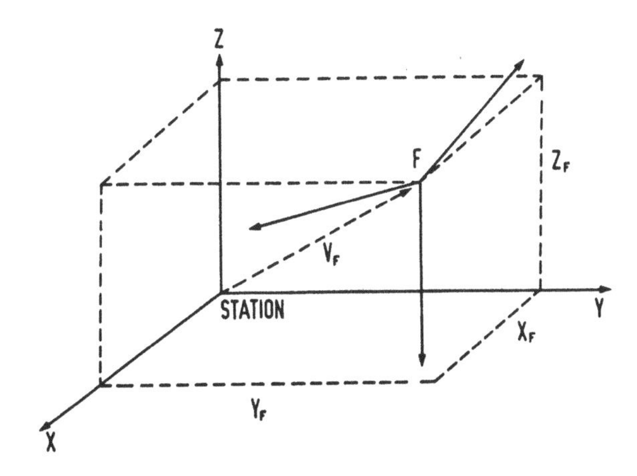

Fig. 7.2. A frame

$$R^{-1} = \text{Rot}\,(N, -\text{theta}) = \text{Rot}\,(-N, \text{theta}) \tag{3}$$

and

$$R\,(N, \text{theta}) \times \text{Rot}\,(\text{N}, \text{theta}') = \text{Rot}\,(\text{N}, \text{theta} + \text{theta}') \tag{4}$$

The composition of rotations is expressed through multiplication on the left. For instance, the rotation

$$\text{Rot}\,(Z, 180) \times \text{Rot}\,(X, -90)$$

is obtained by a rotation about the main X axis with a magnitude of -90 degrees followed by a rotation about the main Z axis with magnitude of 180 degrees.

The multiplication of rotations is not commutative; this means that the order of multiplication will affect the result.

In the same way the frames are composed through multiplication on the left. The product of two frames is a frame, and composition of frames is equivalent to a change of the reference system.

$$F_1 \times F_2 = (R_1 \times R_2, R_1 \times V_2 + V_1) \tag{5}$$

For instance, given the frame

$$A = (R_A, V_A)$$

and the relative position of the frame F with respect to A

$$F_A = (R_{FA}, V_{FA})$$

we obtain the position of F with respect to STATION

$$F = A \times F_A = (R_A \times R_{FA}, R_A \times V_{FA} + V_A) \tag{7}$$

In Fig. 7.3 we have the frames

$$A = (\text{Rot}\,(Z, 180), (4, 10, 0))$$

$$F_A = (\text{Rot}\,(X, -90), (2, -3, 2))$$

from which we compute

$$\begin{aligned} F = {} & (\text{Rot}\,(Z, 180) \times \text{Rot}\,(X, -90), \\ & \text{Rot}\,(Z, 180) \times (2, -3, 2) + (4, 10, 0)) \\ = {} & (\text{Rot}\,(Z, 180) \times \text{Rot}\,(X, -90, (1, 13, 2)) \end{aligned}$$

In the same way, given a vector W_A expressed with respect to the frame A we may obtain the vector expressed in STATION coordinates through multiplication

$$W = A \times W_A = R_A \times W_A + V_A \tag{7}$$

where R_A and V_A are two components of the frame A.

Given any two frames A and F we may compute F_A, which is the frame F in the reference system A, as

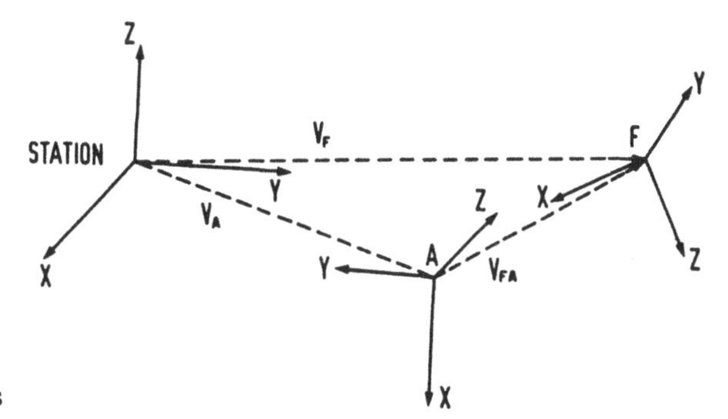

Fig. 7.3. Composition of frames

$$F_A = A^{-1} \times F \tag{8}$$

and given a frame F expressed both in STATION and in A we may compute A as

$$A = F \times F_A^{-1} \tag{9}$$

Both the Eqs. (9) and (10) are derived from (7) and can be explicitely computed by

$$F^{-1} = (R^{-1}, -R^{-1} \times V) \tag{10}$$

A widely used method of representation of frames is based on homogeneous matrices [48].

$$\left[\begin{array}{ccc|c} R_{11} & R_{12} & R_{13} & V_1 \\ R_{21} & R_{22} & R_{23} & V_2 \\ R_{31} & R_{32} & R_{33} & V_3 \\ \hline 0 & 0 & 0 & 1 \end{array}\right] \tag{11}$$

in which the 3×3 part represents the rotation part, and the last column the position in homogeneous coordinates. The elements of the rotation part are the director cosines of the axes with respect to the reference system.

This kind of representation, even though redundant, makes it very simple to do computations, since matrix algebra can be applied. Any rigid body rotation matrix has a determinant value of 1.

There is an interesting property of the rotation matrix

$$R^{-1} = R^{\mathrm{T}} \tag{12}$$

where R^{T} is the transpose of the rotation matrix.

7.2.5 Object Representation in Robot Programming Languages

Let us consider here the problem of how to make this mathematical representation of objects available to the user. We may note that some languages at manipulator or object levels, such as AL [26], have data types specially suited to that, and allow the user to express directly operations on vectors, rotations and frames.

Lower level languages do not have these capabilities and the user has to deal explicitly with equations on matrices.

We may note that the option of making those data types available is becoming more and more podular also in languages running on small systems. VAL [4] has transformations with the same meaning of frames and allows changing the reference system through sequences of

frame1 : frame2 : frame3

whose meaning can be immediately derived by substituting: with the multiplication operator. Even VML [32], an intermediate language designed as a virtual machine language for robotics, has vectors, rotations, and frames as elementary data types.

The change of the reference system, obtained through multiplication on the left of the appropriate frames, can be repeated any number of times. This produces quite interesting effects. For instance we may develop a program in a testing environment and then adapt it to the real situation by changing the reference frame. In a similar way we may use a reference frame for the definition of the grasping position of parts and update the program whenever the fixtures are moved by changing the value of the reference frame. The effect obtained is a rotation and/or a translation of the position where the end effector is moved, without any rewriting of other parts of the program.

Some languages have specific instructions to deal with a change of the reference system in a natural way. For instance VAL has the instruction BASE which offsets and rotates the location of the base of the manipulator by the indicated amount. The effect of the instruction is to modify the computations necessary to convert between joint angles and cartesian coordinates to account for the change of the main reference system.

On the other hand, it can be useful to define a tool to be held in the hand of the arm and to do operations with respect to the tool. The TOOL instruction of VAL does that. The tool dimensions alter what VAL thinks the tool tip is by first adding the rotation to the last joint and then adding the offset to the position of the arm wrist (along the X, Y, and Z axes of the tool coordinate system). It is important to note (although it may be obvious!) that the tool can be an ideal tool. A general solution is available in POINTY [27, 36]. The user can attach a tool to the arm tip and define its position with respect to the hand.

A simple calibration procedure is available to compute the tool transformation. The user moves the arm to any given point, reads the arm position (ARM_1), than attaches the tool to the robot hand, moves the tool tip to the same position, and reads again the arm position (ARM_2). By means of Eq. (9) we obtain

$$\mathrm{TOOL} = \mathrm{ARM}_2^{-1} \times \mathrm{ARM}_1$$

Again we can see how the availability of frames and operations on them at the user level makes it possible to obtain various operations without needing to define all of them in the programming language. For this reason it is quite difficult to justify the decision made in the design of the AL language not to make the operation of inverting a frame available.

One of the problems with frames is the difficulty in defining frames with complex orientations, since it is very difficult to visualize positions and orientations in three dimensional space. One solution is to use the arm as a measuring device in three dimensional space. Although this solution can be considered excellent from many points of view, sometimes it can be easier to determine frames with complex orientations from different pointings.

To this end practical systems have been proposed or implemented. The availability of an alignment frame with marks on the axes at a known distance from the origin makes it easy to compute a frame from four pointings.

In this case the computation of the homogenous matrix can be done by

$$\begin{bmatrix} (x_1 - x_0)/1_x & (x_2 = x_0)/1_y & (x_3 - x_0)/1_z & x_0 \\ (y_1 - y_0)/1_x & (y_2 - y_0)/1_y & (y_3 - y_0)/1_z & y_0 \\ (z_1 - z_0)/1_x & (y_2 - z_0)/1_y & (z_3 - z_0)/1_z & z_0 \\ 0 & 0 & 0 & 1 \end{bmatrix}$$

where we can see that only the vector components of the frames are used to compute the desired frame.

The solution offered by VAL is more similar to the POINTY solution. It is based on the idea of asking the user to point the arm to 3 or 4 points chosen with some constraints.

In VAL the instruction FRAME constructs a frame from four pointings, taken in any arbitrary positions with the constraints that

P_4 is at the origin
P_1 and P_2 individuate a straight line parallel to the X axis
P_3 is on the $X - Y$ plane

All those points should be in the positive part of the three dimensional space.

In POINTY the user has to point the arm at the origin of the frame, on one of the axis and on the plane passing through this axis and another axis. Defaults are for the last two pointings are the Z axis and the X–Z plane.

The user can reach those three points with the hand or with a pointer held in the robot hand. Provided that the tool has been calibrated the program is exactly the same. The frames will be defined as the arm position in the first case or as the tool position in the second case. We call P_1, P_2, and P_3 the three pointings, as illustrated in Fig. 7.4, and we may compute

$$\text{first-axis} = |P_2 - P_1| \tag{15}$$

For the other axes we have one of the two situations

$$\begin{aligned} \text{third-axis} &= \| P_2 - P_1 | * | P_3 - P_1 \| \\ \text{second-axis} &= \text{third-axis} * \text{first-axis} \end{aligned} \tag{16}$$

or

$$\begin{aligned} \text{third-axis} &= \| P_3 - P_1 | * | P_2 - P_1 \| \\ \text{second-axis} &= \text{first-axis} * \text{third-axis} \end{aligned} \tag{17}$$

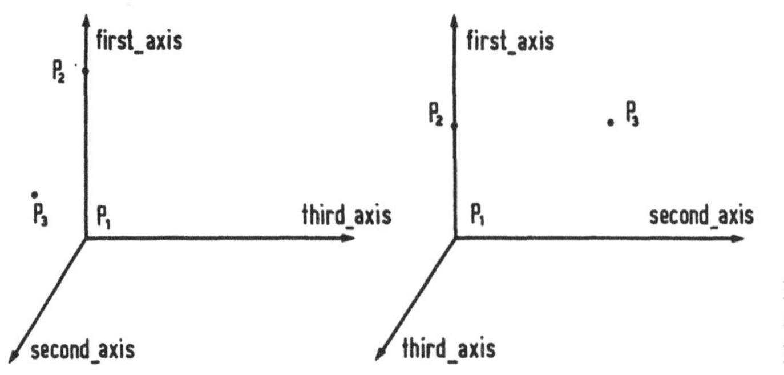

Fig. 7.4. Constructing a frame using three pointings

We see again that having operations on vectors at one's disposal it is quite easy to write any desired sequence of operations. Moreover the availabiltiy of macros or procedures allows recall of sequences of those operations.

7.2.6 At the Object Level: AL and Vision

In the following we will refer to the experimental system for programmable automation designed and developed at Stanford Artificial Intelligence Laboratory [35]. The system is composed of two Stanford Scheinman arms, with six degrees of freedom each, two PUMA arms, a Grinnel Graphic system, and a VS-100 vision module of the Machine Intelligence Corporation.

The software for controlling and programming the system is developed and compiled on a PDP 10, and runs on a devoted PDP 11/45 plus some LSI 11's. The architecture of the system is illustrated in Fig. 7.5.

AL, the language used for programming the system, is at the same time general-purpose and specialized for robotics. It has the structure of the ALGOL language, with the possibility of defining parallel actions and task synchronization. Special data types, such as frames, with appropriate arithmetic instructions, and movement operations, are available to deal with the typical problems of robot programming.

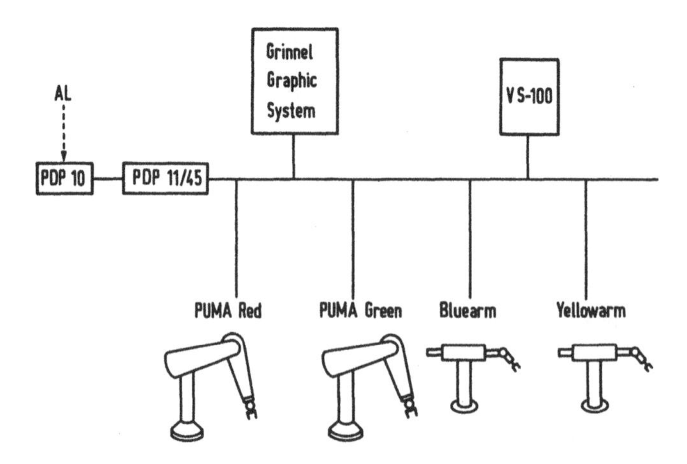

Fig. 7.5. Stanford hand-eye system architecture

Objects are described as a tree of affixed frames, where each frame represents an important feature of the object, as its grasping point or its position. The affixment relation specifies the dependencies between the frames to avoid the book-keeping of modifying their values each time one of them is moved.

Some of the modules are locally programmable. For instance the PUMA's can be programmed through the language VAL, which is available on their controlling LSI 11.

The vision module [40] can be programmed by an unskilled person through a light-pen controlled menu of instructions for teaching and recognizing parts.

The vision system operates in two phases. Initially it thresholds a video gray-level image into a binary image and thereafter works with object silhouettes. Their location is simply the 2-D centroid, while the orientation can be computed with user selectable criteria.

Silhouettes are characterized on the basis of such features as area, perimeter, number of holes, elongation. During the teaching phase objects are shown in different orientations. The system computs statistics on the object's features, which will later be used during the recognition phase.

The availability of AL and of the POINTY programming environment suggested interest in having a single programming system.

The possibility of defining procedures and macros in AL has been used to define operations with the vision system. Only one general procedure for communication through the parallel interface with the VS-100 is needed to handle the dialogue between the manipulation and the vision activities. In a similar way VAL instructions can be sent from AL to the PUMA's.

We intend now to illustrate a complete AL program for quality control. Suppose the robot has assembled some sprinklers. We want to control the quality of the assemblies.

The complete sprinkler has the two stable positions illustrated in Fig. 7.6a and Fig. 7.6b.

Since the insertion of the stem is the most difficult operation it may happen that the robot drops some stems it was not able to insert, together with the assembled sprinklers. As indicated in Fig. 7.6c, the stem has only one stable position.

The vision system should be taught to recognize three models which correspond to the three stable positions of the parts illustrated.

The AL program starts with a declaration part, in which all the variables are declared, and with the initialization. The values of frames have been obtained through a previous POINTY section. In the following BARM indicates the

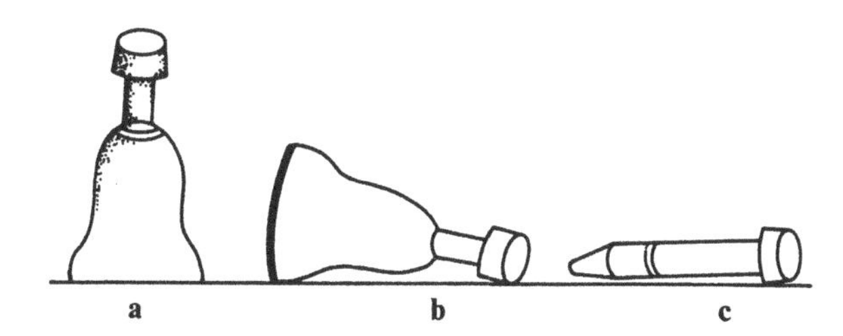

Fig. 7.6a–c. The stable positions of the object

arm and BHAND its hand. The AL procedures used for vision are in the file VISION.AL.

```
REQUIRE SOURCE-FILE "VISION.AL";
FRAME P, SET-DOWN, SET-DOWN-BAD, BLOB;
SCALAR XO, YO, XP, YP, TH, THO, NB NG, PROTOTYPE,
   DISTANCE, DD2;

INTIVISION; (initialize the vision module)
NB := 0; NG := 0; {number of bad and good parts}
SPEED-FACTOR := 2.5;
P := FRAME (ROT (ZHAT, -110 * DEG) * ROT
   (YHAT, 180 * DEG), VECTOR (48.810, 25.510, 7.00);
{position for good parts}
SET-DOWN := FRAME (ROT (ZHAT, 90) * (YHAT, 180),
   VECTOR (24.0, 34.0, 5.00);
{position for bad parts}
SET-DOWN-BAD := FRAME (ROT (ZHAT, 90) * ROT
    (YHAT, 180), VECTOR (24.0, 24.0, 5.00);
MOVE BARM TO BPARK DIRECTLY;
OPEN BHAND TO 2.0;
```

Then we do the calibration between the camera and the robot. By calibration we mean the computation of the transformation between the two reference systems associated with the camera and with the arm.

To do that the arm is moved to a certain position, called P in the program, where a calibration object has been put and the arm position is read. The vision module takes a picture of the same object after the arm has been moved out of the way, and returns x_0, y_0, and θ_0, the coordinates of the center and its orientation. Let us call P the coordinates of the calibration object in the robot system and P_V the coordinates of the same object in the vision system. The position of the vision system in the main reference system, computed from (10), is

$$V = P \times P_V^{-1} \tag{17}$$

After that if the position of any part with respect to the vision system is known and is called OBJECT_V, the position of the same object in the main reference system can be determined using Eq. (7)

$$\begin{aligned} \text{OBJECT} &= V \times \text{OBJECT} \\ &= P \times P_V \quad x\text{-OBJECT}_V \\ &= P \times \text{Rot}\,(z, \theta - \theta_0), \\ &\quad (\text{Rot}\,(z, -\theta_0) \times (x_p - x_0, y_p - y_0, 0)) \end{aligned} \tag{19}$$

where x_0, y_0, and θ_0 have been obtained from the vision system during the determination of P_V, x_p, y_p, and θ are determined during the recognition of OBJECT.

In the program we call

$$\text{BLOB} = (\text{Rot}\,(z, \theta - \theta_0), \text{Rot}\,(z, \theta_0) \times (x_p - x_0, y_p - y_0, 0)) \tag{20}$$

so that the position of the object is given by

$$\text{OBJECT} = P \times \text{BLOB}$$

The first part of the program determines x_0, y_0 and θ_0 needed for the calibration. {we start doing calibration}

7.2.7 Object and Task Levels: Problems

As we have indicated in Sect. 7.2.2, robot programming systems can be classified into different classes. The classification criterium is to differentiate languages according to the amount of specific knowledge that the user should put into programs. The more knowledge is embedded in the programming system, the less detailed will be the specifications of the task on the part of the user.

A task level description can be adopted provided that the appropriate general knowledge is given to the system. The transformation of task level descriptions into executable operations requires the system to be capable of dealing with natural language specifications, of generating plans for actions, of reasoning about geometry, and of computing collision free trajectories.

Geometric reasoning appears to be the central issue for the next generation of robot programming systems. The ability to deal with geometrical models of objects is in fact the major requirement for transforming task descriptions into sequences of operations for the robot. The descriptions of the objects could be expressed in natural language or directly obtained from vision understanding systems.

A few systems currently being developed make use of world models in different ways.

One of the most investigated fields is how to plan collision-free trajectories. A general and complete solution will require too much computational power or time, while solutions based on heuristics, as in AUTOPASS, can be reasonably succes. Neither in RAPT [8] nor in AL is this capability is provided.

Generating trajectories can be approached as a special case of geometric reasoning. Some recent systems, as ACRONYM [49] can be applied to the computation of collision free trajectories although their main purpose is more general.

Another ability is to construct plans of actions from very high-level descriptions. This research direction, developed within artificial intelligence, has attained important objectives in the 70's. The application of plan generation to robots has failed in the past because of the limited capabilities of both planners and robots. Now new emphasis is given to this research and new frameworks, such as expert systems, can help.

In conclusion, many research problems need to be solved in order to obtain robots with a higher degree of flexibility and intelligence. The problem of the choice of a programming language for future robots is premature. A programming language is only a tool for expressing concepts and, obviously, it cannot solve problems not yet solved.

7.2.8 Conclusions

We have presented modern trends in robot programming and we have proposed a classification of robot programming systems. We have indicated the unresolved problems for the next generation of programming languages for robots.

Integrated systems composed of at least two modules, one devoted to manipulation and the other to vision, are appearing on the market. Most of them are based on similar ideas and offer similar results. The present limitations of those systems arise mainly from the vision module because of the constraints on the scenes that can be analyzed. Despite those shortcomings we can forecast that more and more integrated systems will be available in the near future for use in the factory.

Further developments are expected in the integration of different capabilities into a system, provided with an adequate programming language running on the distributed computing system.

7.3 Programming a Vision System

We present LIVIA, a language for writing application programs for industrial vision systems. LIVIA is a Pascal-like language with specific instructions for vision. The objects are identified by the values of their invariant features. Operations on the objects can be done using elementary set theory. The recognition of objects in a data base and the selection of any of their features is easily programmed.

7.3.1 Introduction

The importance of providing robots with more sophisticated vision sensors is widely recognized. Visual feedback for manipulation and assembly operations is highly desirable to increase reliability and to reduce the constraints on the positioning of parts. Visual inspection of manufactured products to determine if they meet their standards plays an important role in industrial automation [50, 51, 52].

The availability of industrial vision systems and their integration into the manufacturing process indicate a growing interest by industry in automatic vision [40, 53].

Flexibility, programmability, and reliability are the key issues for vision systems, although cost and processing time are still important. Portability will become more and more important as soon as software packages become available for a variety of applications.

There is an increasing need for systems that can be easily modified to adapt to new situations, such as changes in lightning conditions, or different classes of parts, or different inspection procedures, or new ways of specifying models.

What is needed is a language in which one can specify the control structures that relate the various functions to be performed on the image as well as a language in which one can specify models of objects. A language and a programming environment specifically designed for industrial vision systems will dramatically reduce the software costs, opening new application areas.

For many aspects we can foresee the same development for vision software we have seen in languages for controlling robots. From simple systems based on "teaching by training", sophisticated programming languages have been developed that make it easier to program robots for complex tasks [27]. Unfortunately not too many languages are available for vision. Recent proposals do not seem to make much progress in this area [54].

The declining cost of computation equipment together with the increasing cost of labor are rapidly pushing the designers of manufacturing systems into new directions. Various functions that were previously separated are integrated into a single distributed system.

A modular system for industrial automation, MODIAC, has been designed in Italy by academic and industrial groups as part of a national effort to advance programmable automation. The field of application of MODIAC ranges from the control of continuous processes to manufacturing processes. As part of the MODIAC system we have designed a general purpose vision system for industrial applications, GYPSY [54].

In this chapter we present a language, LIVIA, designed for programming the GYPSY system. We give an overview of the system and its locigal architecture. Then we describe in more detail the feature of LIVIA. Examples of programs follow.

7.3.2 A Vision System for Industrial Applications

The main task we had in mind in developing GYPSY (a Growing but still Young Portable SYstem) was to develop a flexible system which can be ready for effective factory use in a short time, but which allows the easy incorporation of extensions.

We have set up a general framework and some tools, which allow us to configure the system, to verify the module interfaces, and to write application programs. We have first developed a collection of modules to form a skeleton system. We have incorporated modules to handle several applications and we are in the process of adding more.

GYPSY is programmable at different levels:

- user level. The final user of the system is not interested in the internal aspects; he wants a simple way to communicate what he wants to do. A programming language, LIVIA, is available for writing application programs. A simple interactive system based on menus is available for the teaching phase;
- application level. Application dependent packages can be written for specific fields of application of such as arc welding, or manipulation of particular types of objects. For instance, we are writing a package for the computation of features of 3D objects obtained by projecting a taper light beam [55];
- system level. Any of the basic components of the system can be modified either by changing some code or by adding new functions. Since the system is programmed in high level language using well-defined interfaces it is easy to put modules together to form a system. A simple but powerful operating system can be used to link the various modules and to check the consistency of the interfaces.

The main use of GYPSY is currently for identification of objects to be assembled by a robot. Object images are analyzed during the teaching phase to acquire their features and to build a model. The model is a set of features precisely describing the object. More complex models based on graphs can be constructed for objects with many holes allowing use of the maximal clique program [56].

In the operation phase the images are analyzed to compare their features with the models. A large set of characteristics is computed, scenes with many objects are analyzed, objects inside objects are detected. Even capsized parts can be recognized.

The present implementation of GYPSY is based on the use of a GE TN-2200 solid state TV camera with 128 × 128 pixels and 256 grey levels. The camera is connected through a DMA interface to a DEC MINC-11 computer with 28K words of memory and a LSI 11/02 CPU. A video terminal DEC VT 105 is used for interaction with the user, and a color monitor ISC 8001 provides a graphic display of the image and the results.

The programs are written in OMSI Pascal, except the camera acquisition routine which is in MACRO-11. RT-11 is the standard operating system. A PDP-11/34 can be used for the development of programs. The implementation of GYPSY on MODIAC will be done as soon as the complete hardware and the basic software are availble. We do not expect any difficulties because of the way in which GYPSY has been designed.

7.3.3 Logical Organization of GYPSY

From a logical point of view GYPSY is divided into two levels, a low level process and a high level process, Fig. 7.7. The low level process, connected to the camera, performs the basic computations on the image. The high level process executes application programs and communicates with higher level processes. This organization has been designed to take advantage of the multiprocessing capabilities of the system that will be used for its final implementation.

The lower level process includes threshold selection, binarization, run-length coding, sequential extraction of features, recognition, teaching by showing, management of a library of models, and other modules for image processing [57, 58, 59, 60].

Other modules contain screen display routines which are dependent on the specific hardware. Modules that depend on specific applications can be inserted here. For instance we have one module using the maximal cliques program which we use to recognize objects not completely in the field of view [56].

The user can interact through a terminal on which a menu of commands is available to start various programs. An operating system allows him to reconfigure the system by adding or changing any of the modules. System level programming takes place mainly at this level.

Permanent storage space for programs and libraries of models is available on floppy disks.

The higher level process is mainly used for programming the low level process. It includes the compiler for the LIVIA language, the interpreter of the intermediate code, editor and general purpose software.

```
                          Supervisor
----------------------------------------------------------------------
|                                                                    |
| LIVIA      Editor   Interpreter of  Teaching  Application  Linker |
| compiler            intermediate    module    dependent            |
|                     code                      modules              |
|                                                                    |
|                     Intermediate                                   |
|                     code                                           |
|                                                                    |
----------------------------------------------------------------------
                     Communication handler
----------------------------------------------------------------------
|                                                                    |
| Image        Run     Threshold  Features    Recognition Application|
| acquisition  length             extraction              dependent  |
|              code                                       modules    |
|                                                                    |
| Image          Grey levels     Display     Library                 |
| preprocessing  histogram       routines    management              |
----------------------------------------------------------------------
```

Fig. 7.7. The logical organization of GYPSY

The supervisor allows the user to select how to operate the system. Standard modes of operations are teaching-by-showing, developing LIVIA programs, executing LIVIA programs, and creating a new collection of modules on the low level process. The consistency of the module interfaces is checked at this point. As result of the linking process a table is generated containing the addresses and the names of the routines, so that the LIVIA compiler can generate the appropriate references.

Application dependent programs can be added. If they require specific modules on the low level system the user is responsible for adding the appropriate routines. A second terminal is available for entering LIVIA programs.

Other processes can communicate with the vision machine both directly by feeding commands through the communication handler or through the high level process by sending commands in LIVIA to be compiled and interpreted or by sending commands already compiled to be interpreted.

Communication with the robot mainly takes place by sending commands directly to the communication handler or, for more complex operations, through LIVIA programs. In many cases the robot asks the vision system directly to take a picture and to return the identity, position, and orientation of the object in view. The responsibility for transforming the coordinates of the vision sytem into the coordinates of the robot is given to the robot program [61].

7.3.4 LIVIA: The User Programming Language

We have designed a language, LIVIA, for programming GYPSY. LIVIA is a Pascal-like language with special instructions for vision. We have decided to base most of the operations on elementary set theory.

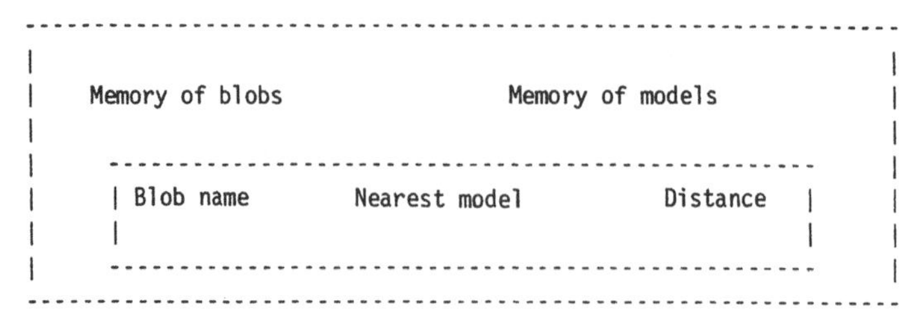

Fig. 7.8. The abstract model of the set

LIVIA has two data types, numbers and sets. Numbers can be integer or real. Sets represent groups of objects. Each object is described by a set of feature values. A list of those features is given in Sect. 7.3.6. Some features are position dependent, some are position independent. The last ones characterize the objects and constitute what we call an object model.

The model is usually constructed with a teaching-by-showing process by showing various times the same object in different positions and orientations. Data from CD data bases could be used to obtain the model. Various libraries of models can be stored on disks. Each library of models can contain any number of elements.

The abstract model that we are using for sets is illustrated in Fig. 7.8.

Each set comprises three component parts. The memory of blobs contains all the objects identified in the image with their feature values. The memory of models contains all the models that should be used. The correspondence table is constructed connecting the blobs with the models. The correspondence is generated by a classification algorithm. Each blob is connected to the nearest model. The value of the distance considered acceptable for recognition can be set by the user.

The function DISTANCE returns the distance of the indicated blob from the corresponding model. NEAREST returns a set constituted by the model nearest to the indicated blob. Usin them the user can write different classification algorithms.

Each time a picture is taken the system constructs a set of the blobs seen by the camera. We call this set TEL. We may construct a set of all the elements that can be found in any of the available libraries of models. We call this set MEM. All the other sets are created from these two sets.

Using the set operations we can intersect the sets TEL and MEM, or find any subset with specific features, such as a certain number of holes or a certain range of perimeter.

Operations available on sets include elementary set operations (union, intersection, and set difference), selection of elements with specific features, and extraction of single elements from any set. These single elements are considered as sets with a single element, so that the same operations can be applied to them.

The operation INTER computes the intersection of two sets, i.e. the set of elements that are members of both the sets. The correspondence table is used to figure out what objects are in both the sets. The operation UNION computes the union of two sets, i.e. the set of elements that are members of either one of the two

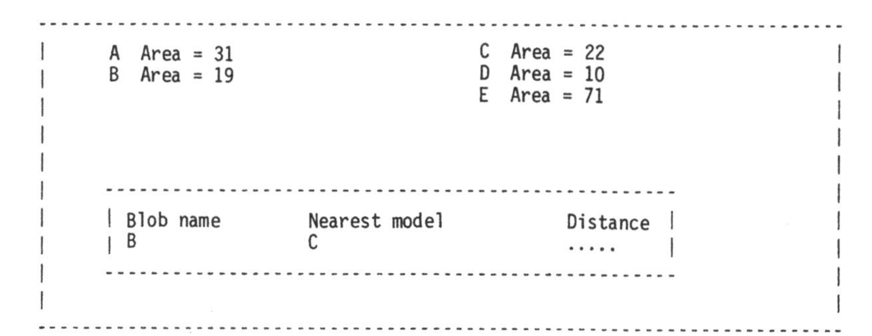

Fig. 7.9. An example of a set

sets. DIFF computes the set difference, i.e. the set of elements that are members of the first set and not members of the second. These operations can be freely intermixed in the same instruction.

Operators are available to select objects with specific features. Selections can be applied to any set to construct subsets of elements satisfying the desired properties.

The user has the option of selecting elements using either the feature values of the models or those of the blobs. Different selectors can be used according to the desired values. Since it is possible to have in the set only blobs or only models, and since the user should not be aware of that, there is a priority scheme. Each of the selectors is applied to its specific class (blobs or models) before and then, if its class is empty, to the other class.

We will explain this mechanism by an example. Let SCREWS be a set containing the blobs A and B, and the model C, D and E. B and C are connected in the correspondence table. The set is illustrated in Fig. 7.9, assuming that only the Area feature is used.

The instruction MEDIUMSCREW:= SCREWS * AREA < 20 inserts into the set MEDIUMSCREW the blob B and the models C and D. If we want to use the values of the models for the selection we should write the instruction MEDIUMSCREW:= SCREWS * AREAMD < 20. In this case the set contains only the model D. After that the selectors AREA and AREAMD will return the same results.

The control structures available in LIVIA are the same as in Pascal (do..while, repeat..until, if..then..else, go to). I/O is performed exactly as in Pascal. Procedures are not yet available.

There are a few specific instructions for vision operations.

TAKEAPICTURE takes a picture, computes all the features of the objects and construct the set TEL. The previous content of TEL and of any set derived from it are cleared.

TAKEMEM (name) constructs the set MEM inserting into it all the models present in the named library. The previous content of MEM is cleared.

SETAREA (val) sets the minimum area (in pixels) that will be considered by the system. In this way the user can decide not to consider objects that are too small or that can be little spots.

SETTHRESHOLD (val) sets the threshold used for binarization.

SETDISTANCE (val) sets the distance that is considered acceptable by the classification algorithm.

The various elements of the set can be selected by using INIT to start at the beginning of the set, and NEXT to go to the next element in the set. The predicate ENDSET tells whether the set has been completely examined. ELEM returns the current element of the set.

Once a particular element of the set has been identified is is possible to extract each one of its feature values by means of extraction operators. The extraction operators are the same as the selectors. When they are used as extractors they return a numeric value.

7.3.5 Examples of LIVIA Programs

Some examples of short programs are illustrated here.

The first problem is to select in the library GROUP 2 the objects that have an area greater than 200 pixels and more than one hole. NAMEINT is an extractor which returns the integer identifying the object, while NAMEEXT returns its symbolic name.

```
PROGRAM WRITE;
VAR NUM: NAME;
  SET  AA;
BEGIN
  TAKEMEM ("GROUP 2");
  AA:=MEM * AREA>200 * NHOLES >=1;
  PRINT ("NAMES OF MODELS WITH AREA >200 AND NHOLES>1");
PRINTELN;
INIT (AA);
NEXT (AA);
WHILE NOT ENDSET (AA) DO BEGIN
  NAME:=NAMEINT (ELEM (AA));
  PRINT (NAMEEXT (NAME)); PRINTELN;
  NEXT (AA)
  END;
END.
```

The next program takes a picture, selects objects with specific features, and recognizes them using a library. The intersection between TEL and MEM contains the recognized objects. We then use two different extractors for the same feature, since we want to compare the value of the blob with the value of the model. AREA returns the area of the object and AREAMD is the area from the corresponding model.

```
PROGRAM SELECT;
VAR NUM: NAME, AR, ARMD;
  SET: BB;
BEGIN
TAKEAPICTURE;
TAKEMEM ("GROUP 1");
BB:=TEL INTER MEM;
INIT (BB);
NEXT (BB);
PRINT ("RECOGNIZED OBJECTS"); PRINTELN (1);
PRINT ("NAMES AREA OBJECTS SEEN AREA MODELS");
PRINTELN;
WHILE NOT ENDSET (BB) DO BEGIN
  NAME:=NAMEINT (ELEM (BB));
  PRINT (NAMEEXT (NAME));
  AR:=AREA ELEM (BB));
  ARMD:=AREAMD (ELEM (BB));
  PRINT (" ",AR," ",ARMD);
  PRINTELN;
  NEXT (BB);
  END;
END.
```

The last program is used to inspect parts. Objects not recognized and without holes are discarded. Their position is printed so that the robot could be used to remove them. The set UNKNOWN is constituted by the objects that are not recognized and that have no holes. In fact the intersection between TEL and MEM is the set of recognized objects. The difference between TEL and the set of recognized objects constitutes the set of non recognized objects. Among this set the objects that do not have holes are then selected using the selector HOLES. The coordinates of their centers of gravity are then printed.

```
PROGRAM INSPECTION;
VAR NUM: COORDX, COORDY;
  SET: UNKNOWN;
BEGIN
  TAKEMEM ("KNOWN");
  SETAREA (200);
  TAKEAPICTURE;
  UNKNOWN:=(TEL DIFF (TEL INTER MEM)) * HOLES=0;
  INIT (UNKNOWN);
  NEXT (UNKNOWN);
  WHILE NOT ENDSET (UNKNOWN) DO BEGIN
    COORDX:=XBAR (ELEM (UNKNOWN));
    COORDY:=YBAR (ELEM (UNKNOWN));
    PRINT ("DEFECTIVE PART IN", COORDX," ",COORDY);
    NEXT (UNKNOWN);
    END;
END.
```

7.3.6 Additional Position-Independent Features for Blobs and Models

Position independent features computed both for the blobs and for the models. For each of them there are two selectors available, one used when the feature value of the blob is desired, the second for the feature value of the model:

Perim: the perimeter of the blob, computed with the distance method
Area: the area of the blob (in pixels)
DMin: the minimum distance from the centroid to the perimeter
DMax: the maximum distance from the centroid to the perimeter
Holes: the number of holes
HArea: the total area of all the holes (in pixels)
AvHArea: the average area of the holes (in pixels)
AvHDist: the average distance of holes from the blob centroid
Compact: the ratio of the square of the perimeter to the area of the blob
FrHDist: the distance from the farthest hole to the blob centroid
NrHDist: the distance from the nearest hole to the blob centroid
HoHoDist: the distance between the centroids of the farthest hole and the nearest hole
MinMaxDif: the difference between MaxAng and MinAng. It may be used to test whether the object is capsized
FrHNrHDif: the difference betwen FrHAng and NrHAng. It may be used to test whether the object is capsized

Features available only in the models:

NameInt: the integer identifier of the model
Nameext: the string of characters associated with each model. It represents its external name

Position dependent features computed for the blobs:

Name: the identifier of the blob
MomX: the first moment with respect to X
MomY: the first moment with respect to Y
XBar: the X coordinate of the blob centroid
YBar: the Y coordinate of the blob centroid
Dad: the identifier of the blob which contains the blob examined
Sons: the number of blobs interior to this blob
FrHName: the name of the hole farthest from the blob centroid
NrHName: the name of the hole nearest to the blob centroid
MinAng: the angle between the X axis and the line passing through the blob centroid and the point on the blob perimeter closest to the centroid
MaxAng: the angle between the X axis and the line passing through the blob centroid and the point on the blob perimeter farthest from the centroid
FrHAng: the angle between the X axis and the line passing through the blob centroid and the hole centroid (FrHName) farthest from the blob centroid

NrHAng: the angle between the X axis and the line passing through the blob centroid and the hole centroid (NrHName) nearest to the blob centroid

XMin: the X coordinate of the point on the blob perimeter closest to the blob centroid

YMin: the Y coordinate of the same point

XMax: the X coordinate of the point on the blob perimeter farthest from the blob centroid

YMax: the Y coordinate of the same point

7.4 Towards Automatic Error Recovery in Robot Programs

Unexpected events can cause the failure of apparently "correct" robot programs. The interaction with the real world and its unpredictability make the problem of error recovery in robot programming specially important. The goal of this section is to present a general framework in which the activity of error recovery can be automated. This is accomplished by introducing a monitor program which identifies the appearance of any error and attemps to correct that error. The correction is done using a knowledge base, where the knowledge that the user has about error identification and correction is expressed in symbolic form. An inference mechanism allows extension of this knowledge base for use in complex and unanticipated situations.

7.4.1 Introduction

Robots are being used in a wide variety of applications. To operate successfully they should be able to handle unexpected events. A more intelligent perception of the robot environment is needed. The capability of making decisions in response to external conditions should be improved. This should also result in greater safety for the operating personnel and the equipment installed in the vicinity of the robot.

With current robot programming languages [2], one can recover from failures caused by arm errors only by using ad hoc error recovery procedures. In writing and debugging manipulation programs, users must depend on their experience, intuition, and common sense to decide what errors to watch for.

Errors in robot programs are difficult to identify because of their unpredictability. The same program can work well hundreds of times and then stop because of a minimal variation in size of one part or because of a little spot of oil on it. Moreover since the programming is done on-line [27] the robot must be used for large amounts of time to check new programs before they can be reasonably used in production.

The problem of recovering after an error has not yet been fully addressed. To do this the system needs to have a knowledge of how the world in which the robot is operating is structured [63].

The problem of dealing with errors has been approached in various ways and with different objectives in plan generation research. Systems such as NOAH [64], and HACKER [65], tried to solve errors arising during the planning. The TROPIC system [66] has a similar mechanism for failure correction. These approaches have not been applied to real robot tasks.

The system closest to our solution is presented in [67]. Here a practical system for analysing failures and their causes, and for replanning the recovery activity was designed. Its main limitation derives from the extensive use of plan formation as the basis for constructing robot programs, and on the decision to check only the preconditions of the actions. In this way an error may be discovered later than when it appeared.

The problem of error recovery plays an important role in industrial robotics. The possibility of using robots unattended, such as during the night, requires at least a reasonable solution of this problem. Strategies to fulfill safety requirements in the case of failures of the robot are important too.

7.4.2 A Method for Automatic Error Recovery

This section presents a general framework for automating the error recovery activity. This is accomplished by an intelligent monitoring system running concurrently with the robot program. Every time an error arises the appropriate recovery procedure is detected using information extracted from a knowledge base [68]. The knowledge base contains rules about correction activities and about interpretation of sensor data.

To detect what happened and to identify the recovery action the system should know the effect on the world of each of the instructions of the program. Some form of dynamic model of the robot environment and the ability to interprete information gathered by sensors are also needed [34].

The general scheme is

```
                               WHAT HAPPENED?
                               (sensor
                     error     interpretation
                     detected  rules)
                     -------->
program execution
                     <--------
                     recovery  WHAT TO DO?
                     actions   (recovery
                               rules)
```

We examine in more detail the recovery method. We start by defining the dynamic model of the world, and the semantics of the robot programming language. Then we present the organization of the knowledge base.

7.4.2.1 Dynamic Model

An initial model of the world is constructed from the declarations present in the program, and data from sensors.

For each instruction InitialModel is the model valid before the execution of the instruction.

ExpectedModel is the model expected to be valid after the execution of the same instruction. It is obtained from the InitialModel and the postconditions of the instruction.

ExpectedModel can contain conditional expressions since postconditions can be expressed with conditional parts. For instance the instruction that closes the hand can be used either to grab an object or to close the hand. The sensors in the finger can identify the situation at run time.

Let CurrentModel be the currently valid model.

It should be obvious that if there are no errors before executing an instruction CurrentModel is the same as InitialModel. If there are no errors after executing the instruction CurrentModel is the same as ExpectedModel.

Example

At the beginning of the program the InitialModel can be:

```
Arm = ParkPosition
HandOpening = X
if TouchSensorTriggered
   then ObjectHeld,
        ObjectSize = HandOpening,
        ObjectPickedUpAt = Arm
   else ClearHand.
```

7.4.2.2 Semantics

We describe the semantics of the language in a STRIPS-like form [69].

Each instruction has an associated list of preconditions and postconditions. The preconditions express what should be true before executing the instruction, the postconditions express how to modify the current model after the execution of the action. They are expressed in term of additions (ADD), deletions (DEL), and updating (UPD) to the model.

Examples

We consider a small subset of AL isntructions [35].

```
MOVE ARM TO frame
prec:
post: UPD: arm = frame;

OPEN HAND TO d
prec:
post: if ObjectHeld
         then ADD: ClearHand
              DEL: ObjectHeld,
                   ObjectSize = X,
                   ObjectPickedUpAt = Y
      UPD: Opening = d
```

```
CLOSE HAND TO d
prec: ClearHand
post: UPD: HandOpening = d
      if TouchSensorTriggered
         then ADD: ObjectHeld,
                   ObjectSize = d,
                   ObjectPickedUpAt = Arm
              DEL: ClearHand
```

Note that we consider rigid objects so that after OPEN and before close the hand does not hold anything.

Using postconditions the ExpectedModel and the CurrentModel can be determined. For instance, after a MOVE instruction the ExpectedModel is computed by updating the arm position in the InitialModel, while the CurrentModel is computed by reading the actual arm position.

7.4.2.3 Knowledge Base

We use a knowledge base containing two types of rules, sensor rules (used to interpret the sensor data), and recovery rules (used to produce the recovery).

Sensor rules have the form

```
if D, .. then C
```

where the D's express what we want to know from sensors and C is their "logical" interpretation. This organization allows a certain independence between the raw data from sensors and their interpretation.

The recovery rules have the form

```
to obtain G, .. when S, .. do R, ..
```

where the G's express what we want to achieve, the S's express what we know is true, and the R's are recovery actions.

Examples

SensorRules:

```
if FingerTouchSensorTriggered
   then ObjectHeld

if not FingerTouchSensorTriggered
   then ClearHand
```

RecoveryRules:

If the object is lost during the movement we can recover with

```
to obtain ObjectHeld
   when   ClearHand
   do     Compute NextPickUp;
          GrabObject(NextPickUp,ObjectSize)
```

knowing that

```
if ObjectPickedUpAt=X
   then NextPickUp=X + d
```

If the arm is not in the right place we can use the rules

```
to obtain Arm=Frame2
   when   Arm=Frame1, Dist(Arm,Frame2)<.5
   do     MOVE ARM TO frame2 DIRECTLY

to obtain Arm=Frame2
   when   Arm=Frame1, Dist(Arm,Frame2)>.5
   do     OPEN HAND TO frame2
```

If the hand is too closed

```
to obtain HandOpening=ObjectSize
   when   HandOpening<ObjectSize
   do     OPEN HAND TO ObjectSize
```

7.4.2.4 Recovery Procedure

The recovery procedure is activated by the identification of an error. As we said before, an error is identified every time. CurrentModel at the end of the execution of any instruction is different from ExpectedModel. Knowing the situation in which we are and where we want to be the appropriate error recovery rules can be fired.

We control both the preconditions before executing any instruction and the postconditions at the end. The first check should not be needed since we assume that the program does not have logic errors. We consider it useful as a protective measure.

After the recovery we resume the execution of the original program at the point where it was suspended. The problem of deciding whether to restart it at a different point has not yet been approached.

7.4.3 Concluding Remarks

Although the examples shown are limited, we think we have supported our claim that we have presented a general framework for error recovery in robot programs. Research is under way to write more rules, to introduce strategies in recovery, and to extend our work to complete programming languages. A preliminary implementation is under development.

In our opinion the strong points of our method are:

- It is based on the use of a real robot programming language, not a planning system intended for purposes other than manipulator control;
- The reasoning process used in error recovery is based on information provided by sensors. Any sensor can be incorporated, provided that interpretation rules are available;

- The knowledge base can be easily extended to cover more errors and more recovery procedures;
- the language used to program the robot could be changed, provided that its semantics are supplied in the same form;
- It can be used to recover errors not only for robots but also for more complex automation systems.

7.5 References

1. Kempf, K (1983) Artificial Intelligence Applications in Robotics – A Tutorial. IJCAI 83 Tutorial, Karlsruhe, Germany
2. Bonner, S, Shin, KG (1982) A comparative Study of Robot Languages. IEEE Computer, No 12, pp 82–96
3. Lozano-Perez, T (1983) Robot programming. Proc of the IEEE 17, no 7
4. Schimano, B (1979) VAL: an industrial robot programming and control system. Proc. Programming Languages and methods for industrial robots. IRIA, France
5. CAM-I proposes standards in robot software. The Industrial Robot, pp 252–253 (1982)
6. Biomatik (1983) PASRO – Pascal for robots. Biomatik Co., Freiburg, West Germany
7. Weck, M, Eversheim, E (1981) ROBEX – An off line programming system for industrial robots. Proc. 11th ISIR, Tokyo, Japan, pp 655–662
8. Popplestone, RJ et al. (1980) An interpreter for a language for describing assemblies. Artificial Intelligence, vol 14, pp 79–107
9. Silver, D (1973) The littler robot system. MIT Artificial Intelligence Lab. RP. AIM 273
10. DoD (1980) Reference Manual for the ADA Programming Language. Proposed Standard Document, Dept. of Defense, USA
11. Volz, RA, Mudge, TN, Gal, DA (1983) Using ADA as a programming language for robot-based manufacturing cells. RSD-TR-15-83 Dep. E & C Eng., The University of Michigan, Ann Arbor, Michigan
12. Descotte, T, Latombe, JC (1981) GARI: a problem-solver that plans how to machine mechanical parts. Proc. 7th IJCAI, Vancouver, Canada, pp 766–772
13. Falek, D, Parent, M (1980) An evolutive language for an intelligent robot. The industrial robot, vol 7, no 3, pp 168–171
14. Taylor, RH, Summers, PD, Meyer, JM (1982) AML: A Manufacturing Language. The International Journal of Robotics Research, vol 1, no 3, pp 19–41
15. Verardo, A, Zaccaria, R (1982) Lenny Reference Manual [in Italian]. Internal Report, University of Genova, Genova, Italy
16. Gonzalez, JC, Williams, MH, Aitchison, DE (1984) Evaluation of the effectiveness of Prolog for a CAD application. IEEE CG & A, no 3, pp 67–75
17. Camera, A, Migliardi, GF (1981) Integrating parts inspection and functional control during automatic assembly. Assembly Automation, vol 1, no 2, pp 78–82
18. Donate, G, Camera, A (1980) A high level programming language for a new multiarm assembly robot. Proc. 1st Int. Conf. on Automated Assembly, pp 67–76
19. Latombe, JC, Mazer, E (1981) LM: a high-level programming language for controlling assembly robots. Proc. 11th ISIR, Tokyo, Japan, pp 683–90
20. Mazer, E (1983) Geometric programming of assembly robots. Proc. Advanced Software in Robotics, Liege, Belgium
21. Miribel, JF, Mazer, E (1982) Manuel d'utilisation du langage LM. Research report IMAG, University of Grenoble, France
22. Haurat, A, Thomas, MC (1983) LMAC: a language generator system for the command of industrial robots. Proc. 13th ISIR, Chicago, Illinois, pp 12–69
23. Bach, J (1983) LPR Description. Renault, France (unpublished)
24. Gini, G et al. (1979) A multi-task system for robot-programming. ACM Sigplan Notices, vol 14, no 9

25. Gini, G et al. (1979) MAL: a multi-task system for mechanical assembly. Proc. Programming Methods and Languages for Industrial Robots, IRIA, France
26. Finkel, R et al. (1975) An overview of AL, a programming system for automation. Proc. 4th IJCAI, Tbilisi, USSR
27. Gini, G, Gini, M (1982) Interactive development of object handling programs. Computer Languages, vol 7, no 1
28. Ambler, AP, Popplestone, RJ (1975) Inferring the positions of bodies from specified spatial relationships. Artificial Intelligence 6, pp 157–174
29. Banzano, T, Buronzo, A (1979) SIGLA – Olivetti robot programming language. Proc. Programming Methods and Languages for Industrial Robots, IRIA, France, pp 117–124
30. Salmon, M (1978) SIGLA – The Olivetti SIGMA robot Programming Language. Proc. 86h ISIR, Stuttgart, Germany, pp 358–363
31. Blume, C, Jacob, W (1983) Design of a Structured Robot Language (SRL). Proc. Advanced Software in Robotics, Liege, Belgium
32. Gini, G et al. (1980) Distributed robot programming. Proc. 10th ISIR, Milan, Italy
33. Nitzan, D, Rosen, C (1976) Programmable industrial automation. IEEE Trans. Computer, vol C-25, December 1976, pp 1259–1270
34. Rosen, CA, Nitzan, D, (1977) Use of sensors in programmable automation. IEEE Computer, December 1977, pp 12–23
35. Gini, G, Gini, M (1982) ADA: a language for robot programming? Computers in Industry, vol 3, no 4, pp 253–259
36. Binford, TO, Liu, CR, Gini, G, Gini, M, Glaser, T et al. (1977) Exploratory study of computer integrated assembly systems. Progress Report 4, Stanford Artificial Intelligence Laboratory Memo AIM-285.4, Stanford, California
37. Gini, G, Gini, M (1980) POINTY: a philosophy in robot programming. In: Rembold (ed.) Information control problems in manufacturing technology. Pergamon Press
38. Gini, G, Gini, M (1983) The integration of manipulation and vision for assembly and quality control. International Journal of Production Research, vol 21, no 2, pp 279–292
39. Cassinis, R, Mezzallira, L (1977) A multiprocessor system for the control of an industrial robot. Proc. 7th International Symposium on Industrial Robots, Tokyo, Japan
40. Carlisle, B et al. (1981) The PUMA/VS-100 Robot Vision System. Proc. First International Conference on Robot Vision and Sensory Controls, Stratford-upon-Avon, England, April 1981
41. Ward, MR et al (1979) CONSIGHT: a practical vision-based robot guidance system. Proc. 9th International Symposium on Industrial Robots, Washington, DC, March
42. Bonner, S, Shin, KG (1982) A comparative study of robot languages. Computer, vol 15, no 12
43. Lieberman, LI, Wesley, MA (1977) AUTOPASS: an automatic programming system for computer controlled mechanical assembly. IBM Journal of Research and Development, vol 21, no 4, pp 321–333
44. Latombe, JC (1979) Une analyse structuree d'outils de programmation pour la robotique industrielle. Langages et methodes de programmation des robots industriels, IRIA (Institut de Recherche d'Informatique et d'Automatique), France, pp 5–22
45. Shimano, BE (1979) VAL: a versatile robot programming and control system. Proc. 3rd International Computer Software Applications Conference, Chicago, Ill., November
46. Lozano-Perez, T, Winston, PH (1977) LAMA: a language for automatic mechanical assembly. Proc. 5th International Joint Conference on Artificial Intelligence, Boston, Mass., pp 710–716
47. Dijkstra, EW (1968) Cooperating sequential processes. In F. Gennys (ed): Programming Languages. Academic Press, New York
48. Roberts, LG (1965) Homogeneous matrix representation and manipulation of N-dimensional constructs. Document MS1045, Lincoln Lab, MIT, Cambridge, Mass. May
49. Brooks, RA, Greiner, R, Binford, TO (1979) The ACROYM model-based vision system. Proc. 6th IJCAI, Tokyo, Japan, pp 105–113
50. Dodd, GG, Rossol, L (1979) Computer vision and sensor based robots. Plenum Press, New York

51. Kruger, RP, Thompson, WB (1981) A technical and economic assessment of computer vision for industrial inspection and robotic assembly. Proc. IEEE, vol 69, pp 1524–1538
52. Mundy, JL, Jarvis, JF (1980) Automatic visual inspection. In: Fu, KS (ed): Applications of Pattern Recognition. CRS Press, New York
53. Chin, RT (1982) Machine vision for discrete part handling in industry: a survey. IEEE Industrial Applications of Machine Vision, pp 26–32
54. Gini, G, Gini, M (1982) General purpose vision sensor. Proc. SPIE 26th Annual Technical Symposium, San Diego, Calif.
55. Dai Wei, Gini, M: The use of taper light beam for object recognition. In: Robot Vision, IFS Ltd (to be published)
56. Bolles, RC (1979) Robust feature matching through maximal cliques. SPIE, vol 182, Imaging Applications for Automated Industrial Inspection and Assembly, pp 140–149
57. Agin, GJ (1980) Computer vision systems for industrial inspection and assembly. Computer Magaz, pp 21–31
58. Agrawala, AK, Kulkarni, AV (1977) A sequential approach to the extraction of shape features. Computer Graphics and Image Processing, no 6, pp 538–557
59. Rosenfeld, A, Kak, A (1976) Digital Picture Processing. Academic Press, New York
60. Weska, JS, Nagel, RN, Rosenfeld, A (1974) A threshold selection technique. IEEE Trans. on Computers, vol C-23, pp 1322–1326
61. Gini, G, Gini, M: The integration of manipulation and vision for assembly and quality control. International Journal of Production Research (to be published)
62. Gini, G, Gini, M (1982) A general purpose vision sensor. Proc. 26th SPIE Technical Conference, Sn Diego, Calif
63. Gini, G, Gini, M, Somalvico, M (1981) Deterministic and non deterministic robot programming. Cybernetics and Systems, vol 12, pp 345–362
64. Sacerdoti, E (1977) A structure for plans and behavior. American Elsevier Publ. Company
65. Sussman, GJ (1975) A computer model of skill acquisition. American Elsevier Publ. Company
66. Latombe, JC (1979) Failure processing in a system for designing complex assemblies. Proc. 6th IJCAI, Tokyo, Japan, August 1979, pp 508–513
67. Srinivas, S (1976) Error recovery in a robot system. PhD Thesis, CIT
68. Stefik, M et al. (1982) The organization of expert systems: a tutorial. Artificial Intelligence, vol 18, pp 135–173
69. Fikes, RE, Nilsson, NJ (1981) STRIPS: a new approach to the application of theorem proving of problem solving. Artificial Intelligence, vol 2, pp 189–208

8 Present State and Future Trends in the Development of Programming Languages for Manufacturing

U. Rembold and W. Epple

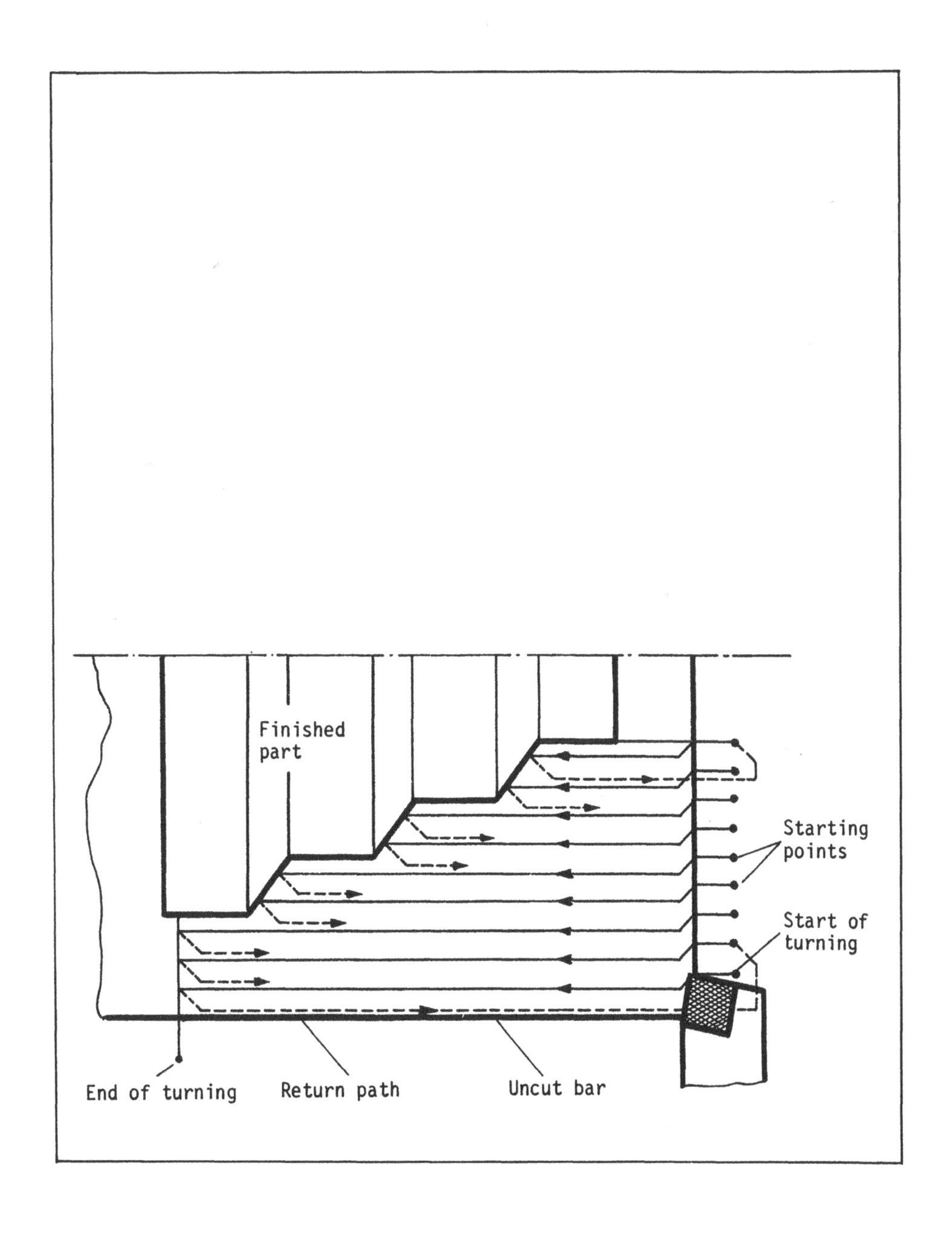

Programming a turning operation

(Courtesy of PDV Funding Agency,
Kerforschungszentrum Karlsruhe GmbH, FRG)

About 25 years ago, there were essentially only assembly languages available. In the course of time, a larger number of higher programming languages were developed. Today they are so numerous that even an expert has difficulty in knowing them all. The same problem exists for manufacturing. There are not only one but several programming languages which have been developed for different manufacturing applications. They can be divided into the following categories:

1. Numerical control of machine tools
2. robot control
3. process control and
4. commercial data processing.

In this section, a description of programming languages and programming-support-tools for each of these categories is given.

The development of programs for numerical control will be simplified by means of easy-to-use independent work stations. In the near future, it will be possible to generate control programs with the aid of graphic representation of a work-piece and by animating the machining process on a display.

In the case of robot control the essential features of the most important programming languages will be discussed. The future of robot programming is characterized by what is called "implicit programming" and the use of expert-systems.

For process control, there exist different real-time languages (e.g. ADA, PEARL, CORAL, RTL/2) which take into consideration parallelism and the problem of interfacing the technical process with the computer. In this area, programming support environments like APSE or UNIX and the use of descriptive languages together with automated tools will simplify program development.

Finally, in the commercial data processing area of a manufacturing organisation, there is wide use of COBOL as the principal programming language. Several methods and tools for program development have been proposed and are in use. In this field, programming will be strongly influenced by new languages like PROLOG and the results of research work on artificial intelligence.

The computers of the 5th generation will process knowledge at a high level and will yield solutions at this level. The dialog with the computer will be held using natural means of communication. Such a system for example could be used to lead a factory. Its expert system contains the status of all orders, information on the manufacturing process and other resources, knowledge about alternative manufacturing methods and the boundaries of the manufacturing environment. This system is capable of learning from past and present operations and can, with the aid of knowledge and deduction rules generate production plans for new products. The communication to describe these products is done via graphical input. Orders for the factory are entered via speech communication or another natural language. With the aid of the expert system, the computer is able to plan production runs and dates and to control the production process.

Although there are many programming languages and program development aids available, the software as well as the planned expert systems used for the control of manufacturing processes are still very complex, and for this reason also very expensive. It will be the combined task of the manufacturing engineer and the computer scientist to conceive user friendly software tools.

In the development of programming languages, or in a broader sense, of tools and languages, a milestone for manufacturing has been reached in two distinct areas, namely factory-information processing and factory automation; many excellent, specialized tools and languages have been developed. There are numerous, independent computer aids for most activities within a factory, e.g. production control, factory data collection, material handling and process control.

Until now, the languages and computers for these applications have been developed independently. For this reason it is very difficult to design a homogeneous computer integrated manufacturing system. The future modernization of factories will be strongly influenced by the development of general purpose programming languages and compatible computing systems. Furthermore, the controls of processing equipment have to be made computer compatible. For the factory of the nineties, expert systems will be available which perform the majority of routine tasks.

8.1 Introduction

Manufacturing creates between 60–80% of the real wealth of the major industrialized countries. For this reason most of these countries have numerous industry and government supported programs to increase manufacturing productivity.

Within the last two decades, many important innovations have been made in the development of new design aids, manufacturing processes, engineering materials and manufacturing systems. The digital computer has contributed to the most significant changes. This so far almost untapped resource has the greatest potential to improve manufacturing productivity compared with any other invention which contributed to the industrial revolution. With conventional production know-how it becomes increasingly difficult and expensive to improve manufacturing processes. The computer offers possibilities for improving the manufacturing technology in many areas. It can directly control production and quality control equipment and adapt these quickly to changing customer orders and new products. The computer makes it possible to evaluate data instantly, to assess the flow of information in the plant and to initiate corrective actions immediately to optimize the manufacturing process. Probably the most important asset of the computer is its capability to integrate the entire manufacturing system. Its ability to make decisions will contribute to the concept and design of flexible manufacturing systems (FMS) which can be reconfigured to changing market requirements and new manufacturing processes.

Unfortunately, the use of the computer has a serious drawback. For the average manufacturing engineer it is very difficult and for the factory worker virtually impossible to communicate with the computer.

8.2 Programming of Machine Tools

8.2.1 The APT Language

APT (Automatically Programmed Tools) is the first high order language for programming NC machine tools. It can be recognized as the origin of many other languages and dialects. The first attempt to conceive this languages dates back to the late fiefties. An organized development effort started in 1961, when the APT long range program was created and contracted to the Illinois Institute of Technology Research Institute (IITRI). Over 130 companies supported the creation of this language through joint funding.

APT uses English-like instructions to describe the geometry of a part. The input statements are of four different types:

- Geometric statements are used to define the part configuration. With these the programmer is able to describe geometric elements such as points, lines, circles, ellipses, planes, cylinders, cones and general conics and quadrics with different surfaces.
- Motion commands which control the path of the cutter along the surface of a workpiece. The repertoire includes start-up and point-to-point instructions, modifiers to change cutter movement direction and methods of describing the cutter.
- Postprocessor commands which control different machine functions such as spindle speed, feedrate, acceleration, deceleration and coolant supply.
- Special control instructions which generate translations, rotations and output listings. There are also possibilities for programming loops, jump instructions and subroutines.

The contour of the part is described by the programmer with the aid of a sequence of instructions. Figure 8.1 shows how the tool is directed to generate the contour of the workpiece with the help of control surfaces. The tool travels along the part and drive surfaces until it encounters the check surfaces. Here its path is changed and the tool follows a newly defined part surface. This process is repeated until the entire part contour has been generated.

A so-called processor is used to transform the part program into a tool path description, Fig. 8.2. The translator compiles the program language into computer executable instructions which are processed by the arithmetic unit. The mathematical calculations needed to generate cutter location coordinates, are done by this unit. It allows the inspection of the actual path the cutter will follow. The output of the processor is a machine tool independent program.

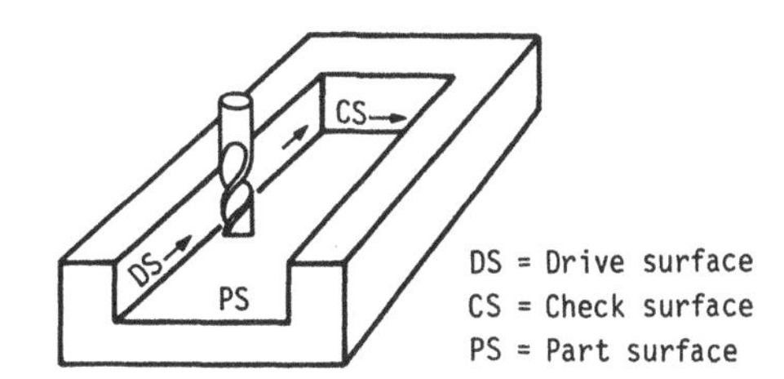

Fig. 8.1. APT principle

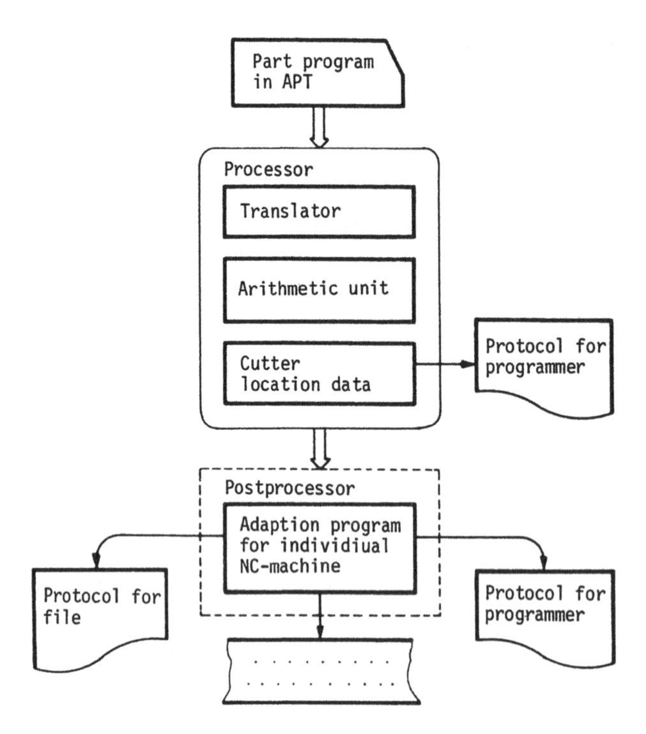

Fig. 8.2. Translation of an APT programm

Since the control unit of a machine tool is of machine specific design, the cutter location data are adapted by a post processor to machine specific code. The program to generate the part contour consists of dimensional and technological data which are necessary to operate the machine tool.

APT was conceived as a generalized program development tool to handle many different machining processes. For this reason the processor is very large and needs a considerable amount of memory space. This fact, however, limited the use of APT to manufacturers who have access to large computers. During the sixties and seventies, attempts were made to simplify the programming system so that it could be handled by smaller computers. This has led to the conception of APT like languages and subsets of APT. As a matter of fact, several minicomputer manufacturers now offer APT program development systems for their equipment. In the following section, the German development EXAPT (EXtended APT) will be discussed.

8.2.2 The EXAPT Programming System

Syntactically EXAPT is identical to APT. In addition to the geometric description of the workpiece, technologic data of the machine tool and workpiece can be considered. The EXAPT system is of modular design and can be operated on smaller computers. The universities of Aachen, Berlin and Stuttgart developed this system during the late sixties. Now it can be purchased from the EXAPT society located at Aachen, Federal Republic of Germany. The fundamental concept of the EXAPT system is shown in Fig. 8.3 [1]. With this system the most important 2 and 3 axis machining operations can be handled. The concept of EXAPT is as follows:

BASIC EXAPT

This module is the basis of all other modules. It can be used similarly to APT to program a tool path to generate a workpiece surface. In addition feeds, speeds and the geometry of tools can be described. The degree of automation which can be obtained is limited.

EXAPT 1

This programming module was devised for boring operations. The following capabilities were added to the BASIC EXAPT features:

- test for tool collision,
- automatic calculation of number of required cutting paths,
- automatic selection of machining parameters,
- selection of tools and
- selection of machining cycles.

EXAPT 1.1

With EXAPT 1.1 boring and milling operations can be programmed. For boring all features of EXAPT 1 are available. For milling additional capabilities are available to test for tool collision and calculation of automatic cutter path segmentation.

EXAPT 2

This module was designed to handle turning and boring operations on lathes. It is possible to describe the contour of the raw workpiece and that of the finished workpiece.
Additions to EXAPT 2 are:

- test for tool collision,
- automatic cutting path segmentation and
- selection of machining parameters.

Figure 8.3 shows how with increasing degree of automation the EXAPT system is being integrated into design. At the present time a strong development effort is being made to close the gap between CAD and CAM and to add the following features, for example for boring:

- selection of machining sequences,
- selection of machine tools and
- selection of fixtures and jigs.

The automation of the programming process for milling and turning is not that easy. However, in the long term, solutions will also be found to tie CAD together with CAM for these machining operations.

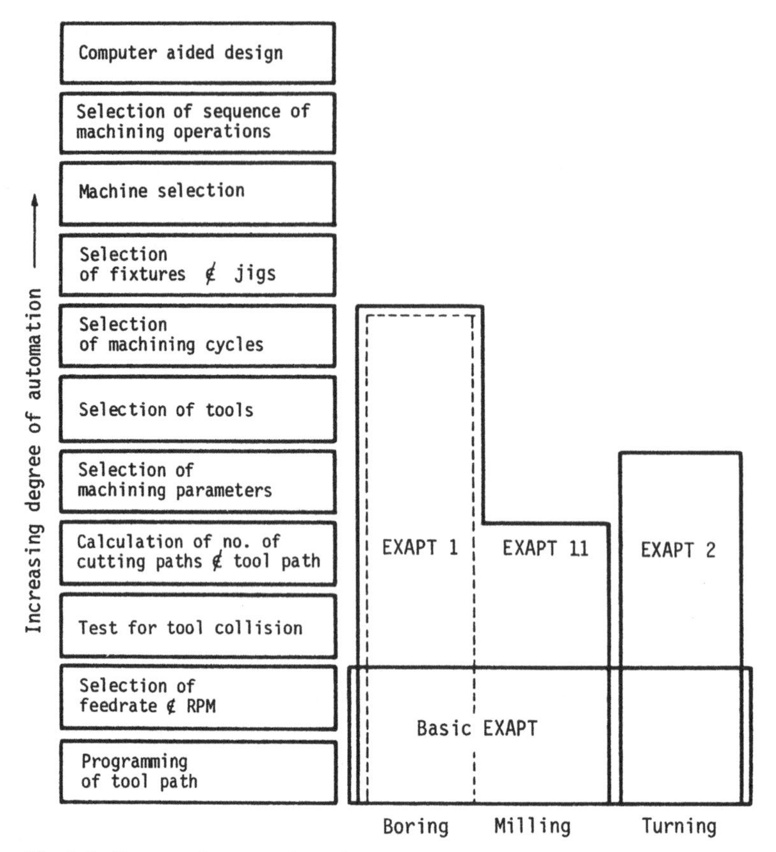

Fig. 8.3. Degree of automation of EXAPT

The process of preparing the NC tape with an EXAPT module is shown in Fig. 8.4 [1]. The part program is entered into the computer and interpreted by the processor. In the first phase the geometric data are processed, as with APT. In the second phase the technologic data are incorporated into the program with the aid of information obtained from the tool, material data and machining files. The output of the processor is a machine independent part program. The subsequent postprocessor serves the same function as that used with APT.

A workpiece to be machined is represented in Fig. 8.5a. The corresponding part program for EXAPT is shown in Fig. 8.5b [2]. The program consists of the following parts:

- *header data* defining the part, machine tool and the coordinate system,
- *geometric data* describing the workpiece,
- *technologic data* describing the machining operations,
- *executive instructions* to direct the machining operation and the
- *end instruction* to complete the machining cycle.

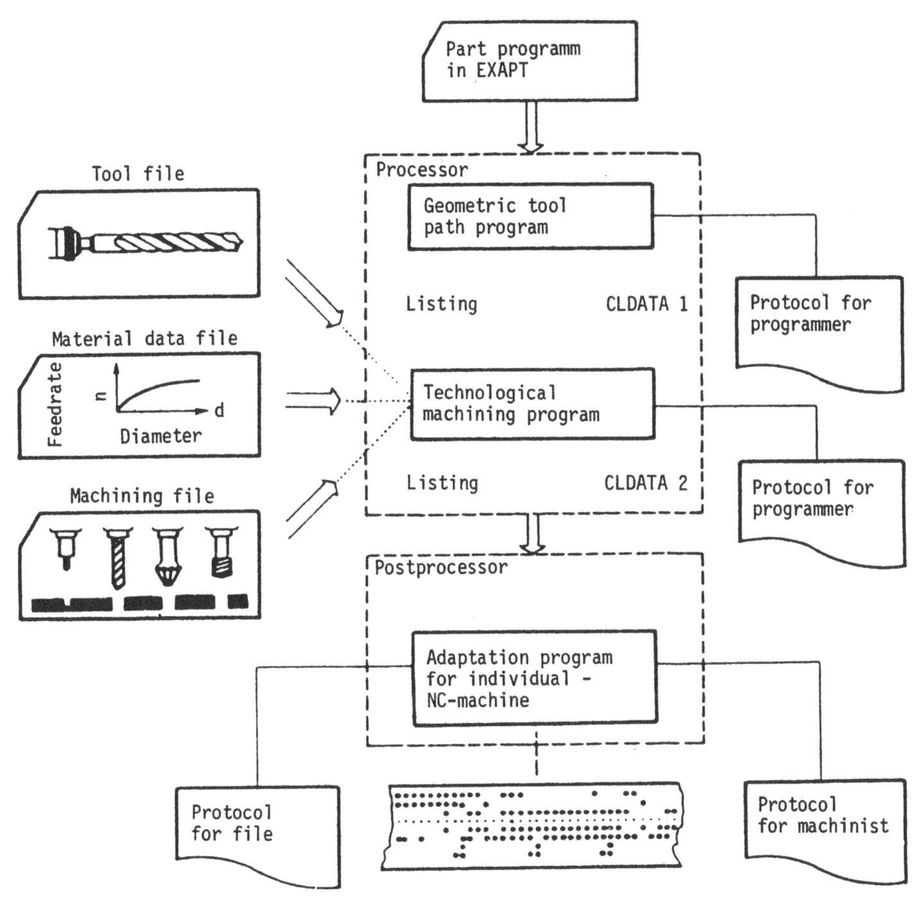

Fig. 8.4. Sequence of programming an N/C machine tool with EXAPT

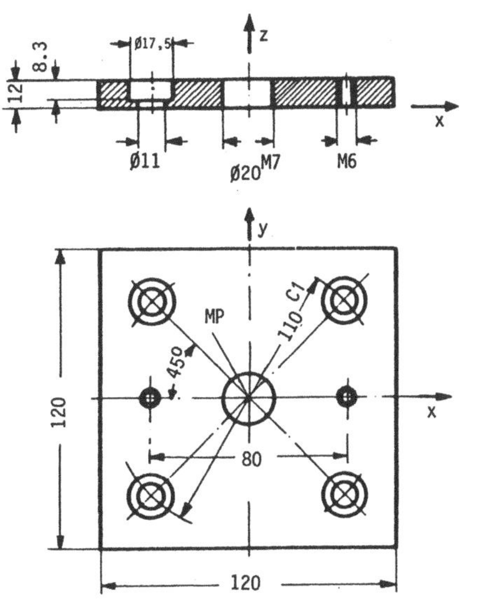

Fig. 8.5a. A workpiece to be machined ▶

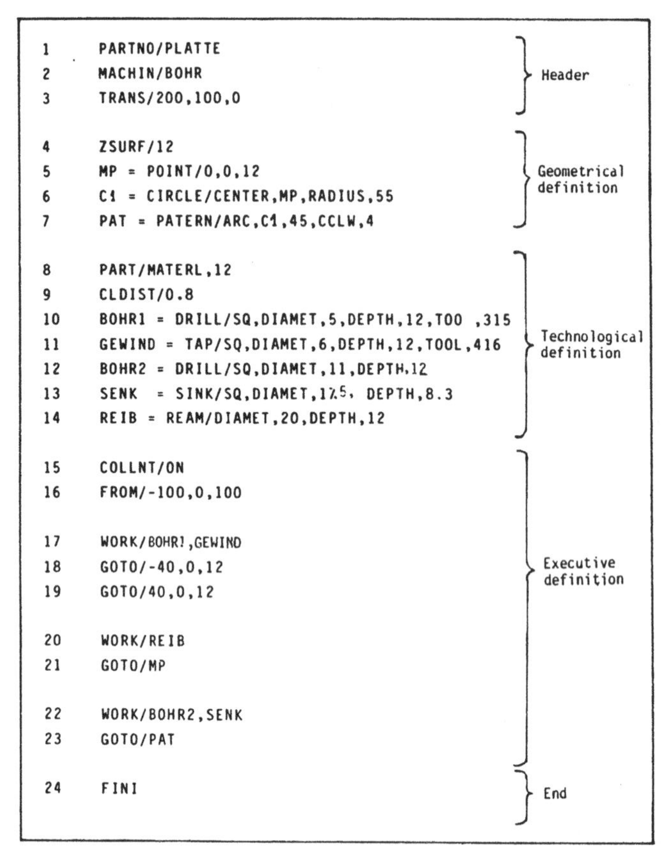

```
1     PARTNO/PLATTE                                        }
2     MACHIN/BOHR                                          } Header
3     TRANS/200,100,0                                      }

4     ZSURF/12                                             }
5     MP = POINT/0,0,12                                    } Geometrical
6     C1 = CIRCLE/CENTER,MP,RADIUS,55                      } definition
7     PAT = PATERN/ARC,C1,45,CCLW,4                        }

8     PART/MATERL,12                                       }
9     CLDIST/0.8                                           }
10    BOHR1 = DRILL/SQ,DIAMET,5,DEPTH,12,TOO ,315          }
11    GEWIND = TAP/SQ,DIAMET,6,DEPTH,12,TOOL,416           } Technological
12    BOHR2 = DRILL/SQ,DIAMET,11,DEPTH,12                  } definition
13    SENK  = SINK/SQ,DIAMET,17.5, DEPTH,8.3               }
14    REIB = REAM/DIAMET,20,DEPTH,12                       }

15    COLLNT/ON                                            }
16    FROM/-100,0,100                                      }
                                                           }
17    WORK/BOHR1,GEWIND                                    }
18    GOTO/-40,0,12                                        } Executive
19    GOTO/40,0,12                                         } definition
                                                           }
20    WORK/REIB                                            }
21    GOTO/MP                                              }
                                                           }
22    WORK/BOHR2,SENK                                      }
23    GOTO/PAT                                             }

24    FINI                                                 } End
```

Fig. 8.5b. EXAPT 1 programm for the workpiece of Fig. 8.5a

8.2.3 Interactive Symbolic Programming

When workpieces are complex, program development by the APT and EXAPT principles may be quite time consuming and cumbersome. For this reason, several machine tool manufacturers have developed symbolic programming facilities. In the simplest case, programming is done via a keyboard directly at the machine tool. Each key contains an instruction, for example "Drill, Countersink, Grid pattern, Keyway etc.". With the help of the keyboard the programmer enters step by step the part program into the computer which converts the instructions into machine code. Obviously this programming method is designed for specific applications such as turning.

More advanced symbolic programming systems use a keyboard in connection with an interactive terminal. The FAPT TURN system, developed by the Siemens Corporation, will be discussed in detail. The program development system consists of a computer, a CRT, a keyboard, a bubble memory and a tape reader/punch combination. The rotational part shown in Fig. 8.6 [3] is taken as an

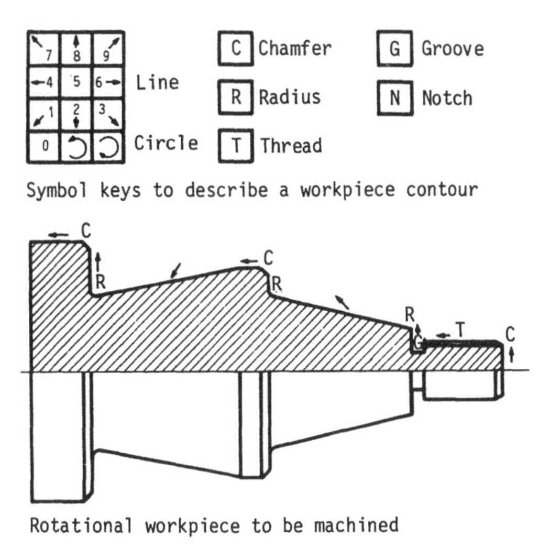

Fig. 8.6. Programming of a rational workpiece with the FAPT system

example to show how programming works. The different steps, which are greatly simplified, are explained below.

1. Initiation of the programming system.
2. Selection of the raw material; there are 17 different choices.
3. Selection of the surface finish.
4. Positioning of the coordinate system, Fig. 8.7.
5. Raw part selection, cylinder, hollow cylinder or a special shape, Fig. 8.6.
6. Generation of the finished part contour, Fig. 8.6 and 8.7. This is done with the help of symbol keys. The sequence of keys to be depressed is shown in the lower part of Fig. 8.6. Basically, lines and circles can be entered. The system will request more information when a line or a circle is indicated. For example dimensions, diameters etc.
7. Generation of the thread by indicating length, pitch and cutting direction.
8. Groove cutting and dimensions.
9. Chamfer and dimension.
10. Programming of mathematical functions such as sin, cos, square root etc.
11. Specification of machine tool reference points, Fig. 8.7.
12. Tool position selection, Fig. 8.7.
13. Determination of tool holder.
14. Determination of cutting parameters, feed and speed.
15. Selection of machining directions.
16. Contour segmentation.

For machining of certain segments it may be necessary to redefine the surface finish. This can be done with the aid of parameters when the surfaces are described.

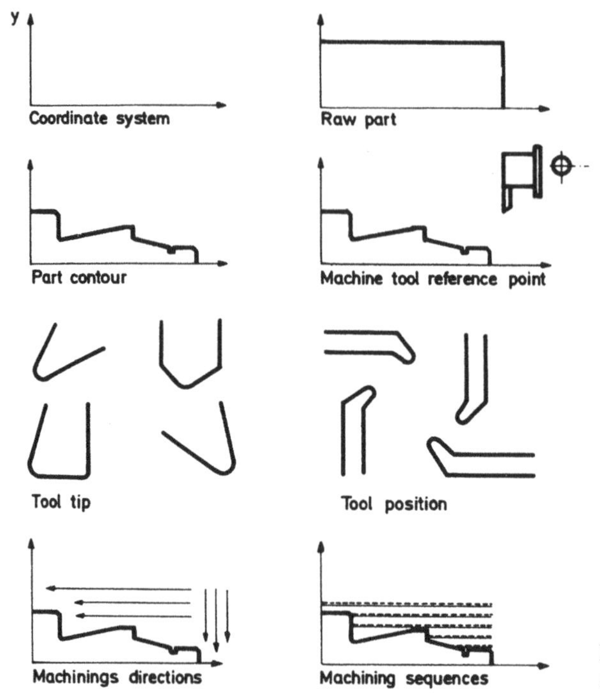

Fig. 8.7. Interactive CRT display of different programming phases

8.2.4 Special Purpose Languages

Many factories produce special purpose parts which have common features. For example the workpiece shown in Fig. 8.8 [4] has a typical design found in sand mold castings. Since it is very difficult to program such a part by hand, a computer is normally used. In this particular case, a manufacturer of precision patterns (Anderson Industries, Muskegon, Mich., USA) introduced a programming system which can easily handle this type of part. The objectives of the programming system are:

- short programming instructions,
- simple syntax,
- no hand computations,
- the operations and ability to handle many different geometries and machining operations,
- the ability to program a wide spectrum of machine tools, e.g., from point-to-point machines to 5-axis continuous path equipment.

The lower half of Fig. 8.8 shows a section of the program which describes the center cavity of the part. There is a 10 field record into which only numbers are entered. The points which are needed to describe the contour are 1, 2, 3, 5, 6, 7, 8. The contour points of the radii and parabolic fillers as well as the coordinates of point 4 are calculated by the programming system.

The known points are encircled and designated as code points. They are enumerated in field 7. They may be defined absolutely or incrementally from a

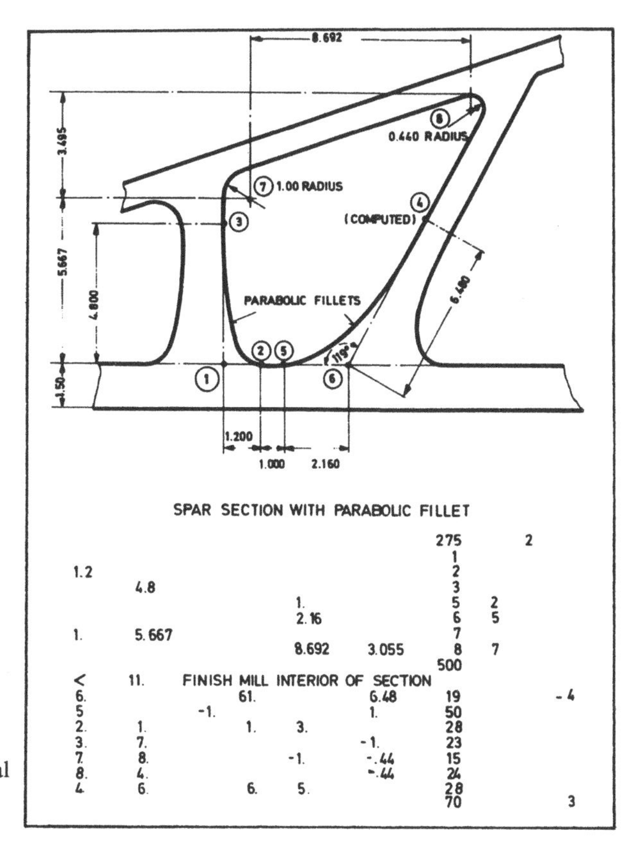

Fig. 8.8. Example of a special purpose NC programming language

previously entered codepoint. For example point 8 is defined incrementally from codepoint 7. The number 500 in field 7 separates the geometric description of the cavity from the milling instructions. First the compiler calculates the coordinates of point 4, see entry 19 in field 7. It is located at a distance of 6.48 from codepoint 6 at a positive angle of 61° from the x-axis. The -4 at the end of line 19 is an instruction to calculate point 4. The 50 in the next line turns on the spindle and coolant. It commands the tool from the home position to its work position at a travel clearance of 1″ above the part. The tool is lowered by 1 inch to contact point 5. The next line has a 2 in field one and a 1 in field two. These numbers define the line at which the parabola is tangent to point 2. Likewise the 1 in field four and the 3 in field 5 define the second line to which the parabola is tangent to codepoint 3. The cutter is moved from point 5 to point 2 and then along the parabola to point 3. The next line entry 23 instructs the cutter to move from point 3 to the beginning of the circular element, defined by points 1 and 2 and the radius originating at point 7. The -1 in this line indicates that the radius of the filler circle is 1″ and that the cutter approaches the circular element from its inside. The next line indicates that the cutter travels from the circular elements whose center is point 7, to the element whose center is point 8. Line 24 is the instruction to bring the cutter from the circular element about point 8 to point 4. Line 28 instructs the cutter to move back to point 6, and the instruction 70 brings the cutter back to its home position.

This coding method might look complex to a reader who is not familiar with it. However, it is of great benefit and easy to use when there are always similar parts to be coded.

8.2.5 Generative Programming by the Machine Tool Control

With the EXAPT language it is possible to calculate the number of cutting paths needed to machine a part. For this purpose feed and speed parameters must be available for different material and tool combinations. For future programming it will be possible to optimize the cutting operation with the aid of intelligent controllers built into the machine tool. Therefore it is necessary to equip the machine tool with sensors, Fig. 8.9. The part program is sent to the control computer of the lathe. It describes the raw part and contour of the workpiece. In addition, preliminary feed and speed conditions are given. The part is chucked and machining is initiated, Fig. 8.10 [5]. First the tool rapidly approaches the raw part at its smallest diameter until a proximity sensor detects the workpiece. At this point, the computer directs the tool drive to assume the normal cutter speed. The tool now engages the workpiece and tries to start machining. A sensor tells the computer that the cut force is too high. The tool is retracted and a new cut attempt

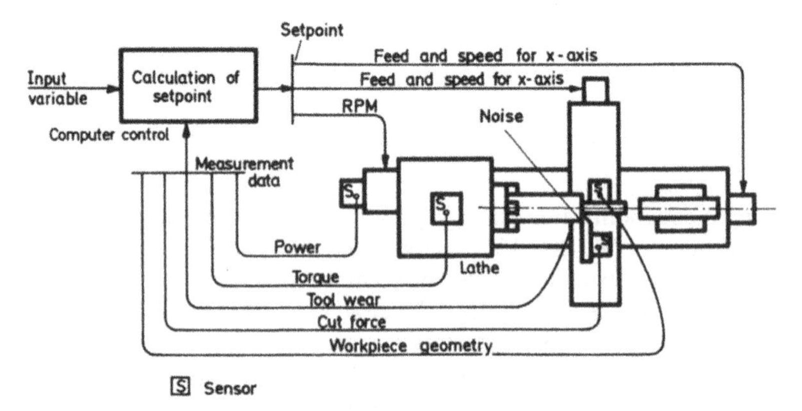

Fig. 8.9. Principle of an adaptive control system for a lathe

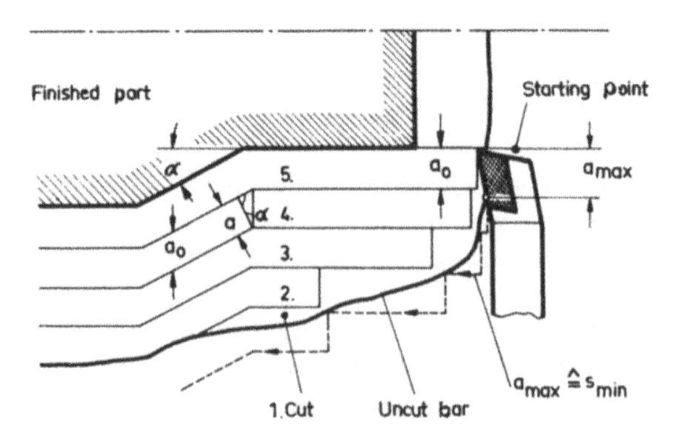

Fig. 8.10. Finished part oriented cut selection with contour parallel cuts

is made at a greater diameter. This procedure is repeated until the permissible cut force is obtained. From here on, the feedback control commences its machining operation and partitions its own cuts. The operation may be done under the supervision of an optimization model. Optimization criteria may be minimum cost or maximum throughput.

8.3 Programming Languages for Robots

Within recent years the industrial robot has matured into a universal tool which can handle efficiently many manufacturing operations. It was used first for simple material handling work and for welding. However, its greatest potential lies in the automation of assembly. For this purpose it must be equipped with sensors and with a comfortable programming system. In order to describe the work of a robot, the language must have conventional constructs, which are also robot specific. With these requirements a programming language for robots usually becomes very complex. Researchers first started to develop simple assembly languages. Valuable experience resulted from their use and led to various concepts of explicit high order languages. At present languages are being developed for implicit programming. With these it will be possible to handle assembly tasks with the help of expert systems.

8.3.1 General Requirements for Programming Languages for Robots [6]

One of the essential features of a computer language for a robot is programming of the trajectory. For this purpose it is necessary that the programmer has the tool to enter the start and end positions of a movement and of the trajectory. In general, it is possible to describe textually any trajectory of a given coordinate system with the aid of a sequence of points. However, it is very difficult for a human operator to visualize this series of points in a 3-dimensional space and to describe the coordinates of each point via a programming language. In order to obtain accurate parameters, the exact location of a point has to be found with the aid of a measurement. This, however, is very time consuming and awkward. In practice, the problem is resolved by leading the robot's effector through its desired path and by reading the coordinates of the corresponding arm joints at predetermined points along the trajectory. The parameters of these points are entered into the robot memory. The robot path is then reconstructed by the compiler with the aid of an interpolation algorithm. In addition to the parameters of the trajectory, the control system of the robot must have information about the orientation of the effector. This parameter can also be entered into the program by the teach-by-showing method.

Modern explicit programming languages for robots use the "frame" concept. With this method, a spacial point is described by a position vector with respect to a standard coordinate system. The orientation of the gripper is described by a rotation. Thus a frame consists of a position vector and an orientation, Fig. 8.11. When the effector is described by a position and an orientation, the following convention may be useful, Fig. 12:

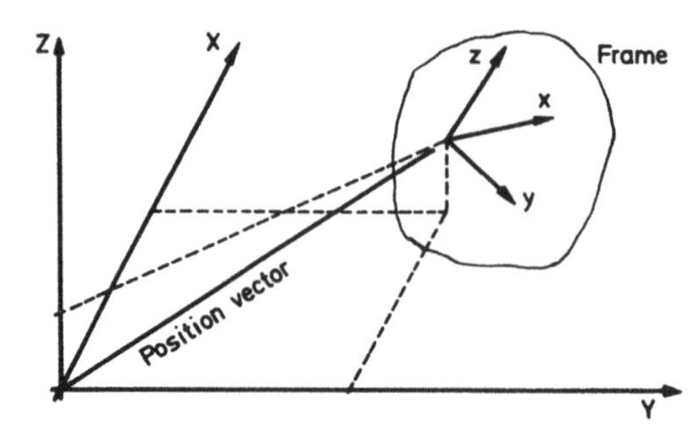

Fig. 8.11. Geometric position of a frame

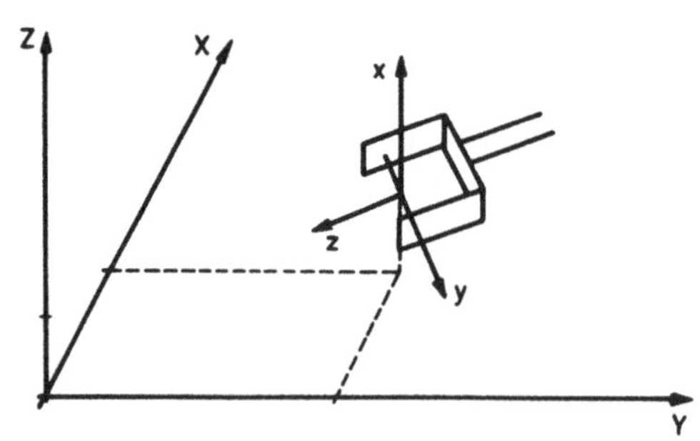

Fig. 8.12. Geometric position and orientation of a gripper

- the endpoint of the position vector is located in the middle of the center line which connects the two gripper jaws,
- the gripper points into the direction of z-axis of the base coordinate system,
- the y-axis of the effector runs through the gripping points of both jaws.
- the right-handed-screw-rule is applied to orient the x-axis.

When comparing the assembly instructions for a human operator with programming instructions of a computer language, the followig factors are of interest:

- The transfer of the dimensions from the drawing to the actual assembly object is done by the operator by looking at the drawing and by translating this information directly into an action.
- Missing information is supplemented automatically by the operator with the help of his experience. For example, the instruction "Assemble a Flange" suffices to perform the described operation. The operator searches for the flange, places it on the assembly object and inserts the fasteners. Exact information about the position of the insertion holes is not necessary.
- The operator automatically performs a sensor controlled positioning operation. For example if he enters a screw into a hole, he uses several sensors. Course positioning is done with the aid of vision and fine positioning under the guidance of the touch sensors of his fingers. In case the thread of the screw does not engage with that of the hole, he instinctively takes corrective action.
- Missing assembly elements or tasks may be automatically supplemented for by the operator. For example the insertion of a screw implies that a screwdriver has to be used and that the fastening process has to be done according to a given sequence of operations.

- Fixturing needed during assembly may be done automatically by the operator without explicit instructions from the drawing.

It can readily be seen that many of the above mentioned tasks have to be programmed explicitly if a robot is to perform the assembly. In order to obtain a fast and flexible programming system for robots, the following components should be provided:

- Installation of sensor for vision, force-torque, slip, proximity etc. into the robot.
- Control of the robot by a runtime system which is able to adapt itself to online changes during assembly.
- The compiler to translate the programmed work cycle should have a component which can automatically generate missing information.

These criteria imply that the robot should be controlled by a computer and that a higher programming language be available.

To support this statement, several programming methods for industrial robots will be compared.

8.3.2 Programming Methods for Robots

All programming methods which will be discussed here, need a facility to program trajectories. From the user point of view, the languages can be classified as follows:

8.3.2.1 Manual Programming

With this method the endpoints of the trajectories are set by hand. They are either hard wired or set through a plug board. Other functions are not provided. For this reason no control computer is needed.

8.3.2.2 Programming with the Help of the Robot's Brake System

For programming purposes the brakes of the robot axes are disengaged and the end effector is brought into the desired work position and orientation by the operator. By depressing a function key, the coordinates of the individual axes are entered into the computer memory. In order to reproduce the desired trajectory, the coordinates are read back from memory and the computer calculates the required motion parameters for the axes drives. For this programming method, the robot must be provided with position decoders.

8.3.2.3 Sequential Optical or Tactile Programming

With this method the effector is lead by the operator through its trajectory. A guide handle with tactile or optical sensors is attached to the arm. A computer polls these sensors at request or continuously and reads the coordinates of all joints along the travelled effector path into the memory. With the aid of these sensor data the robot is capable of reproducing movement along the trajectory.

8.3.2.4 Master-Slave Programming

For this programming method a second, often small master robot, is used to follow the desired work path. The movement is recorded via sensors and stored in memory. The coordinates are then transferred to the slave robot. Thus it is possible to program positions and trajectories.

8.3.2.5 Teach-In-Method

A teach-in pendum is used to direct the robot along its trajectory. At strategic points, the coordinates of the robot joints are polled and entered into the computer memory. This is initiated by the actuation of a push button. The pendum has control buttons for each robot joint. With these it is possible to lead the robot joint through any practical movements and to point the effector in any desired directions. More advanced teach-in systems allow the movement of the effector along the x, y and z-axes of a cartesian coordinate system. Additional features implemented in the pendum are:

- control of velocity motion,
- control of time motion,
- simple program branching,
- setting of counters and
- miscellaneous functions (control mode, weight compensation, single step operation).

8.3.2.6 Textual Programming

These programming methods are similar to those used by data processing. Operations and data are described with the help of character strings. The user enters the sequence of instructions which describes the movement of the robot into the computer. The information is translated by the compiler into machine code. This can be done off-line independently of the robot. However, it is not possible to program positions and orientations. These parameters have to be entered on-line by a teach-in method. For this reason no pure textual programming is possible.

8.3.2.7 Acoustic Programming

With this method the program is entered via voice communication into the memory of the robot. Thus the programmer is freed from the cumbersome process of writing down the instructions on paper. However, he still has to learn the vocabulary and the rules of the acoustic programming language. Programming by natural languages, including its syntax and semantics, will be discussed later.

8.3.2.8 Design Considerations for a High Order Language

The teach-in method and the textual programming language do not only allow the description of spacial movements, they also permit branching, looping and the use of subroutines. At present, the most frequently used programming method in industry is teach-in. However, modern robots are increasingly being employed to handle complex tasks such as assembly work. Here, teach-in may become very cumbersome and there is an ongoing trend towards the use of higher order languages. Several robot manufacturers offer a combination of textual programming and teach-in. A spacial point can be entered by defining it and by assigning a name to it. The name is used by the program. It is also possible to lead the effector to the spacial point and to enter an instruction by the teach-in mode with the aid of a function key into the control computer of the robot. This procedure is supported by an editor. The system enters the text of a motion instruction into the program. It is parametrized with a defined speed and the position of the point. With this method it is possible to program a complete sequence of movements and to obtain readable software. The program can also easily be changed.

In the future, however, the textual method will predominate for programming of assembly work in the factory. It offers the following advantages:

- the program is readable by programmers and other users,
- the program can easily be changed and expanded also by other programmers,
- the program may have variables, during programming only a data type is assigned to it and not its value,
- the program can be stored in readable form, which is important for documentation and
- the program can be written off-line without the availability of a robot, no definition of the operating points is necessary.

In principle, the instruction set of a programming language for robots is similar to that of the teach-in method. Thus for each basic symbol of the language, a function key may be provided and installed in a key board. For simple assembly tasks this programming method may be adequate. However, textual programming in combination with teach-in is needed when the following conditions exist:

- complex assembly work is done which requires frequent program branching and subroutine calls,
- when sensor signals of the robot have to be processed,
- when external data have to be accessed, e.g., from a data file of a manufacturing system and
- when an internal world model is used and its data are manipulated.

In addition, textual programming permits the implementation of special strategies which assure an orderly and safe assembly. The following operations may be included:

- Measurement of a position during program execution: For example if the exact height of a workpiece is unknown, the effector can be lowered slowly until it touches the object surface. A force sensor records this event and this position of the workpiece is then entered into the computer with the help of an instruction.
- Supervision of an operation: The success of an assembly is monitored by sensors which watch for misalignment, breakage or missing parts. For example, the insertion of a pin into a hole can be monitored with a fource torque sensor. During insertion the effector may move slightly in direction of the axes perpen-

dicular to the centerline of the pin. A force signal in either direction will indicate that the pin is surrounded by its mating hole, otherwise no hole is present. In the latter case, the robot will be alerted and a corrective move can be made.
- Search strategies: During bolting or mating operations, the workpiece may not be in its exact position. In this case the robot may try to locate the center of a hole by a search operation. For example, it may try to target the object by a spirally shaped search procedure.

For the following reasons, the number of such special programming instructions should be kept to a minimum

- to reduce the assembly time of the object,
- to save memory space,
- to reduce compilation time and
- to reduce the number of programming errors.

8.3.3 A Survey of Existing Programming Languages

There are numerous programming languages under development. A ranking of different programming languages is shown in Fig. 8.13 [6, 7]. The majority of these are explicit assembly or compiler languages. Here, every movement has to be described explicitly by the programmer. The number of statements to program

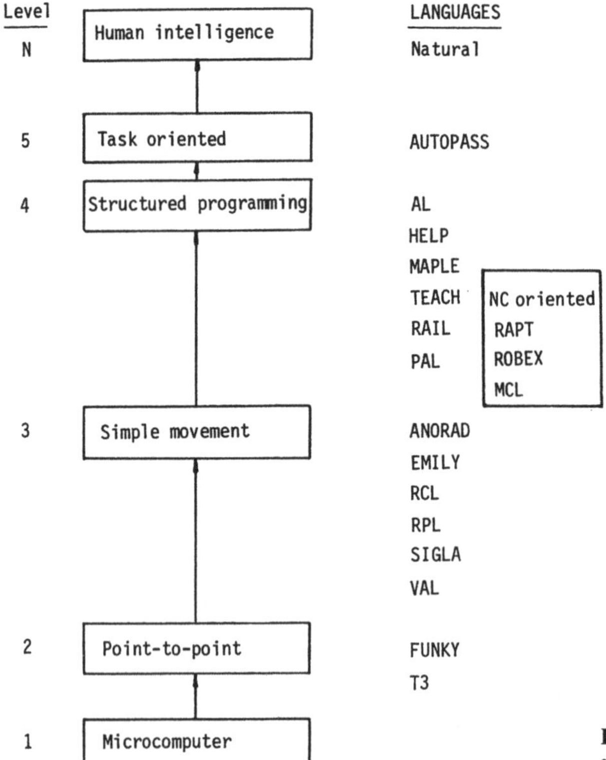

Fig. 8.13. Ranking of different robot languages

Teach-in programming
Control structure
Subroutines
Nested loops
Data types
Comments
Trajectory calculation
Effector commands
Tool commands
Parallel operation
Process peripherals
Force-torque sensors
Touch sensors
Approach sensors
Visions systems

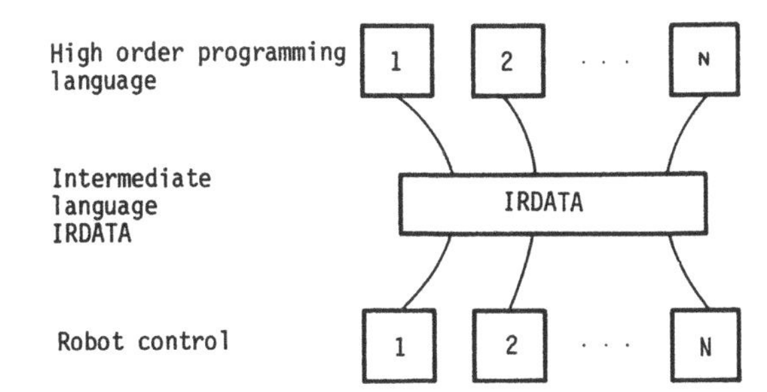

Fig. 8.15. The use of an intermediate language

◀ **Fig. 8.14.** Features of programming languages for robots

an assembly task will be quite numerous, even for a simple problem. Presently, there is only one implicit programming language known, and that is AUTOPASS. It has limited capabilities to describe simple assembly primitives. A typical instruction would be "Pick up a Bolt, Insert it into a Hole". There are three languages which are based on the NC-language concept. They may be used in connection with loading and unloading of NC-machine tools. In this case the programmer only has to be familiar with the NC-language concept. The NC-type languages, however, become very complex when variables have to be defined and when sensor data derived from moving objects have to be processed. Several of the listed languages are conceived for universal applications, others are designed for a specific type of robot.

Figure 8.14 [7] shows desired features of programming languages for robots. In addition to the constructs of conventional languages, there should be several robot-specific ones. For example, typical data types are vector, frame, rotation and translation. It should also be possible to describe to the robot an effector trajectory and how to handle the synchronisation of the work of several arms. The robot must be able to operate the effector and the work tools under program control. In addition, there must be language constructs available which can handle sensor signals to which the robot is able to react.

The many languages currently available suggest that the output of a compiler is a standard intermediate code, Fig. 8.15. In this case the robot manufacturer has to organise his control system in such a manner that the interface with the controller accepts the intermediate code. Thus it is possible to use different languages for different robots via a standardized interface.

8.3.4 Concepts for New Programming Languages

At present a considerable amount of research work is being done to develop new programming languages and systems for robots. Two different development trends can be observed:

1. Making the individual robot autonomous. This gives it the capability to adapt itsèlf to the work environment, to make own decisions and to take actions to handle unforeseeable situations. In other words, the robot will be provided with an amount of limited intelligence which is needed to perform its assembly task.
2. The entire manufacturing process will be completely automated to eliminate the necessity for human intervention. The robot becomes an integral part of a manufacturing facility which is supervised by a hierarchy of computers. The description of the workpieces and that of the assembly is automatically generated from the design process, and transferred from the CAD-system to the control computer of the robot. There is no need to program the robot directly and to describe to it the geometric shape of the object, its surface description, gripping position etc. This information is all available from a central CAD data base.

Both of these developments will have applications in different industries:

- The autonomous and flexible robot, equipped with sensors and a routine system, will be used by small and medium size companies. This system can be adapted and reprogrammed quickly to new production runs. It may make its own decisions with the aid of an expert system, Fig. 8.16. The drawback of this device is its complexity and the unavoidably high development effort.
- Systems with several robots, where each unit performs a specific task, will find their place in mass production. They will be the domain of large companies. Such an integrated manufacturing system will include machine tools, robots, material handling, peripherals, conveyors etc., Fig. 8.17 [8]. The investment cost will be quite high. With increasing product diversity the intelligent robot may also be installed in flexible manufacturing systems which can handle many product variants.

Independently of these two developments, several high order programming languages have been developed to solve assembly tasks. Most of them use the frame concept. Their use, however, requires that the programmer has the capability to

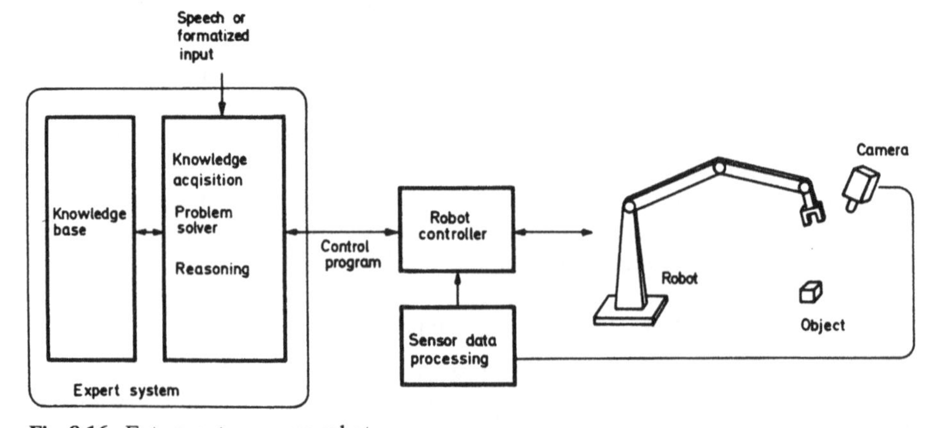

Fig. 8.16. Future autonomous robot

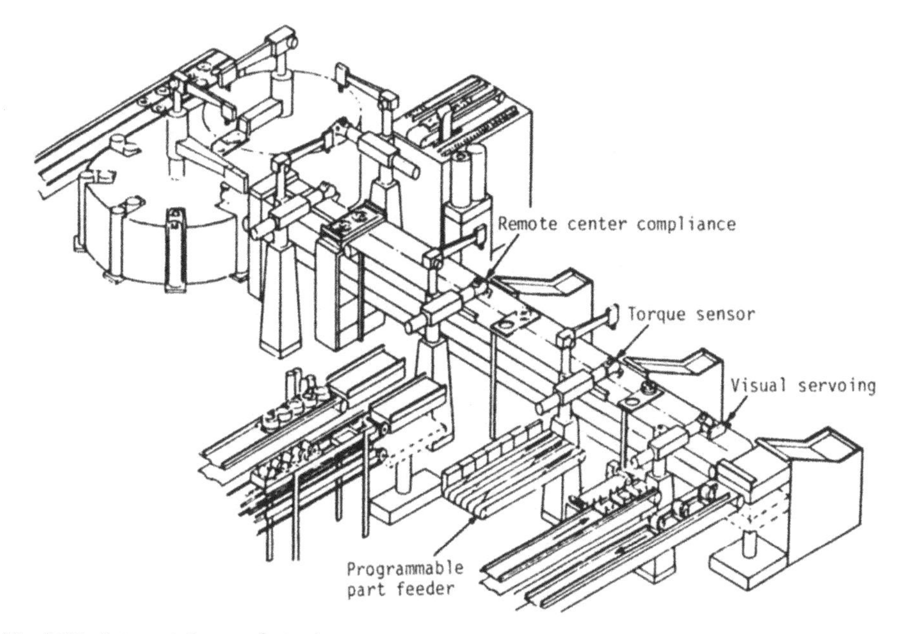

Fig. 8.17. Integrated manufacturing system

view the assembly object in a three dimensional space. For example he must be able to visualize the permissable path of an effector in moving from one point in the assembly space to another. Typical features of a high order programming system will be explained for the SRL language developed by the University of Karlsruhe.

Based on an extensive study of the VAL [9] and AL [10] systems and the result of a comparison of several existing languages for industrial robots, a new language SRL was developed. The frame concept with its geometric data types and geometric operators was taken from AL. Experience obtained from an AL-implementation led to the conception of the following new features:

1. Generalization of the geometric data types "vector, rotation and frame" with the help of the structured data types of PASCAL.
2. A time compound statement to distinguish between a compound statement for syntactical reasons, a block and a sequence of statements to be executed sequentially.
3. A general structure for parallel, cyclic or delayed execution of parts of the program.
4. Input-output to digital or analog ports and sensors.
5. Specification of system components like robots, sensors or interrupts.
6. General sensor interface.
7. Several move statements for different kinds of interpolation.

Most of the points mentioned above were derived from the fact that AL was designed to be hardware in dependent. Therefore, only in the MOVE statement does AL use measuring data from a force and torque sensor.

To overcome hardware dependence and to support structured and selfdocumenting programming SRL includes the above stated language constructs. As a new facility, SRL has an interface to a general world model at program run time. The world model can contain data about objects and its attributes, like workpieces, fixtures, robots, frames and trajectories.

Another fundamental feature of SRL is the language PASCAL. The data concept and file management are taken from PASCAL because it gives the user a very flexible and problem oriented data structure.

The goal in developing SRL is the design of a language which can easily be learned and adapted for further developments and applications. It will also provide an interface between future planning modules and the "traditional" programming system. A planning module will be used to generate SRL statements from a task (goal) oriented specification. This will replace explicit programming for every action, Fig. 8.18. Therefore, SRL has to be well structured and universal in nature, and it has to include all features of robot programming and process control. The standard data types and the RECORDs of structured data types, defined by the programmer are from PASCAL. New data types have been added to improve handling of synchronization between the program and external events. Predefined records can be used for geometrical computation needed for robot moves. The standard data types of SRL are:

```
INTEGER   ________ }
REAL               }
BOOLEAN            }  from PASCAL
CHAR      ________ }
VECTOR             }
ROTATION           }  from AL
FRAME     ________ }
SEMAPHOR           }  for synchronisation
SYSFLAG   ________ }
```

Using a frame in SRL notation, the programmer describes textually, the cartesian position and orientation of the robot effector. It consists of a position vector and a rotation. The rotation can be defined by one or several rotations and it is stored as a 3 by 3 rotation matrix. The user has access to the matrix.

A semaphor is used for synchronisation and queueing of tasks within a program. The system flag SYSFLAG is introduced to synchronize programs. The programmer has no direct access to the data type SEMAPHOR or SYSFLAG but he can use the statements SIGNAL and WAIT to handle them. SRL includes the structured data types ARRAY, RECORD and FILE of PASCAL. Furthermore the programmer can define his own problem oriented data types as in PASCAL by enumeration and subrange. Pointers are also included in SRL. With respect to data types, the programmer can write records and their components in any expression as is done in PASCAL.

Example:

```
save-distance := posvector.x + tolerance;
```

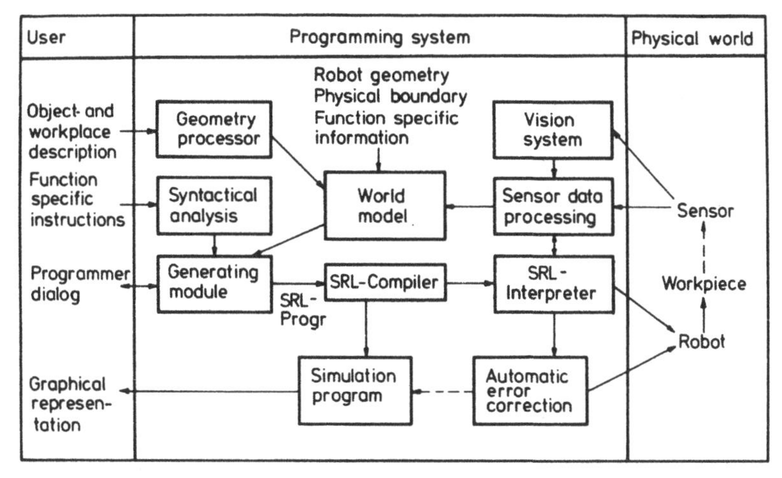

Fig. 8.18. An advanced programming system based on SRL

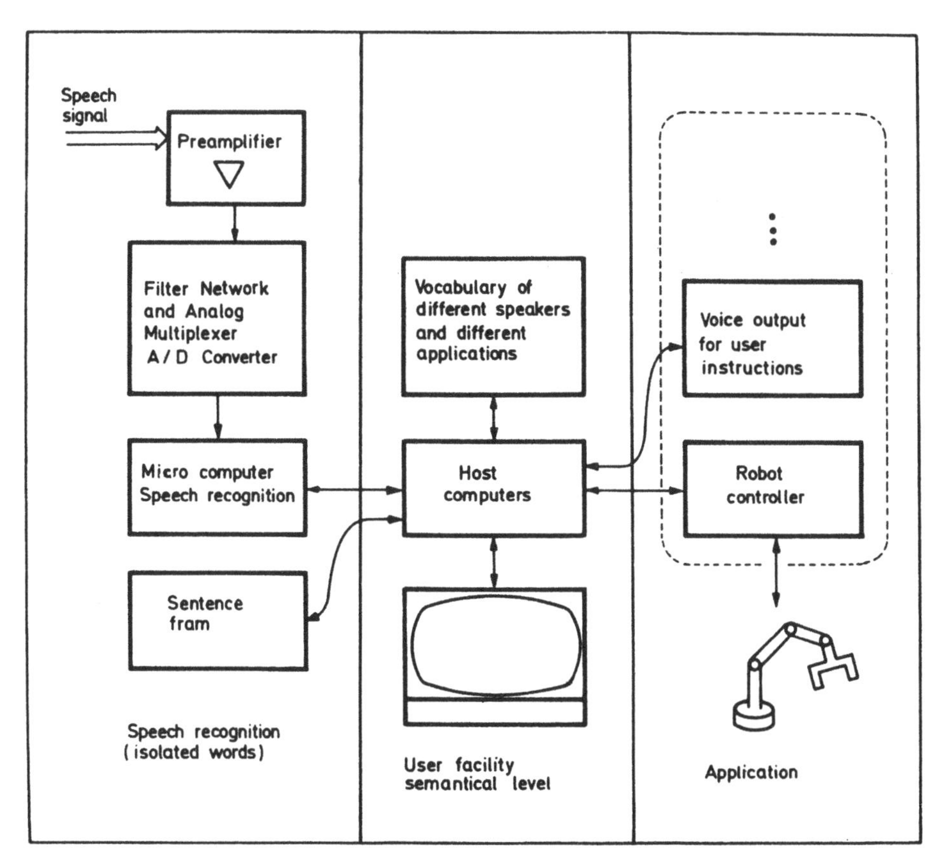

Fig. 8.19. Natural language programming system

8.3.5 Programming with a Natural Language

Natural language programming systems are being developed to assist untrained personnel in teaching the desired movements of a task to a robot, Fig. 8.19. Because of the complexity of a natural language, its entire vocabulary cannot be used. Usually simple syntax and semantics are selected which allow the task of the robot to be described by a quasi natural language. The programming system performs the syntactic and semantic analysis of the speech input, it extracts the pertinent information and takes a plausibility check. Thereafter a corresponding formal robot program is executed. These systems need large memory and require long execution times.

8.3.6 Implicit Programming Languages

Systems to program robots automatically have been under development since the seventies. Here the programmer does not need to formulate explicitly every instruction of a task, e.g. "MOVE ARM TO POS. 1". However, he gives task-oriented instructions, e.g. "FASTEN FLANGE WITH 4 BOLTS". The system tries to interpret this instruction and plans its execution. The system searches in its library for different operators which will perform the required robot actions. Starting with the initialization state, each succeeding state is planned until the finish state has been reached. The result of this search is a sequence of operators and states which can be visualized as an operation plan. This plan is equivalent to a program obtained from a programming language. A system capable of automatically setting up such a plan is called a problem solver. The individual steps to assemble the plan may be as follows:

First the flange has to be recognized. Then the effector picks up the flange and places it on the mating part. Now, a check is made with a sensor for proper alignment. The next steps are to fasten a screw driver to the effector, to locate a bolt, to pick it up and to insert it into a bolt hole of the flange. In order to ensure the correct tightness, a torque sensor supervises the fastening operation. Thereafter, the other 3 bolts are inserted. In the last step the presence of all 4 bolts is verified by a vision system.

8.3.7 Programming Aids

In addition to the language a powerful programming system must be available, consisting of several software packages and of a low cost program development computer. Figure 8.20 shows a comprehensive programming system for assembly robots.

The user describes the object and the workplace to the robot with the help of an application oriented language. This information is processed by a geometry processor and entered into a world model. Likewise the movement of the robot is functionally described by implicit instructions, and a syntactical analysis is performed. This program is combined with information from the world model. The result is sent to the SRL-compiler via a generating model. It is also possible to communicate interactively with the SRL-compiler to enter or edit instructions.

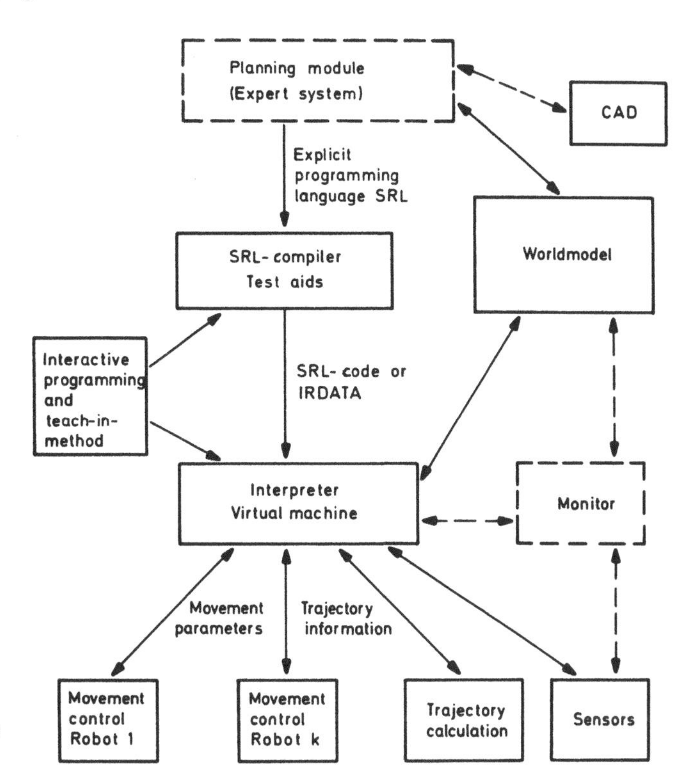

Fig. 8.20. Hierarchical programming structure for industrial robots

The output of the SRL-compiler, in the form of interpretative code, is loaded down to the control computer of the robot. Sensor signals from the robot can be brought back to the sensor data processing module. Where an object or a workpiece has changed its position, this module will send instructions to the world model to update it. The same information is needed by the SRL-interpreter to correct the movements of the effector. A simulation program is also available which allows the programmer to display graphically the work environment of the robot, to check its movements and to detect possible collisions.

The graphical emulation system is part of the programming system or of the real-time controller, Fig. 8.21 [11]. In the early stage of the robot design, its kinematic attributes (joints, links, end-effectors) and the assembly cell as well as its environment can be described on a graphic display. Trajectory planning, the interpolation in cartesian coordinates and the corresponding coordinate transformation can be tested and optimized. By adding a program for the simulation of the robot design the response of the axis motor drives and their control can be traced. This allows evaluation of the dynamics of the robot. For debugging of assembly programs, the emulated robot is interfaced with the programming system which defines multiple moving tasks. With an offline program test facility, the workpiece and the robot components can be emulated without the risk of collision. When it is certain that all assembly sequences are performed without

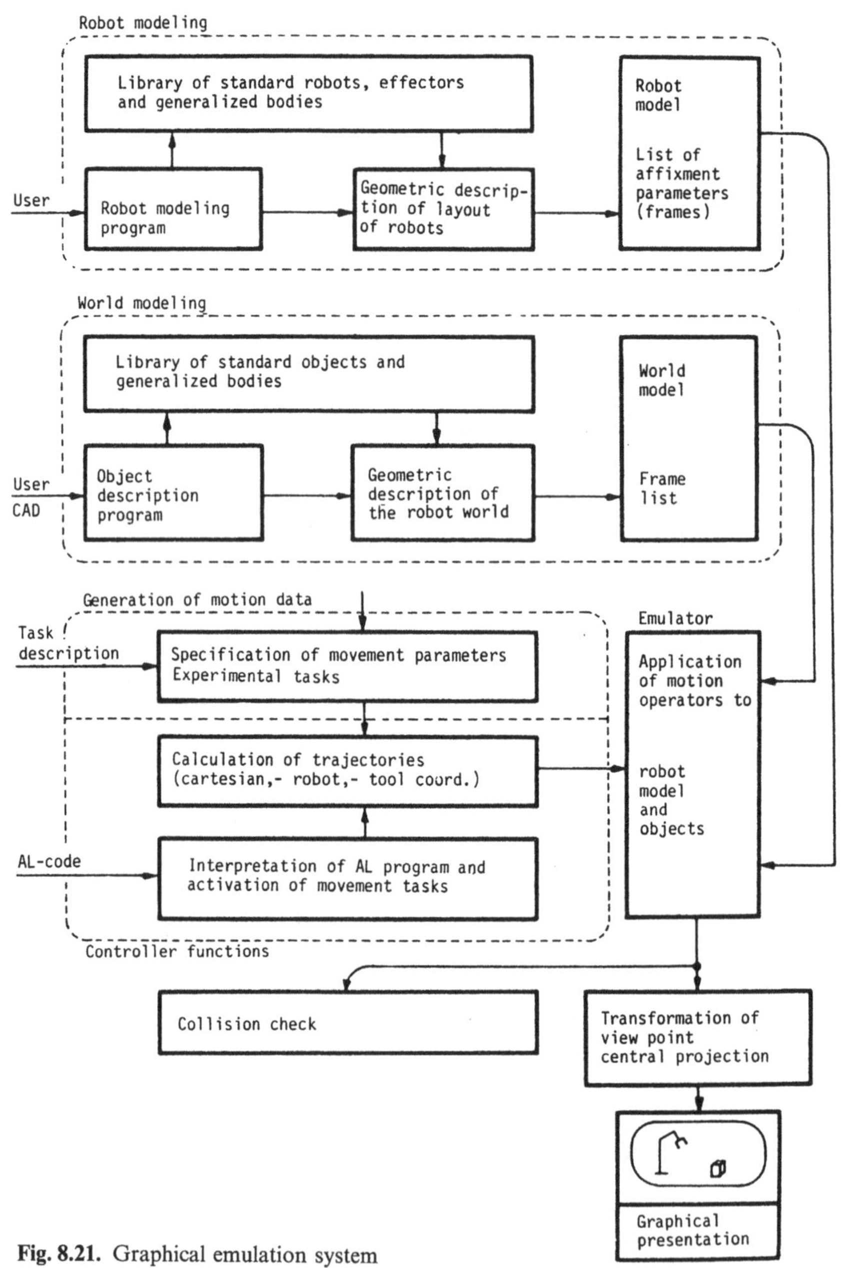

Fig. 8.21. Graphical emulation system

conflict, the program can be transferred to the robot control computer for execution in real-time to move the mechanical manipulator. Verification of the assembly can be performed at this stage. Fig. 8.22 and 8.23 show how a robot is constructed from basic components with the aid of a simulator.

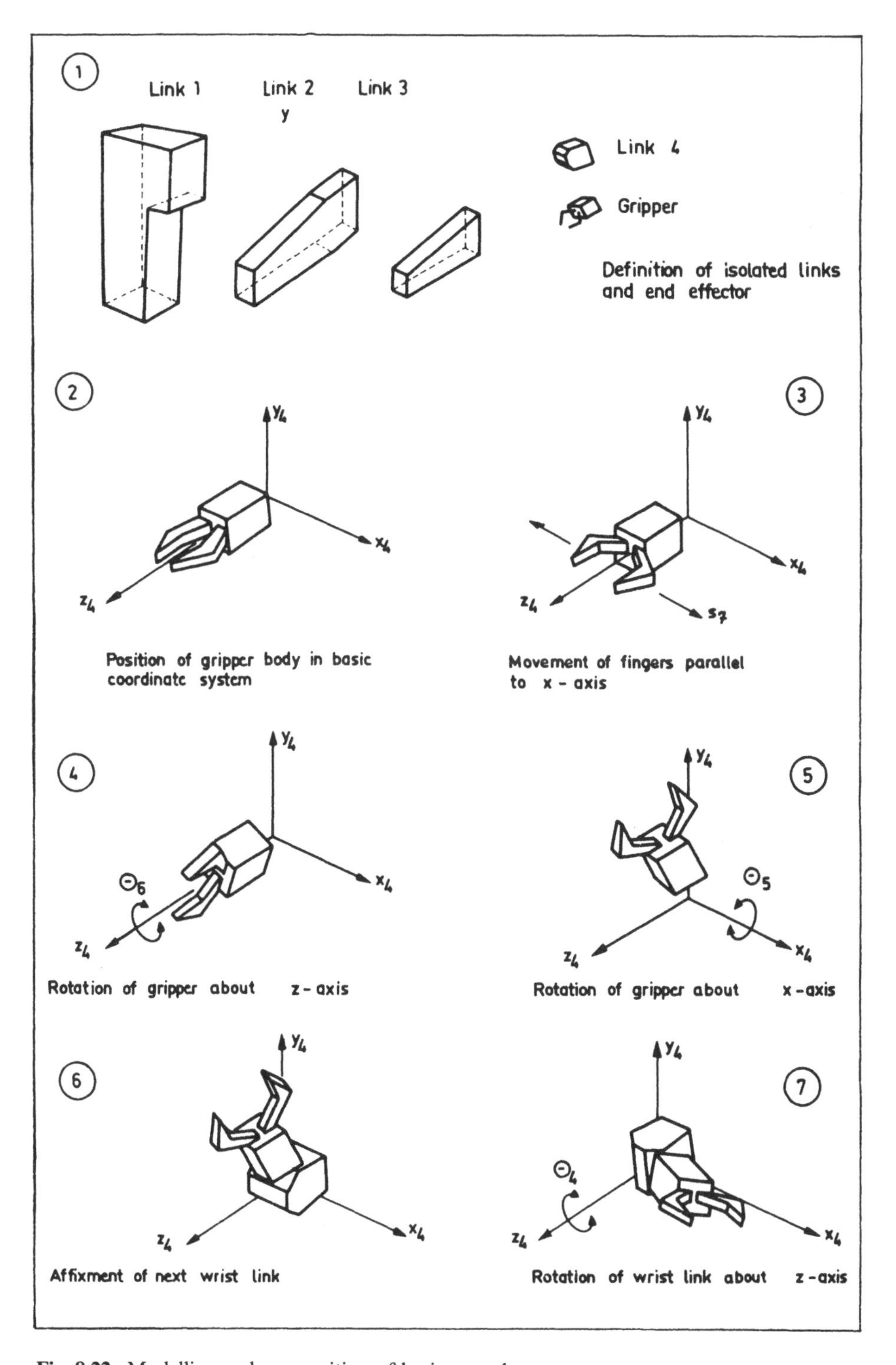

Fig. 8.22. Modelling and composition of basic arm elements

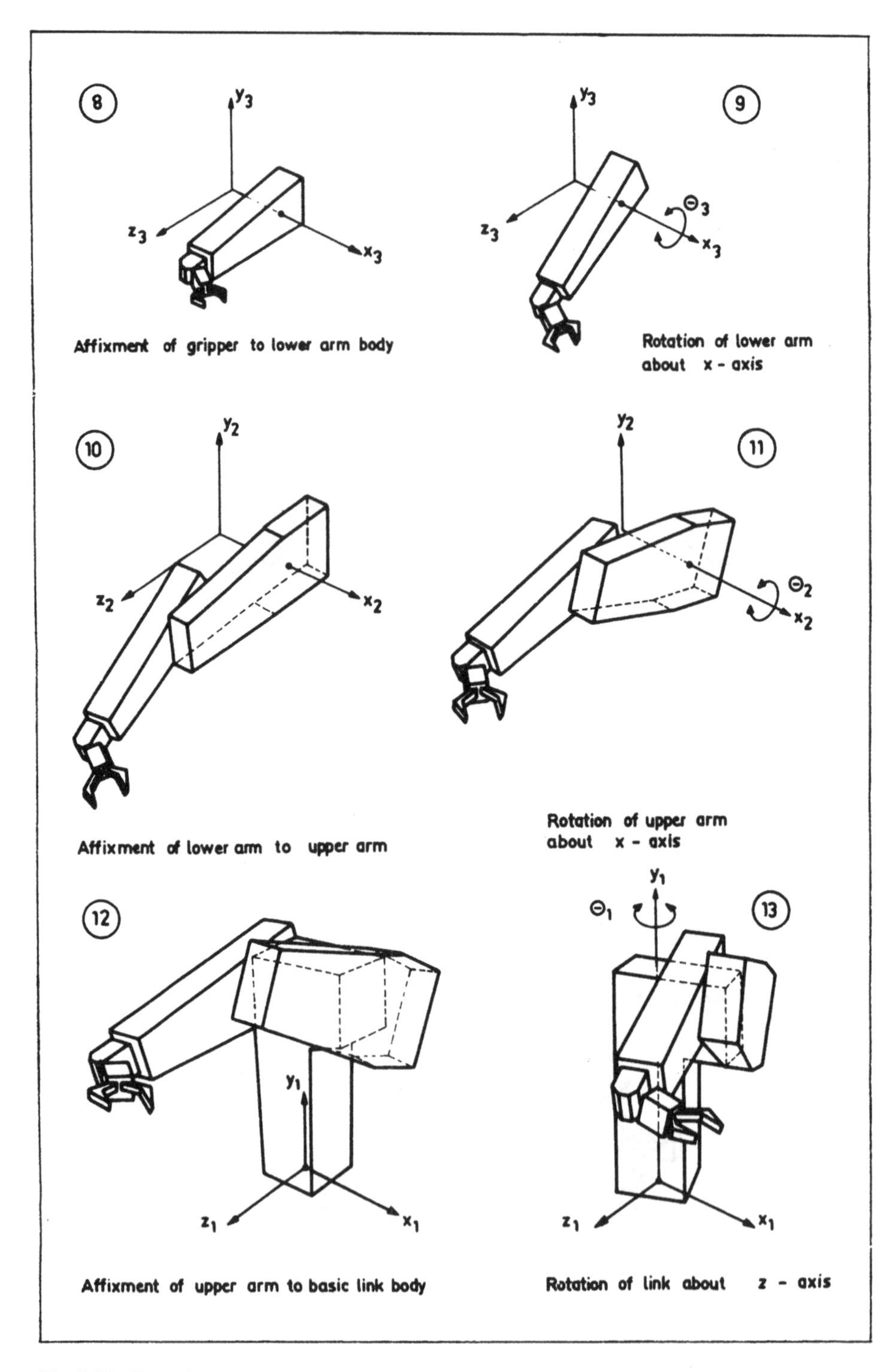

Fig. 8.23. Succesive composition of the PUMA 600

8.4 Process Control

Languages for process control – also named real-time programming languages – are characterized as follows:

1. They permit the description of parallelism. Depending on the structure of technical processes, there are usually several activities which have to be controlled concurrently. Process control languages therefore contain concepts for the definition of tasks.
2. They allow the formulation of time expressions and the interconnection of time and activities.
3. They enable the synchronization and communication between different tasks. The programmer can define which task uses which data from other tasks.
4. Some of the existing languages allow the description of the interface between the technical process and the computer. Usually they possess more data types than commercial data processing languages.
5. They support the reaction to spontaneous events in the technical process.
6. Some of them assist structured programming. Thus they facilitate the implementation of reuseable process control software.

Current languages for process control were developed either as extensions of existing scientific programming languages or as completely new languages. Process-FORTRAN, Process-BASIC are of the first type, PEARL (Process and Experiment Automation Realtime Language), RTL/2 and ADA represent the second type.

8.4.1 Extensions of Existing Programming Languages

Process-FORTRAN is an example of the extension of a language. It emerged from FORTRAN IV. Special features were included by adding new procedures. They can be activated by

```
CALL <Procedure Name> (<Parameters>).
```

These procedures permit:

- manipulation of single bits or a group of bits,
- scheduling of different tasks and
- input and output of measurement and control information to and from the technical process.

The bit-manipulation is very useful in process control because it is often necessary to process a binary coded state information. Usually a set of such state information is represented in a single word, and bit-manipulation-procedures are necessary to extract the desired state information, e.g. the state of a (binary) switch.

Procedures for task scheduling include the definition of different priorities for different tasks, and the specification of which tasks depend on which events (of the technical process). It is possible to activate a task after a predefined time period or to bring it into an inactive state.

The procedures for input and output of measurement and control information permit the transfer of data to and from interfaces. They may contain digital/analog or analog/digital convertors.

8.4.2 PEARL – A Process and Experiment Automatic Realtime Language

PEARL is a language which was developed as a system implementation language for process control application. A PEARL-program consists of several modules which can be compiled separately. Thus it supports the modular decomposition of complex programs. Each module consists of a system part and a problem part.

The *system part* contains the description of the hardware configuration on which the program has to run. All connections between the process control computer and the peripheral devices must be described in this part. These devices are either standard peripherals or disks, displays or special process control equipment. Furthermore it is possible to assign names to these peripherals. They are used later on in the problem part for input- and output operations.

The assigned names represent the logical connection between system part and problem part. Therefore it is possible to adapt a process control system implemented in PEARL easily to a new environment. Only the system part has to be modified to be compatible with the new environment.

The *problem part* contains the algorithms which solve the desired process function. Procedures, tasks and data are components of this part. Tasks enable the independent reaction to different events in the technical process. Each task can be activated either by an external or a time event. Furthermore it is possible to change the state of a task through the functions ACTIVATE, TERMINATE, CONTINUE and SUSPEND. In addition to the data types used by scientific languages, PEARL is equipped with data types for

- time expressions (clock and duration)
- string handling and
- input and output.

The available control structures are similar to those of other higher programming languages.

PEARL offers various facilities to solve the synchronization problem. The logical synchronization comprises synchronization of independent tasks and the activation of actions by events. Both operations can be described by means of semaphor variables. These variables are manipulated by REQUEST and RELEASE operations. The time-synchronization can be described by new data types (clock and duration) and new functions. The following statements give an example:

```
AT 10:30:00  SUSPEND TASK-1;
AT 10:00:00  EVERY 10 SEC UNTIL 11:00 ACTIVATE TASK-2;
AT 12:00:00  ACTIVATE TASK-3;
AFTER 10 MIN RESUME;
```

8.4.3 ADA

ADA is a programming language which includes many facilities offered by classical high-level languages such as PASCAL. It also has facilities often found in specialized languages. ADA is suitable for programming embedded computer systems. This language has the usual control structures of high-level languages.

It offers the following features:

- possibility of defining types and subprograms,
- support of the module concept,
- separate compilation and
- real-time programming with facilities to model parallel tasks, including exception handling.

Furthermore, systems programming is possible.

An ADA program consists of several units. A unit is either a subprogram (procedure or function), a package, a task, a generic subprogram or a generic package. A *unit* consists of a specification and a body. The specification contains the information which is visible to other units. It represents the interface of a unit. The body comprises the implementation details. All parts described within a body do not have to be visible to other units. The body is usually composed of a declarative part which defines the logical entities to be used in the unit and a sequence of statements which specifies the execution of the unit.

A *subprogram* is either a procedure or a function. Its specification part contains the name of a procedure or a function and a description of parameters. Every parameter must be characterized by its name and type. Additionally it must be specified whether it is an input, an output or an input and output parameter.

A *package* is the basic unit for defining a collection of logically related entities. Examples of such entities are data, types, a collection of related subprograms, a set of type declarations and associated operations. The specification of a package contains all components which are visible to all other units, unless they are specified as private.

A *task* unit is the basic unit for defining a task whose sequence of actions may be executed in parallel with those of other tasks. The specification of a task is composed of the name of the task and optionally a list of entries. An entry can be called by other tasks. The calling of an entry resembles the invocation of a procedure or a function. If the specification part of a task contains an entry, then the body must have a corresponding (the name must be the same) accept statement.

A generic unit (*generic subprogram* or *generic package*) is a template which may be parameterized and from which corresponding subprograms or packages can be obtained. A generic unit cannot be called, it can only be instantiated. The result of an instantiation is an instance which can be called. The generic feature enhances a modular approach because the same algorithm can be used for different types of data.

In ADA the rendezvous-technique is used to describe the synchronization between independent tasks. A rendezvous is realized by an entry call inside the calling task and the corresponding accept statement in another task. The following example represents a rendezvous between the tasks t1 and t2:

```
task tl is
  entry ALERT (MESSAGE: in INFO);
end tl

task body tl is
begin
  loop
    .
    .
    accept ALERT (MESSAGE: in INFO) do
           .
           .
    end ALERT;
    .
    .
    end loop;
end tl;

task body t2 is
MESS: INFO
begin
    .
    .
    tl. ALERT (MESS);
    .
    .
end t2;
```

t1 and t2 are independent tasks. The rendezvous mechanism is as follows: the first task to initiate the rendezvous (represented by accept ALERT in t1 and t1.ALERT in t2) waits for its partner. Then the calling task t2 is delayed until t1 has finished the exchange part (end ALERT). When the exchange is completed, each task continues its execution independently. The rendezvous technique is the only mechanism in ADA to describe the logical synchronization. It is also used to couple an event in the technical process with an action of the process control system.

For time synchronization, ADA offers two predefined data types, the duration and the time. It also has a delay statement which inhibits further execution of the task for a specified time.

In ADA there exist no statements which facilitate information transfer to and from the technical process. The problem can be solved by implementing new I/O packages. However, this requires a thorough knowledge of ADA.

ADA as well as PEARL (and other languages not mentioned) enable the implementation of real-time systems. PEARL offers a wider range of real-time functions, e.g. process I/O and task scheduling, whereas the strengths of ADA are its strong typing and its package concept which encapsulates data and the routines that modify these data.

8.4.4 Tools for the Development of Process Control Systems

The need to use tools for software development, especially for requirements specification, is no longer disputed. One of these tools, a high level software development environment, called SARS will be described [12].

The software development tools of the SARS system presented here make it possible to capture requirements in a form the user can understand, and to analyze them for completeness and consistency. In contrast to similar computer assisted systems, SARS offers the advantage of graphical input. Since it is well known that human beings can assimilate complex information much more easily when it is presented in a graphical form; this is a significant step in assisting the user.

Specifying a system is not a straightforward and well structured process. Roughly speaking, it is an iteration of three activities:

1. The collection of information about the system to be automated and its environment.
2. The structuring and storing of the collected information.
3. The analysis and improvement of the stored information.

SARS is a high level software development environment for process control systems. It stands for *S*ystem for *A*pplication Oriented *R*equirements *S*pecification. It consists of various media and a series of tools. These are the following principal components (Fig. 8.24):

- SARS Information System,
- MARS (*M*ethod for *A*pplication Oriented *R*equirements *S*pecification),
- LARS (*L*anguage for *A*pplication Oriented *R*equirements *S*pecification),

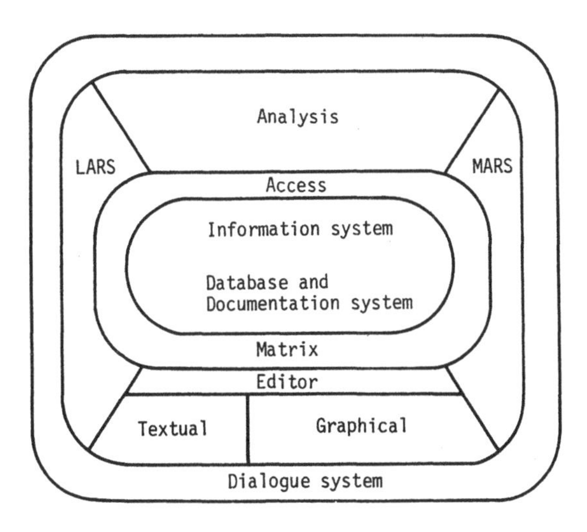

Fig. 8.24. Components of SARS

- Textual and Graphical Editor
- Dialogue System.

SARS Information System

The first step of a specification is a collection of ideas and informal requirements. The specification by means of SARS will start with the step by step transformation of these informal descriptions into a formal notation (LARS). The basis of informal requirements are ideas and instructions of the customers. The SARS specification also contains reports of project members (e.g. minutes, reports to the customer and notes from the system analysts).

The SARS Information System stores information about the project. Among them are:

- databooks,
- deadlines,
- project management information,
- personal data,
- first solutions,
- known methods and aids,
- references to information and
- data sheets.

Due to the iterative nature of a formal specification process there are usually several solutions (versions). Therefore alternate solutions are stored in a central database which is available to the user.

The nucleus of SARS is an information system that manages all data of the specification work station. The descriptive media of SARS are:

- formal language texts (LARS) and
- graphics.

There is a common data management system, which can be accessed by all software components, and which controlls all elements of the database. Underlying this database is a relational database, in which the specification is stored. The database is accessed through an access matrix (Fig. 8.25). It represents the user's logical view of the database. One column of this matrix is reserved for descriptions of the entire process to be automated. The other columns are generated by the project implementer. Their number depends on the number of subprocesses and are named according to the subprocesses.

The rows of the access matrix contain the basic components of the LARS language. This part of the access matrix cannot be changed by the user. To allow the user to store additional information like reports, notes and so on, further lines can be created which represent "private" classifications.

Pictures and tables represent the formal language. The graphic work station consists of a graphic display and a digitizer. Graphical representation aids fast information transfer to the user. The interrelations between LARS components can be seen immediately. Formal graphic input is syntactically analyzed and translated into the internal language representation. The actual transformation between LARS text and LARS graphics will be done by the system.

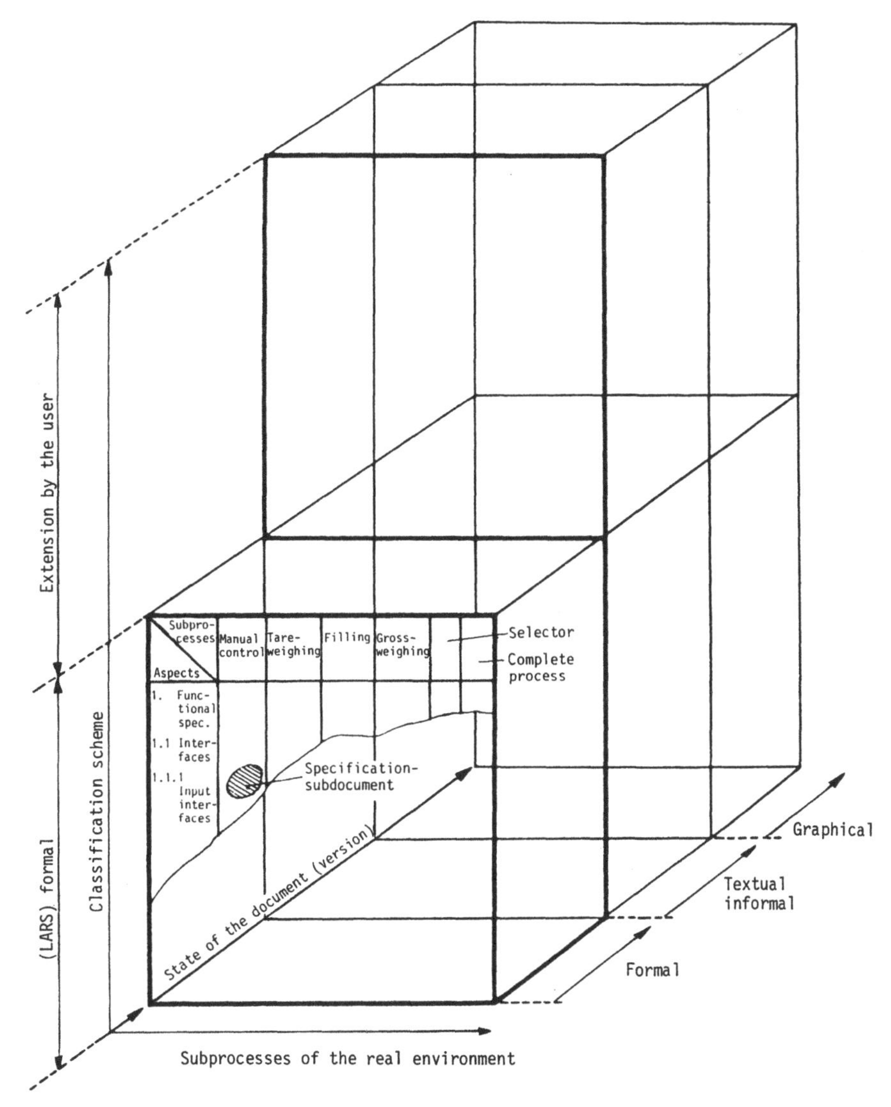

Fig. 8.25. Access matrix

The common data management for all descriptive media makes selected retrieval possible which is necessary to produce reports for different user groups.

The MARS Method

This method was the conceptual point of departure in the creation of SARS. The following two considerations characterize the method:

1. Every automation system to be developed – or more concretely, the software for that system-reflects the environment associated with the system. Thus, in

the office domain, for instance, the existing or future organizational structure is as much a binding aspect of a requirements analysis as is in process control the existing or planned structure of a technical process.
2. Every action to be performed by a behaviorel unit like a computer can be described with the Stimulus-Response model. Either it is evident that a causal event (stimulus) will have as an effect the reaction to that event (response), or an attempt is made with the aid of an analysis to trace a certain effect (response) back to an identifiable cause (stimulus). A simple example: On the birthday of the operator (stimulus: a particular date becomes valid) the computer is to print out a Marilyn Monroe poster (response: the printout).

Both principles determine the methodology of the requirement specification. The "requirement specification" defines all of the necessary stimulus-response (SR) functions according to the subprocesses. Although these SR-objects are still independent and neutral with respect to the later implementation, they can nevertheless be understood as autonomous computing processes. The SR function

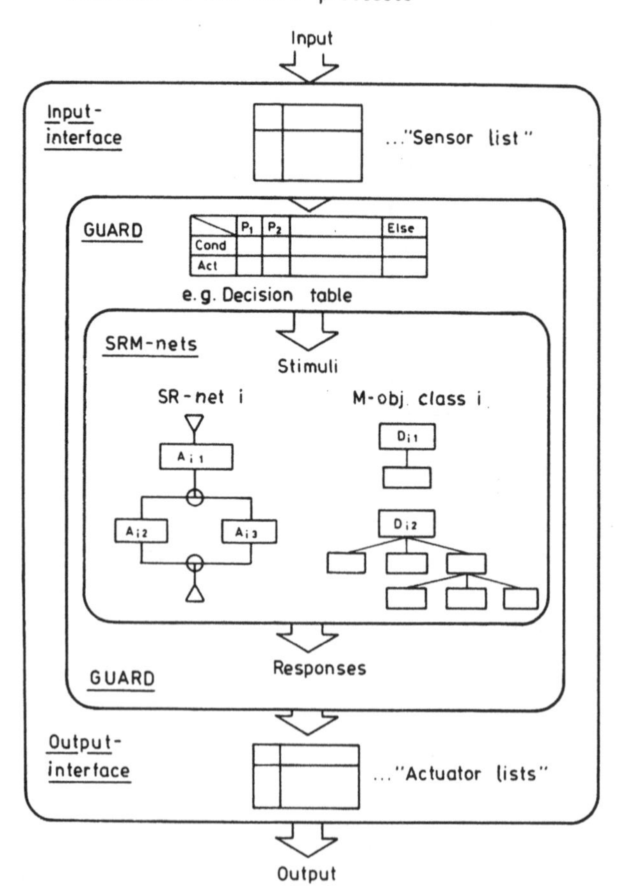

Fig. 8.26. LARS model

blocks are refined in further steps to nets. The syntax of these (Stimulus-Response-Memory) nets corresponds approximately to R-nets, which were published.

The LARS Language

The primary components of LARS are interfaces, SRM-Nets and Guards, Fig. 8.26. Through the Interfaces the user can model input and output signals in a way a non computer scientist is accustomed to. SRM-Nets are stimulus-response-memory nets, which can be regarded as basic functional components. Data objects are data and data structures which are used by the nets. The Guard has been introduced as a supervisory constituent, which controls the flow of input and output signals and data. It determines which data objects are transfered to which SRM-Nets and which nets must be activated or have to be synchronized.

The LARS language allows:

- a formal description of SRM-Nets and the data used by them (Net LARS),
- the definition of the interfaces between the control system and the external world (Interface LARS) and
- the description of the logical connections between the interfaces and the SRM-Nets (Guard LARS).

There exists in SARS a compiler for LARS, with the help of which it is possible to analyze requirements spezifications formulated in LARS for completeness and consistency. SARS contains two *editors*. One manipulates the textual representation, and the other manipulates the graphical representation.

Special characteristics of the SARS system are the graphical representation of information, the formal definition of the specification language and the harmonious integration of the various tools by means of a dialogue system.

8.5 Commercial Data Processing

Currently there are many different programming languages available which are used for commercial data processing, such as Algol, Basic, Cobol, Fortran, PL/1, Pascal, and RPG. All these languages provide concepts to describe the flow of control within a program. Usually they include conditional statements (IF THEN ELSE) and loops (a repeated sequence of statements). The language most frequently used for commercial data processing is Cobol. The features of this language will be described briefly:

Cobol

One reason for the widespread use of Cobol is its orientation towards the English language. All instructions are coded using English words. A Cobol program consists of four divisions namely:

- identification division,
- environment division,

- data division and
- procedure division.

The *identification division* serves to identify the program. It contains the program name and some organizational information, like author, installation, date and remarks.

The *environment division* assigns the input and output files to specific devices. It is divided into a configuration section and an input-output section. The configuration section supplies data concerning the computer on which the Cobol program will be compiled and executed. The input-output section comprises information concerning the specific devices used in the program. This is the only section which will change significantly if the program is to be run on different computers.

The *data division* consists of a file section and a working storage section. It describes in detail the field designation of the input and output file. Each field has a corresponding picture clause denoting the size and type of data. If the program requires constants or work areas, they are also contained in this section.

The *procedure division* contains all the instructions to be executed by the computer. It is detailed into paragraphs. Each paragraph defines an independent routine. The instruction set is comparable with that of an assembly language. But in contrast to assembly language, the word-symbols are easier to understand because they were chosen in accordance with natural English. From the scientific point of view Cobol is a poor language. It supports neither the module concept nor structured programming. The flow of control is realized by means of the harmful GOTO statement. There is no way to implement abstract data types. Nevertheless in practice there is a widespread use of Cobol.

In the near future Cobol will be replaced by languages like Pascal. It is very likely that languages currently used primarily in scientific and advanced industrial environments e.g. LISP and PROLOG will have a broader application. PROLOG will be the language of the Japanese fifth generation computers.

8.6 Future Trends

Currently the automation of the factory (material, tools, machines) and factory information processing is kept largely independent of each other. In the future both of these activities must be developed in parallel. Otherwise, it will be very difficult, if not impossible, to link together different computers running under different operating systems and programmed with different languages. It must also be possible to interface the manufacturing units with the computer. The need to improve productivity and quality is well recognized. But to be really successful in the modernization of manufacturing, or in a broader sense in the modernization of an entire factory, the following requirements have to be fullfilled:

- The mechanical, electronic and computer software worlds must be brought together. It is necessary to improve the communication between mechanical, electronics and software people.

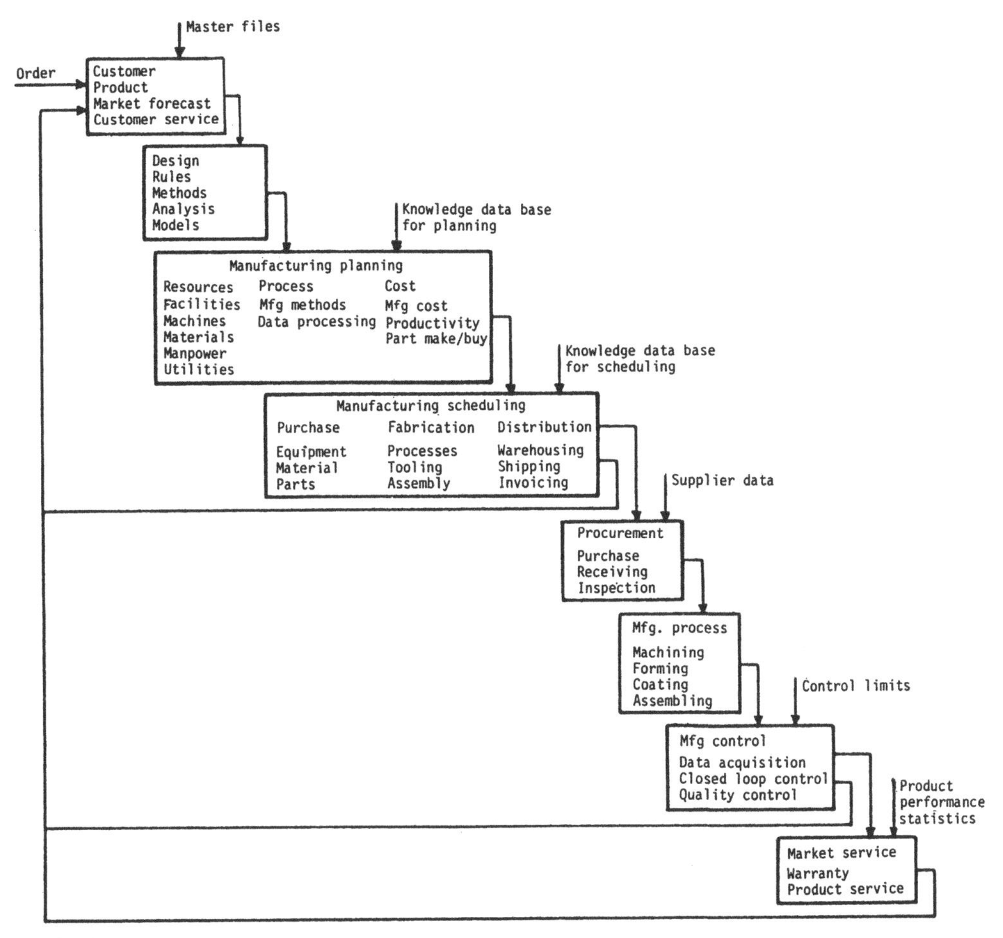

Fig. 8.27. The manufacturing activities

- Before an existing manufacturing process is automated there should be a thorough analysis. Single subprocesses should not be automated in isolation, however, the process should be studied, and if necessary modified in order to arrive at a global and homogenous solution.

An important part of such a solution will be realized by software. A programming language like ADA, which can be adapted to different applications by the definition of suitable packages can essentially contribute to a homogeneous solution.

At the beginning of the nineties expert systems will be used to perform many routine tasks of a manufacturing organization. The communication with the computer will be done with graphical, voice or fill-in-the-blanks programming aids. The computer will have decision rules to plan and control the manufacturing process. It will be able to learn from past and present operations and will plan with the entrance of an order all manufacturing activities. A typical integrated manufacturing system is shown in Fig. 8.27. The order may be processed as follows:

- Upon entry of an order to expert system determines if the item is a current product, a variant or new. The order from a current product passes design and directly enters manufacturing planning. For a new product or a variant the engineering department must furnish design data and manufacturing documents. Expert systems will be available which contain design rules, mathematical models and analysis tools. Depending on the degree of automation the design is done either completely automatically or with some manual assistance.
- In manufacturing planning the material, processes and process sequences are selected. The planning module will have information about the entire plant resources, manufacturing methods and fabrication alternatives. A cost analysis will be made as well as a make or buy decision.
- The product order and the manufacturing plan activates production scheduling. There are long and short term schedules which are used to meet the manufacturing due dates. The new order will be queued into these schedules. A knowledge data base contains information on all orders, available inventory and the status of the manufacturing equipment of the plant. Equipment scheduling will be done with the aid of group technolgoy and Operations Research tools. In the case of problems with due dates or manufacturing equipment an alternative manufacturing schedule will be conceived.
- At the next stage tooling is activated and procurement issues orders for parts and raw material from outside vendors. When the order is released for manufacturing the control computers and data collection devices will supervise the fabrication process. There will be mechanisms to evaluate the capability of processes, to automatically set up test limits and to detect faulty equipment.
- The final activities are quality control testing and distribution of the product. Quality control data from the process and warranty information from the customer will be important feedback information for the expert system to maintain an efficient and a high quality product.

An expert system which can autonomously solve many of these manufacturing functions may be conceived as follows:

- It should understand the problem description and the requirement specification.
- It must be able to synthesize processing procedures.
- It must be capable of optimizing between the machine system functions and the processing procedures.
- It must be able to synthesize responses based on the outputs from the machine system.
- It must understand verbal and graphic instructions via natural means of communication.

An expert system with these features must have following capabilities:

- It must understand the language used at the man-machine interface.
- It must know solutions to the problems to be solved.
- It must know the machine system.

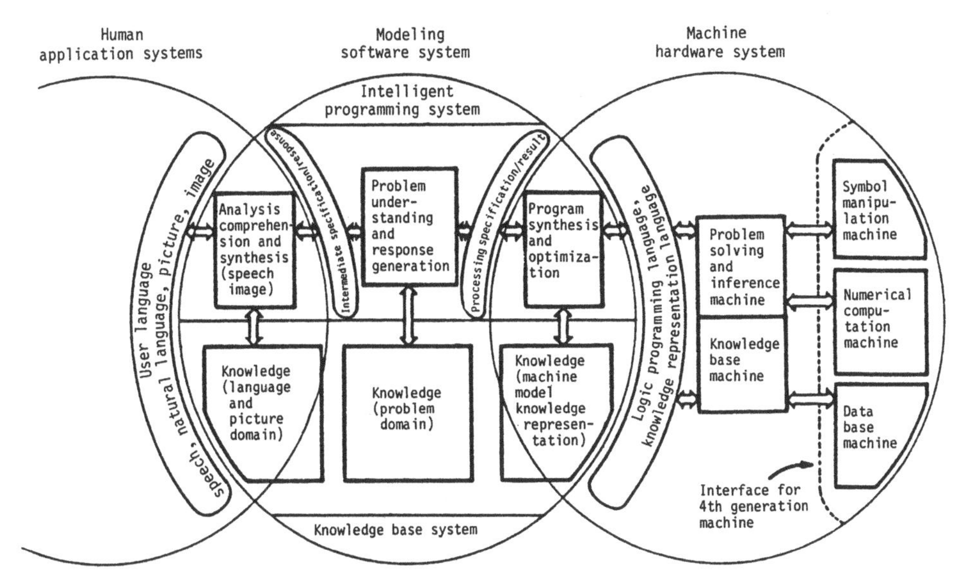

Fig. 8.28. The 5th generation computer concept

Figre 8.28 [14] shows the concept of 5th generation computers. It consists of application software, system software and computer hardware. The core of this computer is the system and application software. With the help of these it is possible to instruct the hardware, for example, to optimize a manufacturing plant for a given product spectrum.

8.7 References

1. EXAPT-NC-programmiersystem. EXAPT-Verein, Aachen, Germany
2. Langebartels, R (1969) Programmierung von numerisch gesteuerten Werkzeugmaschinen mit den Fertigungstechnisch orientierten Programmiersprachen APT und EXAPT. VDI-Z Bd III, Nr 12
3. Programmierplätze für NC-Bearbeitungsprogramme: System P-D/P-F SYMBOLIC FAPT TURN. Siemens Corporation, Munich
4. NC Machining by the Numbers. Manufacturing Engineering, May 1979
5. Leonards, F et al. (1976) Prozeßlenksysteme für die Drehbearbeitung. PDV Berichte, Gesellschaft für Kernforschung mbH Karlsruhe, Aug. 1976
6. Rembold, U, Blume, C, Dillmann, R, Mörtel, K (1981) Technische Anforderungen an zukünftige Montageroboter, Part 4. VDI-Zeitschrift Nr 21, Nov. 1
7. Bonner, S, Kang, G (1982) A Comparative Study of Robot Languages. Computer, Dec.
8. The Advent of Adaptable Programmable Assembly Systems. Manufacturing Engineering, April 1979
9. Blume, C (1980) VAL – a Robot Control System of Unimation. University of Karlsruhe (unpublished)
10. Mujtaba, S, Goldmann, R (1979) AL User's Manual. Stanford University
11. Dillmann, R (1983) Graphical Emulation System for Robot Design and Program Testing. Conference Proceedings of 13th ISIR/Robots 7, July 1983, Chicago, Ill.

12. Epple, W, Koch, G (1983) Specification of Process Control Systems with SARS. Proc. IFIP TC-2 W.C. on System Description Methodologies, Kecskemet, Hungary, 23–27 May, 1983, 6. David (cd). North-Holland Publ. Co.
13. Alford, MW: A Requirements Engineering Methodology for Real Time Processing Requirements. IEEE Trans. on Software Engineering, vol SE-3, pp 60–69
14. Motorola, I, et al. (1981) Challenge for Knowledge Information Processing Systems. Proceedings of the International Conference on Fifth Generation Computer Systems, Oct. 19–22, 1981, Japan

9 Quality Assurance and Machine Vision for Inspection

P. Levi

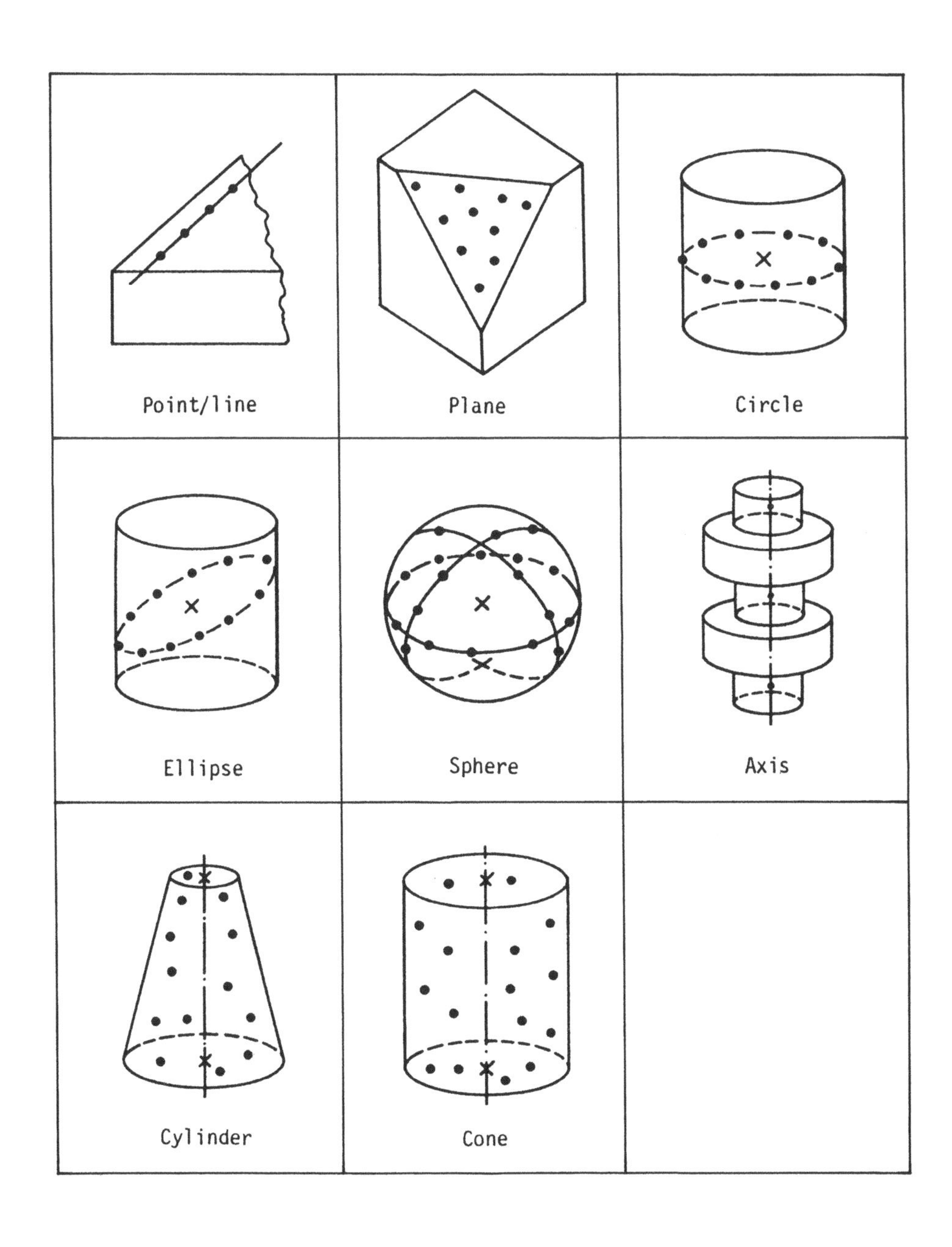

Measuring various solids

9.1 Introduction

Quality control (QC) has been recognized for some time as a crucial point in a manufacturing system. This activity determines the deviations of the quality of actually produced workpieces from the specified product quality. The quality standards are set by the market (customer, standard, industrial acceptance), by the legislator (producer liability, authorities, safety provisions for workers), and last but not the least by the enterprise (failure rates, costs, profit, image). The increasing complexity of products and installations, as well as the rising quality judgement of the customers accentuate the importance of the product quality. Quality control has to be performed at many different levels of a manufacturing process. Quality control testing of the finished product is only a part of it.

Reliable and efficient quality control must be performed as an entity which includes all phases of the life cycle of a product: planning, purchasing, manufacturing, acceptance testing and the use of the product. This entity is called *quality assurance* (QA). It integrates all quality control activities into one system. The origin of product defects can be attributed to the following production phases:

40% planning and development cycle
40% purchase cycle
20% manufacturing cycle

This fact emphasizes once more the need for a comprehensive quality engineering approach.

Quality assurance is usually structured in a hierarchical fashion, which is similar to other architectures of organizational activities. At each level, the control activities to be performed, as well as the data to be collected and evaluated, differ considerably. In future manufacturing facilities, hierarchical computer systems will be employed to supervize an integrated quality assurance system. The ever increasing complexity of the products, the large model diversification and the increasing complexity of the manufacturing process will need the use of these distributed systems, otherwise quality assurance will be the bottleneck of the enterprise. To date, not all the building blocks for a quality assurance system are available. Methods have to be devised to integrate quality assurance procedures directly into the fabrication process to gain more flexibility in manufacturing. A considerable amount of research is yet to be performed to arrive at an optimized distributed data base in which data manipulations and communications are reduced to a minimum. Updating of files presents another problem. If updating is not properly done the designed system can be subject to severe difficulties, when a computer or other important parts of the system fail.

The hierarchical structure of a QA system is influenced by the configuration of the local network acquiring operational data. It also depends on the equipment configuration conceived for the different quality control levels. It is also influenced by the problem-oriented languages used for programming of the production systems. Depending on the desired degree of automation the equipment configuration for QA can be very different. There are three measuring methods used to enter parameters into the measuring system: manual input, contact measurement and non-contact measurement. The last type of measurement plays an ever-increasing role in the automation of visual inspection and robot vision.

This contribution is divided into two parts. The first section is devoted to the specification of quality assurance, to its functions and to the implementation of an computer integrated quality assurance (CIQ) system. The second part is concerned with machine vision techniques for automatic inspection with laser based measurements and inspection, and last but not the least with robot vision for parts recognition and inspection.

9.2 Quality Assurance: Functions, Problems and Realizations

9.2.1 Quality Assurance Functions

Quality is defined by the sum of all attributes and characteristics of a product or an activity which contribute to their ability to perform specified functions.

This definition of quality should be extended by three additions.

(a) The quality is determined by the correctness of the planning cycles and the workmanship which contributes to the quality of the product.
(b) A product is any kind of a device, a raw material, a part or a draft or design. An activity is any kind of service, but is also a sequence of operational steps of a technique or a process.
(c) The specified functions include security-measures, pollution control and an appropriate use of tools. These functions are determined by the application of the product or by the goal of an activity. Furthermore they must be feasible.

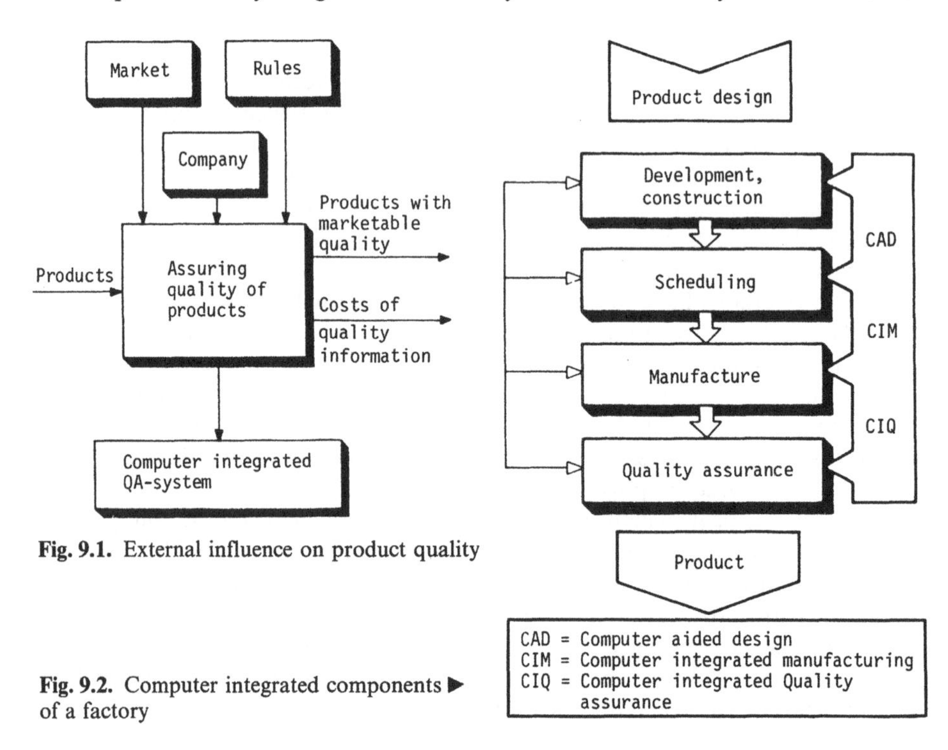

Fig. 9.1. External influence on product quality

Fig. 9.2. Computer integrated components ▶ of a factory

Quality assurance is a regulative process to measure quality performance, in order to compare it with standards and, corrective actions are initiated if necessary. The QA methods of an enterprise are influenced by the market, national and international standards, governmental regulations and the company's objectives (Fig. 9.1). Adherence to the standards which become a contractual matter between the manufacturer and the customer is the aim of a quality assurance system.

As quality assurance systems integrate computer controlled test methods, they should be equipped with a flexible and fast data acquisition system and based on a consistent information system. In order to be efficient QA system should be interfaced with the other computer integrated tools of a factory: CAD and CIM. Figure 9.2 sketches this connection.

Defects may be caused by any active part of the manufacturing system. For this reason quality assurance must be considered as a complete system which is integrated into the organizational concept of a company. Such a quality assurance system can be considered as a complex, adaptive control loop [1]. When the cause of a defect and its location is known, measures for its correction or possible elimination should be initiated. Corrective actions can be performed directly at the manufacturing process or may reach far back into the product design.

An integrated QA system operates as follows (Fig. 9.3):

- The manufacturing equipment constitutes the process to be controlled and it is part of the control loop. Disturbances influencing the process derive from: men, material, machines and methods.
- The local data aquisition system for quality control determines the actual value of the controlled variables.
- The controlled variables are compared with set points obtained from predefined quality specifications or standards.
- For the correction of any deviation from a quality standard a new manipulated variable is calculated and if necessary, a corrective action is performed to adjust the manufacturing process.
- The function of the controller is performed by manufacturing planning and scheduling.

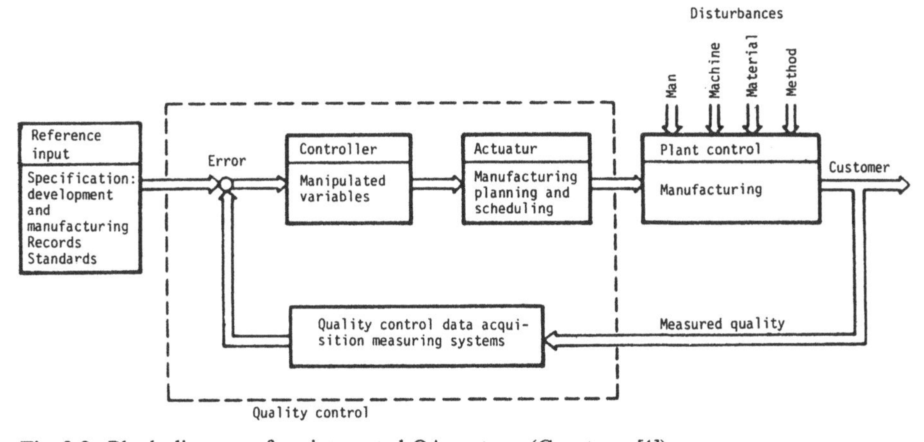

Fig. 9.3. Block diagram of an integrated QA system. (Courtesy: [1])

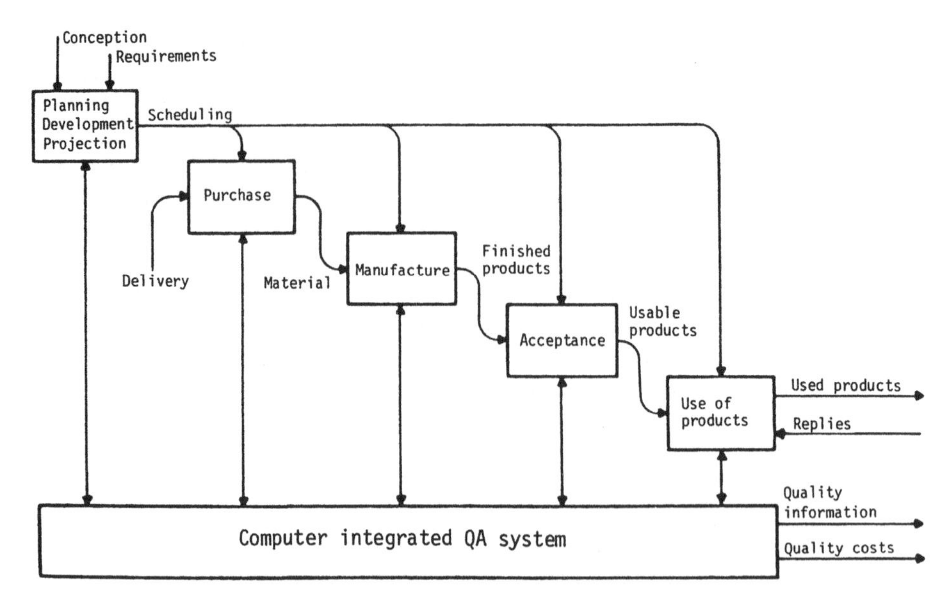

Fig. 9.4. QA function distribution within a company

With the help of the sensors of a QA system all properties of the product and its performance are identified. This can be done at many places in a factory, such as receiving inspection, at the locations where parts originate in the manufacturing process, at subassembly stations and during the final acceptance test of the finished products. The distribution of the QA functions in manufacturing is shown in Fig. 9.4.

Quality control must be done during every production cycle of the product. Five different types of quality definition can be specified:

(1) *design quality*: planning, development, layout, design
(2) *supply quality*: purchasing, material reception
(3) *manufacturing quality*: manufacture, assembly
(4) *delivery quality*: acceptance, final test
(5) *user quality*: sale, user installations, use of the products

When a product has a defect, the following questions are of interest: type of defect, place of occurance, seriousness of defect and cost of defect. The problem may lie in any of the five production cycles of a product. To improve the quality the following three important functions must be performed.

(1) *Quality planning*: goal, techniques
(2) *Quality loop*: control loop for defect location, defect analysis and corrective actions
(3) *Quality cost*: minimizing the product cost (error protection, test, reject).

The principle of a control loop and a quality control loop are shown in Fig. 9.3 and Fig. 9.5 respectively.

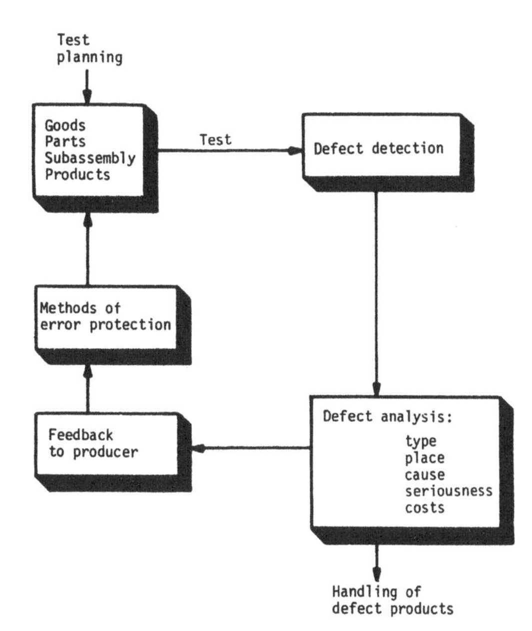

Fig. 9.5. Quality control loop for production cycles

This control mechanism itself is again subdivided into five parts: measurement of quality characteristics, comparison of the nominal and actual value of quality characteristics, defect detection and feedback, defect analysis and initiation of activities for error protection.

9.2.2 Design of a Computer Integrated QA System

Local hierarchical networks have been in use for quite sometime. However, most systems are tailored to very specific applications. Quality assurance procedures which are performed in different factories are not the same even in similar operations. Similar to other manufacturing activities, a very systematic approach must first be made to find common building blocks which form the quality assurance operations. The current trend towards model proliferation makes it necessary to change production runs, often several times a day. In the automobile industry for example, one can hardly talk about mass production any more. Cars are customer-ordered and different parameters have to be checked with each car coming from the production line. Thus the quality assurance system has to respond dynamically to changes in production [4].

While designing the data base management system for QA the different requirements of the different hierarchical levels have to be taken into consideration. General and common data should be kept in a master file in the executive computer. In order to eliminate bottlenecks, frequent data transfers over long distances, as required for example in data acquisition systems, should be avoided. On the other hand, the local quality control operators at the end terminals, residing at the lowest hierarchical level, need very specific information for an acceptance test to be performed. These data should be stored in a decentralized manner.

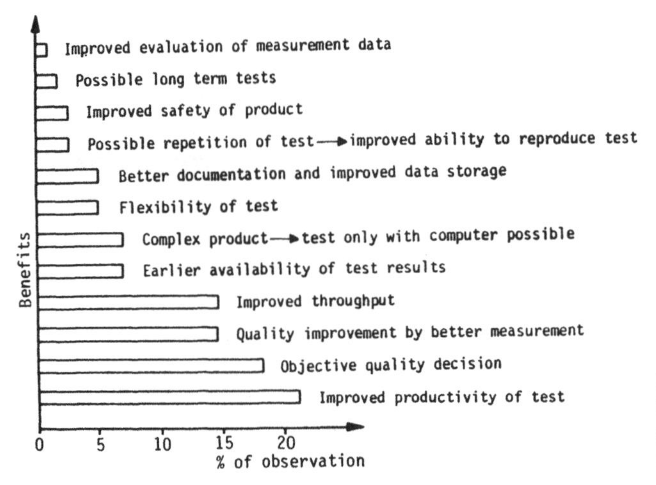

Fig. 9.6. Frequency distribution of benefits of a computer controlled test system. (Courtesy: [1])

The data structure and the design of the distributed data base [5] must be specified. The question is whether a relational model is appropriate [6], or a hierarchical or a network model is more suitable [7]. Protection mechanisms and the integrity of the data bases have to be resolved [8]. Furthermore, from the user's point of view, the types and the dynamics of information retrieval and the queries must be defined.

A data acquisition system is an important part of a quality assurance system. It is the basic building block for the information management system and it is necessary for the evaluation of test data. The architecture of this system defines the structure of the computerized quality assurance installation. The data collected by a data acquisition system are called (factory) operational data. These are data which are generated and used in the course of the manufacturing process. They are acquired at their origin (manual/automatic) and at the instant they are created.

When computer controlled test methods are introduced it may not be good practice to duplicate conventional tests. These tests were usually developed to perform limited repetitive functions. Adaptive decision making, as well as large data storage capabilities, are not available in the measuring equipment. The computer offers the possibility of conceiving new and improved tests which are impossible to perform with conventional equipment. Thus, whenever a computer is introduced a possible redesign of the test method and test apparatus should be investigated.

A systematically and well developed computerized test system has many benefits. Figure 9.6 depicts results obtained from an industry survey in which approximately 100 users were asked about the benefits they had experienced with this equipment. The most important benefits are improved test productivity, more objective quality decisions, better and improved measurements and improved throughput.

Depending on the desired degree of automation, quality assurance activities can be supported by different, local computer configurations (Fig. 9.7).

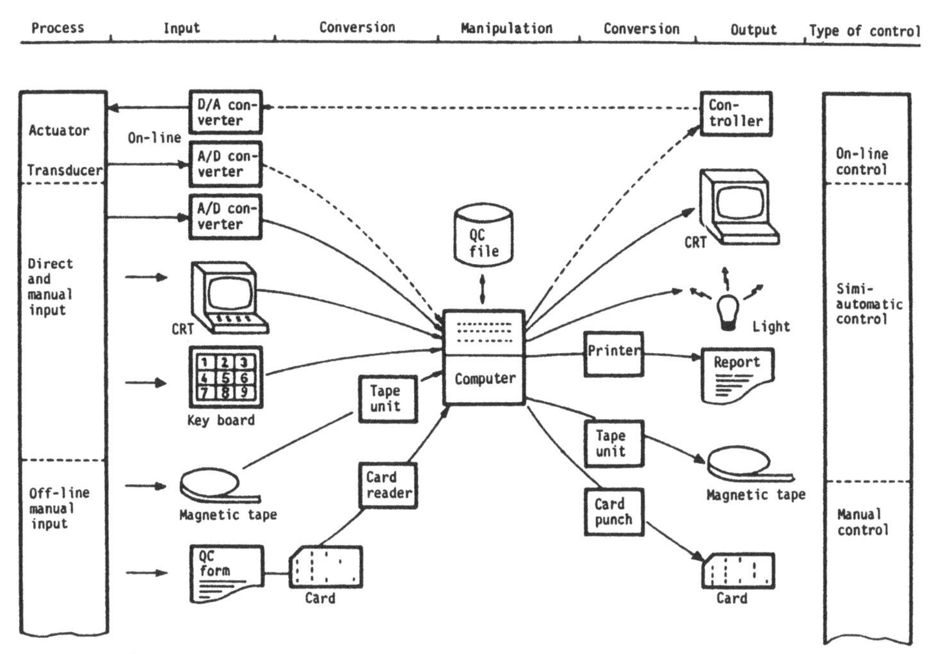

Fig. 9.7. Different equipment configurations for a quality assurance station. (Courtesy: [1])

In a simple application quality control, observations are recorded by an inspector on a QA form. The contents of this form are transferred to a punched card and read into the computer. Data are evaluated and stored on a central QA file. The operator is provided with a report, generated with the help of a printer. The operator will act upon the report and will control the process manually.

The other extreme is the installation of a completely automatic quality assurance system which senses process parameter changes with a transducer. An actuator performs the desired process corrections and in this way generates the feedback. In general, this extreme solution is very expensive. For this reason, in practical installations often a less sophisticated quality assurance system with some manual interactions is selected. The manual interactions, however, may be a source of problems. In addition, in quality assurance there are many activities which cannot be completely automated, since there is no adequate sensing equipment available, i.e. the detection of scratches or impurities on a coated surface.

The performance of a computer integrated quality test is a rather complex process and does involve different steps. They are the recognition of the product, positioning of the sensor, acquisition of test parameters, evaluation of test results, calculation of a quality index and the output of test data. If all these operations are done completely automatically, the programming language must have capabilities for controlling the test equipment, supervizing the input/output of data and performing statistical calculations. The first two groups of these functions require instructions which can easily manipulate individual bits; whereas, statistical calculations require a language which can readily do mathematical calcula-

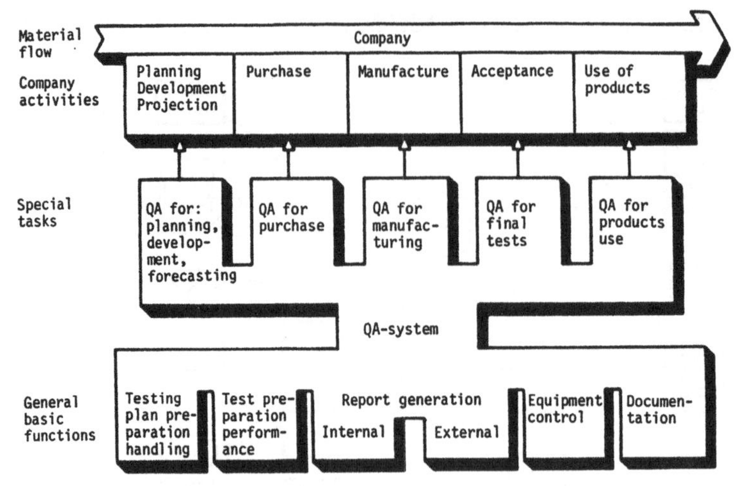

Fig. 9.8. Planning modules of a CIQ system

tions. Therefore, the selection of the programming language is determined by factors such as the type of test, the size of the available computers, execution time of the programs, software portability, flexibility, etc.

The language spectrum offered to meet these requirements ranges from machine oriented languages (e.g. assembler) to problem-oriented languages which are user-friendly and have several aids such as user guidance, menus to select commands and pictorial data on a monitor.

To summarize the introductory discussion, consistent planning and installation of a computer integrated quality assurance system must take the following points into consideration:

(1) Planning and scheduling of QA
(2) Software assurance and hardware reliability
(3) Architecture of the local network for QA
(4) Information management
(5) Data acquisition system
(6) QA methods and test procedures
(7) Equipment configuration for QA measurements
(8) Computer language for test application

The planning and the scheduling of a quality assurance system must offer the following five basic functions (user interface): creation of testing plans, performance of testing plans, reporting, equipment control and documentation (Fig. 9.8).

More details relating to point 1 are stressed elsewhere in this book. Point 2 will also not be discussed in this chapter. However, it should be stated, that software assurance is an important field of software engineering and has many similarities with quality assurance [9]. The remaining points will be described in more detail. The information management problem will be described in connection with data acquisition systems.

9.2.3 Hierarchical Computer Systems for Quality Assurance

A hierarchical control system for a plant can be divided into three parts: planning, control, and acquisition and preprocessing of data (Fig. 9.9).

In analogy to Fig. 9.9 there are three levels of computers employed in advanced quality assurance systems [10]. Typical tasks to be performed by each level can be outlined as follows.

At the executive level, general administrative quality control functions are performed to supervise the quality acceptance of a product by a customer and to compare it with the quality planned by the firm. Special functions at this level are:

- Monitoring of quality for each product and component
- Calculation of cost of quality for each product and component
- Administration of a master file
- Monitoring of reliability data for competing products.

The computer at the next lower level (host computer) monitors the functional integrity of the quality control equipment, performs statistical calculations and aids programming of the computers at the lowest level. These functions are listed below.

- Coordination of measuring devices
- Functional testing of measuring equipment
- Reporting system errors
- Automatic change of test procedures
- Storage of test data
- Automatic trend analysis
- Optimization of test parameters

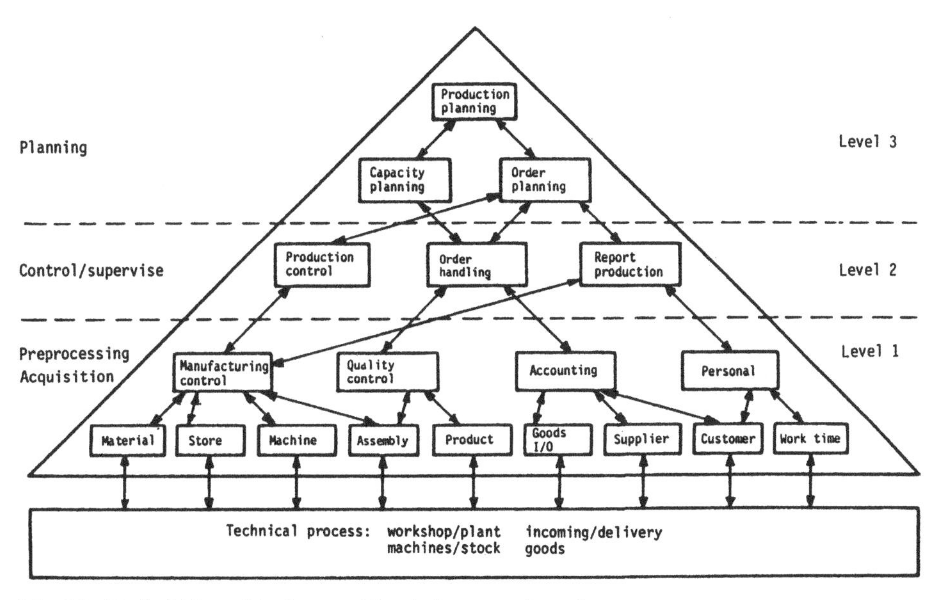

Fig. 9.9. Typical hierarchical control levels for manufacturing

- Calculation of quality indices
- Comparison of test results obtained from parallel tests
- Recognition of errors of different test personel
- Accumulation of statistical data
- Automatic determination of number of tests to be performed
- Location of manufacturing difficulties.

At the lowest level the quality control function is directly concerned with the product. For this reason it is almost impossible to conceive a universal computer system at this level, since the requirements vary considerably. Fundamentally, there are four different functions to be performed to control a test system. These are:

- System control functions
- I/O of operator and control data
- Operation and control of the test program
- Processing of measurement data.

All four functions can be performed efficiently by microprocessors. They add to the flexibility of the test system. The test program can be quickly adapted to accomodate the test object and the instruments. With local processing of the test data the results are immediately available and a higher throughput is obtained.

In general, the benefits of a hierarchical system (also of a local network) can be listed as follows.

(1) *Function sharing*. The entire technical process is too complex and too vast to be controlled by one computer. The process must be divided into subprocesses, which are handled by dedicated computers. In practice, sharing is realized by resource sharing, data sharing and it is often extented to load sharing.
(2) The assignment of different functions to different computers decreases the logical *complexity* of the system.
(3) The *reliability* of the hardware and software components increases. For example, the operating systems for the dedicated machines are smaller, simpler and their structure is more understandable.
(4) More *flexibility* is obtained by an increased adaptability of the whole system.

The communication between different computers in such a network is done at the process (task) level. A data transport system defines the tools for the interprocess communication and synchronization. These tools must be realized by the distributed operating system. It integrates synchronization concepts which are tailored to the networks [11]. The topology of a distributed system can be very sophisticated. The data transfer between tiers is handled in a master/slave mode.

9.2.4 Architecture of a Data Acquisition System

Structure

The basic function of a data acquisition system is the collection of operational data and the processing of these data. The system collects data about production,

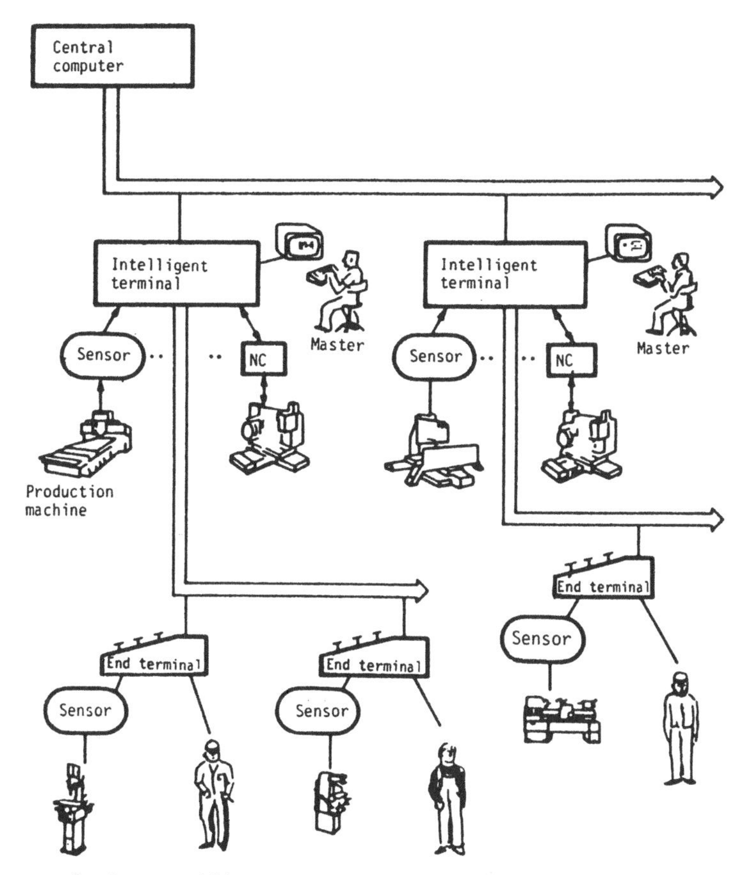

Fig. 9.10. Architecture of a data acquisition system

storage, assembly and quality control. It also communicates with product engineering, production scheduling and control. Typically, the architecture of a data acquisition system forms a hierarchy of interconnected terminals with different intelligence, as shown in Fig. 9.10 [12]. In general, advanced systems use three tiers.

Level 1

At the lowest level operational factory data are acquired by end terminals. Their task is to collect data from keyboards, punched cards, badge readers, document readers, magnetic code carriers and input devices for digital signals. The keyboard has three types of keys: alphanumeric, enabling and function keys. The output of data from this terminal is visual information via a screen (display) or via illuminated function lights or digital signals. If requested, the output can also be obtained on a printer. Simple peripherals are used at the end terminals. Some end terminals can be installed as stand-alone acquisition devices. However, their usefulness is questionable since they only collect data and do not use the capability of the computer to perform calculation.

Level 2

Usually, several end terminals are connected to an intelligent terminal at the next level of the hierarchy. It obtains data needed for issuing instructions to the operators at the end terminals. It handles and preprocesses the raw data from the lower tier and stores them in a local memory or in a remote file at the central computer. In contrast to the end terminals, intelligent terminals can communicate directly with each other (cross communication). The peripherals connected to these terminals usually have complex functions such as dialog or printing capabilities.

The intelligent terminal at this level may also be a full grown computer. In this case it may serve as a management information system operating in open loop.

Level 3

Finally, all intelligent terminals are connected to the administrative computer or the master. Its task is to control management, the principle processing of data and their long-term storage; it also monitors the activities of all party lines. This master function can be transferred to a predetermined intelligent terminal if the master station is not available. Connection to a large planning computer is also possible. Smaller systems can operate without intelligent terminals; in this case the end terminals are connected directly to the central computer.

A data acquisition installation controlled by a master may be configured as a factory control system. In this case the manufacturing processes are supervised in a close loop control mode.

The data communication link topology is of non-switched connection. The intelligent terminal obtains messages from the end terminals by polling, and in turn, the central computer polls the intelligent terminals. Two types of buses connect the terminals:

- a fast bus which connects the intelligent terminals with the center; it operates at about 200 kBaud.
- a slow bus, which connects the end terminals with the intelligent terminals; it operates at about 10 kBaud.

Both may be of the PDV bus design. The activity on the party line is controlled by the PDV-data communication procedure [13].

Transaction Language

The dialog with the end terminal and that with the intelligent terminal will differ considerably. Operator communication at the end terminal will be done via a display. It allows formulation of tasks and system communication such as instructions to the operator at the end terminal or requests for shift reports. The dialog of the operator with the end terminal is characterized by a rather rigid procedure issued by the intelligent terminal. This dialog is referred to as operator guidance. A problem-oriented language for manufacturing control including commands for terminal tests and for operator guidance is used to conduct the dialog between the master computer and the operator. The language can handle in-process control and human/computer communications. It is based on a transaction concept [14] which provides preplanned instruction sequences for a machine or an operator. They are of logical design. A transaction is a complete operation to collect operational data from a manufacturing machine, to process the data in the end terminal and to transmit it to the intelligent terminal. In order to reduce the error rate, the commands of this language are of simple structure.

A distinction is made between two elementary types of transactions depending on whether they are of internal and external nature. The *internal* transactions are used for the control of the transaction interpreter, which is located at the end terminal. Examples of this kind of transaction are the reset of an end terminal and the revocation of a transaction. The *external* transactions are further subdivided into two classes: the mask and I/O transactions.

The *mask* transactions are used for the man/machine dialog via a display terminal. The image shown on the screen is divided into output and input fields. The output fields are used for guiding the operator at the end terminal and cannot be changed. The input fields can be filled in at random. Mask transactions are employed for the generation of forms [15].

The *I/O* transactions activate peripherals connected to an end terminal. This can be done in a strict sequential or in parallel mode (default). The parallel I/O transactions include concurrent operations of several devices determined by one transaction or the use of one device by several transactions. Twelve I/O commands have been identified. The most important (8 commands) are listed below.

input Initiation of the data collection from input devices. The devices are specified by their full names.

move This command allows transfer of data from input devices to output devices.

check This command offers the possibility of verifying the input data. It is always combined with an input command. If the check result is true, then the input data is transferred to the private buffer of the input device. If the result is not true, then this part of the transaction is marked as erroneous and the transaction is continued by a case statement.

output This command is similar to the input command. Data received from the intelligent terminal are placed on output.

collect All input data of a transaction are transferred to the intelligent terminal, initiating the transaction.

buffer This command is used to collect all input data of a transaction which have not yet been checked and to send them to an intelligent terminal.

exclusive This command allows a transaction to request exclusively peripheral devices (input, output), later it releases them after use.

sequence This command constrains all different operations of a single transaction to a strict sequential execution.

Data Handling Schedule and Data Structures

A data acquisition system, e.g., the one used for QA, collects information from different local subsystems (stations) and must contain a distributed data base which can be accessed by different users, such as manufacturing or assembly personnel. Many basic functions are very similar for both systems. These are:

- Dialog oriented data manipulations, e.g. form generation
- List controlled data acquisition (the process of taking information from an order list)
- Order processing
- Graphical protocols.

The solution of the problems can be done in a similar fashion in both applications. Thus three levels of a software control hierarchy can be implemented on the intelligent terminals (operating systems). The highest level is the command tier. It receives and decodes orders coming from a dialog device or from the control computer and initiates the appropriate actions. The next lower tier is defined by the user programs. It guides all terminals according to a unique schedule. Here, the semantics of the acquired operational data is ignored. The data evaluation program represents the lowest tier in this control hierarchy. It is activated if a transaction is prepared, and is specific to every application.

The handling of operational data coming in from factory floor is performed by a four level approach (Fig. 9.11).

The rate of data arriving from the end terminals is very high. For this reason they cannot be sorted instantly according to their origin and information contents. This is why, as a first step, data is stored temporarily in an accumulation file. In the following sorting process (compression, distribution) the input data are transmitted in a sorted fashion to a permanent working file. Operator queries transfer these data into temporary report files. Finally a printer, a screen or a similar output device, is used for the output of the sorted data.

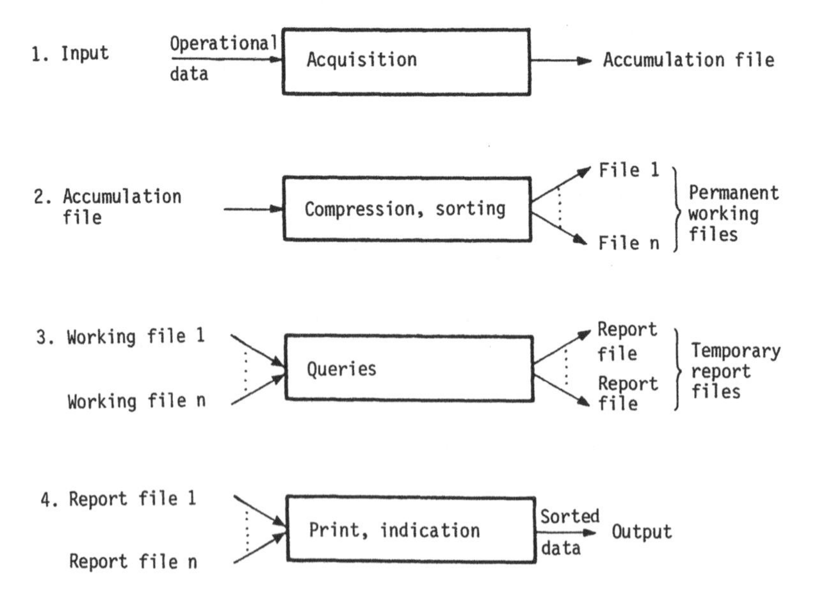

Fig. 9.11. Data handling schedule

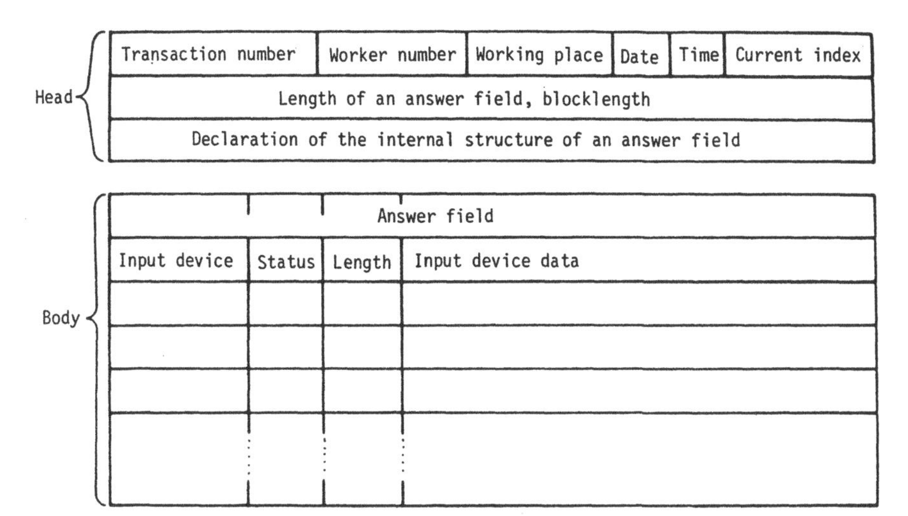

Fig. 9.12. Data structure of a permanent working file

Every transaction is divided into a header and a body. In addition, the data structure which is constructed on an end terminal as a response to the request of an intelligent terminal shows the same form as that of a transaction. The response header characterizes the type of response and identifies the transaction. The response body is built up from the usable data of the response and has four parts which describe the input device, the status (result of a transaction), the length of the data and the data themselves. A permanent data structure is located in the intelligent terminals (permanent working files) to aid storage of the responses of the supervized end terminals. It is also subdivided into two parts: a header and a body (Fig. 9.12).

The header carries the identification and the control information which defines a unique assignment of the response data. The body saves the response data which is evaluated by a special program.

The Operating System

A data acquisition system is typically a distributed system. For this reason, two operating systems for the two terminal types have to be implemented. The basic requirements for the structure of these two operating systems are the same. Both systems must meet real time requirements and have to support multitasking in order to be able to handle asynchronous events. These two requirements can be satisfied by dividing the operation systems into a kernel and a shell. The same kernel can be used for the end terminals and the intelligent terminals. The shell of an end terminal accomplishes the following basic tasks: scheduling of the peripherals, communication, buffer management for messages, transaction interpretation, and clock management for wake-up requests. The shell of the intelligent terminal has, in addition, a dialog device management, a volume management and a virtual file system. The control hierarchy mentioned before (e.g. user guidance) is implemented on top of this shell.

9.2.5 Quality Assurance Methods

When a QA system is designed an effort should be made to divide the product spectrum into similar groups or variants, and to design for each of these groups a universal test system. Typical groups are motors, switches, thermostats, actuators etc. These universal test system should be self sufficient so that they can be used by different departments. They also have an advantage when they are connected to a computer network. When the main computer fails the individual test systems can still be operated.

The computer can be programmed so that the measuring system is checked at different time intervals. Thus a failure or an inaccurate measurement can be detected early. Likewise it may be of advantage to calibrate all transducers before a test is initiated. The computer can also do automatic correction of test results, i.e. compensation for temperature drifts.

The configuration of the measuring system and the type of test to be performed depends on the product and on the rate at which it is tested. For this reason, the test concept and the test system must be integrated into the manufacturing process. The different steps necessary to perform the test are as follows:

- Recognition of the product
- Positioning of the sensor
- Acquisition of the test parameter
- Evaluation of the test result
- Calculation of a quality index
- Output of the test result.

The degree of mechanization of these individual components depends on the desired degree of automation for the entire system. For a simple product with a low production rate it may only be necessary to perform automatic data acquisition. High production rates may require that all functions are completely automated. In accordance with these two extremes there will be different requirements for every test system.

Automation of the recognition function can be done with the help of light barriers, binary coding or machine vision. These aids are also of importance to robots. For many measurements, accurate contact positioning of the sensor on the test object is necessary, i.e. for temperature or vibration measurements. This may be an awkward task on moving objects as they are commonly encountered on conveyors. There may be very complex fixtures necessary for positioning. To circumvent this problem, non-contact sensors are frequently used.

Acquisition of test parameters, the evaluation of the test results, the calculation of the quality index and the output of the test results should be done by the computer. These functions can be performed by it quickly and very efficiently. Automatic test systems are the more efficient the more they are automated. Most errors during testing are introduced by man. For this reason a high degree of automation may be desireable for a test system.

9.2.6 Measuring Methods for Quality Assurance

9.2.6.1 Contact Measurement

In order to obtain readings of high accuracy the contact measuring method is preferred. Typically temperature, vibration, dimension, pressure etc. are measured with this method. Dimensional measurements play a predominant role in part manufacturing. For this purpose, different computer controlled measuring machines are available.

Figure 9.13 shows a two column universal measuring machine of the Carl Zeiss Company, measuring an engine block [16]. The measuring table, all guideway elements and the crossbar are made of quality granite for high dimensional stability. A universal 3-D probe measuring head is attached via low friction air bearings to the crossbar. It can be moved under program control along the surfaces of a workpiece. During this movement the dimensions of the workpiece are recorded with the help of a probe to an induction or optical measuring system. A computer records these measurements and calculates the dimensions, in reference to a given coordinate system. Depending on the application and the size of

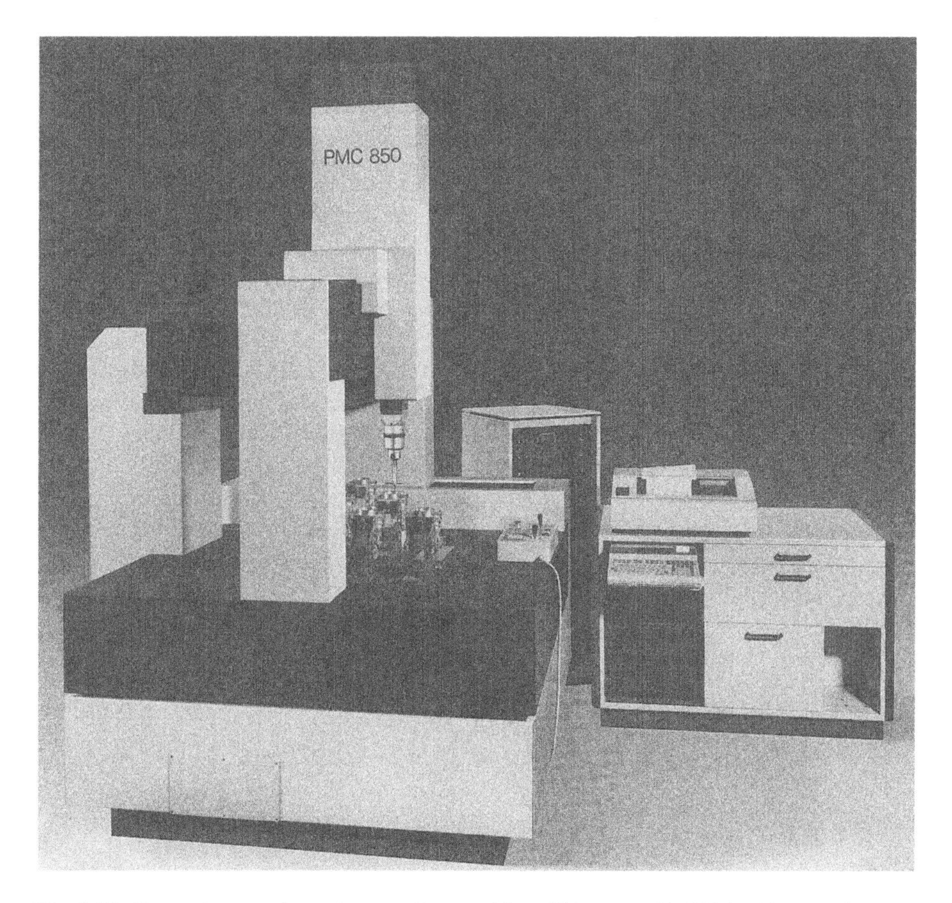

Fig. 9.13. Two column universal measuring machine. (Courtesy: Carl Zeiss Company)

Fig. 9.14. Measuring a gear. (Courtesy: Carl Zeiss Company)

the workpiece, there are different types of measuring machines available. A cantilever type machine can cover a large workspace. In principle, the measuring head and recording mechanism can be the same as that of the above-discussed machine. A wide spectrum of workpieces can be handled with this type of machine. Figure 9.14 shows an application, in which the dimensional accuracy of a gear is defined.

The gauging mechanism which measures the travel of the measuring head in either of the 3 main axes may operate on inductive, optical or mechanical measuring principles. Figure 9.15 shows the operation of an optical measuring system. The movement of the measuring head is recorded by light pulses which are generated when a light beam is interrupted by the graduation marks on a glass scale attached to the head. The accuracy of present glass scales is 0.1 microns.

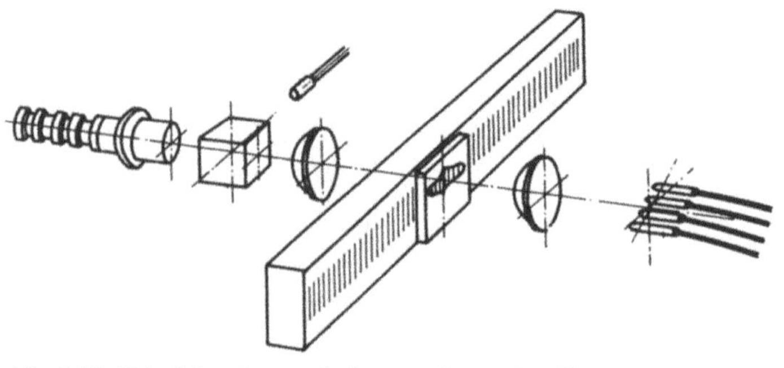

Fig. 9.15. Principle of an optical measuring scale. (Courtesy: Carl Zeiss Company)

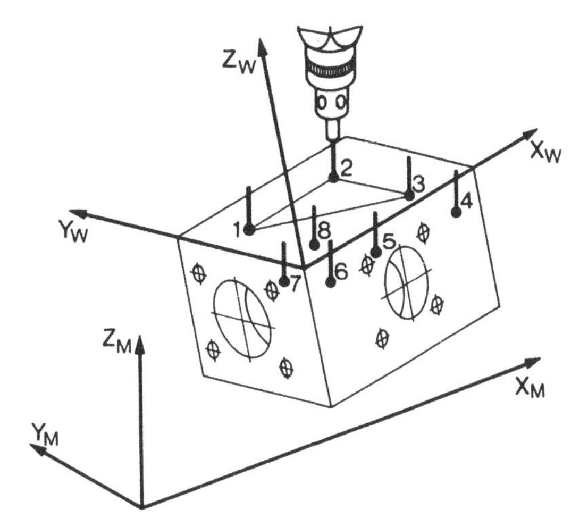

Fig. 9.16. 3-D measuring principle. (Courtesy: Carl Zeiss Company)

With this measuring principle a reference cartesian machine coordinate system is used with respect to which a probe head moves. The spatial reference point of the probe head may be the center of the probe tip (Fig. 9.16).

The illustration shows a machine coordinate system x_m, y_m, z_m and the workpiece coordinate system x_w, y_w, z_w. The test object can be placed at any location and orientation on the table. In this way the two coordinate systems do not have to coincide. The exact position of the workpiece is determined by scanning several points of its surface. For example, the surface normal is determined by the measurements 1, 2 and 3. The direction of the x-axis is defined by the points 4, 5 and 6 and that of the y-axis by point 7. With this measuring procedure the computer compensates all measurements for the offset between these two coordinate systems.

There are different types of probes used in the measuring head. Theoretically the diameter of the probe should be infinitely small in order to be able to reproduce the actual dimensions of the workpiece. Such probes, however, cannot be manufactured. Actual mechanical probes will have shapes, which are shown in Fig. 9.17. The computer of the measuring machine is programmed to compensate for any offset due to the finite dimensions of the probe tip.

The measuring probe is connected to a one, two or three dimensional measuring system. For universal measurements the probe head may contain several probes extending into different directions. The calibration of the probe is done with the help of a high precision spherical standard. For this purpose, the diameters of all probe tips and their position with regard to the machine coordinate system are determined and made known to the computer. During actual measurements the workpiece is scanned by the probe tip, and the measuring head is thereby displaced (Fig. 9.18).

The machine control system tries to move the measuring system back to its zero point. At this location the coordinates of the measuring machines are entered into the computer. Since the dimension of the probe tip and the tip position is

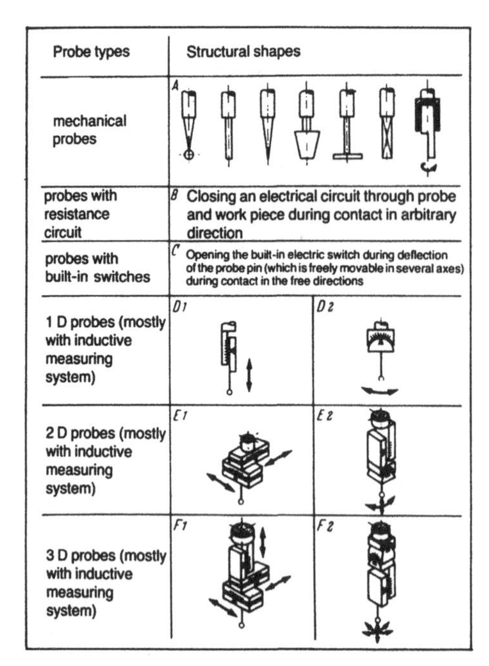

Probe types	Structural shapes	
mechanical probes	A	
probes with resistance circuit	B Closing an electrical circuit through probe and work piece during contact in arbitrary direction	
probes with built-in switches	C Opening the built-in electric switch during deflection of the probe pin (which is freely movable in several axes) during contact in the free directions	
1 D probes (mostly with inductive measuring system)	D1	D2
2 D probes (mostly with inductive measuring system)	E1	E2
3 D probes (mostly with inductive measuring system)	F1	F2

Fig. 9.17. Various structural shapes of a mechanical probe system.
(Courtesy: Carl Zeiss Company)

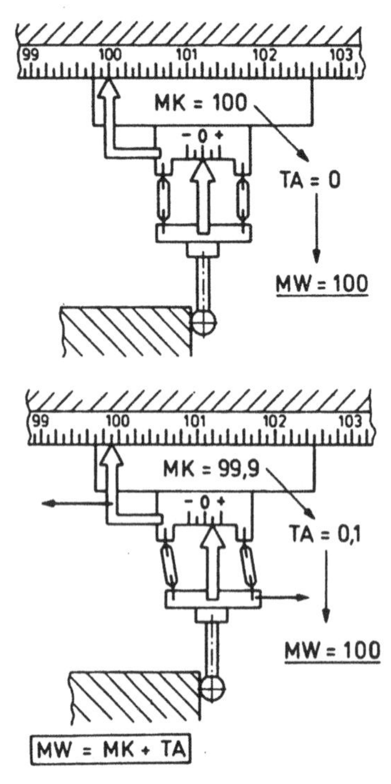

Fig. 9.18. Measuring the displacement of a probe tip.
(Courtesy: Carl Zeiss Company)

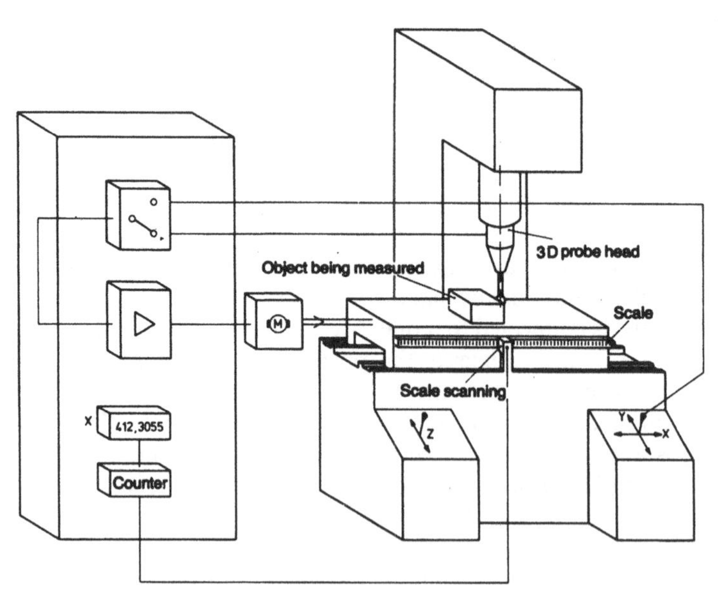

Fig. 9.19. Principle of position control by manual machine operation. (Courtesy: Carl Zeiss Company)

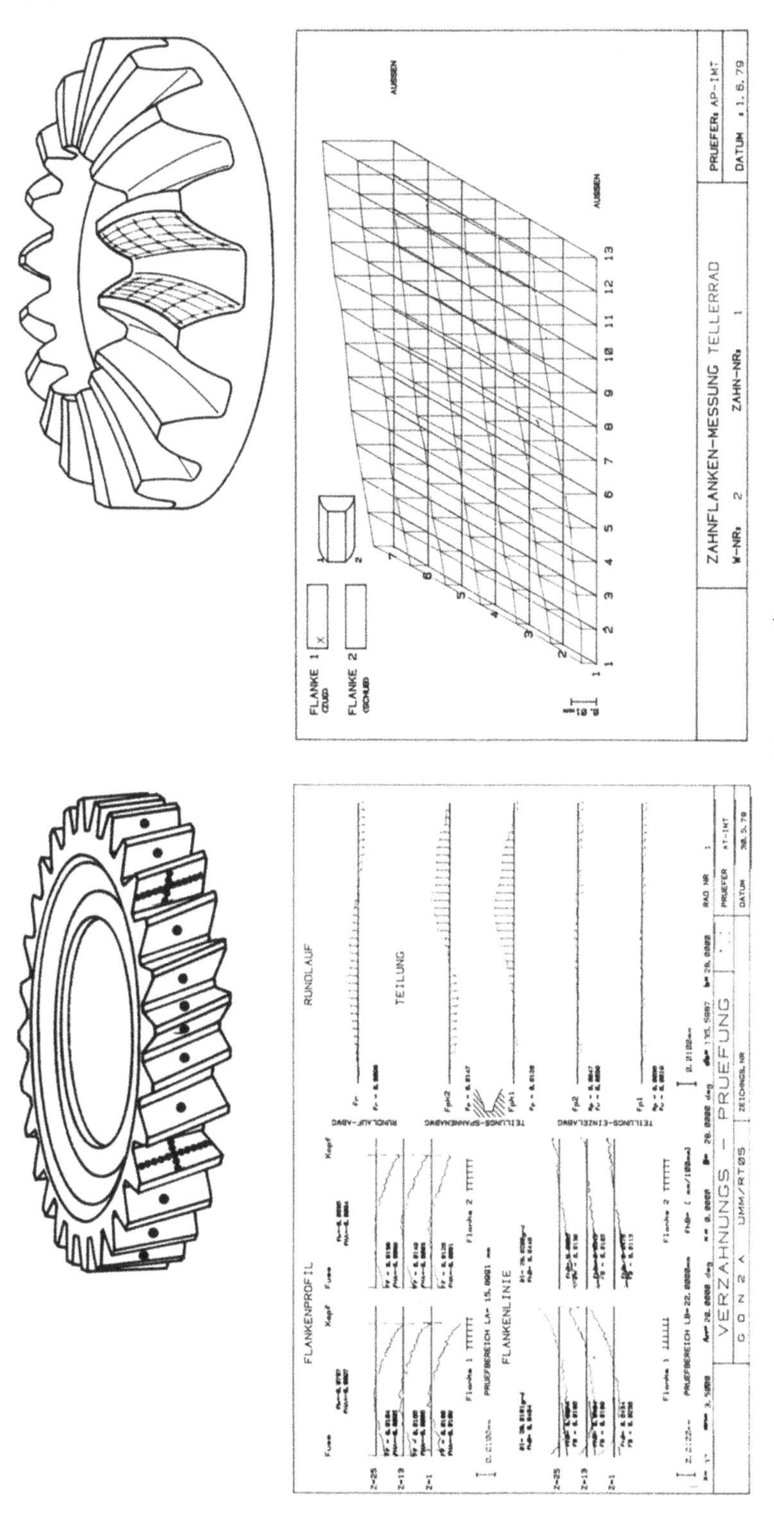

Fig. 9.20. Measuring the contour of the teeth of a gear. (Courtesy: Carl Zeiss Company)

known to the computer from the calibration procedure, the computer can calculate the actual position of the workpiece contour in reference to the origin of a predefined coordinate system. A simplified control loop of the coordinate measuring machine is shown in Fig. 9.19. In this case the probe is brought into contact with the workpiece under manual control with the help of X, Y and Z position control sticks. After contact with the object the hand control becomes inactive and the measurements can be taken.

With a coordinate measuring machine it is also very easy to measure contours of cams and irregular surfaces. The object to be measured is rotated about an axis, i.e., with the help of an indexing table. Measurements can be taken at different angular increments and compared against a mathematical curve describing the contour of the object or against a curve obtained from a master cylinder. This type of test is very time-consuming when done with conventional measuring devices. Under computer control the test time may be reduced by several orders of magnitudes.

Hard operation of the machines is very time consuming when complex workpieces are measured. Figure 9.20 shows the example of a curved gear surface. For such tasks CNC controlled measuring machines are available. The probe is guided by a program which describes the contour of the workpiece along its surface. The actual measured coordinates are compared with those of the described contour and the deviations are determined. The results may be plotted for visual inspection.

9.2.6.2 Non-Contact Measurements

Most measurements require positioning of a probe on the measuring object. This procedure is often very cumbersome and time consuming. In order to increase the productivity of the measuring process, many noncontact methods have been developed. In general such instruments available today have low accuracy and are very complicated. In future, there is a need to develop low cost remote measuring devices with high accuracy. Advanced laser technology may be used for dimensional measurements and for the investigation of the surface integrity of parts. Figure 9.21 shows the principle of a laser telemetric system to measure online the diameter of a rod. The laser light beam strikes a rotating mirror and is projected as a ligth sheet with the help of a collimating lens towards the object to be measured. A second lens collects the residual light which passes the object. The intensity of this light is recorded by a photodetector. With this laser principle, an accuracy of 0.0005 mm can be obtained.

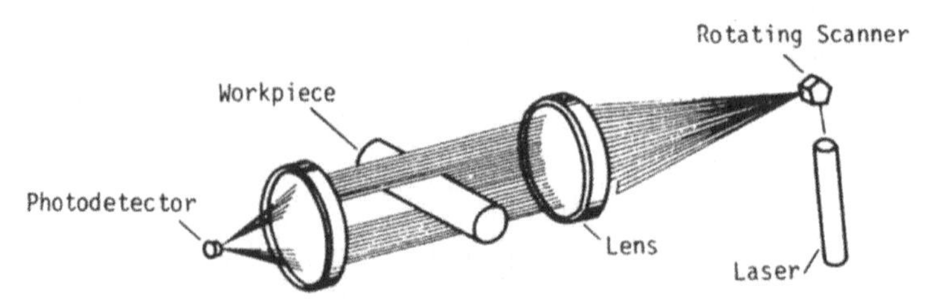

Fig. 9.21. Thickness measurement by a laser. (Courtesy: [1])

Another example of a non-contact dimensional measuring system uses a linear photodiode array. The object to be measured must be clearly distinguishable from its background. For this purpose it is illuminated either by directly shining light or by a translucent backplane. The light emitted from the illuminated objects is collected by a lens and projected on the photodiode array. To distinguish the object from its background, a threshold is set automatically or by hand. The image from the photodiode array is digitized and stored for further processing. Such arrays are presently available with up to 4096 photodiodes. For this reason the accuracy of this measuring device depends on the number of pixels of the array and the distance at which the object is located. Through a discrimination procedure the edges of the metal strip are located by the camera and the pixels of the photodiode array showing these edges are marked by the computer of the camera. Now the computer counts the number of pixels located between these two marks and this number is proportional to the width of the strip. Then the unknown object to be measured is viewed, and again the number of pixels between the detected edge marks are counted. The actual dimension is proportional to the number of pixels and can be calculated with the help of the information obtained from the standard by simple arithmetic.

The use of the computer for both of the above discussed methods is advantageous. It will retrieve the information about the stored image from memory and calculate, with the help of the known threshold, the distance between two edges of the object. Such computerized devices are capable of performing over 500 measurements per second.

9.2.6.3 Manual Input

There are many quality control parameters which cannot be directly measured or which are very difficult to acquire with presently available instruments. These parameters have to be recorded by an operator and entered into the computer by means of a keyboard or other manual device. Typical defects of this type are scratches on painted surfaces, bent sheet metal parts and many others. In the past these defects have mainly been recorded on paper and thereafter entered into the computer. When data acquisition systems became available, passive data entry terminals were used. With the development of low cost intelligent terminals manual parameter entry can be greatly facilitated. The terminal will be able to guide the operator through his quality control procedure, carry out validity checks on data and immediately perform process capability calculations. The results of these calculations can be made visible on a low cost display device. Thus at the workstation, it will be possible to observe at the work station process drifts, wear of measuring devices, operator interaction with the process, performance trends between different shifts etc. The intelligent terminal can be connected to the factory bus system and will thus be able to transfer its quality control data to the host computer. The manual input stations for quality control are usually identical to those used in data acquisition systems.

Voice communication with the computer is another enhancement of a data acquisition terminal. The operator is able to call out the defects to the terminal. It in turn recognizes the instruction and records the defect entries in a file. To date, these devices have a limited vocabulary and are very selective with regard to the master's voice by which they were trained. In case of a shift change the voice recognition system has to be familiar with the voice of its new master. Also physical stresses may change the tone of a voice and cause a recognition problem.

9.2.7 Computer Languages for Test Applications

Machine oriented languages are mainly used on small computers, such as microcomputers, which do sequential control and simple arithmetic. This group of languages is machine dependent and requires good programming skills. The macro and block languages are more user-oriented and may reduce programming effort and time considerably. The machine-oriented languages offer a major advantage when memory cost has to be minimized and fast execution times are of importance. Their great disadvantage is that they are cumbersome to use for complex arithmetic operations. With the help of the system programming languages more abstract formulation of the problem to be solved is possible. Thus they are more suitable for the implementation of complexs algorithms. They allow conception of clearly formulated, well-structured programming modules which are also self explanatory. Their major disadvantage is that they do not support programming of I/O operations and real time requirements.

The high order real time languages have been specifically designed for computer process communication. They support real time computer process interaction, task communication, simple input/output programming and easy implementation of complex mathematical operations. Their main application area is process control. For that reason they are of more generalized design and may lead to unwieldy programs when used for test applications. However, they are easy to learn and easy to use and will render satisfactory solutions for most test installations. Their main disadvantages are slower execution times compared with the machine and system languages, and they also have large memory requirements. The extended scientific languages are very popular, as FORTRAN or BASIC are well-known to engineers. For real time applications they have been supplemented with subroutine calls to support computer process (input/output) communication, task scheduling and to perform manipulation of bits and bit patterns. These languages are best suited for less sophisticated test systems where the real time requirements are not very stringent. They also occupy large memory space and have slow execution times.

There are several languages which use APT type constructs [17] for measuring applications. They may be of advantage when the program which is used to manufacture a workpiece on a NC or CNC machine is similar to that which measures the part. The program sends the probe along the surface of the workpiece contour and directs the measuring machine to take dimensional measurements at consecutive contact points. These points may be predetermined or automatically selected according to the results obtained from previous measurements. The APT language for this application has to be an extended version of the basic APT to include measuring functions. Some typical functions are as follows:

- Coordinate transformation
- Coordinate selection, displacement correction, shift of workpiece position, conversion of spatial axes

- Selection of stepping increments of indexing table
- Selection of number of contact points along a circle, ellipse, plane, cone etc.
- Mean value calculation
- Calibration of probe tips with calibration sphere and precise definition of their position including the tip geometry
- Graphical output of measurement results and deviation from the drawing
- Determination of statistically significant number of measurements.

Manufacturers of coordinate measuring machines supply such programming aids with their product. There are usually several levels of programming comfort the user may select from.

Problem-oriented languages consist of program modules which can be freely combined, depending on the application, by user guidance and menus. Therefore, they simplify the use of a test system considerably. Typical modules for test applications are:

- Control of test cyclus
- Supply of test limits
- Measurement data acquisition
- Parameter limit control
- Plausibility control of measured parameter
- Trend analysis
- Digital filtering
- Conversion from physical to technical values
- Output of control information
- Graphic representation of data.

9.2.8 Implementation of a QA Computer System

In mass production it is often necessary to test a product while it is moving on a conveyor or an assembly line. If several tests are to be performed on the same unit at different test stations all data pertaining to this unit must be positively stored in the same quality file. This may lead to very difficult synchronisation problems between test stations. Synchronization may be complicated by start and stop operations of the assembly line or by swinging hangers of a chain conveyor. When the product to be tested is moving on an assembly line, difficulties may be encountered in transmitting the test data to the computer. With a long test track and many tests to be done it becomes impractical to connect instruments and data acquisition terminals to a stationary central computer because of the signal communication lines which have to be pulled along with the product. Depending on the type of test to be performed there are different solutions to this particular problem. They are:

- Instant test. Here the performance of the test unit is time invariant. In this case it is only necessary to synchronize the storage of performance parameters taken at the different test stations.
- Transient test. The test unit may have a typical pull-down characteristic. Tests are taken at different times and different locations. It is the task of the supervisory computer to reconstruct the pull-down curve.

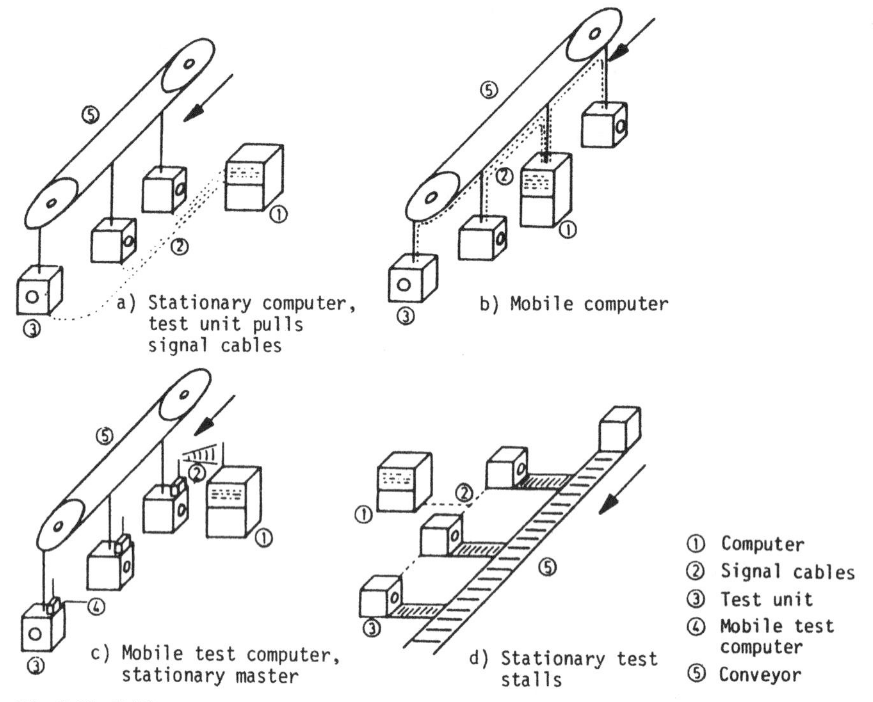

Fig. 9.22. Different test configurations for moving production

- Test at steady state condition. In this case the unit has to be constantly monitored until the performance variables have reached steady state condition. Thereafter the reading is taken.

Figure 9.22 shows different test configurations for handling moving line tests. The main problems are encountered when the test signals are sent to the computer.

The easiest solution is to connect the sensors directly to the computer. This would require pulling the test lines along with the moving unit (Fig. 9.22a). In most cases this is impossible. The next solution would be the use of sliding contacts to transmit the signals from the moving line. But these signals are usually very weak, therefore they are falsified by poor or bouncing contacts.

Figure 9.22b shows a solution where the computer travels together with the test units on the conveyor system. In this case only the power to the computer has to be transmitted via sliding contacts. If the computer is provided with adequate power buffering, short interrupts of the power by the sliding contacts can be eliminated.

Figure 9.22c shows a test set-up whereby each test unit is accompanied by its own battery powered data acquisition computer. At the end of the test track the central computer receives the test data by radio or infrared communication. The evaluation of the test data will be done by this more powerful computer. Upon the completion of the test the data acquisition computer is placed on a new test unit.

Figure 9.22d shows a test system where the test units are pulled off from the main product stream and monitored in stationary position. This solution renders the best test data. However, it is expensive and interrupts the flow line process.

9.3 Machine Vision: Inspection Techniques, Mensuration and Robotics

9.3.1 Visual Inspection Tasks

Machine vision is a discipline related to artificial intelligence. If the sensory constraints are appropriately simplified, it provides a powerful tool for important applications to automated inspections. Here, inspection means any visual process which is performed to measure the quality of individual parts or to monitor an ongoing manufacturing process. Often QA must be automated, because visual inspections by humans may be influenced by operator fatigue, lack of motivation and different experience levels.

Machine vision for inspection can be classified into two groups: precision measurements and qualitative inspections including measurements [18]. This classification and its subdivisions are shown in Fig. 9.23.

Inspection methods requiring highly quantitative mensuration are, for example, in-process optical gauging for NC, tool wear measurement (e.g., by holography) and analysis (e.g., form factors) of metallurgical specimens. Qualitative in-

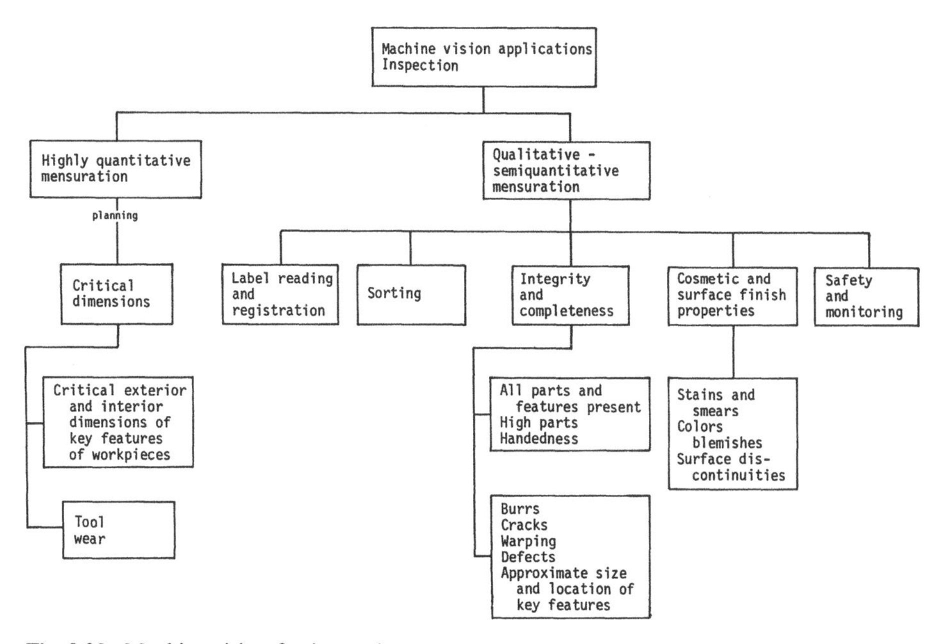

Fig. 9.23. Machine vision for inspection

spection with semi quantative mensuration is aimed at emulating an operator, who visually inspects a workpiece or assembly without the aid of measuring device. Label reading and registration is used for the classification of characters printed on labels or on bottles, cans, cartons, etc. Sorting includes packing of one type of workpiece in a orderly fashion in a container, the selection of a specific workpiece from a random mix of parts, or counting of number of workpieces for inventory purposes.

The integrity and completeness subclass is the most important one of the whole group. Here, individual workpieces are examined for many types of defects incurred during production and handling; they include burrs, fractures, surfaces, defects on printed wiring boards and on silicon integrated circuits, etc. The size and location of key features, such as drilled holes can be measured for conformity to specification. Inspection for completeness is required to ensure that subassemblies and assemblies have all parts in the right places. This inspection method is usually quite different from the other types mentioned and requires the use of a range finder and three dimensional models [19].

Deviation from a specified colour, blemishes, dirt, stains, smears, runners, etc. are all defects found on the surfaces of many finished goods. These defects are of cosmetic rather than of functional nature, they are aesthetically unacceptable and affect the marketability of the product.

Safety and monitoring may be needed for modern robot arms moving at high speed and with powerful forces. Therefore, they can be dangerous to human workers and destructive to machinery (safety and monitoring).

The input peripherals for these inspection tasks are cameras (vidicon, CCD, etc.), linear arrays, OCR (Optical Character Recognition) devices, microscopes and optical scanners [20]. The resolution of cameras available today is usually 256×256 pixels and may go as high as 512×512 pixels. When the scene to be scanned is in continuous linear motion (e.g. constantly moving conveyor) some economy can be achieved by using a linear array instead of a two dimensional array. Linear arrays are available with resolutions of up to 2048 elements. Laser scanners can not only monitor the productions of paper, cloth, sheet, plastic etc., they are also very useful as range finders.

Vision systems are classified as binary or gray-scale devices. Most commercial vision systems only process binary images. However, many defects can only be identified by gray-scale (e.g. 256 values) or colour sensors.

A further task of inspection is to check the dimensionality of images. Label reading and the registration of properties of finished surfaces can usually be done by two dimensional images. Sorting has similar proplems to "bin picking" and inspection for integrety and completeness. This all require additional distance information and 3-D representation. This three dimensional information is also necessary, when a robot is needed for inspection. The robot can be trained to pick up an object, to show it to the camera for identification [21] and to place the object after a good/bad decision in an appropriate bin (e.g., accept, reject).

9.3.2 Machine Vision Techniques for Inspection

The various methods for designing a machine vision system can be grouped into three major categories: template-matching, the decision-theoretical approach and the syntactical (structural) approach. Each of these approaches can be divided into three parts. First, the data from the imaging devices must be preprocessed; secondly the representations must be selected and the feature extraction for simple constituents has to be performed. In the last step the classification or the structural analysis (complex pattern) has to be done [22].

9.3.2.1 Template Matching

With this method a set of templates or prototypes, one for each pattern class, is stored in memory for QA. The input pattern is compared with the templates of each class. The classification is based on a comparison between the template and the recorded image (e.g. correlation). The defect free master image I of the product (reference image) can be generated by the computer (synthetical image) or by photographing a defect free product. I is a two dimensional image. The numerical value of each pixel represents the gray level of the pixel. During operation, the product to be inspected is placed under the "camera" to generate a real image R. R is aligned with I and their gray values are compared pixel by pixel.

This pixel by pixel method is often used in the inspection of printed circuit boards, integrated circuit masks and bank check readers. The major advantage of this approach is its simplicity and high speed. A fundamental limitation is the need to align the two images perfectly. Usually one image is rotated by software before matching. Then the comparison can be performed. However, rotation is very time consuming. Template matching systems will not be able to handle the rotation in real time unless faster microcomputers become available. Another shortcoming of this method is the inflexibility in defining the inspection criteria, since information beyond the gray value of a pixel cannot be easily implemented. Each pixel is processed independently of its neighbours, therefore contextual information contained in the image is ignored.

9.3.2.2 Decision-Theoretic Approach

With the decision-theoretic approach (feature inspection) a pattern is represented by a set of N measured features classifying a pattern class. Therefore, storage of the master image is not necessary.

The designation of each sample is usually done in terms of N-dimensional conditional probability density functions. The classification (decision making) is based on statistical decision rules.

The template-matching approach can be regarded as a special case of the decision-theoretical approach. Each pattern is represented by a feature vector (N-dimensional), and the classification is performed by a simple similarity criterion like a correlation.

Applications of this approach include character recognition, remote sensing, target detection and identification, failure analysis, machine parts recognition and inspection with automated manufacturing processes, etc. These examples demonstrate that this method is a most popular approach. It is also used in vision systems and in robot control.

Segmentation is a process which breaks up a sensed image into its constituent parts or objects. The algorithms are based either on the discontinuity or similarity method. The first one is edge detection and the principal approach of the second category is region growing [23]. Edge detection is usually done by the gradient techniques. This approach effectively generates an outline of the workpiece and its internal holes. Region growing techniques are applicable in situations where objects cannot be differentiated from each other, or from their background by edge detection or thresholding.

Feature extraction is done by segmentation techniques which generate descriptions. They should be independent of object location and orientation, and must contain enough discriminatory information to separate uniquely one object from another. Object features which are needed to support the work of industrial computer are based on shape and intensity. Size and shape features are: area, length, compactness, radius vectors, the count of holes, center of gravity, moments of inertia, center of a bounding rectangle, etc. Intensity features are reflectivity, texture, porosity, surface-imperfections (cracks, scratches, dirt), etc. The aim of feature extraction is to describe an object for the purpose of classification (medium level vision) or analysis (high level vision). These representations define object models by which the parts can be identified in an industrial scene. They can be used for learning purposes to build the model by selecting features from sample workpieces.

Classification of objects or defects is done with the help of decision (discriminant) functions. Given several prototype representations (models) and the image of an unknown part, the problem is to determine, with the aid of the decision functions, to which object class the part belongs. Two popular approaches to this, are the nearest neighbor method and the use of a binary decision tree.

9.3.2.3 Syntactical Approach

In the syntactical approach a formal grammar is developed to generate elements of a language which defines a pattern class. An automate is developed to recognize this language. The language can consist of strings of primitives and relational operators (e.g. directed line segments) or of higher order data structures, such as trees or graphs. The decision making process is a syntax analysis.

It is characteristic of this approach, that no comparison between the workpiece image to be identified and the "master model" is needed [25]. Here, the phrase "master model" (or the defect free product) rather than "master image" is used because the master image may not exist at all. This method relies only on knowledge of localized generic properties and it transforms the properties into a set of generally applicable rules. A small window is moved over the whole sensed image. At each location placement, only the window area is investigated. If the current window violates the set of rules, a defect is registered.

The template-matching approach can also be regarded as a special case of the syntactical approach. In such a case, each pattern is represented by a string (or graph) of primitives and the decision making process is based on a similarity or distance measure between two strings (or two graphs).

Applications of the syntactic approach include character recognition, wave form analysis, automatic inspection, finger print classification, machine part recognition and remote sensing.

9.3.3 Automated Microscopic Material Testing

The input devices used for this application are the scanning electron microscope, the laser microscope or the TV-camera [26], [27]. The tested object is two dimensional; it is prepared (contrast, sectional view) and mounted on a specimen holder. The probe is illuminated by a well-designed light-source (e.g., back light, directional light). The resolution is usually 256 × 256 pixels. The original grayscale image is transformed into a binary image. Typical applications of this image "analysis" system are evaluations of shapes and patterns (e.g., ceramic specimens, medical/biological tissues) and quantitative measurements (e.g., diameters of grains and pores, surface computations, mesh size of precision grids).

The most frequently used inspection techniques are template-matching for the recognition of shapes, and the decision-theoretical approach for the extraction of features and the classification of material. Usually, an interactive command language is available to facilitate the proper use of the image operation.

The recognition of the shape (texture classification) of the objects in a microscopic image is usually done with the aid of a structuring element using a set of basic image operations [28]. Structuring elements are points, lines, triangles, squares, hexagons, etc. These structuring elements can be freely selected and are used as templates. The matching of these templates is performed by basic image operations which are defined by set operations. Typical operations are:

dilation: $p' = p \oplus t$
erosion: $p' = p \ominus t$
ouverture: $p' = (p \ominus t) \oplus t$
fermeture: $p' = (p \oplus t) \ominus t$

The original image is given by p, the transformed image is defined by p', t is the template and $\oplus$ or $\ominus$ are the Minkowsky addition or subtraction [29]. The mode of action of these two operations is depicted in Fig. 9.24.

To perform the operation of dilation the oriented template t is directed along the perimeter of the image p. The dilated image p' is determined by the union of

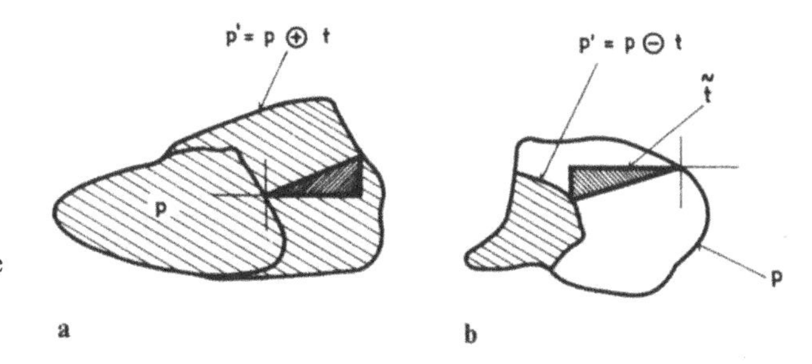

Fig. 9.24. Operation of the Minkowsky addition (**a**) and subtraction (**b**)

p and the set of points which are generated by traversing the obect contour of the template (Fig. 9.24a). The erosion is the inverse operation of dilation (Fig. 9.24b), where *t* is the reflected template.

The dilation operation can be used to merge several distributed objects in *p* into one object (*thickening*). The erosion separates "bridges" – between different objects in *p* and eliminates small elongated parts of an object (*shrinking*). The ouverture operation eliminates from *p* all elements which are smaller than *t* (*opening*). The inverse operation of fermeturing eliminates for *p* all elements which are smaller than *t* (*closing*). These four operations can be used to classify the shape of the individual components (e.g. pores in a sintered substance) and for high quantative measurements for quality assignments. The last point may be illustrated by the inspection of a precision square grid. The structuring element (template) is a square. The ouverture operation with this template demonstrates, that all sieve holes are actually squares. The nominal value for the individual grid distances may be for example 20 microns. Templates with variable edge length (e.g. 20, 18, 17 microns, etc.) are then needed to measure the mesh width of the grid and to divide the parameters into different classes.

A completely different technique for detecting defects is the filtering approach. A filter can be constructed from a set of simple operations for pairs of image points (e.g. addition, substraction, multiplication, linear filtering, etc.). The application ranges are defined by so called receptive fields [30]. These receptive fields are realized by hardware. The parallel image operations are obtained by processing the original image according to a pipelining principle. Non-linear low-pass filters are for example suited to detect spots and grains. Non-linear highpass filters are used to advantage for the detection of thin lines and peaks. Contour filters can be constructed for the recognition of areal parameters.

9.3.4 Laser Based Measurements and Inspections

9.3.4.1 Quantitative Mensuration

Lasers are ideal light sources for distance measurements and for material testing. With the aid of holography it is possible to extend the use of the classical interferometric technique from specular objects to diffuse reflecting test objects.

Optical Gauging

Over the last decade the servo control technique for numerical machine tools has advanced to a point where the limiting factor in producing the workpiece is not the machine drive but the real-time measurement to verify accuracy of the produced dimensions. Usually, the workpiece itself is not part of the control loop. For this reason tool wear, thermal distortion and backlash of the lead screw is outside the measurung loop. Laser optical techniques offer the possibility of measuring the dimension of the actual workpiece as close as possible to its contact point with the cutting tool. Thus all errors are within the control loop of the machine tool. The purpose of the next section is to describe a combined optical servo system for in-process gauging for numerical machine tools [31].

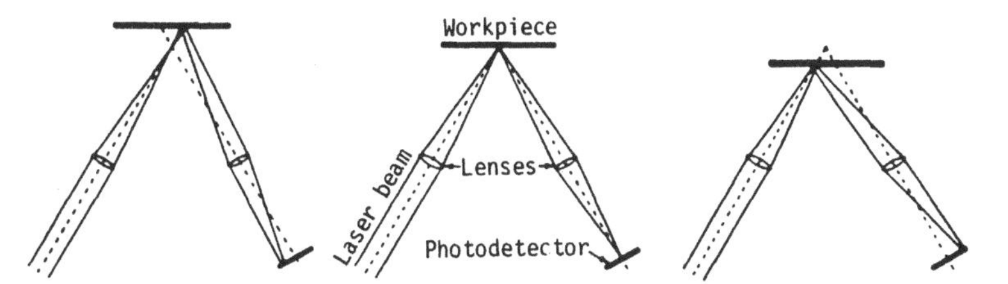

Fig. 9.25. Geometric principle of optical measurement

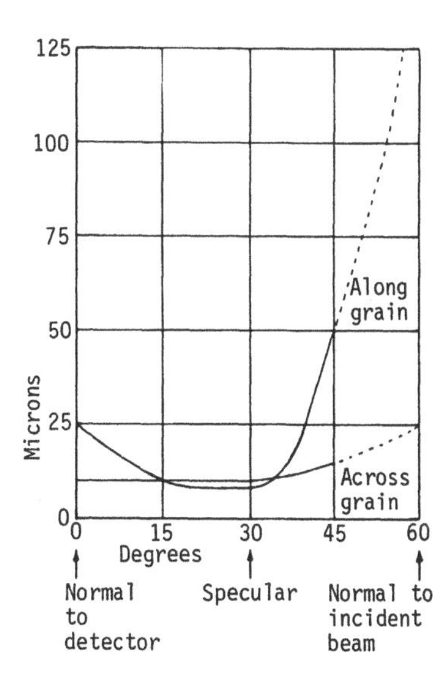

Fig. 9.26. Distance resolution for various orientations and two different surface grains of the workpiece surface

The distance measurement is a geometric approach, where an incident laser beam is focused on the workpiece and the image is analyzed with a position sensitive photodetector. Interferometric techniques are not appropriate here, because they give relative readings and require precision reflectors. For this purpose direct absolute measurements of usually non-specular surfaces are required. The gauging principle is illustrated in Fig. 9.25.

The incident laser beam (spot diameter 12 microns) and the centerline of a position sensitive photo-detector are focused at the common intersection point. The position sensitive photo-detector determines whether the focal point is in front of, behind, or at the workpiece surface. A servo control loop drives the optical assembly with precision actuators to maintain the focal point at the workpiece. The distance is determined by triangulation. The precision of the optical system for a distance of 20 centimeters is shown in Fig. 9.26.

"Specular" reflection gives the best results (10 microns). The maximum precision is obtained along the grain, but good precision over a wide range of angles is obtained across the grain. The precision degrades rapidly when the workpiece surface is turned toward the incident beam. Good results are achieved, when the workpiece is turned towards the photo-detector. It is obvious, that the direction of the surface grain relative to the plane of the incident beam and the optical axis of the photo-detector affects the results of the measurements. This is due to the fact, that the reflected light tends to scatter over a much wider angle across the grain than along the grain.

Non-destructive Testing

The most important use of holography in industry is in non-destructive testing (NDT), [32]. It connects interferometric techniques with holographic methods. Small deformations, a fraction of a micrometer, are induced on the surface of the test object by pressure, heat, ultrasound, vibration, etc. The deformation is detected by comparing the interference of the hologram pattern before stressing with that after stressing (double exposure technique). The hologram pattern reveals the location of defects such as poor bonds in laminated composite materials, or cracks and fractures. It can also be used to measure tool wear. Holographic non-destructive testing is very useful, when the surface of an object changes due to stresses caused by a defect. Often, a defect manifests itself as an irregular interference pattern (fringe pattern). For example, a set of closed rings is deformed (Fig. 9.27), or a set of equally spaced dark and light rows are disturbed (Fig. 9.28).

Another, very attractive approach to perform stress analysis is the technique of specklegrams [33]. Here, the same photographic film is directly (without reference beam) exposed twice (without and with application of stresses). This pattern is called a specklegram and it reveals the deformations in a similar way to the hologram. If a laser beam is directed at this specklegram a diffraction pattern (halo) is visible on the screen (Fourier transformation) as equally spaced interference lines (Young's interference lines). These lines are normal to the stress direction. Their distance is inversely proportional to the deformation modulus. The spatial frequencies of these lines are proportional to the amount of deformation.

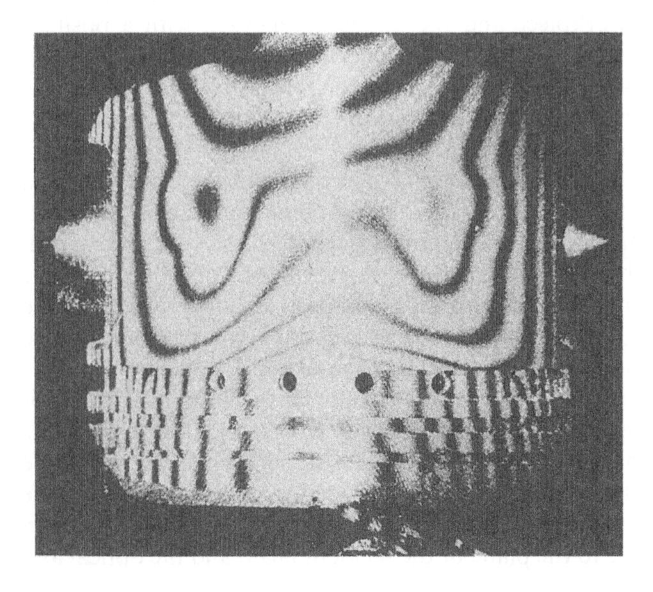

Fig. 9.27. Holographic stress analysis by NDT. Increased internal pressure in a polyethylene bottle

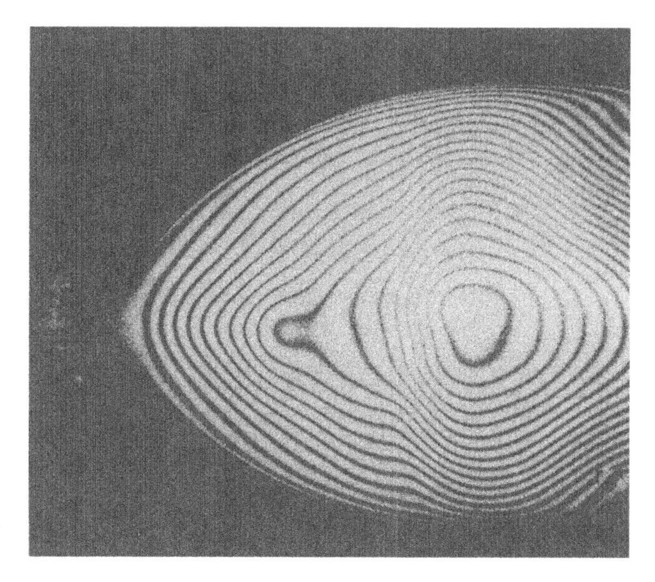

Fig. 9.28. Thermal deformations of a heated motor part

9.3.4.2 Semi-Quantitative Mensuration (Scanner)

Holographic Laser Scanner

The most widespread application of laser scanners is reading of the bar code that is attached to most supermarked goods. There are even scanners which can rotate around a product. The checker does not have to place the bar code in any particular position over the window as he usually must do. With such scanners the two dimensional scanning mode can be obtained by rotary motions of two deflecting mirrors or by one rotation of a holographic disc (Fig. 9.29).

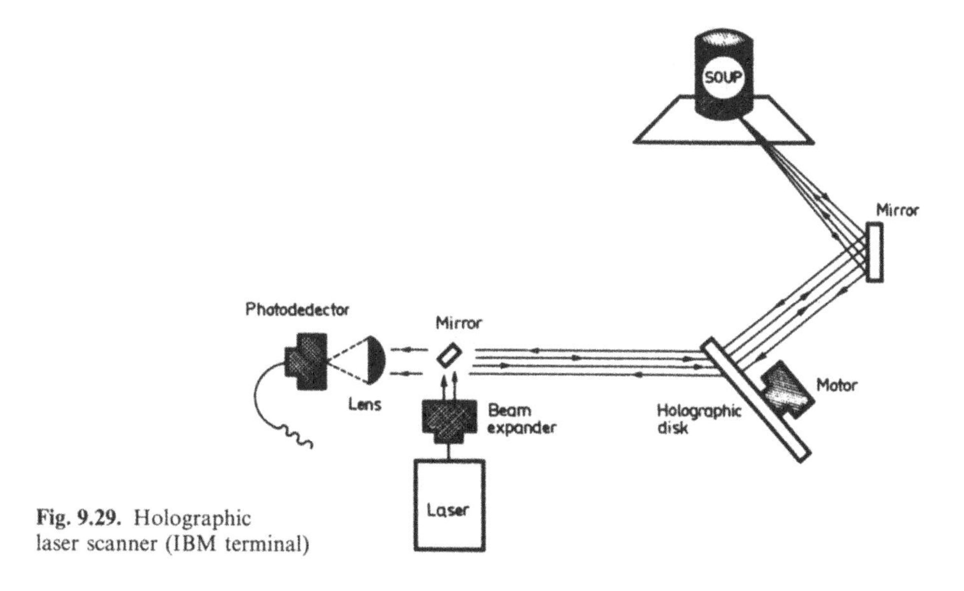

Fig. 9.29. Holographic laser scanner (IBM terminal)

The scanner has a disk which holds a series of mini reflection holograms. As the disk rotates on a spindle a low power laser beam passes through it. Each hologram on the disk diffracts the beam to a different location, forming a continuous pattern of laser lines that passes through the window and encircles the product. Such a scanner is reliable (insensitive to shock disturbance) and inexpensive.

Flight Time Laser Scanner

The three different measurement techniques used to obtain distances for industrial robots (from 10 cm up to 10 m) are based on the phase shift, the time of flight and the triangulation principle. All of these use laser scanners. However, they have severe restrictions. These are the large dynamic range of the reflected beam (up to 100 dB), the low signal-to-noise ratios (e.g. due to photon noise) and sometimes the slow pixel scanning rates (e.g. 500 msec per pixel).

Laser scanners which are based on phase shift analysis (indirect flight time measurement) are described in the reference [34]. This subject will not be further discussed. The only fact to be mentioned is that the maximum pixel rate which can be obtained by this kind of scanner is about 25 msec/pixel. Modern laser scanners are controlled by computers and are not now used as off-line devices, as was done for example in reference [34].

The most ambitious approach to measure distances is the direct measurement of the flight time. Distance and intensity data can be obtained in about one microsecond. However, for a distance of 1 meter the flight time of light is 3.3 nanoseconds. This means that a device must be used whose time resolution is 16.5 picoseconds, in order to get a distance resolution of 5 mm. The only technique which can meet these requirements is the coincidence measurements (time spectroscopy) which is used in nuclear science and in high energy physics. Time-to-amplitude convertors (TAC's) are used to measure the time relationship between a start and stop signal [35]. The timing resolution of these instruments is defined by the full width half maximum (FWHM) of the timing spectrum these can discriminate. To date, it is possible to obtain with a TAC a timing resolution of about 10 picoseconds [36]. Figure 9.30 indicates a block diagram of a flight time laser scanner.

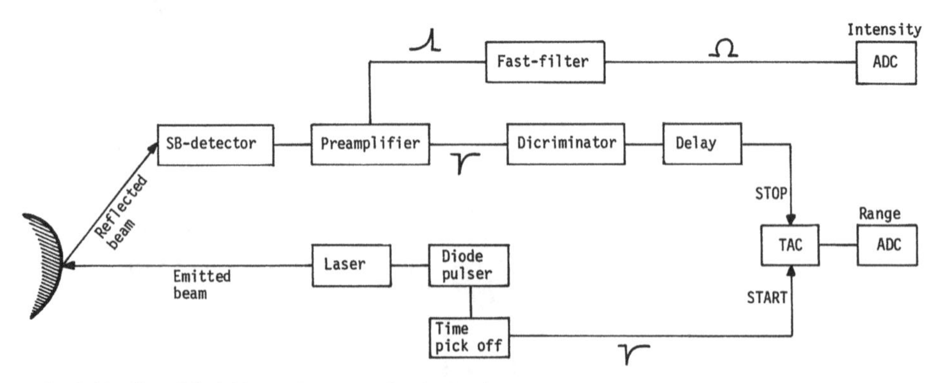

Fig. 9.30. Simplified block diagram of a flight time laser scanner

A laser diode (e.g. pulse laser diode, 50 mW) shines light on an object. The laser diode pulser is used to generate the starting signal. At the same time a time-pick-off element produces a logic pulse which is to a great extent independent of the shape and amplitude variations of the input pulse. The output signal of this trigger initiates the TAC. The reflected beam from the object impinges on a silicon surface barrier detector (SBD). The barrier eliminates time jitters as they exist in conventional photomultiplier tubes. The output signal of this detector is preamplified and divided into two signals. A fast filter amplifier intensifies and forms the shape of the "intensity" signal. This new shape (flat roof) is more suited for an analog-digital conversion (8 bits) than the original shape form (peak). The final signal is used for the intensity measurement.

The signal for determining the distance is discriminated (constant fraction discriminator) and delayed. The delay is necessary to allow the TAC to return to the second receiving mode after it has processed the start.
It defines the time signal to stop the TAC. The output of the TAC is precisely (time resolution ≤ 10 psec) correlated to the time interval defined by the start and stop signal. The TAC output is converted to a digital value (12 bits).

The most difficult problem with this kind of laser scanner is the high experimental dynamic range of about $10^5:1$. The timing resolution is very sensitive to the intensity (energy) of the reflected beam. The more energy is deposited in the reflected beam the better is the resolution. Furthermore, in nuclear experiments the dynamic range is limited to about 100:1. This means that the main building blocks of a time flight measuring system, the preamplifier, the constant fraction discriminator and the TAC, must operate efficiently if the reflected intensity is to attain its maximum and if the dynamic range is to approach zero. However, these two assumptions can not be taken for granted in a harsh industrial environment. First results which have been obtained in an experimental set-up demonstrate, that realistic timing resolutions for direct time of flight laser scanners are about 5 nanoseconds (distance resolution of 1.5 meters). Therefore, only very dedicated applications (e.g. constant specular reflection, constant temperature) are candidates for distance measurements by this approach.

Triangulation Laser Scanner

Figure 9.31 depicts the principle operation of a triangulation laser scanner. One camera of the stereoscopic approach is substituted by a laser source.

The control unit of the scanner can also be a multiprocessor system. For example, in reference [37], 64 parallel operating module-processors control the laser and process the data. The range data can be calculated with the help of the horizontal deflection angle δ or the vertical deflection angle γ of the emitted beam, the basis line length d, the focal length f_0 of the camera lens, and the picture plane coordinate x or y which correspond to the angles α and β of the reflected beam.

By the use of parallel projection (calculation of the distance h between the baseline and the point p') only two of the four deflection angles are needed (e.g. β and δ). This distance h can be calculated as follows:

$$h = \frac{d}{\operatorname{ctg} \delta + x/f_0}.$$

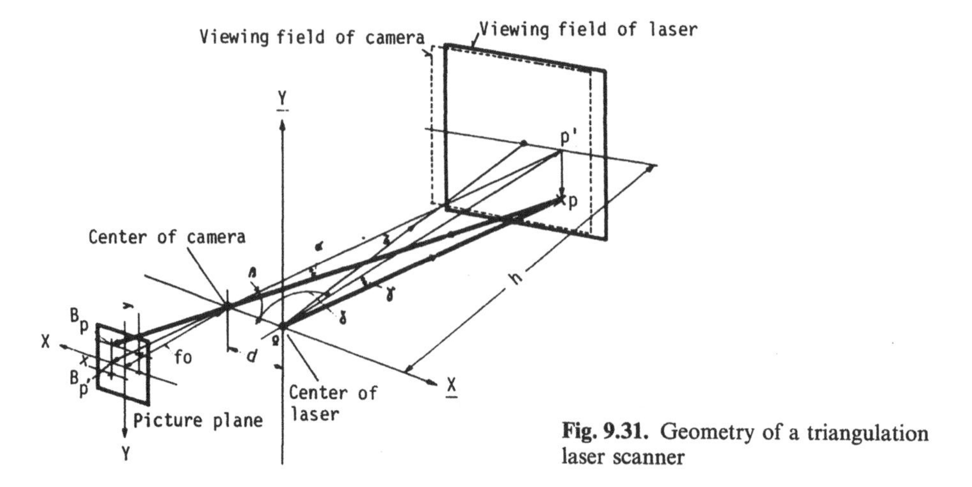

Fig. 9.31. Geometry of a triangulation laser scanner

The distance resolution $\Delta h(\Delta h(x, \delta) = h(x + x_{min}, \delta + \delta_{min}) - h(x, \delta))$ can therefore be defined in a linear approximation as

$$\Delta h = (h^2/d)\, x(-\, x_{min}/f_0 + f_0\, \delta_{min}/\sin^2 \delta).$$

Here, x_{min} gives the minimum detectable x-distance in the image plane and δ_{min} defines the minimum horizontal deflection angle which can be measured. The resolution is optimal when the distance is small (h^2), the base line is long (d), δ is equal to 90° (normal to object surface), and when x_{min} and δ_{min} are as small as possible.

To cite the result of an experiment, it can be stated that such a system renders a very good resolution of 0.25 mm at a distance of 1 meter. The time for the distance evaluation for every pixel can be decreased to about 0.5 microseconds (e.g., CCD-linear array, hardware multiplication)

Often, this type of laser scanner is connected with the structured light approach (e.g., line projection, grid projection) [38]. However, this approach operates with parts which are presented and preoriented. This renders a unique assignment between every projected object point and the corresponding image point. Thus the distance can be calculated. If there is no such unique assignment, this approach can only be used to classify different object classes. In this case absolute distance measurements are not possible.

9.3.5 Synthetic Images for Defect Classification

Low level vision must determine where the intensity (irradiance) values of an image are coming from. The following parameters are responsible for these values.

(1) surface geometry,
(2) spatial surface organization (texture),
(3) illumination (source position),

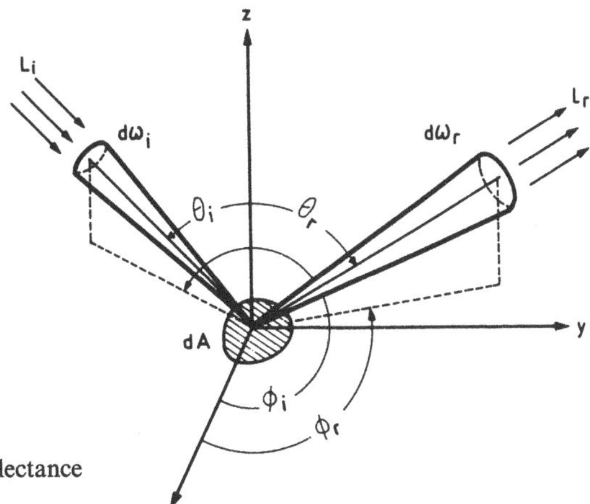

Fig. 9.32. Geometry of differential reflectance functions

(4) viewpoint (receiver position),
(5) orientation,
(6) discontinuities.

The reflectance map technique of Horn [39] renews the old radiometric knowledge of reflectance functions and relates the points (1)–(6) to the intensity changes in an image. However, this approach is usually restricted to perfect diffuse surfaces and to the assumption that the image projection is orthogonal. This assumption implies that the viewing direction is constant and that the observer is at a great distance (all rays from the object are parallel).

These two restrictions are too rigid and can be bypassed by the use of differential reflectance functions [14]. Figure 9.32 depicts the geometrical definitions for these bi-directional reflectance functions.

The differential reflectance function is defined by

$$\mathrm{d}\varrho(\theta_\mathrm{i}, \phi_\mathrm{i}; \theta_\mathrm{r}, \phi_\mathrm{r}) = \frac{L_\mathrm{r}(\theta_\mathrm{r}, \phi_\mathrm{r}) \cos\theta_\mathrm{r} \mathrm{d}\omega_\mathrm{r}}{E_\mathrm{i}(\theta_\mathrm{i}, \phi_\mathrm{i})}.$$

Here, L_r means the reflected radiance and E_i is the incident irradiance.

Defects change the reflectance properties of a surface. They absorb, deflect, scatter and alter the polarization of the incident laser beam. Figure 9.33 a shows a polar diagram of an object with specular reflectance; Fig. 9.33 b presents the modifications of the differential reflectance function by a scratched mirror.

Synthetical images can be constructed, if the following three points are known:

(1) differential reflectance function(s)
(2) surface gradients
(3) incident and reflected angles (source and view points).

The synthetical images are then produced in two steps (perspective projection): (a) measurement of the surface gradient and the incident and reflected angle,

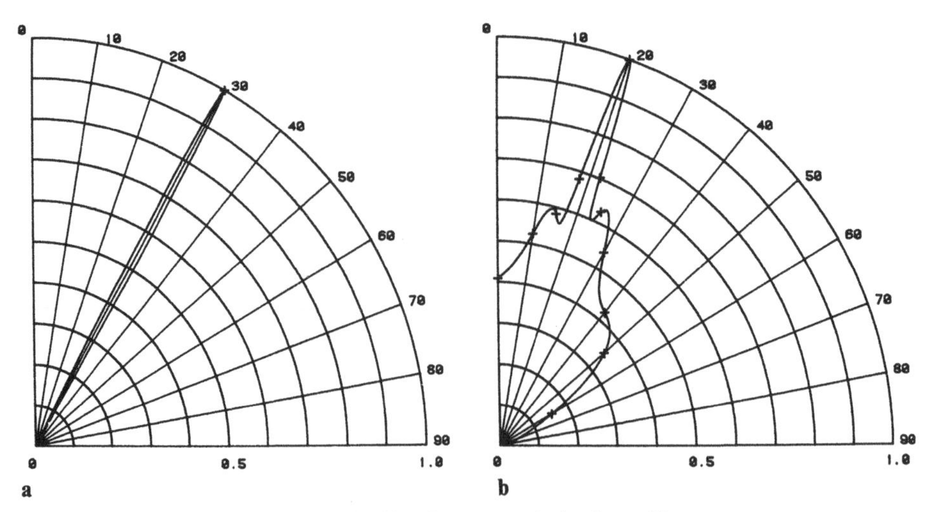

Fig. 9.33. $d\varrho/d\omega_r$ for a mirror (**a**), $d\varrho/d\omega_r$ for a scratched mirror (**b**)

(b) calculation of the image intensity with the aid of point (a) and the differential reflectance functions.

The measurements of points (1) to (3) can be performed with a laser scanner which measures distance and intensity (e.g. triangulation scanner). The classification of the defect is then obtained by comparison (e.g., difference image) of a synthetic image (ideal surface) with the real image of a surface containing the defects. The benefit of this approach is that this comparison can be done directly without the application of time consuming coordinate transformations (e.g., rotations) required in the stereographic approach. There is no correspondence problem, since the real and the synthetic image are in complete alignment.

The defect classification by the use of synthetic images is mostly suited for workpieces which are not oriented in a fixed position. Examples are goods which are hanging on hooks in batch production of refrigerators, washing machines, etc. Typical defects which can be detected by this approach are rust spots (pits), fissures, scratches, dents, etc. Furthermore, it is possible to detect holes, tears, and contaminations in flat material. The observance distance typically is 1 meter and the laser diameter 1.5 mm. The detection of defects is strongly influenced by the distance resolution of the laser scanner and by the extent of the defecct. If the defect is very small (about 0.5 mm or less), then it cannot be detected by this approach.

9.3.6 Robot Vision for Recognition and Sorting

9.3.6.1 Interfacing of a Vision System with an Assembly Robot

Often different randomly oriented parts are moved by a conveyor. The vision system has to recognize the parts, their location and orientation. It advises the robot to pick up the parts and to place them in a defined position on a work bench. With a sensor controlled robot, handling of a broad spectrum of workpieces may be possible eliminating the necessity for special tools. This section describes results obtained from an experimental set-up where a robot, a vision system and a conveyor were interconnected to sort parts.

Vision System

A vision system which assists an assembly robot is expected to fulfill the following tasks:

- Identification
- Determination of positions and angular orientations
- Positioning, fitting and sorting
- Set-up or adjustment of tools
- Measurement of dimensions, distances and angles
- Extraction and identification of specific parameters
- Visual testing for error and completeness.

A set of features is derived from each pattern (decision-theoretic approach) and is stored as a reference model in a microcomputer. The first feature extraction is performed during the teach-in mode, in which the system is taught the identity of the object by showing the workpiece to the camera. The most important features are position and orientation. The position is defined by the two dimensional center of gravity. The angular orientation is determined by a modified polar coding method.

Another feature which is important for the object recognition is its area. In order to increase the classification accuracy, circular segments about the center of gravity of the workpiece projection are placed over the image. The number of picture elements having segments in common with the image serve as a second classifier. This feature vector is independent of the angular orientation of the object.

When the recognition is completed, all features are available which are necessary to control grasping for an assembly robot. They are:

- Object identification
- Object position
- Object orientation.

The optical sensor of the recognition system described below [40] is an 4096 photodiode array which is located in a camera housing together with the scanning unit and a signal amplifier. This sensor consists of a 64 × 64 square matrix and its picture field can be described in cartesian coordinates. Modern sensors have up to 512 × 512 pixels. The picture, which is scanned by the camera, is digitized and transformed into a binary image. Figure 9.34 shows the architecture of the vision system. The optical sensor is coupled to a microcomputer TI 990/4 via special hardware processors.

Assembly Robot

The handling system connected to the camera is a Siemens industrial robot (Fig. 9.35). Its mechanical design has four degrees of freedom: a vertical lift axis, a swivel axis, a horizontal transfer axis and a gripper rotation axis. All axes of the equipment are operated by DC pancake motors.

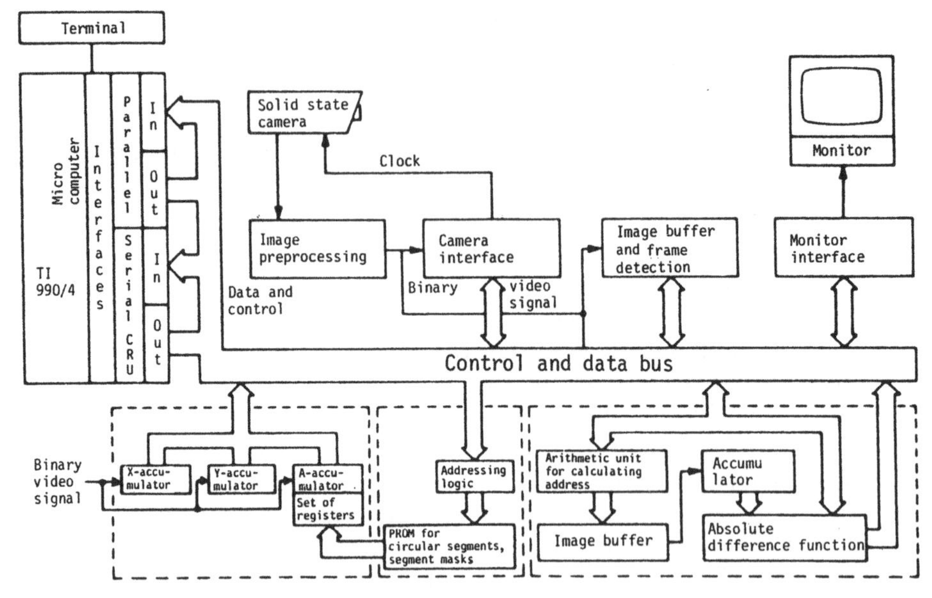

Fig. 9.34. Block diagram of the vision system

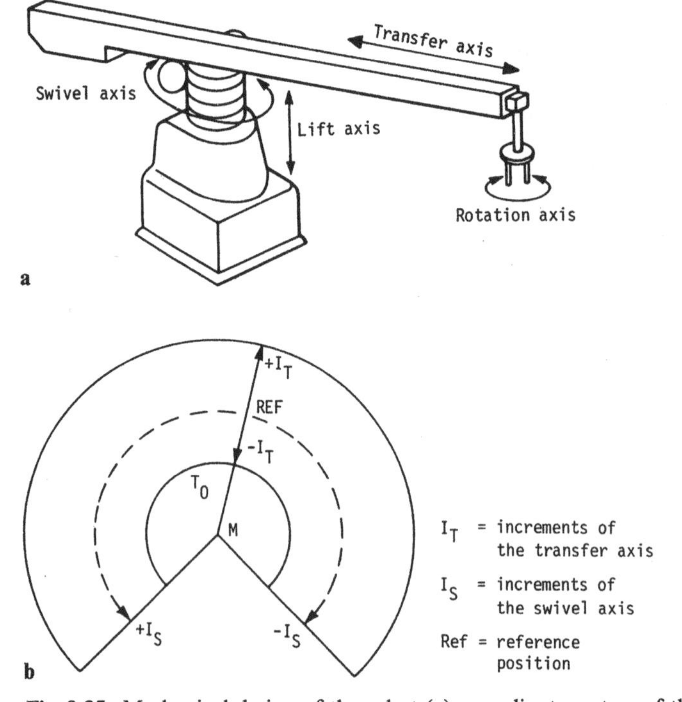

Fig. 9.35. Mechanical design of the robot (a), coordinate system of the robot (b)

A Siemens minicomputer PR 310 serves as the control unit. This unit has interfaces for the four axes of the robot and the input/output devices. These devices are used to transfer the control signals between the control unit and the operator panel, or the peripherals (floppy disk, control computer of the vision sensor).

The function of the control unit can be divided into the following subfunctions: processing of program information, processing of manually-entered information, action processing, handling of control signals, control of the operator panel, and last but not the least, the synchronization of the just mentioned subfunctions. The majority of these subfunctions are realized in software modules.

Interfacing the Vision System with the Robot

Three aspects must be considered if a visual sensor is coupled with an industrial robot: the physical interface, the data transfer protocol and the coordinate transformations. To date there is no standardization of this problems. The designer of such a system relies therefore on his own intuition. However, he is limited by the support of the sensor system and of the control unit of the manipulator. In this case, the physical interface is realized by a V. 24 interface.

For the sensor-robot system the following concept has been developed. For each workpiece known to to the sensor system, an action program is assigned by means of a program number. The grasp position of the workpiece, which may not correspond to the center of gravity, is calculated. This is done by bringing the gripper under hand control to the grasping position and by transferring the coordinates of this position to the sensor computer. From these coordinate axes the grasp position is calculated with reference to the sensor system. With this information the positional values are determined which are needed for the calculation of the grasp position in the work mode.

During the work mode, the control unit waits for the transfer of the grasp coordinates. After a workpiece is identified by the sensors, first the grasp position has to be calculated with reference to the sensor system, and then this grasp position is transferred to the robot's coordinate system. The coordinate axes obtained by this transformation are transferred to the control unit together with

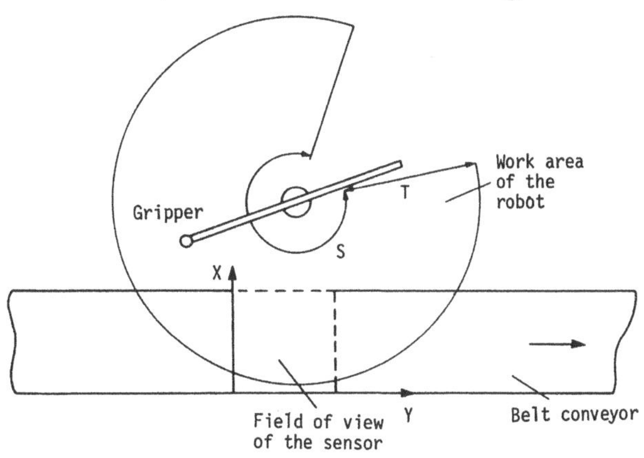

Fig. 9.36. Geometry of the robot-vision system

the program number of the action program assigned to the workpiece in the teach-in phase. The action program, modified by the new coordinate axes, can then be processed by the robot control.

The geometric set-up of the robot-vision system is demonstrated in Fig. 9.36.

The (visual) sensor coordinate system is described by X and Y, the robot coordinates by T and S.

Calibration

The robot vision system must be calibrated before performing the calculation and transformation of the grasp position from the vision system to the robot. The robot constants such as the distance between the centre of the gripper and the rotation centre of the swivel can be measured directly. The parameters of the interfaced sensor-robot system are:

F_T Number of horizontal transfer axis increment units (I_T) per pixel units.

S_0 The angular offset of the swivel axis when it is located parallel to the sensor X-axis.

X_0, Y_0 Coordinates of the centre of rotation of the swivel axis with regard to the sensor system.

To measure these parameters the sensor and the robot are brought to a fixed position. Assisted by the robot, a disc is moved between the three points P_1, P_2 and P_3 (Fig. 9.37).

A disc is usually chosen because of the following advantages. The possibility of an error while calculating the centre of gravity is small. There are no geometric changes when it is rotated. The grasp position can easily be defined as the centre of gravity of the disc.

The coordinate values of the robot system (S_1, T_1), (S_2, T_2) and (S_3, T_3) are measured for each point. Since the center of gravity of the disc and the grasp position are identical, the grasp position with regard to the sensor system (X_1, Y_1), (X_2, Y_2), (X_3, Y_3)) can be obtained by calculating the centre of gravity at the points P_1, P_2 and P_3. Thus a fixed assignment between sensor and robot coordinates is obtained and the four parameters mentioned above can be calculated.

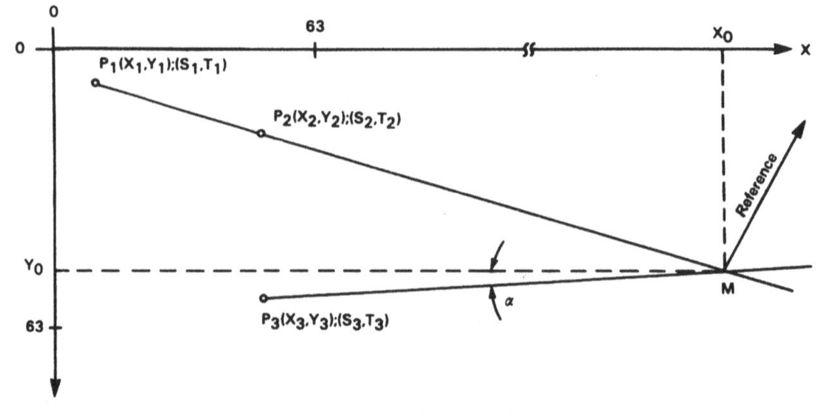

Fig. 9.37. Calibration of the sensor and robot coordinate systems

Coordinate Transformation During the Working Mode

In the work mode, besides the workpiece identification, the following tasks have to be performed:

- Calculation of the center of gravity $P_A(X_p, Y_p)$ of the workpiece
- Calculation of the rotational position of the workpiece
- Calculation of the grasp position G with regard to the sensor $G_A(X_G, Y_G)$
- Transformation of the coordinate of the grasp position to the coordinates of the robot system.

Figure 9.38 shows the interrelation of the different coordinates and the main system parameters during the working mode. The index L describes the values obtained during the teach-in mode. The values obtained in the working mode are marked by the subscript A.

The workpiece orientation γ_A is simply given by

$$\gamma_A = \gamma_L + \delta .$$

Here, γ_L is the workpiece orientation in the teach-in mode and δ is the rotational angle of the workpiece. With the aid of γ_A the grasp position $G_A(X_G, Y_G)$ can be calculated. Now the grasp position G_A in reference to the sensor system has to be transferred to the robot coordinate system. For this reason the position of the four robot axes must be calculated. The values of these parameters are then transferred to the robot control unit. Now the assembly robot is able to move to the calculated grasp position.

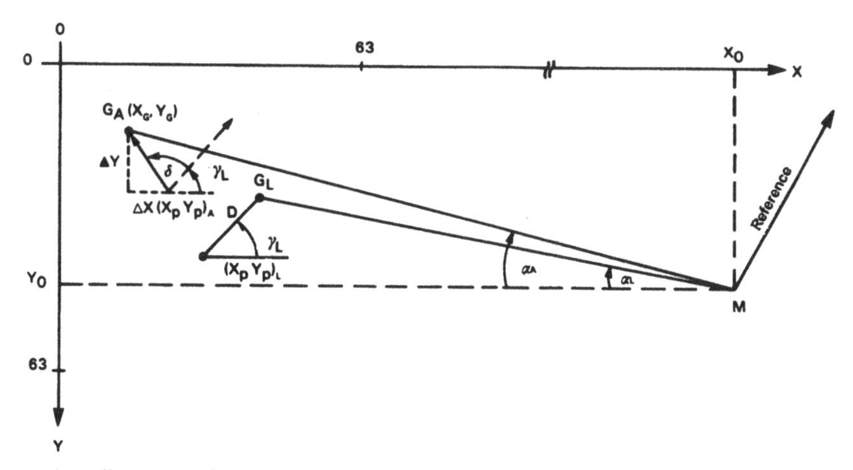

Fig. 9.38. Coordinate transformation during the working mode

9.3.6.2 Sorting of Castings

When using a robot in a foundry it becomes necessary to look for methods to distinguish part information from "noise". For example, in a foundry finishing room parts are not located in accurate fixtures. They may be scattered on the conveyor belts as they are discharged from tumble or cleaning machines or from

Fig. 9.39. Two robots used for sorting castings

grinding operations. Finishing room personnel stand alongside these belts, inspect the parts and segregate the good pieces from scrap. Downstream from inspection, the good parts are removed from the belt and placed on the shipping container. The remaining parts go to scrap.

An automatic method is required to recognize the different castings (e.g. 6) on the belt and to locate good parts. The average line rate can be about 1200 pieces/h.

To solve such a problem a vision system must be coupled with several robots. Figure 9.39 shows two of three robots, sitting adjacent to a process belt [41]. These three robots are coupled to one vision system. Each robot has a single gripper capable of picking up any of the (six) castings. The tracking of a robot is two dimensional.

Visual sensing is done with structured light (Fig. 9.40). Each lens focuses the light onto the belt forming a line across the width of the belt. About 2 meters above the belt a 256 cell linear diode array camera is mounted. When a part passes the line of light, the beams from the two sources are deflected out of the camera's field of view. One line is split up into two lines which are visible on the castings.

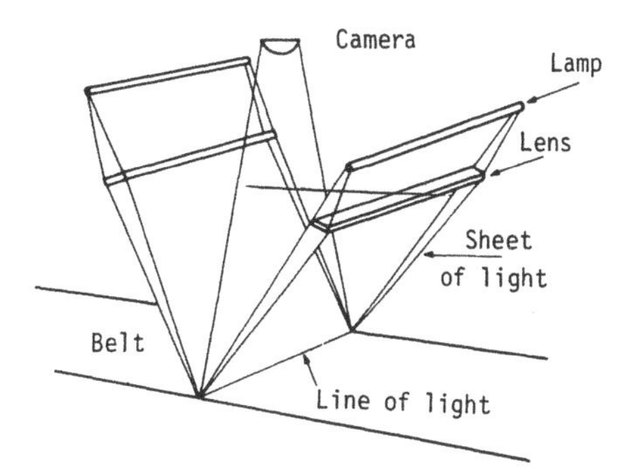

Fig. 9.40. Two linear sources projection sheets of light at 15° on either side of a camera

If a single light sheet is used instead of the two sheets a tall part would obstruct the light beam before the part is under the camera. The result is a false indication of the size and orientation of the part on the belt. The second sheet of light overcomes this problem.

Displacement of the light sheet indicates when a part is present. This information is then used to locate and identify a part. By segmentation (edge detection) the parts are separated. Once all the information from an object has been accumulated the vision software constructs the image and generates the part area, maximum and minimum part dimensions, number of holes and several other statistical characteristics. It then compares each of these features with the content of a previously taught library of parts, looking for a statistical fit (correlation). If a fit is found, then the system calculates the angular position of the part on the belt and locates the gripping point to be used by the robot. Coordinate transformation is performed and the robot picks the part off the moving belt and places it in the proper container.

A dispatching routine runs an algorithm which determines which casting will be picked up by which robot. As parts are identified by the vision system their positional and pick-up coordinates are placed into a queue. As a robot becomes available to pick-up castings the information is removed from the queue and transmitted to the robot. Only those robots which tell the vision system that they are available are considered. Their availability is examined on a continuous basis. The system is always aware of which parts are within the reach of a robot, and if for some reason a robot cannot reach a part the missed part will be assigned to the next robot downstream.

If a robot malfunctions, it will flag this to the vision system which will do an orderly shutdown of the data link and transfer the part load to the functional machines. In practice it has been found that two robots are able to handle a normal part flow of 1200/h. The third robot sits waiting to be called upon if the part rate increases or a malfunction of one of the upstream robots is detected.

9.4 References

1. Rembold, U, Blume, C, Dillmann, R (1985) Computer Integrated Manufacturing Technology and Systems. Marcel Dekker Publishing Co, New York Basel
2. Strohbach, K (1982) Rechnergestützte Qualitätssicherung. In: Fachseminar 3 zur FATEMA '82, Kernforschungszentrum Karlsruhe
3. Thierauf, RJ (1979) Distributed Processing Systems. Prentice-Hall, New Jersey
4. Warnecke, HJ (1983) Production Systems with Automated Information and Material Flow-State of the Art and Applications. In: Warman, EA (ed). North Holland, Amsterdam, pp 23–40
5. Paulus, TJ (1982) High Speed Timing Electronics and Applications of Hybrid Electronics in Nuclear Instrumentation. Electronic Devision, ORTEC, Oak Ridge
6. Codd, EF (1982) Relational Database: A Practical Foundation for Productivity. CACM, vol 25, no 2, pp 109–117
7. Lockemann, PC, Mayr, HC (1978) Rechner-gestützte Informationssysteme. Springer, Berlin Heidelberg New York
8. Bittner, J (1983) Data Independance in CAD/CAM Data Bases. In: Warman, EA (ed) CAPE '83. North Holland, Amsterdam, pp 573–587
9. Boehm, BW (1981) Software Engineering Economics. Prentice Hall, New Jersey
10. Rembold, U (1979) Prozeß- und Mikrorechnersysteme. Oldenbourg, Munich
11. Levi, P (1981) Betriebssysteme für Realzeitanwendungen. Data Kontext, Cologne
12. Levi, P (1979) The System Architecture of Data Acquisition Networks. Euromicro Journal, vol 5, no 6, pp 350–357
13. Buxmeyer, E et al. (1981) Serielles Bussystem für industrielle Anwendungen unter Echtzeitbedingungen (PDV-Bus), KFK-PDV 150. VDI-Berichte Nr. 399, pp 13–17
14. Levi, P (1983) Transaction Language Characteristics and User/Computer Interface in Manufacturing Systems. Microprocessors and Microsystems, vol 7, no 1, pp 3–17
15. Tsichritzis, D (1982) Form Management. CACM, vol 25, no 7, pp 453–478
16. Herzog, K (1980) Zeiss Multi-Coordinate Metrology, Hardware-Software Application. Reprint from Zeiss Information 91. Carl Zeiss Company, Oberkochen, W. Germany
17. APT 360 (1980) Reference Data. McDonnell Douglas, USA
18. Rosen, CA (1979) Machine Vision and Robotics: Industrial Requirements. In: Dodd, G, Rossol, L (eds) Computer Vision and Sensor-Based Robots. Plenum Press, New York, pp 3–20
19. Jarvis, JF (1982) Research Directions in Industrial Machine Vision: A Workshop Summary. Computer, Dec 1982, pp 55–61
20. Nagy, G (1983) Optical Scanning Digitizer. Computer, May 1983, pp 13–24
21. Kelly, RB (1982) Pose Refinement Vision. Proc. of the 2nd International Conference on Robot Vision and Sensory Control, pp 379–388
22. Niemann, H (1981) Pattern Analysis. Springer, Berlin Heidelberg New York
23. Gonzales, RC, Safabakhsh, R (1982) Computer Vision Techniques for Industrial Applications and Robot Control. Computer, Dec. 1982, pp 17–32
24. ´Agin, GJ (1980) Computer Vision Systems for Industrial Inspection and Assembly. Computer, May 1980, pp 11–20
25. Fu, KS (1982) Pattern Recognition for Automated Visual Inspection. Computer, Dec. 1982, pp 34–40
26. Vollath, D (1982) Das Bildanalysesystem PACOS. Praktische Metallographie, Bd 19, Heft 1, pp 7–23, Heft 2, pp 94–103
27. Rohde, A (1982) Application of an Image Analyser in Quality Control. Proc. of the 2nd International Conference on Robot Vision and Sensory Control. pp 53–61
28. Wasmund, H (1980) Automatische Produktionskontrolle und Materialprüfung. In: Kazmierczak, H (ed) Erfassung und maschinelle Verarbeitung von Bilddaten. Springer, Berlin Heidelberg New York, pp 245–256
29. Nawrath, RF et al. (1980) Visual Inspection and Quality Control by TAS. Proc. of the 5th International Conference on Automated Inspection and Product Control, pp 115–124
30. Giebel, H (1983) Nichtlineares Filtersystem zur Bildvorverarbeitung. BMFT-FB-DV83-001

31. Mergler, HW (1978) In-Process Optical Gauging for Numerical Machine Tool. 6th NFS Grantees Conference on Production Research and Industrial Automation. West Lafayette, India
32. Caulfield, HJ (1982) Holography: A Reassessment. IEEE Spectrum, Aug. 1982, pp 39–45
33. Bruhn, H, Felske, A (1981) Schnelle und automatische Bildanalyse von Specklegrammen mit Hilfe der FFT für Spannungsmessungen. VDI-Berichte 399:13–17
34. Nitzan, D et al. (1977) The Measurements and Use of Registered Reflectance and Range Data in Scene Analysis. Proc. of IEEE, vol 65, no 2, pp 206–220
35. Knoll, GF (1979) Radiation Detection and Measurement. John Wiley & Sons, New York
36. Levi, P, Weirich, E (1983) Differential Reflectance Functions and Their Use for Surface Identification. Proceedings of 13th ISIR/ROBOTS 7 Symposium, pp 17.61–17.77
37. Martini, P, Nehr, G (1983) MOBIP: Ein modulares Bildverarbeitungssystem mit Parallelrechner. Elektronische Rechenanlagen, 25. Jahrgang, Heft 2, pp 55–65
38. Bolles, RC (1981) Three-Dimensional Locating of Industrial Parts. Eight NFS Grantees Conference on Production Research and Technology, Stanford, California
39. Horn, BK, Sjoberg, RW (1979) Calculating the Reflectance Map. Applied Optics, vol 18, no 11, pp 1770–1779
40. Nehr, G, Martini, P (1983) The Coupling of a Workpiece Recognition System with an Industrial Robot. In: Pugh, A (ed) Robot Vision. Springer, New York, pp 83–96
41. Baumann, RD, Wilmshurst, DA (1983) Vision System Sorts Castings at General Motors in Canada. In: Pugh, A (ed) Robot Vision. Springer, New York, pp 255–266

10 Production Control and Information Systems

A.-W. Scheer

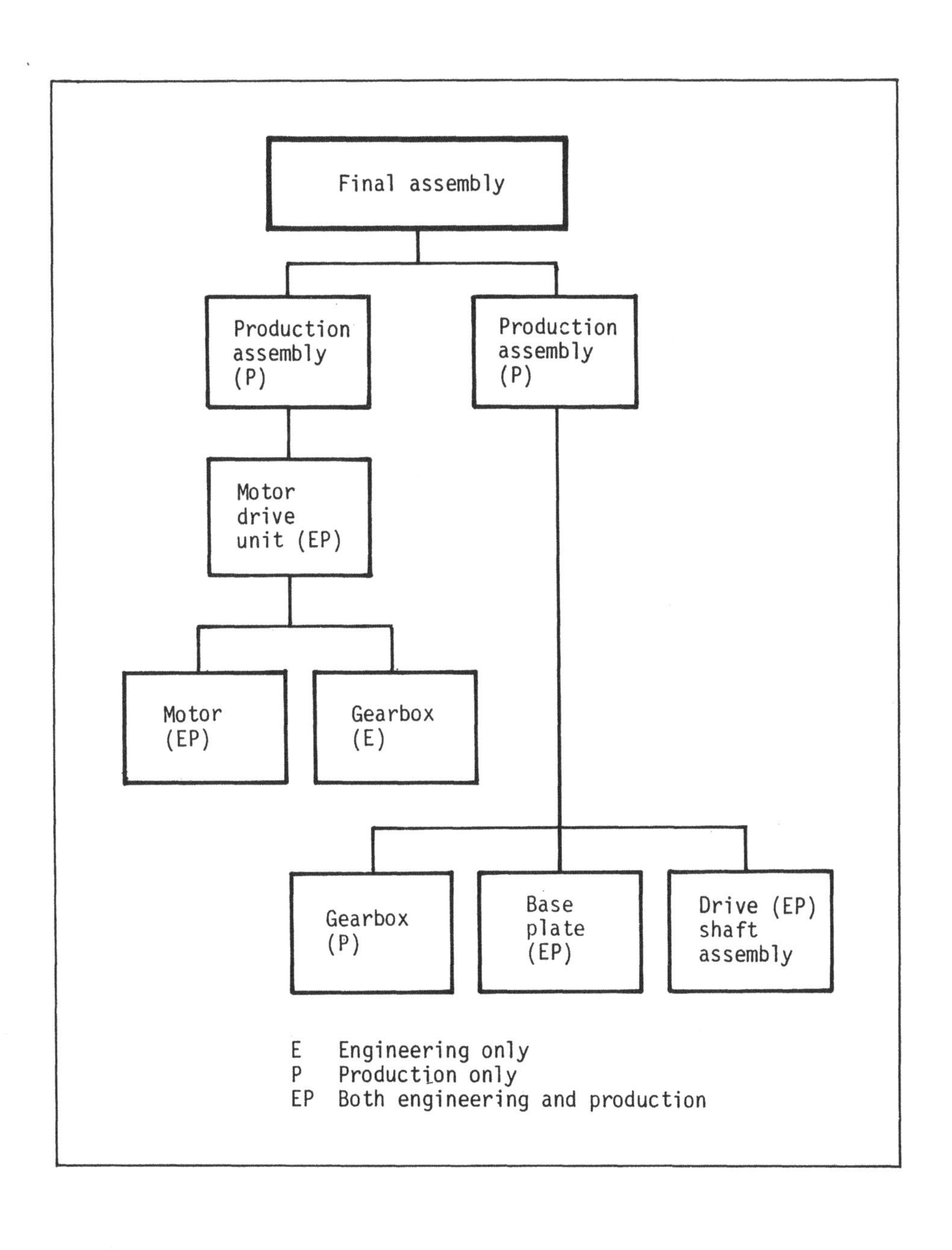

Data structure of a product

(Courtesy of International Business Machine Corporation, White Plains, New York, USA)

10.1 Strategies for the Selection of Software Packages in Production

Within recent years the commercial availability and implementation of software packages for production planning and control have suddenly increased in the Federal Republic of Germany (see Fig. 10.1). One major reason for this development is their favorable cost in comparison with software developed in-house. The quality of these software packages has also increased.

Empirical studies have shown that approximately 75% of the total cost of a software system arises after the development phase, i.e. in the maintenance phase. A typical cost structure for software systems is shown in Fig. 10.2. The utilization of standard software can lead to cost decreases as compared with inhouse developments, especially for the implementation and test phases. Also, considerable

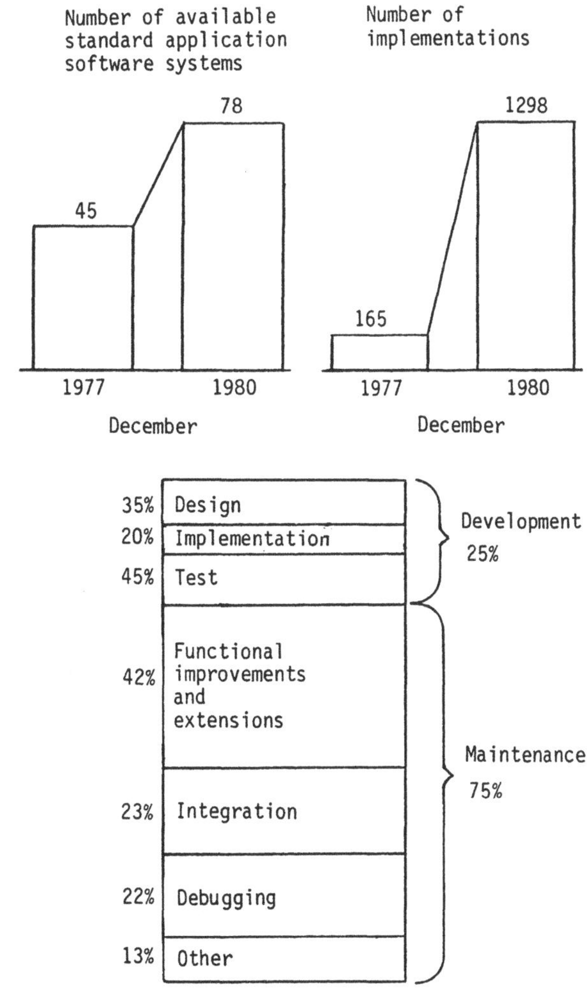

Fig. 10.1. Number of available software systems and implementations for production planning in West Germany

Fig. 10.2. Cost relation for self developed application software

cost decreases can be realized for the improvement and expansion of software since standard packages are maintained by the producer. This allows the user to obtain modifications and extensions at relatively low cost, even after the completed purchase of a software package. If a system is bought which is already integrated within a software family, the user also saves the cost of adapting the existing software environment.

These factors lead to the current situation where the cost of purchasing a software package amounts to only 10 to 15% of the cost of software developed in-house. It must be noted, however, that in judging the profitability of a software package the necessary integration of the software into the existing organization must be taken into account. An additional cost equal to the purchasing cost of the entire software package can often be incurred. Nevertheless, the effect remains the same, that the purchase of a software system is more cost effective than in-house development.

During the process of selecting a software package to be purchased, careful attention must be paid to the question of whether or not the requirements of the enterprise can be fullfilled by the standard package.

One further item to note in dealing with profitability is that, while the cost of a production planning and control system can be determined quite exactly, difficulty may arise in estimating the utility of such a system. In contrast to a typical production investment, e.g. the purchase of machinery, the causal chain between the output of an information system and the actual change in the cash flow of the

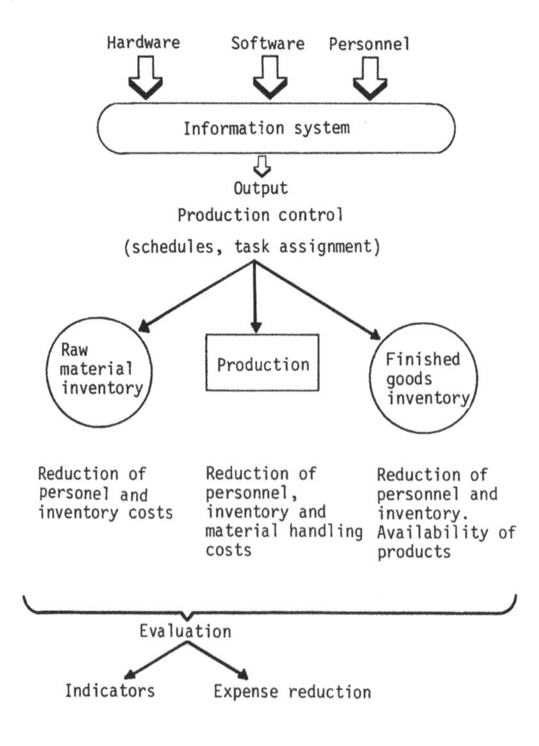

Fig. 10.3. Task of the factory information system

enterprise is significantly longer. The output of a production planning and control system consists solely of information, which can effect a decrease in inventory, an increase in throughput etc. This only happens after human directives have translated the information into a concrete action (Fig. 10.3).

Advantages of standard software include:

- standard solutions are tried and tested
- high degree of availability
- low personnel requirements
- reliability
- time-saving in development and maintenance.

Advantages of in-house development include:

- exact fit of the software to the problem
- integration into the existing problem solution
- less overhead in terms of additional functions
- more readily modified

The following steps are involved in the selection of standard software systems:

1. *Determination of functional requirements.* After the relevant departments of the enterprise have thoroughly studied the planning philosophy embodied in a production planning and control system, the requirements which are typical for the enterprise are catalogued. The question of whether the enterprise is oriented more toward customer-centered piece rate production or toward mass production is of the utmost importance for the selection of a standard software system. It can be seen that, while planning systems developed in the US are generally oriented towards mass production, the situation in the Federal Republic of Germany in, say mechanical engineering, is characterized by close contact with the customer and his special wishes. These factors must be taken in account when designing a production system.
2. *Analysis of the current state of the EDP-technical and commercial environments.* Extremely important for the selection of a standard software system is the underlying hardware configuration. If a decision concerning the hardware configuration is dependent upon the selection of a software system, then the number of alternatives to be considered is significantly larger than in cases where the hardware system has already been decided upon. Further applications to be covered, such as cost accounting and payroll, may also serve to narrow down the selection process. For this reason it is of extreme importance that the prospective environment is studied, so that from the very beginning the required data interfaces are taken into account.
3. *Conception of a preliminary requirement profile.* The catalogue of the requirement forms the basis upon which a preliminary requirement profile may be constructed. This serves to allow a preliminary selection of those software systems which are to be more carefully considered during later evaluation.
4. *Cataloguing of systems to be considered.* For a preliminary selection the number of EDP-systems initially considered should be as large as possible. Sources of

information in Germany are the ISIS-Catalogue as well as numerous publications in technical journals.

5. *Preliminary selection.* Using the requirement profile constructed under 3. above, an evaluation of the various systems may be made based on the information provided by the system producers. Systems which do not fulfil important requirements can be eliminated by use of specific criteria. For example, the question of whether data management is to be accomplished as a data base system or with conventional data files, or whether a real-time dialogue is provided, may be examined at this point.
6. *Development of a detailed catalogue of criteria.* Since the amount of knowledge available to the members of the selection team from the relevant production department and from the EDP department increases during the selection process, a more refined catalogue of criteria may be compiled for the final selection.
7. *Evaluation.* Based on detailed criteria, which may include, for example the lengths of key data fields, a more thorough evaluation of the remaining systems may be carried out.
8. *Inspection of reference installations.* The evaluation may be further verified by a visit of reference installations, and in particular by a discussion with the users about their experience.
9. *Cost-utility analysis and final decision.* Based on the information compiled so far, a cost-benefit analysis can be made, the results of which embody the final decision proposal. A decision tree may be used for the selection of a standard software system.

Important criteria which must be considered in the selection of a software package are derived both from EDP-experience obtained from handling of the system and from the business functions of the software. From the EDP-point of view of particular importance are the degree of dialogue support, type of data management, and the modularity.

It must be emphasized that, in addition to the management of fundamental data, interactive decision-making systems and simulation systems all require dialogue support. This in particular applies to capacity planning and job scheduling. Batch-oriented solutions have delivered only unsatisfactory results, due to the great complexity of the planning effort involved. At present data base systems are preferred for data management.

A high degree of modularity makes it possible to put together those functions which are suited to the organizational structure of the individual enterprise. At the same time it allows easier interfacing of adjacent areas. With respect to the business functions of the system, the degree of integration of the operational procedures and the orientation of the system with the planning structure as a whole are of importance. The integration of operational procedures is increasingly important in a successful automation endeavour. Redundant data entry can be avoided and a high availability of information can be achieved.

Important for the overall planning structure is the orientation towards mass or piece rate production. A further distinction is that of whether a single endproduct is made up from many parts, or whether many end-products are produced from few raw components. In the former case the management of the bill of materials is of primary importance, while in the latter case it is the timing of the production lines which stands in the foreground.

Many software packages for production planning and control are oriented towards the management of the bill of materials and associated information, in other words, towards the structure of the machinery industry; while the consumer goods industry can expect little support from such systems.

Due to the ongoing development of microcomputers and of standard software, it can be expected in the near future that suitable standard software will also be available to smaller enterprises.

10.2 Data Management Requirements for Production Control

The primary data for a production planning and control system consist of the bill of materials, work description and equipment information. There is usually a large quantity of complex data required. For this reason, special data management systems (bill of materials processor) were employed very early in the manufacturing sector; whereas general data base systems evolved from such systems at a later date. In the following, a data base schema for manufacturing data will be developed based on a network model.

10.2.1 Development of a Data Base Scheme for Primary Data

10.2.1.1 Bill of Materials

The assembly of end-products may be presented graphically by the use of product trees (Figs. 10.4a and b). In such product trees, however, redundancies may

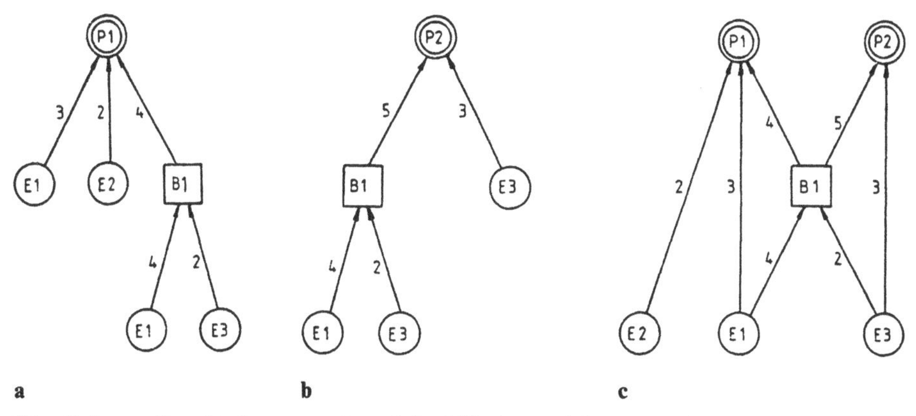

Fig. 10.4a–c. Graphical presentation of the bill of materials: **a**, **b** with possible redundancy, **c** without redundancy

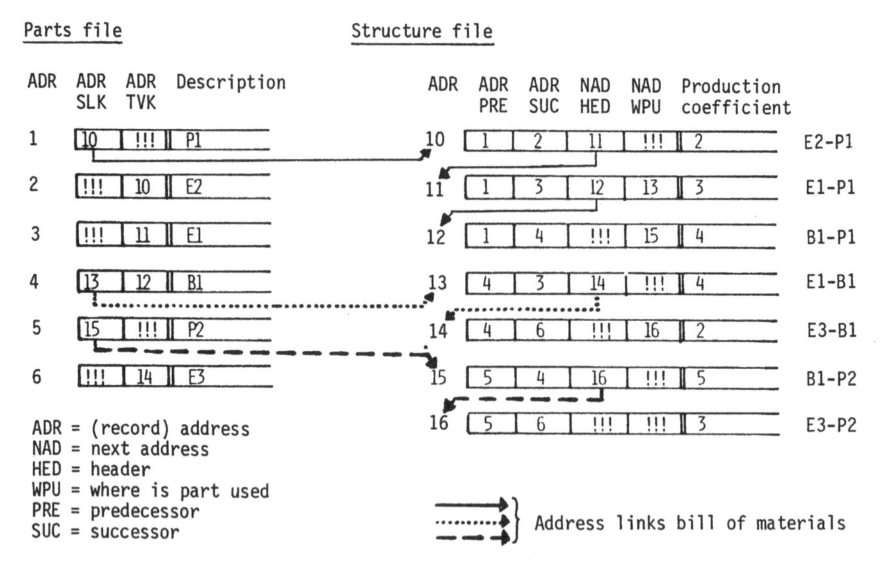

Fig. 10.5. Principle of presenting a part in the bill of materials

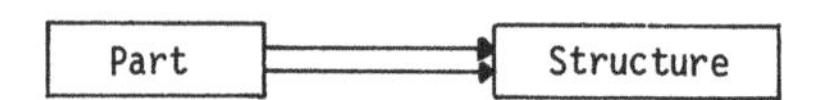

Fig. 10.6. Links within a bill of materials

occur, in the case shown the parts E1 and B1. In order to avoid these redundancies, the product structures may be combined to form a special graph (Fig. 10.4c). In it each part and each arc are represented only once. In order to avoid redundancies in an EDP-supported data management system, and to reduce memory size and updating costs, the data storage for the bill of materials will follow the principle embodied in the special graph. For each part one record in the parts file, and for each arc one record is created in the product structure file. The two record types are connected together by means of set relations (Fig. 10.5). Thus the many-to-many relation of the entity type "part" is reduced to two $l{:}n$-relations. From the bill of materials view the records for all arcs for a part point to that part. Whereas from the view of the part usage the records for all arcs point away from that part. In Fig. 10.6 the address links for the product tree of Fig. 10.4 are shown. From each part record two address pointers are directed to the structure records. Each structure record contains the addresses of the following links and these of the preceeding and succeeding part.

In addition to the part number and part name, typical fields in the part record, are sales price, purchase price, lead times, supplier, ordering quantity, etc. In the product structure record the fields designate the production coefficient, expiration data and information on variants.

10.2.1.2 Work Descriptions

For each part produced in-house there may exist one or more work descriptions, in which production instructions are maintained. Different work descriptions exist in cases where a part can be produced by more than one method or process. This is the case when, for example, small orders are produced in a machine shop and large orders are produced on a production line.

Each work description may contain several operations. A work piece routing is a complete step within a work description. Sometimes a routing can be done through more than one machine. In this case the cost and production times are usually different. If a routing is seen in terms of its header information only, then the actual material required can be assigned in corresponding parallel routing records. This record type contains the actual planning information such as lead time, transition time, splitting factor, etc. Since there are *l*:*n* possible relations for a part, the work description, routing and parallel routing, the data base schema has the structure shown in Fig. 10.7.

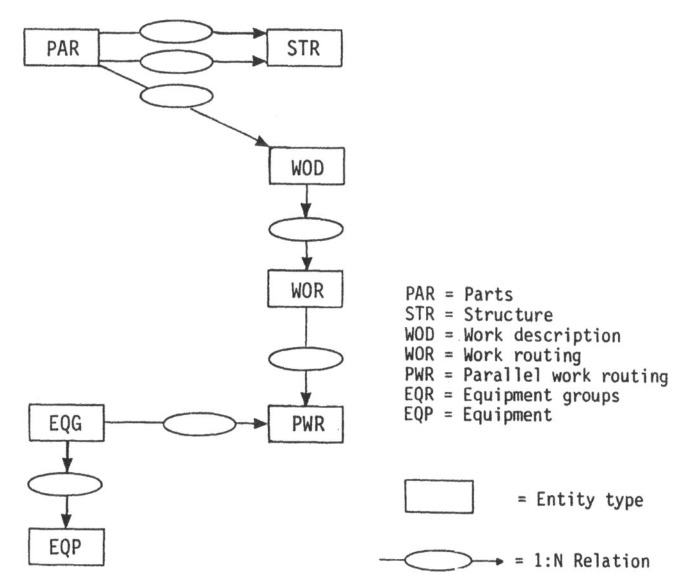

Fig. 10.7. Structure of a data base for production

10.2.1.3 Manufacturing Equipment

For each equipment class, in which similar materials (for planning purposes) are combined, one record of the type equipment class is created, and the information on the individual aggregates are maintained in a record type equipment. Since several pieces of manufacturing equipment may belong to a given equipment group, a *l*:*n* relation may exist.

On each equipment group several work steps may be performed (workpiece routing), but only one (parallel) work step can be performed on one equipment class. Thus a $l:n$ relation exists between the equipment class and the parallel workpiece routing. The completed data base schema, shown in Fig. 10.7 contains all structures required for the management of primary production data.

10.2.2 Special Cases of Variant Production

Many industrial sectors, in particular machinery and automobile production, have products where, through slight modification of an end-product, a great number of similar but variant products may be produced. A number of methods have been developed for the management of the required information (Figs. 10.8a–d). For each method a different number of part or structure records

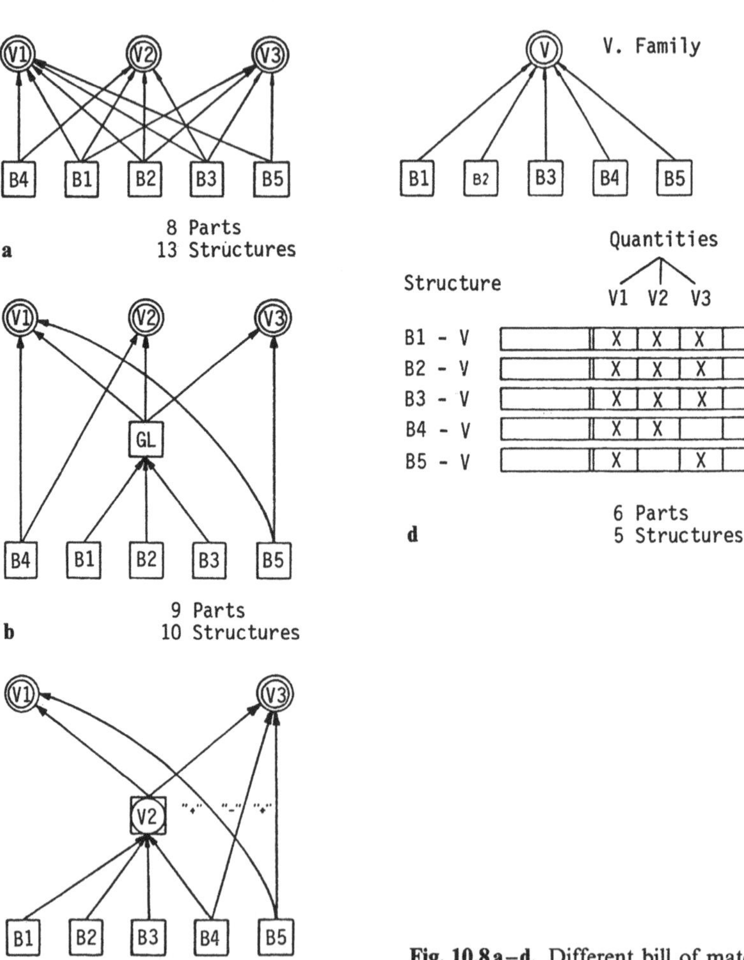

Fig. 10.8a–d. Different bill of materials: **a** storage of variants, **b** uniform parts, **c** plus-minus, **d** multiple

is required for the representation of the three variants V 1, V 2 and V 3 which are made up of the five assembly parts B 1 to B 5.

The most expensive storage form is that shown in Fig. 10.8 a. In this method one record is created for each part, and also for each relevant structural relationship.

In the so-called uniform parts-bill of materials (Fig. 10.8 b), the parts B 1, B 2 and B 3 are grouped into a fictitious assembly group GL, so that these parts enter into each of the variants as an entity. The number of structure records required by this method is smaller than that of Fig. 10.8 a.

In the so-called plus-minus bill of materials (Fig. 10.8 c), one variant serves as a pattern; deviations from this pattern of the other variants are marked in additional fields in the structure records.

In the multiple bill of materials (Fig. 10.8 d) only a single ficticious family of parts is defined, and all assembly groups are linked to it. Markers for a given variant are created and are placed in the structure records.

10.2.3 Data Management with Software Packages

Four different forms of data management can be found in standard commercial software.

10.2.3.1 Conventional File Management

Some software packages utilize traditional forms of file management such as Indirect Sequential Access Method (ISAM) or Virtual Sequential Access Method (VSAM). This has, however, certain limitations with respect to the management of complex data structures, e.g. of the bill of materials. On the other hand, these software packages are highly portable compared with those which employ data base systems.

10.2.3.2 Specialized Data Base Systems

Some vendors offer software containing their own complex file management system which has some of the features of a data base system. The advantage of such a system is that it is highly specialized towards the kind of data structures required in production. However, only limited extensions are possible, so that any integration into available management or technical problem areas is limited.

10.2.3.3 General Data Base Systems

Most hardware producers and large software houses build their software packages with the help of a general data base system, e.g. COPICS (IBM), UNIS (Univac), MIACS (Honeywell Bull) and PM/MM 3000 (Hewlett-Packard).

The advantage of this form of data organization lies in the generalized utilization of a powerful data base system and in the relative ease in integrating such a system with existing problem areas. A further advantage is the utility of having a unified recovery system for the data management of the entire enterprise.

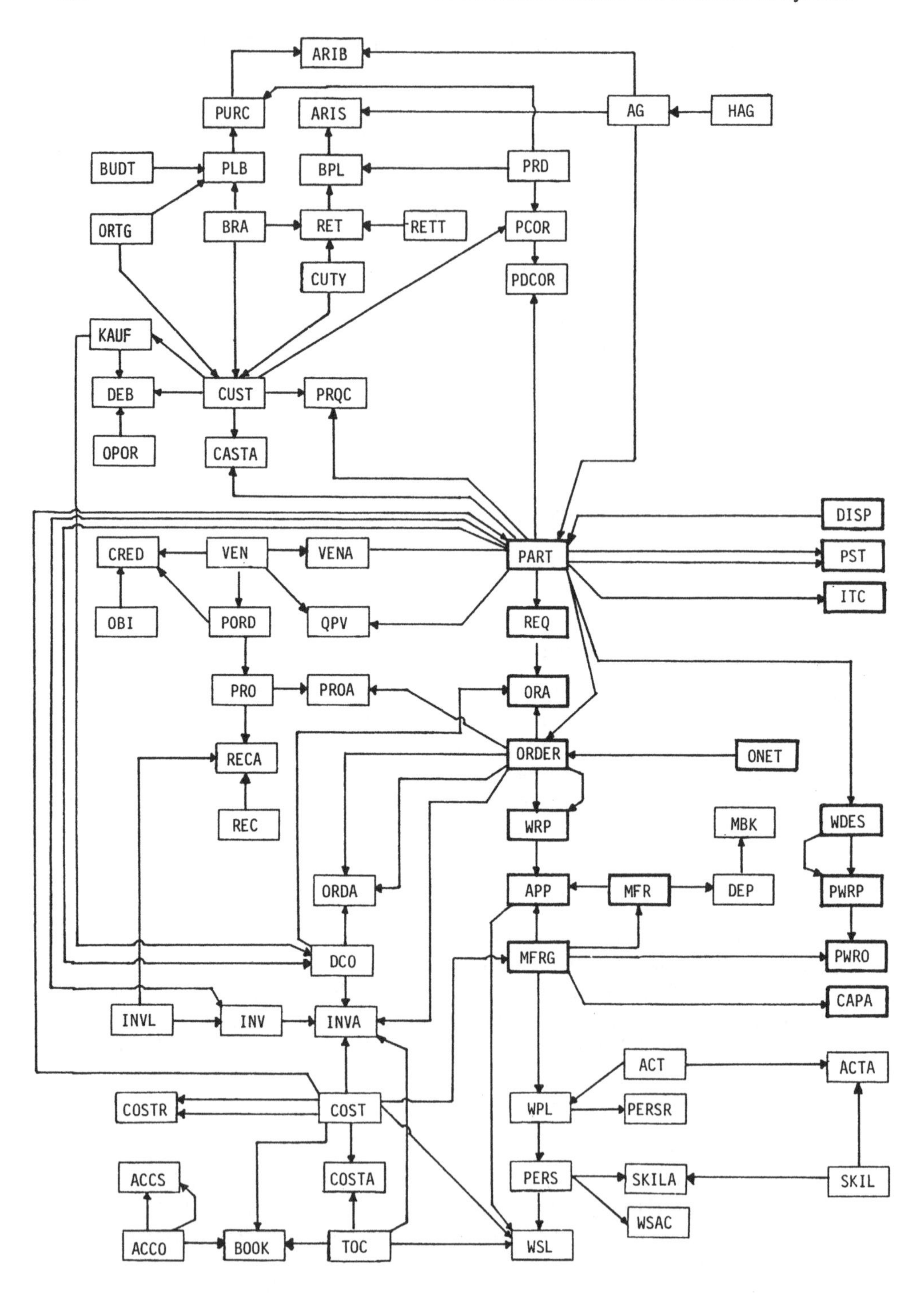

Fig. 10.9. Company data base scheme (explantation see opposite page)

ACCS Account structure
ACT Activity
ACTA Activity assignment
APP Actual parallel process
ARG Article group
ARIB Article related information for planned budget
ARIS Article related information for planned sales
BOOK Book-keeping entry
BPL Budget plan
BRA Branch
BUDT Budget type
CAPA Capacity assignment
CORD Customer order
COST Cost
COSTA Cost assignment
CRED Creditor entry
CUST Customer
CUTY Customer type
DCO Detailed customer order
DEB Debitor book-keeping
DEP Depreciation
DISP Disposition level
INAC Inventory accounting
INV Inventory
INVL Inventory location
ITC Inventory control
MBK Maintenance book-keeping
MFR Manufacturing resources
MFRG Manufacturing resources group
MPC Main product category
ONET Order network
OPBI Open bills
OPOR Open orders
ORA Order-requirements assignment
ORDA Order assignment
ORDER Order (purchase order/ production order)
PART Parts
PCOR Past customer order
PDCOR Past detailed customer order
PERS Personnel
PERSR Personnel requirements
PLB Planned budget
PORD Placed order
PRD Period
PRO Part related order
PROA Part related order assignment
PRQC Part related quotation of customer
PST Product structure
PURC Purchasing (planned budget)
PWRP Parallel work routing plan
QPV Quotation of part vendor
REC Receiving
RECA Receiving assignment
REQ Requirements
RET Retailer
RETT Type of retailer
SOCI Size of city
SKIL Skill
SKILA Personnel skill assignment
SUSTA Part/customer assignment
TOCK Type of cost
VEN Vendor
VENA Part/vendor assignment
WDES Work description
WPL Workplace
WRO Work routing
WRP Work routing plan
WSAC Wage and salary accounting
WSL Wage slip

Fig. 10.9 (continued). Explanation of abbreviations

An integrated data base design for the entire enterprise also promotes a clear arrangement of all data structures and the avoidance of redundancies across all application areas.

10.2.3.4 System-Independent Data Bases

Since the portability of standard software based on a general data base system is limited, some software houses now employ data base independent interfaces. The interfaces Compatible Data Base Interface (KDBS) and Compatible Data Communications Interface (KDCS) were developed in Germany. The Input/Output commands are separate from the programs so that it is possible to link the software core to different data base systems without modification. Only the interface between the I/O commands and the actual data base system needs to be developed for each case.

A disadvantage of this procedure is that the full power of the underlying data base system is no longer available, since the software must be limited to those functions which are common to different data base systems.

10.2.4 Future Developments in Data Management

The employment of general data base systems for the management of primary production data and for the foundation of production planning and control is increasing. This development emphasizes that primary production data play a central role in all areas of an enterprise. For instance, the bill of materials, work description and manufacturing equipment data are needed for product unit costing. Also online costing requires that the cost accounting program can directly access production data. This is the reason why software packages contain routines for product cost calculations. Thus a common data base leads to the integration of functionally different applications (cost analysis, production planning and control).

Generally it can be noted that data bases are becoming more important for information systems. The increasing availability of higher programming languages allows individual departments to call up their own information from existing data bases. Very little programming knowledge is required. Such possibilities, however, only exist when the data bases have been structured very flexibly. For this reason, more attention must be paid to the structure of the data base than has been done in the past.

This fact has also been verified by empirical studies. In one investigation of a CAD system it was found that, prior to the installation of the system, the user mainly stressed the importance of complex software functions; soon after the system was installed the possibility of utilizing the data base more widely was seriously considered, and after several years of use the importance of the data base became apparent.

Data management is important for many adjacent production areas. This applies not only to cost accounting but also to payroll. For this work records contain standard times which are used to compute wages. Also, the bill of materials contains unit quantities which can be added up and used for performance checks.

11.1 Introduction

Computer integrated manufacturing systems provide important economic advantages over less flexible and less controllable forms of manufacturing. Increased *flexibility* is particularly advantageous in the short run for small-batch production of mixed parts on the same system, and for longer run changes in product specifications and volume requirements. Increased *controllability* is particularly advantageous in sychronizing the flow of parts and tools over a greater part of the total production cycle, in maintaining greater and more uniform dimensional accuracy, and in stabilizing production during design and schedule changes. These advantages are realized in the form of lower direct labor requirements, lower in-process inventories, greater machine utilization, shorter lead times, and improved quality. On the other hand, such advantages require larger initial capital investments, more intensive use of such facilities, training in new labor skills, organizational changes, and the use of new planning and control methods.

To evaluate the economic benefit of CIM systems, simulation methods like GCMS [1], Q-GERT [3], and stochastic network models like CAN-Q [2] can be used to estimate how the output capacity of a manufacturing system changes as the various components of the system are increased or improved for a known cost. This data can be used to develop *cost vs. output curves* for a given part or product mix. By comparing these curves for different types of production systems, it is possible to identify break-even points where a particular type of system is economically preferred over other systems for particular products or product-mixes. By varying the design and cost parameters, it is possible to study the effects of changes in economic conditions and changes in the design and cost of hardware.

To demonstrate this approach, a comparison is made between two process plans for the same prismatic part. The one plan, called LINE, represents a production line, with six different stations each manned by an operator. The second process plan, called FMS, uses a multi-function machining center, a computer-controlled transporter, and off-line load-unload stations in a flexible manufacturing system. A convenient notation for describing the process plans is developed and used to define the two process plans. Realistic estimates of the production capacities for plans are obtained by successive application of Solberg's [2] CAN-Q network analysis program to estimate the output for a minimal system, identify the bottleneck station, and then increase that station's capacity so as to increase production. In this way, it is possible to estimate the capital and labor requirements for a given output capacity. The results indicate that capital and labor increase in an approximately linear fashion with output capacity.

A notation is developed for showing how the capacity expansion data can be used to evaluate payback criteria, to make rate of return analysis, and to determine how the cost per part is related to output capacity and system design. These calculations take into account the cost of labor, the cost of capital, and the effect of taxes on computing the net rate of return after taxes. The advantage of FMS over LINE is shown to be sensitive to the capital-labor cost ratio and to the confirmed customer orders. The relationship between marketing and production, in particular with respect to master production scheduling, will be discussed later, in the section on the interface between production control and marketing.

10.3.1.2 Material Requirement Planning

In this stage, the master production requirements are divided, with the help of the bill of materials into requirements for assembly groups and parts to be purchased externally. For the gross-net calculation both stock on hand and projected stock increases are taken into consideration.

Material requirement planning uses the method of disposition levels to avoid redundant processing of individual parts. The disposition level is the length of the longest path from all leaf nodes (externally purchased parts) to the part in question, plus 1. The length of a path is equal to the number of arcs in the path.

To determine the amount of material needed, production lots are formed from the net requirements. Production planning systems offer different approaches to optimize this procedure. The well-know Andler-formula assumes a static requirement curve. This, however, is true only in very few cases.

The sliding economic lot size and the part period methods allow a dynamic requirement curve to be used. The formulas are based on heuristic methods. They do not always lead to an optimal result.

By using dynamic optimization (Wagner-Whitin algorithm), the optimal solution for the dynamic requirements within a given planning period can be determined. This approach has so far hardly ever been considered in PPC systems.

The formulae needed for optimizing have many insufficiencies. There is also significant critizism to be made concerning the procedures employed to determine the optimal lot size. One such criticism is that the lots are formed for each part separately, whereas the parts actually exist as system components. Thus interdependencies must be taken into consideration. Numerous approaches in optimizing lot sizes in multi-echelon production processes are known from the literature. These approaches lead, however, to complicated optimization procedures which can only be performed using mixed integer optimization or complex heuristic calculation methods.

The results of material requirement planning are production orders for the parts produced internally and purchase orders for parts purchased externally. The purchase orders are placed by the purchasing department. The buyer can use this information to determine the most favorable supplier, to check the composition of the orders, to bargain for discounts, and to make the final order.

The production orders are the input to the next planning stage, which is capacity planning.

The document needed for requirement planning is the bill of materials. To perform the gross-net calculation an inventory control has to be carried out for all end products and production components, which are contained in the special graph. Since the bill of materials and an inventory control operation are basic activities of data processing. EDP-supported requirement planning may be a first step in the implementation of a PPC system.

One danger, however, lies in the fact that the relative simplicity of this planning step causes many firms to develop and program their own planning functions. The structural definition of the bill of materials imposes certain restrictions on the succeeding planning stages, and in particular on the use of software packages for the more complicated planning operations.

10.3.1.3 Capacity Planning

In the capacity planning phase the production orders are modified to include time data for the work description and processing steps. For this procedure the primary data on the work description, work steps and equipment must be available. This often represents a real barrier within the planning model, since a careful study is required for the compilation of these data. In addition, these planning functions are significantly more complex than those of requirement planning, thus they place greater demands on the quality of the firm's planning. After time data have been added to the orders, an exact throughput schedule can be made. The results of this planning stage are the capacity plans for all equipment groups. The bottlenecks and idle times of each process are determined. The more modern EDP systems allow interactive capacity planning of orders and the display of capacity summaries on the screen (Fig. 10.10).

When bottlenecks become visible they can be corrected by means which do not alter the work schedule. These include the introduction of overtime or the selection of alternative manufacturing routes.

Difficult problems may arise, however, in the re-timing of operations. Since operations are all interconnected within a network of orders, an isolated change in scheduling of a single operation is impossible. Instead, entire order processing sequences must be changed. This can lead to new bottlenecks for other equipment groups, where none had existed before.

Batch oriented systems to minimize idle capacity have often led to long runtimes and unsatisfactory results. Of particular importance to this project stage is the need to keep a relatively detailed schedule for the work steps. Such a schedule

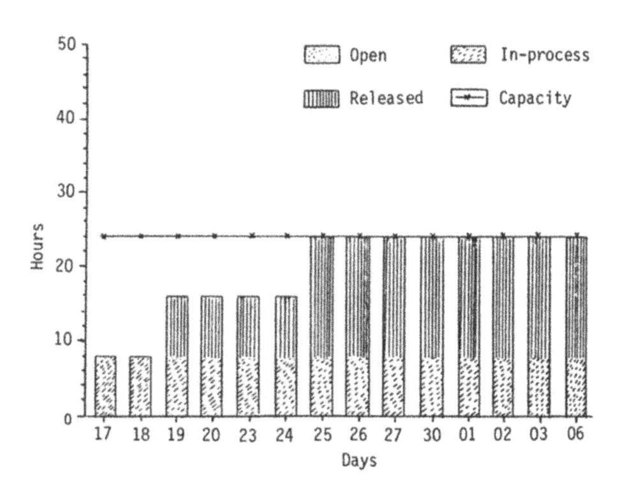

Fig. 10.10. Typical capacity plan for a work center

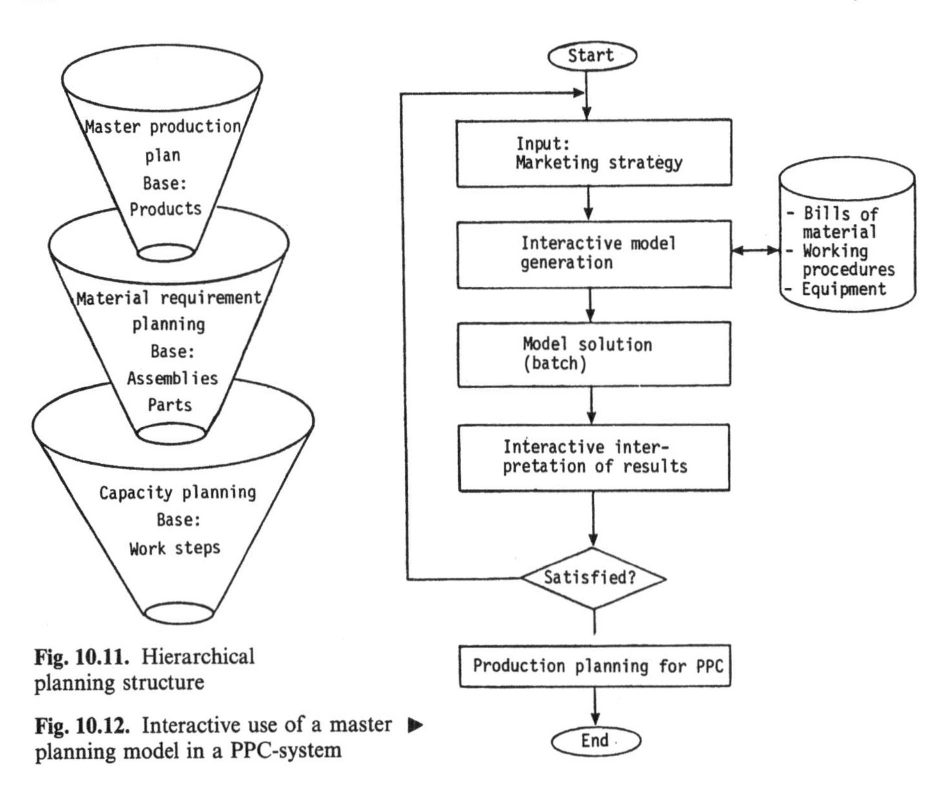

Fig. 10.11. Hierarchical planning structure

Fig. 10.12. Interactive use of a master planning model in a PPC-system ▶

may encounter problems when put into practice. In particular, the assumed lead times are seldom realized in practice, since variations in lead times arise with queuing situations.

For this reason a man-machine dialogue is used by the modern capacity planning systems. When unexpected events occur, such as a defective production machine or the processing of an urgent customer order, they can be solved quickly via dialogue.

One special problem of this planning stage is that the complexity of planning becomes very great due to the numerous interdependencies among the work steps to be scheduled. In addition the quantity of parts to be processed can be very high. Figure 10.11 demonstrates this situation using the funnel concept. Whereas the first planning stage (master production scheduling) only has to deal with a small number of end products, the number of items to be processed increases drastically during the material requirement planning stage, where the parts are itemized in the bill of materials. In capacity planning work steps are added to the model, thus further increasing the number of discrete entries.

At the author's institute a system is under development for master production scheduling which can handle existing capacity limitations. Thus the following stages, especially the capacity planning stage, may be based on quantitative solutions. Central to this model is a linear programming approach which processes the primary production data with the help of a matrix generator, i.e. the bills of materials, work descriptions and data about manufacturing equipment (Fig. 10.12).

10.3.1.4 Job Shop Control

For job shop control the orders are arranged so that they can be processed by the available production equipment in an optimal sequence. First the components and tools are selected for the given planning period to fill the production order on time. After a check the orders are released and placed in optimal sequence, according to their priority level.

The heuristic procedures for determining the optimal priority levels are similar to those used in capacity planning. They are applied to the orders, while in the job shop control stage they are applied to work steps. Typical criteria are first in – first out and shortest operation times. It has been shown by simulation studies that the shortest operation time method leads to a relatively fast throughput, and has minimal variation in processing times.

10.3.1.5 Data Collection

After the detailed planning has been completed, the individual work steps are performed in the manufacturing facility. For this the order processing papers must be made available. In the data collection stage, the relevant data on processing stages are fed back into the planning system. The information collected are on:

Machinery,
Personnel and
Material.

In order to reduce potential error of individual workers the identity of each work step is printed directly on the operation paper in machine readable form. The most important data included are part number, work description number and work step number, as well as a description of the actual operating activity. Thus the worker needs only add up information on quantity and time. This information is automatically linked to the proper order file.

Machine readable data collection devices generally use magnetic code or bar code. Different manufacturers follow different coding philosophies. It appears that at the present time the bar code is most often used.

The data collection phase closes the planning circle. Since job shop control constantly requires an up-date data base a very close interaction exists between job shop control and data collection.

10.3.2 Implementation Strategies

Besides the straightforward implementation strategy mentioned before, which follows the same order as the planning stages, other practices are used in manufacturing. For example, some firms introduce data collection system without coordinating them with the production planning functions, which precede the data collection phase. The argument that exact and up-to-date information on

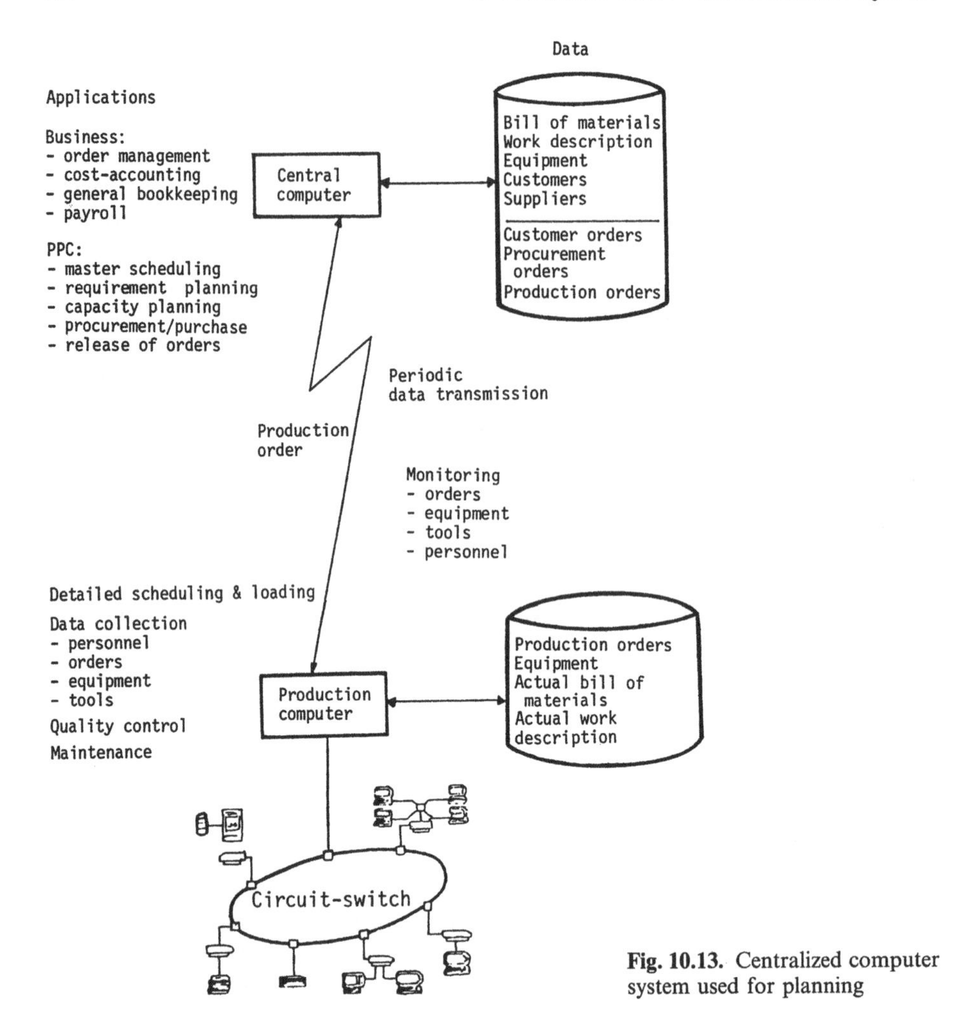

Fig. 10.13. Centralized computer system used for planning

production quality and quantity allows effective handling of unexpected disturbances may suffice for many production managers. This function alone may justify the financial investment in installing a data collection system. However, such implementation strategies lead to great difficulties if the data collection system is to be linked at a later date to the integrated production planning system. These difficulties have to do with software and hardware. Since many different process control and data collection functions are often involved in production, process control computers are frequently employed for process monitoring. Thus real-time data processing plays an important role. If a hardware system is installed for isolated functions, e.g. quality control, then the standard software available for additional data collection and production control functions is limited. For this reason it is recommended that a coordinated implementation stategy for production planning and control, and for data collection will be followed.

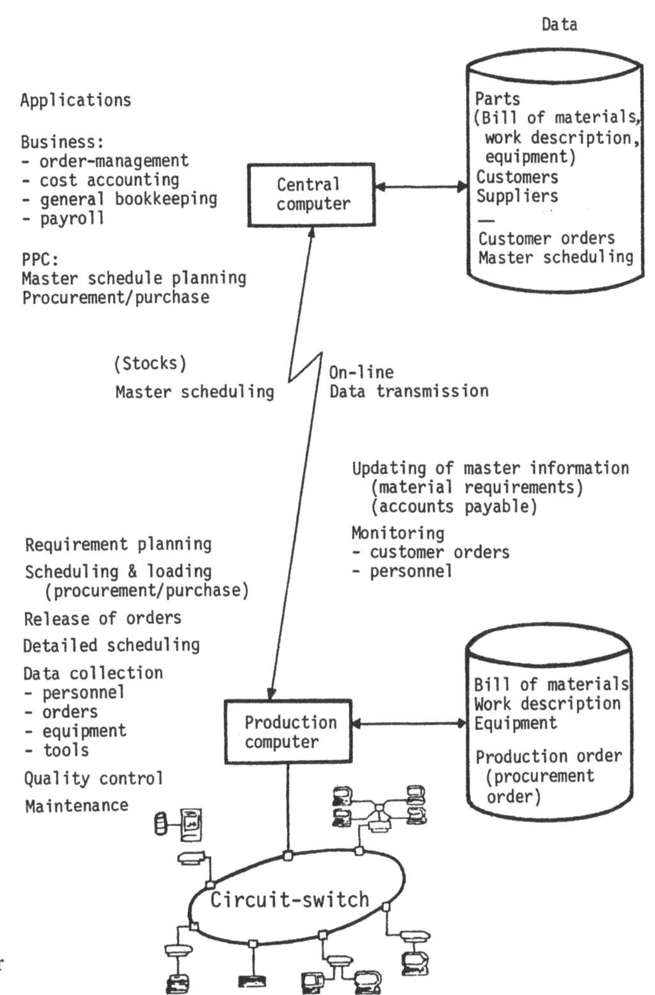

Fig. 10.14. Decentralized computer system used for planning

At present, two different computer strategies are used for hierarchically structuring the planning functions (Figs. 10.13, 10.14). In the central processing system all stages from production planning to order release are carried out on the central commercial computer of the firm, while production control and data collection are run on dedicated computers.

In the decentralized processing system the entire production planning function is run on the production computer, while master production scheduling is carried out on the central computer.

In the first case less updating between the computers is required, both for processing of released orders and for feedback control. Two updating cycles generally suffice per day. In the second case a strong link is necessary between the computers. This can only be obtained by an online connection.

For standard software systems the first alternative is preferred because linking of computers at the transaction level, especially computers which are running under different operating systems and have different data base systems, still remains a difficult technical problem.

10.4 The Interface Between CAD and Production Control

The most significant interface between a CAD-system and production planning and control systems are the data used. CAD-systems generally do not only process drawings, but also access the bill of materials. This applies both to existing parts which are combined to make new parts and to newly designed parts. When making a new design it should be possible to generate the bill of materials directly from the drawing. The most important information required for the compilation of a bill of materials are contained directly in the drawing.

Similar to the connection between CAD and the bill of materials there also exists a close relationship between computer aided manufacturing (CAM) and the work description.

In many organizations there exists the potential danger that CAD/CAM will be installed in isolation and that it will not be fully integrated with the production planning system. This is due to the fact that CAD/CAM software and hardware are produced by highly specialized firms which are not fully familiar with the entire scope of planning. Thus often specialized components for CAD/CAM systems are acquired which are not compatible. The user should consider this potential problem very carefully when an integrated primary data management system for CAD/CAM and for planning is being implemented.

It should be noted at this point that there is also a close connection between computer aided design and certain management functions. It is known that approximately 70% of the cost of a product is determined during design for example with respect to the materials used and the production methods chosen. It is therefore necessary to make data on the cost of raw materials and production methods available to the design engineer. This requires that direct access from the design engineer's EDP system to the cost information must be possible. The designer must also be able to operate the software systems involved.

At present there are only partially integrated CAD/CAM and production planning systems available. There are systems which provide the design engineer with two different terminals, one for graphics support and one for access to the primary production data. Other systems are more integrated, allowing the user to access both areas from one terminal, even through different software systems are involved. A truly direct connection will only be possible when data for both CAD applications and planning functions are maintained in the same data base and are accessed through the same user interface.

10.5 The Interface Between Production Control and Marketing

The EDP activities for production planning and control begin with master production planning. The necessary input information is provided by the sales forecast of the marketing department. This forecast is often quite inaccurate. If this is the case then the results of all the subsequent planning stages will also be inaccurate. This usually leads to a situation where the inventory of both production components and end-products must be maintained at a high level in order to be able to absorb the inaccuracies of the marketing forecasts.

Also unscheduled urgent production orders which are due to unexpectedly high sales can greatly disturb the orderly operation of the production process.

The high interest rates incurred during recent years have led to attempts to reduce inventory of both components and end-products. For this reason production systems emphasizing a minimal in-process inventory are being widely discussed, including the Japanese Kanban-principle.

An other important interface between marketing and production is that of order handling.

10.5.1 Master Production Scheduling

The sales forecast is frequently based on manual estimates made by the marketing department. A large number of statistical forecasting techniques exist to support sales forecast. However, their application has not yet been realized sufficiently. One reason for this is that personnel involved in scheduling have little knowledge of these statistical forecasting techniques.

Another reason that the use of forecasting methods has often failed is that wrong models are applied to the data sets used for the forecast.

Figure 10.15 shows the four typical shapes of sales curves. The well known exponential smoothing (type one) method is applicable to cases of static demand. Frequently, however, this method is applied to the prognosis of sales data which are influenced by seasonal effects or trends. Such an application leads to significant errors.

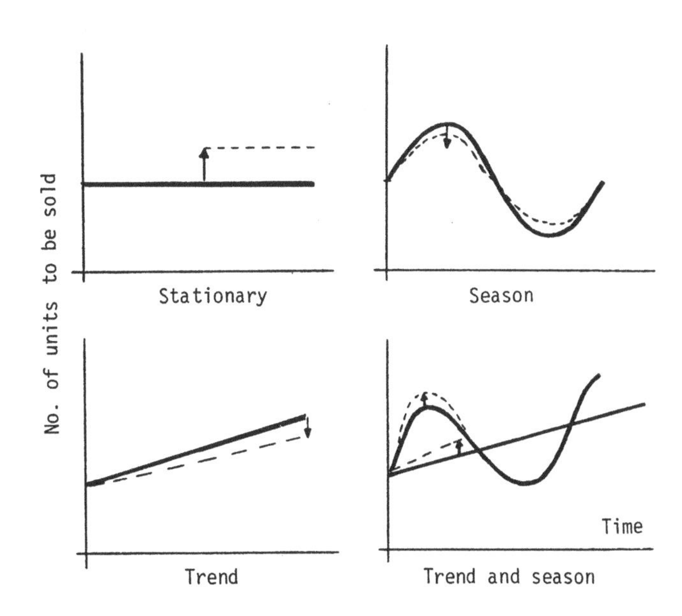

Fig. 10.15. Different forecasting models

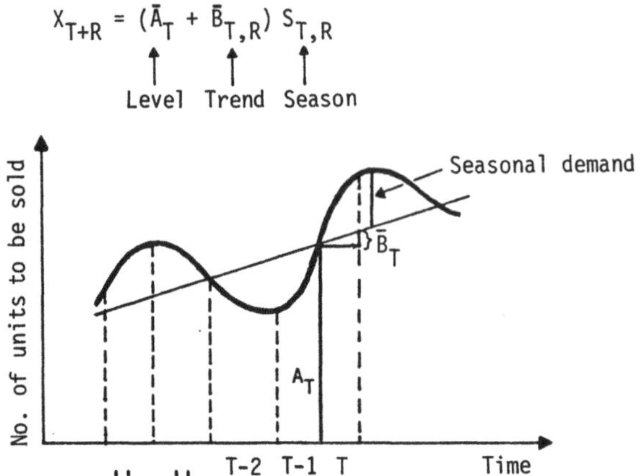

Fig. 10.16. Exponential smoothing for seasonal demand

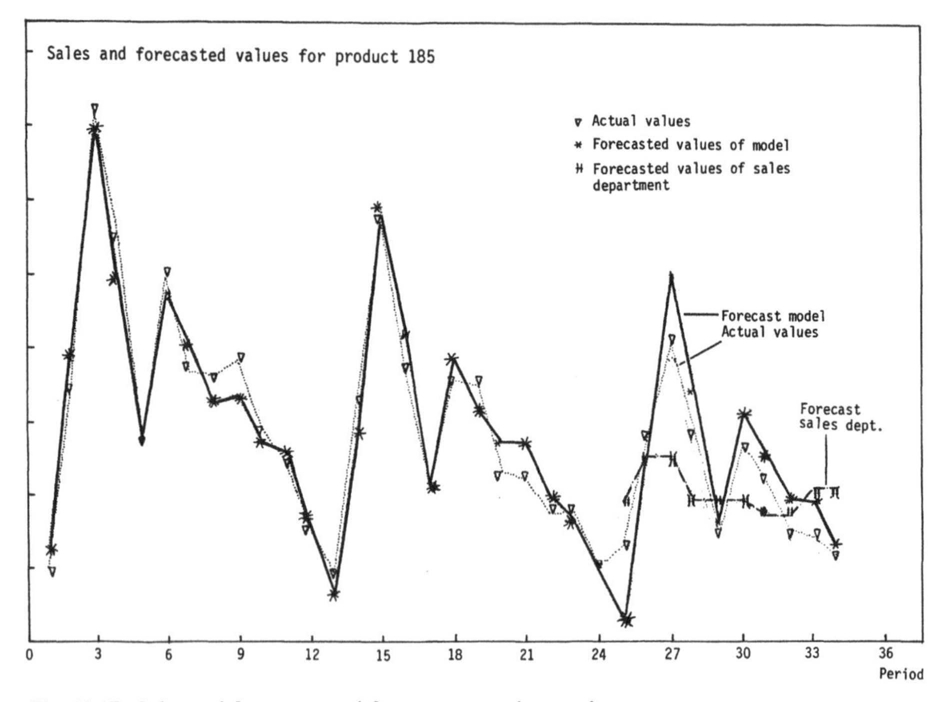

Fig. 10.17. Sales and forecast trend for an automotive product

Since many sales curves are influenced by seasonal effects, more complex forecasting techniques should be employed. Many production planning and control systems offer the Winters model, which takes both trend and seasonal effects into account. This model is shown in Fig. 10.16. Each of the factors "trend, level, and season" is analyzed separately by means of the exponential smoothing method. This method proved to be quite successful in practical applications. Figure 10.17 shows a curve representing the sales of a vendor in the automobile industry. The monthly values show seasonal effects. The dotted lines represent the actual sales. Based on 24 months data, the third year is forecast. The solid line shows the sales forecast for the model, while the dashed line represents the marketing departments forecast. It can be seen that the marketing prognosis is not able to make a sufficiently accurate prediction of the course of the sales curve. The inventory buffer needed to compensate for the inaccuracies is quite substantial.

In more modern logistics models, a coordination procedure is carried out at the beginning of the production planning process among the participating departments marketing, purchasing, production and logistics. In this coordination effort, the sales forecast made with the support of a forecasting system is discussed with the marketing department. Thereafter, the sales forecast is given to the production department to set up a production program within the current capacity limitations. At the same time the availability of the required raw material is checked to ensure that everything is scheduled on time.

With this coordination the interface between marketing and production is assured, and it will have a beneficial effect on the results of the production planning and control system.

10.5.2 Order Handling

When a new customer order is accepted, the current inventory of the endproduct must be known. Thus the customer can be informed about the immediate availability of the product. If the product is not currently in stock then the sales department must have access to the short-term and medium-term production plans, in order to be able to determine delivery times. This requires that the sales department has an up-to-date knowledge of production data. For this reason an online connection between the commercial computer and the production data is necessary. Where the customer enquires about the state of an order which is being processed. This can mean a significant competitive advantage, especially in economically tight situations.

Because of the close relationship between the two departments an integrated EDP-model can take all the aforementioned interdependencies into account.

10.6 Factors Influencing the Acceptance of Production Control Software

Many firms have reservations about employing standard software. This is particularly true of large enterprises. The reservation is partially due to the fact that large enterprises are less willing to adapt their operations to standard software systems than small or medium-sized firms. The latter firms are not in a position to produce complex software in-house. This is due to a lack of know-how and the unavailability of funds to finance such projects. Thus, the firms are more willing to accept a necessary re-structure of their organization. Large firms are in general in the position to do their own complex software development since they have the necessary resources.

For the success of a software package it is important from the very beginning to achieve a consensus with the relevant departments and the EDP department about the features which the planning model should possess. Thus, when the decision has been made that such a system is to be acquired a project committee should be formed in which all the departments involved, including EDP, are represented. Since often the degree of know-how about the planning process in a firm is marginal additional schooling by means of seminars may be necessary.

The strategic importance of a production planning system and of the decisions relating to the installation of such a system require that active support of the project will be guaranteed by top management. Since in some cases the installation necessitates the purchase of hardware or the implementation of a data base, such strategic decisions can only be made by top management.

It is of importance for the acceptance of a software package that the catelogue of criteria for the selection process be agreed upon by all departments involved. Due to the variety of problems presented by such a project it is impossible for each department to take equal part in the entire selection process. For this reason it is necessary for the project committee to solve all highly specialized problems in cooperation with personnel from the relevant departments.

Project control is absolutely mandatory for the success of the selection process. A project director who has know-how in management, the relevant technical areas and in EDP must coordinate the various implementation tasks. Since a production planning and control system is of interdisciplinary nature there are very few specialists available who have a system wide view. It may therefore require quite an extended preparation phase during which the firm accumulates the necessary know-how in all relevant areas before the system can be successfully selected and installed.

11 Economic Analysis of Computer Integrated Manufacturing Systems

F. Leimkuhler

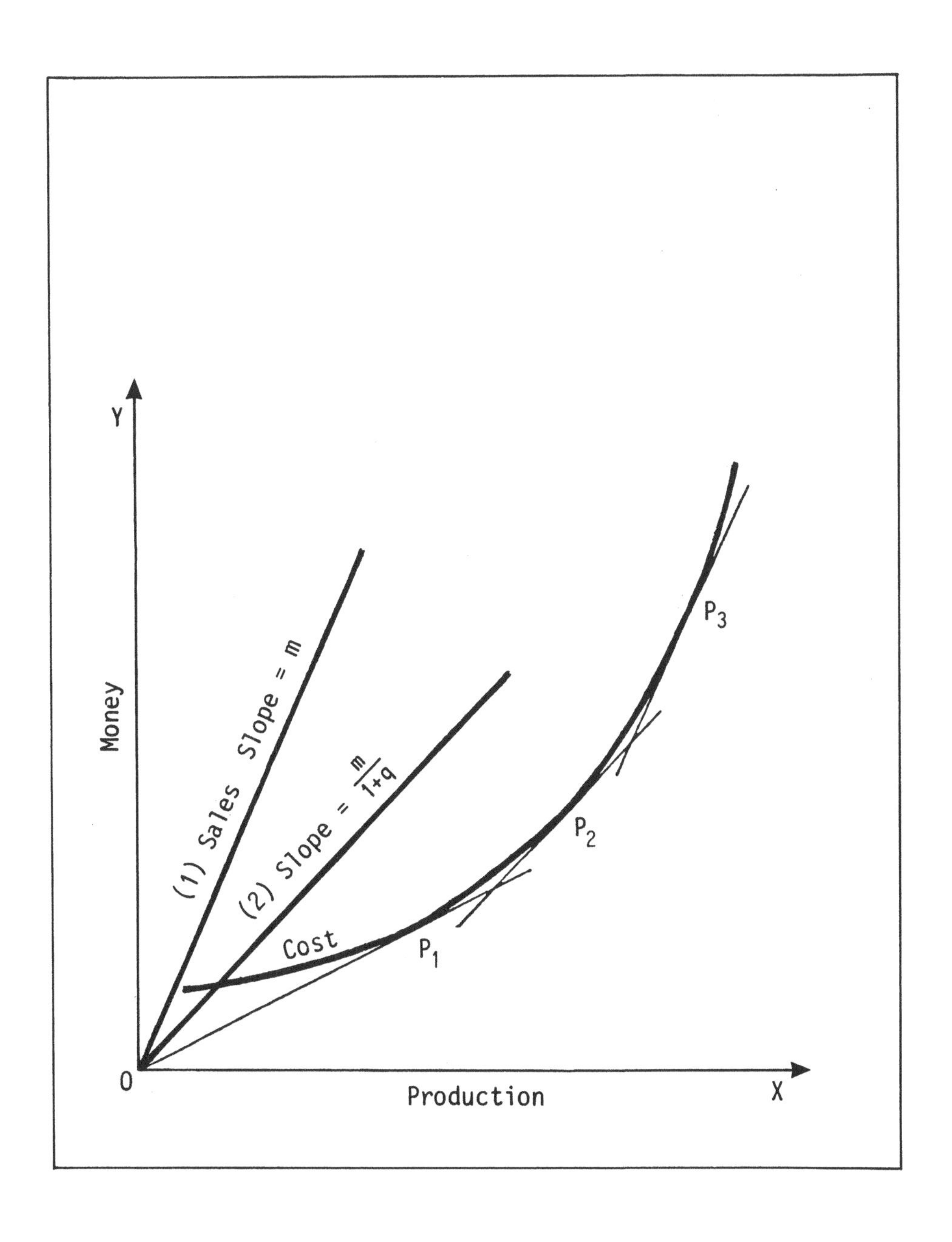

Determination of maximum economic production

(From Max Kurtz, "Engineering Economics for Professional Engineer's Examination" McGraw-Hill Book Company Inc., New York, 1959)

11.1 Introduction

Computer integrated manufacturing systems provide important economic advantages over less flexible and less controllable forms of manufacturing. Increased *flexibility* is particularly advantageous in the short run for small-batch production of mixed parts on the same system, and for longer run changes in product specifications and volume requirements. Increased *controllability* is particularly advantageous in sychronizing the flow of parts and tools over a greater part of the total production cycle, in maintaining greater and more uniform dimensional accuracy, and in stabilizing production during design and schedule changes. These advantages are realized in the form of lower direct labor requirements, lower in-process inventories, greater machine utilization, shorter lead times, and improved quality. On the other hand, such advantages require larger initial capital investments, more intensive use of such facilities, training in new labor skills, organizational changes, and the use of new planning and control methods.

To evaluate the economic benefit of CIM systems, simulation methods like GCMS [1], Q-GERT [3], and stochastic network models like CAN-Q [2] can be used to estimate how the output capacity of a manufacturing system changes as the various components of the system are increased or improved for a known cost. This data can be used to develop *cost vs. output curves* for a given part or product mix. By comparing these curves for different types of production systems, it is possible to identify break-even points where a particular type of system is economically preferred over other systems for particular products or product-mixes. By varying the design and cost parameters, it is possible to study the effects of changes in economic conditions and changes in the design and cost of hardware.

To demonstrate this approach, a comparison is made between two process plans for the same prismatic part. The one plan, called LINE, represents a production line, with six different stations each manned by an operator. The second process plan, called FMS, uses a multi-function machining center, a computer-controlled transporter, and off-line load-unload stations in a flexible manufacturing system. A convenient notation for describing the process plans is developed and used to define the two process plans. Realistic estimates of the production capacities for plans are obtained by successive application of Solberg's [2] CAN-Q network analysis program to estimate the output for a minimal system, identify the bottleneck station, and then increase that station's capacity so as to increase production. In this way, it is possible to estimate the capital and labor requirements for a given output capacity. The results indicate that capital and labor increase in an approximately linear fashion with output capacity.

A notation is developed for showing how the capacity expansion data can be used to evaluate payback criteria, to make rate of return analysis, and to determine how the cost per part is related to output capacity and system design. These calculations take into account the cost of labor, the cost of capital, and the effect of taxes on computing the net rate of return after taxes. The advantage of FMS over LINE is shown to be sensitive to the capital-labor cost ratio and to the intensity with which the system is used, i.e., the number of shifts per day of normal operation.

Comparison of the efficiency of the two systems in using equipment and labor shows that FMS has a relatively high utilization of capital investment at relatively low output rates, while the LINE capital utilization rate gradually approaches that of FMS as output capacity is increased. These results are used to develop formulae for estimating the amount of allowable additional investment in FMS that could be made as a function of the cost of labor, the captial return requirements of the company, the equipment design characteristics of the systems being compared, and their efficiency of operation. These formulae are evaluated for the two process plans, FMS and LINE, and the results are used to derive some general implications for the design and operation of computer-aided manufacturing systems.

11.2 Process Planning

The design and analysis of a manufacturing system begins with the *process plan* which specifies the *operations* to be performed on different *workpieces* at different *workstations* in order to complete one production cycle. The process plan defines the *operation time* t_{ijk} for operation k on piece j at station i, along with the *operation frequency* f_{ijk} with which this operation is performed in any one production cycle. For example, an inspection operation may be performed on every other piece so that the value of f_{ijk} is equal to 0.5. The product $f_{ijk}\,t_{ijk}$ determines the average time to perform operation k on piece j at station i.

The completion of a full production cycle results in the production of one unit or *part* of final product, and this may include operations on several different kinds of workpieces. A workpiece can be a single component, an assembly of different components, or a mix of different components and/or assemblies. It is convenient to let n_j denote the *number of pieces* or items of type j needed for each final unit of product. The sum of the n_j's is the total number of different workpieces that must be handled in order to make a single unit of final product. Furthermore, if P denotes the *production rate of final units or parts* or product, then P_j equal to $n_j P$ represents the *production rate of workpieces* of type j. An alternative way of making these calculations is to do it in terms of the fraction $n_j/\Sigma_j n_j$ of workpieces of type j needed for each unit or part of final product.

The sum of the average operation times for making piece j on station i determines the total average time on station i needed to make one piece of type j, i.e., t_{ij} equals $\Sigma_k t_{ijk} f_{ijk}$ and since n_j pieces of type j are needed per unit of final product, the $t_{ij} n_j$ is the average workload on station i in producing parts of type j for one unit of final product. The total average workload or time on station i for one unit of final product is t_i equal to the sum $\Sigma_j t_{ij} n_j$. The sum of the t_i's, i.e., T equal $\Sigma_i t_i$, is the *total average processing time* for one unit of final product.

Assuming that each operation requires the transfer of material, then since there are n_j pieces of type j needed per unit of final product, there is an average of $n_j f_{ijk}$ transfers needed per unit for operation K on piece j at station i. The total transfer per unit at station i is v_i equal to $\Sigma_j \Sigma_k n_j f_{ijk}$ and the total average number of transfers at all stations in the system for all operations needed in a full production cycle is V equal to $\Sigma_i v_i$. This notation is summarized in Table 11.1. Also shown is the notation used for the cost parameters of a system.

Examples of two processing plans are shown in Tables 11.2 and 11.3. Table 11.2 shows the plan for a prismatic part made on a production line consisting of six different types of workstations. The operator at each station loads

Table 11.1. Notation

T	$\Sigma_i t_i$ = total average process time per final part
t_i	$\Sigma_j t_{ij} n_j$ = total process load per part on station i
t_{ij}	$\Sigma_k t_{ijk} f_{ijk}$ = average time on piece j at station i
t_{ijk}	process time for operation k on piece j at station i
f_{ijk}	frequency per cycle of operation "ijk"
n_j	number of j pieces per final part
v_i	$\Sigma_j \Sigma_k n_j f_{ijk}$ = transfer per part of station i
V	$\Sigma_i v_i$ = transfer per part of final product
K_Q	$K_0 + \bar{K}(Q = Q_0)$ = estimated capital for capacity $Q > Q_0$
L_Q	$L_0 + \bar{L}(Q - Q_0)$ = estimated labor for capacity $Q > Q_0$
K_0	$\Sigma_i k_i$ = capital for a minimum capacity system
L_0	$\Sigma_i L_i$ = labor [integer] for a minimum capacity system
$\bar{K}$	$\Sigma_i k_i t_i$ = average capital per unit of final product
$\bar{L}$	$\Sigma_i L_i t_i$ = average labor per unit of final product
Q_0	smallest value of $1/t_i$ = minimum output capacity
t_i	average processing time on a station of type i
k_i	cost of a type i station
L_i	labor assigned to a type i station
P	production rate in parts per hour

Table 11.2. Line process plan with six machine types

Operation	Sta. 1	Sta. 2	Sta. 3	Sta. 4	Sta. 5.	Sta. 6	Total time
1	14.77						14.77 min/part
2		9.17					9.17 min/part
3			3.22				3.22 min/part
4				23.20			23.20 min/part
5			3.13				3.13 min/part
6					17.23		17.23 min/part
7						15.50	15.50 min/part
Total time	14.77	9.17	6.35	23.20	17.23	15.50	86.22 min/part
Load t_i	0.25	0.15	0.11	0.39	0.29	0.26	$T = 1.44$ h/part
Cost k_i (10^3)	\$250	\$500	\$200	\$500	\$200	\$150	K_0 = \$1,8000,000
$t_i k_i$	62.5	75	44	195	58	39	K = \$448,553/part
Labor l_i	1	1	1	1	1	1	$L_0 = 6$ men
$l_i t_i$	0.25	0.15	0.11	0.39	0.29	0.26	$L = 1.44$ men/part
Note, all $f_{ijk} = 1$ and all $n_j = 1$							$Q_0 = 2.56$ part/h

Table 11.3. FMS process plan with machining centers

Fixture	Load	Transport	Machine	Total time
1	6.0	4.0	17.99	27.99 min/part
2	6.0	4.0	35.50	45.50 min/part
3	6.0	4.0	32.73	42.73 min/part
Total time	18.9	12.0	86.22	126.22 min/part
Load t_i	0.3	0.2	1.44	$T = 1.44$ h/part
Cost k_i (10^3)	\$0	\$500	\$500	$K_0 = \$1{,}000{,}000$
$t_i k_i$	0	100	718.5	$\bar{K} = \$18{,}500$/part
Labor l_i	1	0	0.25	$L_0 = 2$ men
$t_i l_i$	0.3	0	0.359	$L = 0.66$ men/part
Note, all $f_{ijk} = 1$ and all $n_j = 1$				$Q_0 = 0.69$ parts/h

parts, maintains tools, and makes inspections while the processing operations are being executed. Parts move between stations on conveyors. Table 11.3 shows the plan for a flexible, computerized manufacturing system making the same part and consisting of load/unload stations (type $i = 1$), computer controlled transporters (type $i = 2$) and universal machining centers (type $i = 3$). Loading and transport operations are done simultaneously with the machining operations.

11.3 Capacity Analysis Using CAN-Q

Realistic estimates of the performance characteristics of a production system can be obtained from Solberg's [2] CAN-Q Program, which models a system as a stochastic or queueing network. The input data for the program are taken from the process plans and include specification of the types and number of workstations, transporters, and workpieces in the system, along with a listing of the operation sequences, processing times, and cycle frequencies. The program computes expected production rates and average times in the system for each type of workpiece as a function of the total number of pieces in the system. The program identifies the bottleneck station and gives detailed information about the congestion and utilization at each station, showing the sensitivity of the output rate to small changes in the processing times and mixture of workpieces.

For capacity planning purposes, it is necessary to estimate how a system's performance depends on the rate of output. For any given system and product mix, CAN-Q determines both the output rate and the "bottleneck station" which is critical in limiting the rate of output or capacity of the system. By adding one more station of the bottleneck type and re-running the program, the new capacity output rate Q can be determined along with the other measures of system performance. Successive iterations of this procedure make it possible to generate relationships between the output rate and system performance.

Note that in generating capacity relationships by successive iterations of the CAN-Q program, it is necessary to specify the maximum in process inventory of pieces. One convenient way to do this is to set this number as a multiple of the

Table 11.4. CAN-Q estimates[a] of line output rates for part X

System number	Number of machines by type						Output pcs./h	Equipment		Manpower		Machine
	1	2	3	4	5	6		Cost ($ 1000)	Use %	Men	Use %	Use %
1	1	1	1	1	1	1	2.557	1800	64	6	61	61
2	1	1	1	2	1	1	3.319	2300	65	7	68	68
3	1	1	1	2	2	1	2.721	2500	67	8	67	67
4	1	1	1	2	2	2	4.052	2650	70	9	70	70
5	2	1	1	2	2	2	5.146	2900	87	10	74	74
6	2	1	1	3	2	2	6.254	3400	97	11	82	82
7	2	2	1	3	2	2	6.752	3900	89	12	81	81
8	2	2	1	3	3	2	7.306	4100	80	13	81	81
9	2	2	1	3	3	3	7.604	4250	80	14	78	78
10	2	2	1	4	3	3	8.093	4750	76	15	78	78
11	3	2	1	4	3	3	9.281	5100	82	16	83	83
12	3	2	2	4	3	3	10.009	5300	85	17	85	85
13	3	2	2	5	3	3	10.377	5800	80	18	83	83
14	3	2	2	5	4	3	11.372	6000	85	19	86	86
15	3	2	2	5	4	4	11.966	6150	87	20	86	86
								Avg.	80		76	76

[a] Assumes that there are enough workpieces in the system to maintain near maximum output.

Table 11.5. CAN-Q estimates[a] FMS output rates for part X

System Number	Number of stations			Output pcs./h	Equipment		Manpower		Machine
	Load	Trans.	Mach.		Cost ($ 100)	Use %	Men	Use %	Use %
1	1	1	2	1.391	1500	87	2	52	100
2	1	1	3	2.079	2000	85	2	69	100
3	1	1	4	2.720	2500	89	2	90	98
4	2	1	5	3.477	3000	95	3	76	100
5	2	1	6	4.148	3500	85	4	68	99
6	2	1	7	4.692	4000	85	4	77	96
7	2	1	8	4.942	4500	90	4	81	89
8	2	2	8	5.551	5000	91	4	73	100
9	2	2	9	6.161	5500	92	5	81	98
10	3	2	10	6.959	6000	95	6	76	100
11	3	2	11	7.653	6500	96	6	84	100
12	3	2	12	8.336	7000	98	6	92	100
13	3	2	13	8.965	7500	98	7	84	99
14	3	2	14	9.424	8000	96	7	89	97
15	4	2	15	9.915	8500	96	8	82	95
				(parts/h)	Avg.	93		78	98

[a] Assumes enough workpieces in the system to maintain near maximum output.

number of stations and large enough to produce an output level which is fairly close to the maximum possible level of output. In special cases other considerations may govern how this value is chosen.

An example of the use of CAN-Q for capacity planning purposes is shown in Tables 11.4 and 11.5, using the process plans of Tables 11.2 and 11.3. Also shown in these tables are the CAN-Q estimates of how equipment and labor utilization changes as the capacity of the system is increased.

11.4 Capital and Labour Requirements

The basic CAN-Q program can be modified to provide estimates of how the capital and labor requirements change as the production capacity of a system is increased. Additional input data is needed on the acquisition cost of each type of station, k_i, and the number or fraction of workers assigned to each type of station, L_i.

Examples of data of this kind are shown in Tables 11.4 and 11.5 for the process plans of Tables 11.2 und 11.3. These tables show the estimated total capital and labor requirements needed by these systems for different capacity levels. Note that both the capital and labor requirements are based on the need to use an integral number of stations and workers in any particular systems.

A simple and useful approximation of the capital and labor requirements for an automated system can be made by considering two extreme cases of system capacity. A minimum capacity system consists of one station of each type plus the smallest integer number of workers needed to operate such a system. The maximum output rate or capacity of this system is determined by the station with the largest average processing time t_i per unit of output. Specifically, the capacity Q_0 of a minimum system is equal to the smallest value of l/t_i for all station types. For production rates less than or equal to Q, the capital requirement is constant and equal to K_0 which is just equal to the sum $\Sigma_i k_i$, where k_i is the station cost of type i. The labor requirement for a minimum capacity system is equal to the smallest integer L_0 which is greater than or equal to the sum $\Sigma_i L_i$, where L_i is the number (or fraction) of workers assigned to a type i station.

As the production capacity is increased and more stations of each type are added, the processing operations tend to become more balanced and equipment utilization is increased so that the rate at which the investment increases approaches the theoretically optimum rate, $\bar{K}$, which is equal to the sum $\Sigma_i k_i t_i$. In the same way, the rate with which the labor force is increased also approaches the optimum rate $\bar{L}$, which is equal to the sum $\Sigma_i L_i t_i$. By combining these two notions, the following simple linear equations appear to provide good approximations of the capital and labor requirements as a function of the production capacity Q of a system with a given process plan.

$$K_Q = K_0 + \bar{K}(Q - Q_0),\ Q > Q_0 \qquad (1)$$

$$L_Q = L_0 + \bar{L}(Q Q_0),\ Q > Q_0 \qquad (2)$$

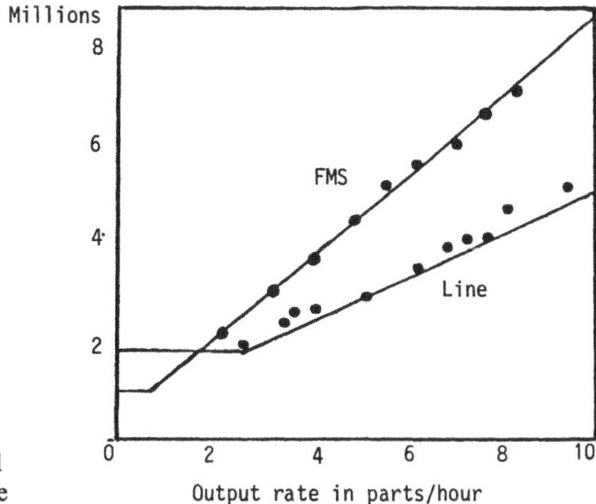

Fig. 11.1. Theoretical (*lines*) and actual (*dots*) capital investment vs. output rate

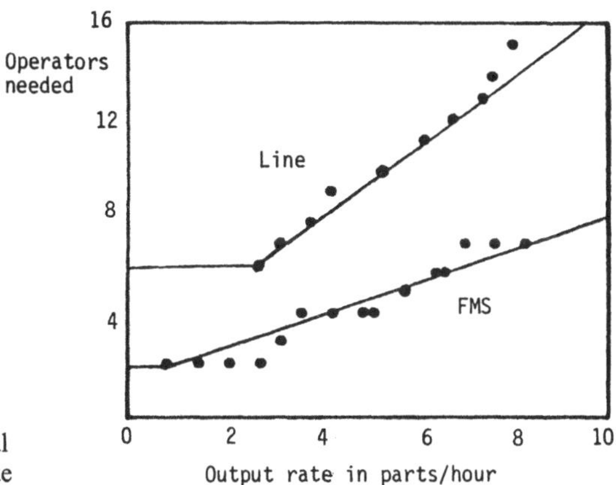

Fig. 11.2. Theoretical (*lines*) and actual (*dots*) labor requirement vs. output rate

Note that in special cases it may be necessary to make some adjustments to these values to account for capital and labor costs that are not included in the values of k_i and L_i.

An example of the application of these equations is shown in Figs. 11.1 and 11.2 which show how the capital and labor requirements change with the production capacity. The lines in these figures are based on Eqs. (1) and (2) above and the data in the process plans of Tables 11.2 and 11.3 according to the following equations:

$$K_Q\,(\text{LINE}) = 1{,}800{,}000 \text{ if } Q \le 2.6 \text{ parts/h} \qquad (3\text{a})$$
$$= 639{,}952 + 448{,}553 \quad Q \text{ if } Q \ge 2.6$$

$$K_Q\,(\text{FMS}) = 1{,}000{,}000 \text{ if } Q \le 0.7 \text{ parts/h} \qquad (3\text{b})$$
$$= 436{,}406 + 818{,}500 \quad Q \text{ if } Q \ge 0.7$$

$$L_Q\,(\text{LINE}) = 6 \text{ if } Q \le 2.6 \text{ parts/h} \qquad (3\text{c})$$
$$= 2.28 + 1.44 \quad Q \text{ if } Q \ge 2.6$$

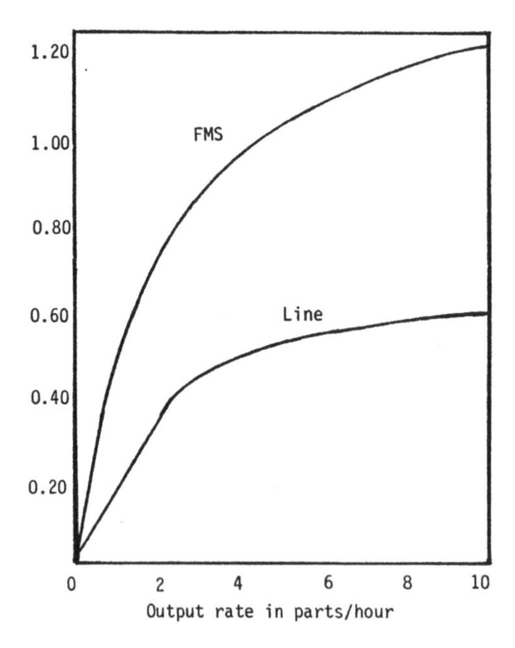

Fig. 11.3. Productivity (parts per manhour) vs. output rate

$$L_Q\,(\text{FMS}) = 2 \text{ if } Q \leq 0.7 \text{ parts/h} \qquad (3\text{d})$$
$$= 1.54 + 0.66 \quad Q \text{ if } Q \geq 0.7$$

The small circles in Figs. 11.1 and 11.2 are the observed capital and labor requirements as computed from the CAN-Q analysis in Tables 11.4 and 11.5.

The labor productivity of the two types of systems can be readily computed from the equations by calculating the ratio of output to workforce for different outpout capacities. For the LINE and FMS type systems, the labor productivity expressed in "parts per manhour" is computed from Eq. (3) and plotted in Fig. 11.3. The curves indicate that the FMS type process plan has a productivity value that is generally twice that of the LINE type plan, and the productivity of both systems increases with the output rate. The LINE curve shows a high rate of increase in productivity for high production rates. The rate of increase in productivity for the FMS type systems is fairly high over a large range of output values.

11.5 Payback, Capital Cost, and Taxes

Payback analysis is widely used in industry to justify equipment investments. If an investment of amout I generates annual earnings of amount J, then the ratio I/J measures the years needed to "pay back" the investment. Payback values of 1 to 4 years are often used for automation projects [4], and even larger values are sometimes used for large projects with long lives. In comparing two systems with capital cost K_1 and K_2, where K_2 is greater than K_1, and with labor requirements

L_2 less than L_1, the selection of the more capital intensive system would have a payback value of

$$I/J = (K_2 - K_1)/(L_1 - L_2)\ 2000\ HS = \text{payback ratio} \tag{4}$$

where H denotes the hourly labor cost rate and S denotes the number of 2000 hour shifts worked per year (assimung there are 2000 h per year per shift).

Payback analysis is closely related to a second popular method of equipment justification, *internal rate of return on investment before tax*, R_b, which satisfies the following capital recovery equation.

$$I/J = [1/R_b][1 - (1 + R_b)^{-Y}] = \text{payback in years} \tag{5}$$
$$B = J/I = R_b/[1 + R_b)^{-Y}] = \text{before tax recovery rate}$$

Here I/B is the payback ratio and Y denotes the number of years in the life of the investment. Because the rate of return method takes into account the life of the project, it is considered to be a more precise measure of investment worth.

Companies often specify their cost of capital in terms of a minimum acceptable rate of return *after taxes*. In his 1967 study of *Numerical Control Justification*, Steffy [5] found that "most companies desire an after-tax return in the neighborhood of 7–10%". A more recent survey by Newnan [6] found that "with a normal level of business risk" the after-tax return for many companies is between 12–15%. For equipment evaluation purposes it is desirable to translate the "after-tax" rate of return into a "before tax" rate, and then into a payback ratio or its reciprocal, which is the before-tax capital recovery factor.

With an after-tax rate of return R_a each dollar of investment must generate an amount of money A each year *after taxes*, where A is related to R_a by the capital recovery function,

$$A = R_a/[1 - (1 + R_a)^{-Y}] = \text{recovery rate after tax} \tag{6}$$

where Y denotes the life of the investment in years. Furthermore, each dollar of investment must generate an amount of money B each year *before taxes*, where B is equal to A plus the taxes on B. If the tax rate is assumed to be 50% and straight-line depreciation is used for a life of Y years, then the taxes on B are equal to $0.5\,(B - 1/Y)$. Therefore, the amount of before-tax annual earnings B for each dollar of investment is defined by:

$$B = A + 0.5\,(B - 1/Y) = 2A - 1/Y = \text{recovery rate before tax} \tag{7}$$

Note that a shorter write-off period Y reduces the capital recovery cost of an investment. In practice, the tax laws make it possible to use effective tax lives which are approximately 2/3 of the nominal life. Therefore, a more realistic estimate of the before-tax capital recovery rate is defined by:

$$B = 2A - 1/(2/3)\,Y = 2A - 3/2\,Y \tag{8}$$

Substituting Eq. (6) into Eq. (8) makes it possible to define the before-tax capital recovery rate B in terms of the after-tax rate of return R_a as follows:

$$B = 2\,R/[1 - (1 + R_a)^{-Y}] - 3/2\,Y \tag{9}$$

Recall that the reciprocal of B is the *payback ratio* I/J required to recover each dollar of investment so as to earn an after-tax rate of return equal to R_a.

In order to determine the cost of output for a production system, it is necessary to derive an estimate of the *capital recovery cost per hour* of system operation, which is equal to the annual capital recovery cost rate B divided by the number of operating hours per year for the system. A useful assumption is that there are approximately 2000 work hours per shift of operation. If S denotes the number of shifts worked, then the hourly capital recovery cost rate G is defined by:

$$G = B/20{,}000\ S \tag{10}$$

As an example of the kinds of numbers determined by Eq. (10), a reasonable assumption about the life and usage of a modern piece of production equipment is YS approximately equal to 30 shift-years, i.e., a machine used on 4 shifts would last about 7.5 years, about 10 years on 3 shifts, and 15 years on 2 shifts. With this assumption, the values of the payback ratio and the hourly capital recovery cost rate that correspond to different values of the after-tax rate of return applied to machines with varying intensities of usage are shown in Table 11.6. The data in Table 11.6 indicate that the payback ratio is much more dependent on the after-tax rate of return than on the intensity of equipment usage. However, the hourly capital cost recovery rate G is highly dependent on the usage of the equipment.

Table 11.6. Hourly cost of capital (10^{-5} \$/h) and payback values (years) for various rates of return and equipment usages

After tax rate of return	$S = 2$ shifts $Y = 15$ years		$S = 3$ shifts $Y = 10$ years		$S = 4$ shifts $Y = 7.5$ years	
	\$/h	Payback	\$/h	Payback	\$/h	Payback
10%	4.08	7.6	2.92	6.2	2.40	5.2
12%	4.84	6.8	3.40	5.7	2.74	4.6
15%	6.35	5.7	4.14	5.0	3.27	3.8
20%	8.20	4.7	5.45	4.2	4.21	3.0
25%	10.46	3.9	6.87	3.6	5.19	2.4
30%	12.80	3.3	8.28	3.1	6.22	2.0

For example, the hourly cost rate is about the same for both a 4-shift machine with R equal to 30% and a 2-shift machine with R equal to 15%. *Doubling the intensity of usage from 2 to 4 shifts will almost cut the hourly cost of equipment in half without making any significant change in the payback ratio.* This result helps to explain why most computerized manufacturing systems are used as intensively as possible. Furthermore, it indicates that a general reduction in the interest rate, which is a critical determinant of the rate of return, and a large increase in the demand for manufactured products would be highly favorable to the development of such systems. Finally, it is noted that payback ratios are not very precise benchmarks for measuring the impact of equipment investments on manufacturing cost.

11.6 Cost Comparisons

In choosing between two production system designs, cost comparisons can be made in the following different ways:

(a) The years to pay back the additional investment in the more capital intensive system (assuming that it is less labor intensive),
(b) the internal rate of return on the added investment before taxes are taken into account,
(c) the rate of return after taxes, and
(d) the average cost per part produced.

The payback method, which is the less precise measure, requires the specification of the hourly cost of labor, the number of shifts, and the hours per shift worked per year. For the rate of return methods, the number of years in the life of the investment project must be specified and for after-tax comparisons, the tax rate and depreciation method must be specified. Finally, in order to estimate the cost per part, it is necessary to specify all of the above factors plus the minimum acceptable rate of return for the investment which is translated into an hourly cost of capital, as discussed above.

Normally, these measures of cost are consistent in indicating if one system costs more than another. The measures with the greater amount of input information are more precise and are able to show how the input factors can change the performance of a system. The most useful and most informative measure is the average cost of producing a part, since it implies all of the other measures. By using the linear equations developed above for estimating the capital and labor requirements for a system, the cost per part can be estimated by the following equation.

$$C = (1/P)(GK_0 + G\bar{K}(Q - Q_0) + HL_0 + H\bar{L}(Q - Q_0)) \tag{11}$$

Here C denotes the cost per part at output rate P and system capacity Q, G is hourly capital cost, H is hourly labor cost, K_0 is system cost with minimum capacity Q_0, $\bar{K}$ is average cost per unit of output capacity, L_0 is labor required for minimum capacity Q_0, and $\bar{L}$ is average labor requirement per unit of output capacity. The value of G can be estimated by using Eq. (10) above.

For the LINE and FMS process plans of Tables 11.2 and 11.3 assuming a \$20 per hour labor cost and 2000 hours per year per shift, the payback ratio for the additional investment in FMS over LINE is given by the following equations for systems with capacity Q parts per hour and S shifts.

$$\begin{aligned} I/J &= (249.22\,Q - 137.12)/S(1 + 1.05\,Q) \quad \text{when} \quad Q \le 2.6 \\ I/J &= (20.46\,Q - 34.09)/S(4.46 - 0.66\,Q) \quad \text{when} \quad 0.7 \le Q \le 2.6 \\ I/J &= \text{payback on investment in FMS over LINE} \end{aligned} \tag{12}$$

These equations are evaluated in Table 11.7 and plotted in Fig. 11.4. They show that the payback ratio increases with capacity, levelling off at higher capacities, and that the ratio is quite sensitive to the number of shifts worked.

Table 11.7. Years to payback (yrs) investment in FMS over line and rate of return before tax (rorb) and after tax (rora) for different output rates (pcs./h) when labor costs $20 per hour

Pcs./h	2 shifts, 15 yrs.			3 shifts, 10 yrs.			4 shifts, 7.5 yrs.		
	yrs.	rorb %	rora %	yrs.	rorb %	rora %	yrs.	rorm %	rora %
2	1.1	91	51	0.7	138	76	0.5	184	100
3	1.6	61	35	1.1	92	52	0.8	172	70
4	2.3	43	25	1.5	65	38	1.2	87	52
5	3.0	34	20	2.0	50	30	1.5	67	40
6	3.6	27	16	2.4	40	25	1.8	55	32
7	4.8	20	13	3.2	29	20	2.4	42	25
8	4.9	19	13	3.3	28	19	2.5	41	24
9	5.0	18	12	3.4	28	18	2.5	40	23
10	5.1	18	12	3.4	28	18	2.6	39	23

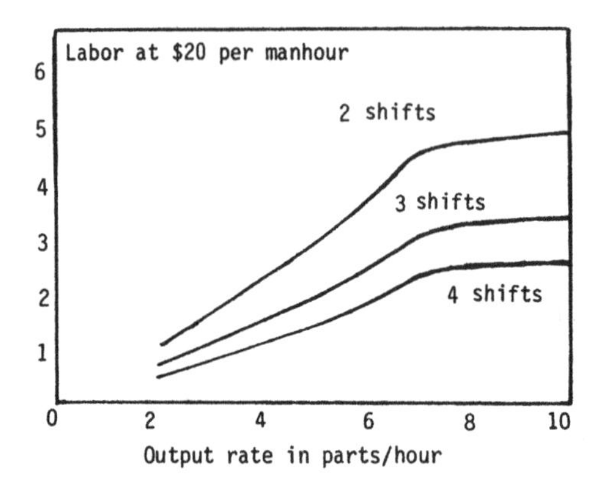

Fig. 11.4. Payback (years) for FMS over LINE vs. output rate

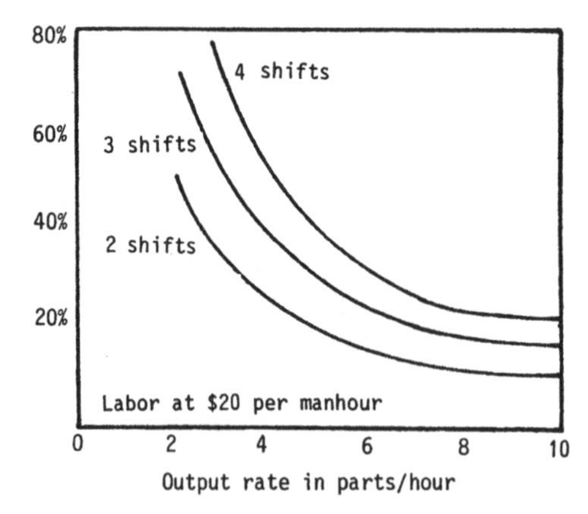

Fig. 11.5. After tax rate of return on FMS over LINE vs. output rate

The before-tax internal rate of return on the additional capital investment required by the FMS process plan over the LINE process plan can be determined by using Eq. (5) and solving for the value of R_b from the appropriate interest tables, or by interpolating the values in the tables. The after-tax rate of return can be found by using Eq. (9). Alternatively, Eq. (7) can be used to determine the after-tax capital recovery rate A as follows:

$$A = 0.5\,[(J/I) + 1/Y] = 0.5\,(B + 1/Y) \tag{13}$$

The value of A determined from Eq. (13) is substituted in Eq. (6) and the after-tax rate of return is found from the interest tables. Values of the before-tax and after-tax rates of return on FMS over LINE are computed in Table 11.7, assuming that the labor rate is $ 20 per hour. It is also assumed that the systems can be used for 15 years at 2 shifts, 10 years at 3 shifts, or 7.5 years at 4 shifts, and that each shift is 2000 hours per year. The after-tax rate of return values are plotted in Fig. 11.5 and are consistent with the payback values of Fig. 11.4, in indicating the economic advantage of using the FMS-type system intensively on a 4 shift basis.

The cost per part for both the FMS and the LINE process plans can be determined using Eqs. (11). The resulting cost formulas are as follows for the LINE process plan of Table 11.2, when the production rate P is greater than 2.6 parts per hour.

$$\begin{aligned} C &= (86.31 + 57.22\,Q)/P = \$/\text{part with 2 shifts} \\ C &= (72.17 + 47.31\,Q)/P = \$/\text{part with 3 shifts} \\ C &= (66.60 + 43.41\,Q)/P = \$/\text{part with 4 shifts} \end{aligned} \tag{14A}$$

The cost formulas for the FMS process plan when the output is greater than 0.7 parts per hour is as follows:

$$\begin{aligned} C &= (58.54 + 65.16\,Q)/P = \$/\text{part with 2 shifts} \\ C &= (48.90 + 47.07\,Q)/P = \$/\text{part with 3 shifts} \\ C &= (45.10 + 39.95\,Q)/P = \$/\text{part with 4 shifts} \end{aligned} \tag{14B}$$

These equations assume a labor cost of $ 20 per hour, 2000 hours per year per shift, a total useful life of 30 shift years, and an after-tax rate of return of 15%

Table 11.8. Cost per part for CMS and LINE with a 15% after tax rate of return and a $ 20.00 per hour labor rate

Output rate/h	2 shifts, 15 yrs.		3 shifts, 10 yrs.		4 shifts, 7.5 yrs.	
	CMS $	LINE $	CMS $	LINE $	CMS $	LINE $
2	94	117	72	97	63	89
3	85	86	63	71	55	66
4	80	79	59	65	51	60
5	77	75	57	62	49	57
6	75	72	55	59	48	55
7	74	70	54	58	47	53
8	73	68	53	56	46	52
9	72	67	53	55	45	51
10	71	66	52	55	44	50

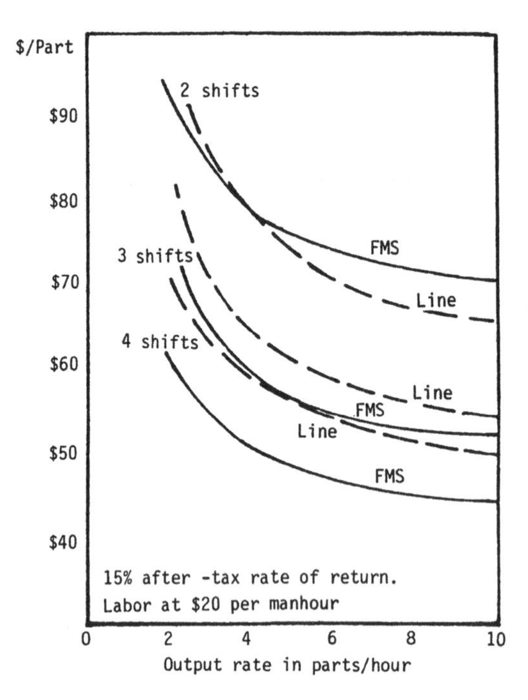

Fig. 11.6. Cost per part vs. output rate

on capital equipment. These equations are evaluated in Table 11.8 for production rates up to 10 parts per hour under the assumption that the system is operated at near maximum capacity, i.e., the capacity Q is equal to the output rate P. The values are plotted in Fig. 11.6 and show the relative advantage of using both the FMS and the LINE system on a 3 or 4 shaft basis. The LINE part cost is greater than the FMS cost with 3 or 4 shifts, but only at lower production rates with 2 shifts.

11.7 System Efficiency

Theoretically, automated production systems can be described by a "fixed proportion" type of production function in which there is a fixed proportional relationship between the input and the output. The optimal capacity expansion pattern for such systems can be represented by a linear function of the inputs. The following equations define the theoretical expansion path in terms of capital and labor inputs.

$$\begin{aligned} K &= (\bar{K}/\bar{L})\, L = \text{expansion path of capital } K \text{ with labor } L \\ \bar{K} &= \Sigma_i K_i t_i = \text{average capital needed per unit of output} \\ \bar{L} &= \Sigma_i L_i T_i = \text{average labor needed per unit of output} \end{aligned} \tag{15}$$

here, K_i denotes capital cost of a type i workstation, L_i is workers assigned to a type i workstation, and T_i is average time per unit of output at a type i station.

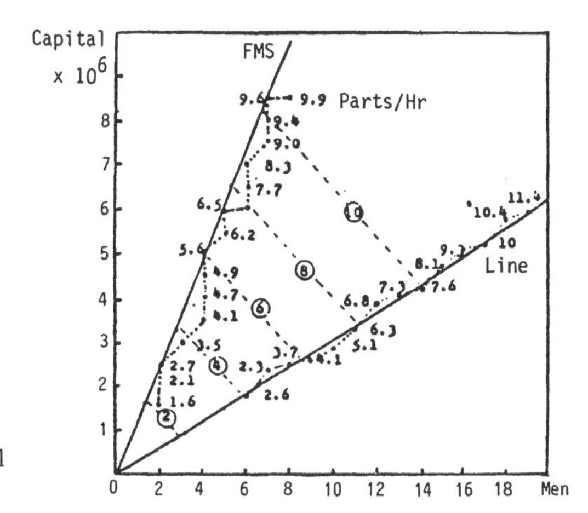

Fig. 11.7. Theoretical (*lines*) and actual (*dots*) output combinations

Numerical values of $\bar{K}$ and $\bar{L}$ are computed in Tables 11.2 and 11.3 for the LINE and FMS process plans. The theoretical expansion paths for these two process plans are plotted in Fig. 11.7 as straight lines. The difference in the slopes of these two lines is indicative of the fact that the LINE process plan is relatively labor intensive and the FMS process plan is relatively capital intensive.

Also shown in Fig. 11.7 as circled numbers, are the theoretical output rates if the systems were operating with 100% efficiency in their use of capital and labor. The dotted paths trace the input-output combinations determined from the CAN-Q analysis as the capacity was expanded. The numbers on these paths are the observed output rates in parts per hour. Generally, it is expected that the theoretical and observed expansion paths will not coincide because of the need to add stations and workers in integral units, and because of losses in efficiency due to congestion delays in the work flow. It is noted that the LINE plan straddles its optimal expansion path and the FMS plan is generally displaced to the right, i.e., the FMS system tends to make less effective use of its labor than its capital resources because of the fractional labor requirements per station.

The theoretical expansion path is useful as a standard for computing the relative efficiency with which a system uses capital and labor. A commonly used measure of system efficiency is the relative utilization of the machine capacity of the system where the actual machining rate is equal to the output capacity of the system, Q, multiplied by the machining time per part, T. It there are N machines in the system, QT/N is the percentage utilization of machine capacity. The estimated utilization of machines by LINE and FMS is computed in Tables 11.4 and 11.5 and is plotted in Fig. 11.8 as as function of the output capacity. The data indicates that FMS has an extremely high rate of machine utilization (over 95%) at almost all output capacity levels, while the LINE utilization is only 65% at low outputs and builds up to about 85% at higher capacity levels.

The high machine efficiency of FMS is not a dominant factor in output cost because of considerable differences in the cost of LINE and FMS machines, and the additional transporter cost for FMS. A comparison can be made of the

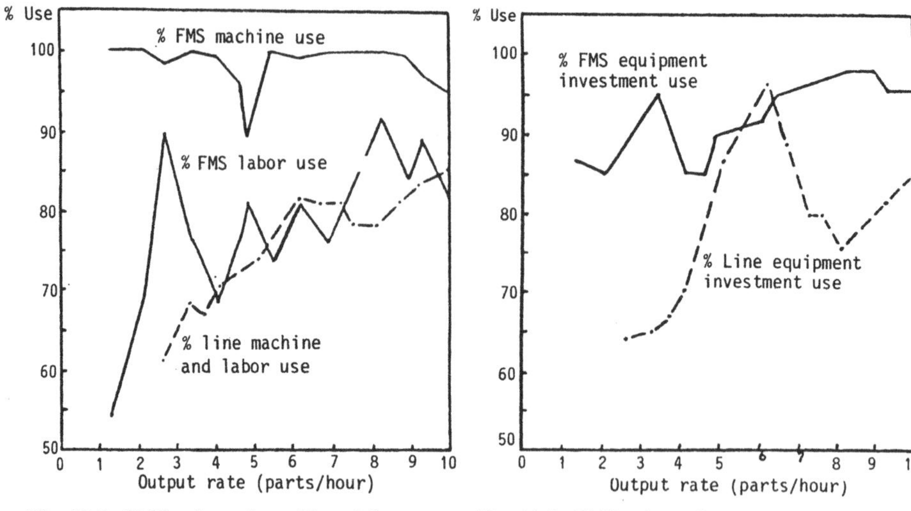

Fig. 11.8. Utilization of machines labor at different output levels

Fig. 11.9. Utilization of equipment investment at various output levels

utilization of the equipment investments in the two systems by computing the amount of equipment investment per part, $\bar{K}$, as in Tables 11.2 und 11.3, multiplying by the output capacity Q, and dividing by actual investment K to get a measure of the relative utilization of the equipment investment, n, where n equals $\bar{K}Q/K$.

Investment efficiences for LINE and FMS are computed in Tables 11.4 and 11.5 and plotted in Fig. 11.9. They show that FMS is generally more efficient than LINE in the use of equipment investment, except at mid-range output rates where there is an unusually favorable combination of LINE equipment. This output interval coincides with a lower than usual FMS efficiency due to the introduction of a second transporter into the system. In general, however, FMS investment efficiency is considerably above that of LINE. This indicates that the relatively high cost of the FMS equipment, along with the hourly cost of capital and labor are critical factors in making FMS output cost competitive with LINE output cost.

Also shown in Tables 11.4 and 11.5 and Fig. 11.8 are the utilization of manpower on FMS and LINE. In the case of the LINE, the manpower and machining utilizations are the same because of the assumption of one man per machine. Although the total manhours per part on FMS is almost half that of LINE, the FMS utilization of labor is not as efficient as its use of machining capacity. The labor use fluctuates considerably and tends to increase as output increases to about an 85% usage rate.

11.8 Justification of Automation Equipment

The justification of automation equipment depends on the amount and value of the labor saved and the value of the increased flexibility and utilization of automated systems. In comparing a capital intensive system with a labor intensive system, the additional capital investment is expected to satisfy the payback relationship.

$$1/B = (K_2 - K_1)/(L_1 - L_2)\ 2000\ HS = \text{payback in years} \quad (16)$$

Here, K_1 denotes capital investment in LINE system, K_2 is capital investment in FMS system, L_1 is labor required by LINE system, L_2 is labor required by FMS system, H is hourly cost of labor, and S is the number of shifts of 2000 hours per year. This relationship can be expressed in terms of the percentage changes in capital and labor as follows:

$$1/B = K' K_1/L' L_1\ 2000\ HS \quad (17)$$

$$K' = (K_2 - K_1)/K_1 = \text{percentage increase in capital}$$

$$L' = (L_1 - L_2)/L_1 = \text{percentage decrease in labor}$$

By rearrangement, the payback equation can be used to determine the amount of additional capital investment that can be justified for a given payback ratio or for a given value of the hourly value of capital, and a given percent decrease in labor.

$$K' = L'(L_1/K_1)\ 2000\ HS(1/B) = L' HL_1/GK_1 = \text{breakeven value}$$

$$G = (1/B)/2000\ S = \text{capital recovery rate per hour} \quad (18)$$

For the LINE process plan of Table 11.2, the capital-labor ratio K_1/L_1 is approximately equal to the theoretical ratio $\bar{K}_1/L_1$, which can be computed from

Table 11.9. Breakeven percentage increases in capital investment or CMS over line

Labor cost $/h	Shifts 1 week	Payback Requirement			
		2 yrs. %	3 yrs. %	4 yrs. %	5 yrs. %
15	2	20	31	41	51
15	3	31	46	61	77
15	4	41	61	82	102
20	2	27	41	54	68
20	3	41	61	82	102
20	4	54	82	109	136
25	2	34	51	68	85
25	3	54	71	102	128
25	4	68	102	136	170

Labor cost $/h	Hourly cost of capital (10^{-5} $/h)				
	$3 %	$4 %	$5 %	$6 %	$7 %
15	85	64	51	43	35
20	113	85	68	57	48
25	142	106	85	71	60

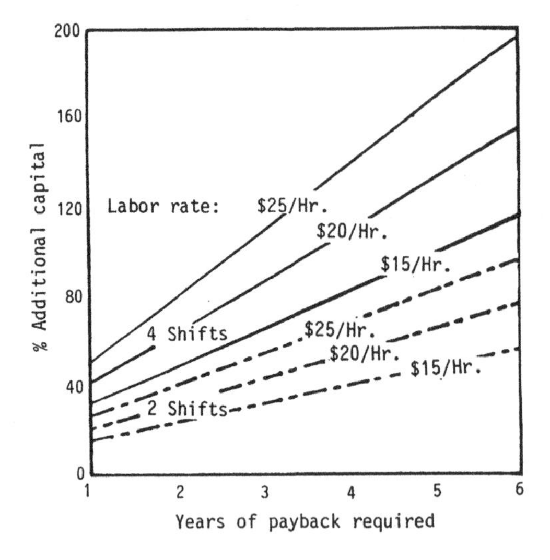

Fig. 11.10. Percent additional capital in FMS justified for different payback rates

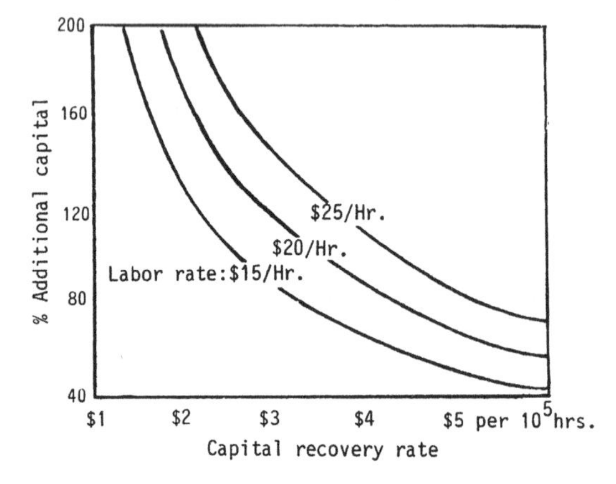

Fig. 11.11. Percent additional capital in FMS justified for different capital recovery rates

Table 11.2 as $\bar{K}_1/\bar{L}_1$ of 311 495 dollars per man, and the FMS plan uses about 54% less labor over the output range studied. Therefore, in designing a FMS system to replace a LINE system, the allowable increase in capital investment is approximately equal to

$$
\begin{aligned}
K' &= (0.54)/311\,495)\ (G/H) = (0.17)\ (10^{-5})\ H/G \\
K' &= (0.54)\ 2000\ HS\ (1/B/311\,495) = (0.0034)\ HS\ (1/B)
\end{aligned}
\tag{19}
$$

Representative values fo these equations are computed in Table 11.9 and plotted in Figs. 11.10 and 11.11. They show that the increased capital investment is highly sensitive to the payback ratio and the intensity with which the equipment is used.

The amount of capital investment in a system depends on the efficiency with which it is used, and the percentage increase in capital needed for a more capital intensive system can be defined in terms of the capital utilization factor u and the minimum capital requirements per part $\bar{K}$ as follows:

$$\begin{aligned} K' &= K_2/K_1 - 1 = [(\bar{K}_2/u_2)/(\bar{K}_1/u_1) - 1 = (\bar{K}' - u')/(1 + u') \\ u &= \bar{K}Q/K = \text{utilization of capital at output } Q \\ u' &= u_2/u_1 - 1 = \text{percentage increase in capital utilization} \\ \bar{K}' &= \bar{K}_2/\bar{K}_1 - 1 = \text{percentage increase in capital/part} \end{aligned} \tag{20}$$

Note that this formulation divides the capital increase into two components that are due to (a) a minimum design requirements $\bar{K}$ and (b) the operating efficiency of the system in using the capital provided.

In a similar way the reduction in labor requirements can be expressed in terms of the minimum labor requirements per part and the efficiency e with which the labor is utilized by the system.

$$\begin{aligned} L' &- (L_1 - L_2)/L_1 = 1 - L_2/L_1 = 1 - e_1\bar{L}_2/e_2\bar{L}_1 \\ \bar{L}' &= (e_1/e_2)\,(e_2/e_1 - \bar{L}_2/\bar{L}_1) = (e' + \bar{L}')/(1 + e') \\ \bar{L}' &= 1 - \bar{L}_2/\bar{L}_1 = \%\text{ decrease in theoretical labor per part} \\ e' &= e_2/e_1 - 1 = \%\text{ change in labor efficiency} \end{aligned} \tag{21}$$

The data on the FMS and the LINE labor efficiency indicates that both systems are about equally labor efficient at a given output rate. This implies that e' is approximately equal to zero and that Eq. (20) can be simplified since L' equals $\bar{L}'$, i.e., actual labor savings are effectively equal to the theoretical labor savings predicted by the minimum design requirements.

With the above definitions and assumptions it is now possible to rewrite Eq. (18) so as to reflect the effects of the capital utilization on the allowable capital investment in automated systems.

$$\begin{aligned} K' &= (\bar{K}' - u')/(1 + u') = \bar{L}'/(\bar{K}_1/\bar{L}_1)\,(G/H) = \%\text{ change} \\ \bar{K}' &= u' + (1 + u')\,\bar{L}'(H\bar{L}_1/G\bar{K}_1)\,(u_1/e_1) \end{aligned} \tag{22}$$

The data on the LINE system indicate that capital utilization u_1 is approximately equal to the labor efficiency e_1 at a given output rate, or the ratio u_1/e_1 is approximately equal to one, which further simplifies Eq. (22) as follows.

$$\bar{K}' = \bar{L}'H\bar{L}_1/GK_1 + u'(1 + \bar{L}'H\bar{L}_1/GK_1) = \%\text{ capital increase} \tag{23}$$

Equation (23) makes it possible to estimate how much additional average capital investment per part can be made as a function of three elements: (a) the minimum design requirements per part, (b) the utilization of capital, and (c) the capital-labor cost ratio.

The utilization of capital equipment in a system depends on its output rate. At higher output rates the utilization increases and approaches an optimum value. An estimate of how the relative utilization of equipment changes with output rate for two systems can be made by using the earlier linear estimates of the capital requirements for production systems.

Table 11.10. Breakeven percentage changes in capital investment for CMS over LINE vs. capacity[a]

Capacity parts/h	% Higher capital use	% Higher capital invest.	
		Breakeven	Actual
	%	%	%
Q = 1	61	154	–
2	37	117	15
3	26	99	46
4	21	91	52
5	17	85	57
6	14	80	61
7	12	77	63
8	11	75	65
9	10	74	67
10	9	72	68

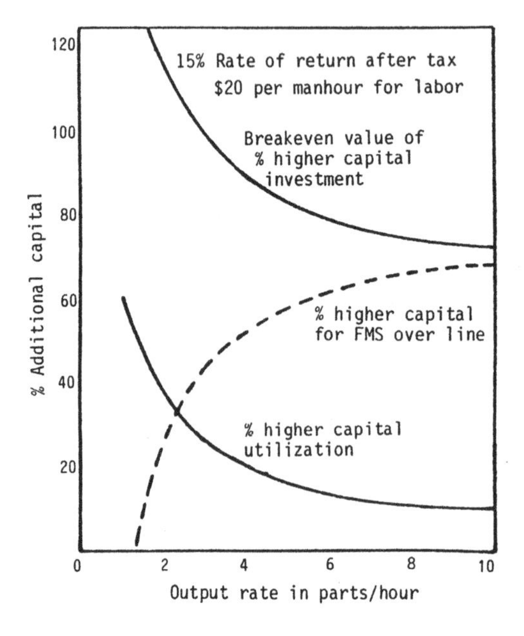

Fig. 11.12. Percent additional capital in FMS justified for different output rates

$$
\begin{aligned}
u' &= (\bar{K}_2 Q/K_2)/(\bar{K}_2 Q/K_1) - 1 = \text{percent increase in capital} \\
u' &= \bar{K}_2[(K_{01} + K_1(Q - Q_{01}))]/\bar{K}_1[(K_{02} + K_2(Q - Q_{02}))] - 1 \\
u' &= (K'_{01} Q_{01} - K'_{02} Q_{02})/(K'_{02} + Q)
\end{aligned}
\tag{24}
$$

$$K'_{01} = K_{01}/Q_{01}\bar{K}_1 - 1 = \$ \text{ extra capital in minimal system}$$

This estimate of how capital utilization changes with output depends on the characteristics of the minimal system, i.e., the system with only one of each type of processing station. The estimation equation, in the third line of the above formulation, depends on the percentage of extra capital needed by such a system K_0' and the output capacity of such a system Q_0 for each of the process plans being compared.

Representative values of the percentage changes in capital utilization when comparing the FMS and LINE system are computed in Table 11.10 using Eq. (24). These values are then used in Eq. (23) to compute the allowable increases in capital investment that can be made for different cost factors. These results are plotted in Fig. 11.12.

This figure compares FMS with LINE showing the percentage increase in capital utilization by FMS, the percentage increase in capital investment required, and the *breakeven percentage* of additional capital allowed with a 15% after-tax rate of return and a $20 per manhour labor cost.

11.9 Summary

The economic justification of computer integrated manufacturing systems depends on many factors as shown in the above analysis and these can be conveniently grouped into three categories as follows:

A. *System Design Factors*
1. Product characteristics and processing requirements which can be defined in terms of the kinds of workpieces and fixturings (k), the sequence of operations (j), and the kinds of workstations used (i).
2. Operation time (t_{ijk}) and frequency (f_{ijk}), and the average station time (t_i) and total time (T) for a unit of finished product.
3. Equipment requirements, cost per station (k_i) and the average capital investment per unit of product ($\bar{K}$).
4. Labor requirements for stations (L_i) and the average labor required per unit of output ($\bar{L}$).
5. Configuration of a minimal size system for making the output, including its cost (K_0), labor force (L_0), and its output capacity (Q_0).

The analysis shows that it is possible to summarize the economic importance of the design factors with linear approximations of the type shown in Figs. 11.1 and 11.2. These curves show that the marginal requirements per unit of capacity are fairly constant for systems above the minimum output level, while the average requirements per unit of capacity fall as the capacity is increased and approach the marginal requirements. A system's economic advantage is improved if (a) the requirements for a minimal system are reduced, or, more specifically, if the average values K_0/Q_0 and L_0/Q_0 can be reduced, and (b) if the capacity Q_0 of a minimal system can be kept small, i.e., the finished product can be made on a relatively small modular system. This was done in the FMS example by using multi-function machining centers instead of special purpose machining stations. However, in this example a relatively expensive material transporter system was used which exaggerated the cost of the minimal system. A less expensive, modular material handling device would be economically advantageous at low output rates.

In both of the example systems used, machining was done with only one tool cutting at a time. The use of multiple-tool operations might have a large effect in reducing operation times. However, this advantage would have to be weighed against the possible loss in flexibility for different product mixes and the higher cost of such hardware.

B. *System Operating Factors*

6. Products and product-mixes determine the average station time requirements per unit of finished product.
7. Production rates (P) relative to the system design capacity (Q) for particular part requirements.
8. Equipment utilization and capital investment utilization (u) which is a function of system capacity and output rate.
9. Labor efficiency (e) which is a function of capacity and output rate.
10. The intensity of system use as measured by the number of shifts (S) worked per week.

The analysis indicates that there is considerable economic advantage in matching the production rate of a flexible system to its design capacity by adjusting the product mix. This ability to vary the product mix on a flexible system makes it possible to operate the system more intensively and get much higher capital recovery rates.

The analysis shows that the flexible system has a much higher machine and capital investment utilization at relatively low capacities than does more specialized LINE-type systems, while the labor efficiency is about the same for both types of systems over the entire output range. At relatively high output rates, the equipment utilization of the specialized LINE system approaches that of the flexible system. Thus at high output rates the added cost of equipment flexibility cannot be justified by increased efficiency. This result appears to help explain why the FMS-type systems that have been built are designed for mid-range capacities and have not been modularly expanded to higher output rates, where dedicated LINE-type systems are economically preferred.

C. *Economic Environmental Factors*

11. Cost of labor (H), taking into account the hours worked per shift-year.
12. Payback requirements (I/J), taking into account the intensity of capital usage in shifts per week.
13. Rate of return before taxes (R_b), taking into account the useful life of the equipment.
14. Taxes and the rate of return after taxes (R_a), taking into account tax credit and depreciation provisions.
15. The cost of capital on an hourly basis (G), which incorporates all of the above factors.

The economic environmental factors can play a decisive role in the economic justification of capital-intensive, labor-saving production systems. A relatively high cost of capital and a relatively low cost of labor make it virtually impossible to justify the replacement of conventional systems with more automated ones. The more intensive use of automated systems on a three or four shift per week basis can do much to reduce the negative effect of high capital rate of return requirements. This result would seem to be consistent with the tendency for existing FMS-type systems to be used on a very intensive basis, even when other production operations are being cut back. It would also be consistent, for example, with the interest of many industries in developing automated systems for around-the-clock operations with un-manned operations on evening shifts.

Changes in the tax laws which allow greater investment credit and faster depreciation write-offs can foster the introduction of automated equipment by reducing the net cost of capital after taxes are taken into account.

In general the analysis shows that while simple payback rules can be used to give consistent measures of the economic advantage of automated systems, considerable care should be given in choosing appropriate payback rules. The break-even values should take into account the other economic factors listed above.

The present study serves as a beginning for a more comprehensive study of the ecomomics of computerized manufacturing systems. The study clarifies many of the more important factors that must be considered and shows they can be combined into relatively simple computational formulae. These formulae need to be tested on a variety of different products, product mixes, and different types of computerized production systems. The tested models and the refinements and corrections that should result from the tests, need to be applied and tested further at three levels of decision-making: at the *design level* where decisions are made about equipment specification and selection, process capabilities, and automated controls; at the *operational level* where decisions are made about production requirements, scheduling, and operating procedures; and at the *strategic level* where decisions are made about the extent and timing of the introduction of automated production facilities.

At the design level, it would seem desireable to put a great deal of effort into seeing how best to incorporate the economic analysis with the physical systems analysis into a total computer-aided design package. In doing this, however, considerable attention should be given to the need for including recent efforts to develop artificially intelligent systems that simulate and make maximum use of "human expertise" and good judgment in design decisions. At the operational level, there is a need to build more comprehensive models of the total operating environment for a particular system, taking into account the medium-run production objectives of a firm, its existing production facilities and other resources, and the prevailing economic climate.

It is at the strategic level that the problems of economic modelling are most difficult and most important, since the decisions at this level bend the decisions at both the design and operational levels. As Skinner [7] has argued, there is a pressing need "to change a corporation's basic management approach in manufacturing from short-term to long-term, from operations to policy, and from tactical to strategic." This change requires the development of more explicit methods and models for planning long-term changes in manufacturing facilities. It appears that there has been relatively little explicitness in the decisions to install all or most of the computerized manufacturing systems in the U.S. to date, since it is commonly reported that these decisions depended on the intuitive judgment of a "gutsy vice president" with enough clout and foresight to commit his company to highly innovative, expensive, and risky new directions in manufacturing technology. While it is certain that such decisions will always depend in a crucial way on expert human judgement in particular situations, there is an important need to be able to rationalize these decisions in an explicit manner in order to translate them into planning policies that can be delegated down the line, and in order to develop coherent patterns of decision making over future planning periods. An even greater challenge is to find ways to make strategic planning fully consistent with design planning and operations planning, i.e., to integrate the dominant man-machine concerns of operations with the dominant user-product concerns of design, and with the dominant economic concerns of strategic planning.

Economic analysis is one part of the overall planning process required for the proper introduction of computer integrated manufacturing methods and systems. Recently, the Charles Stark Draper Laboratory prepared a *Flexible Manufacturing System Handbook* [8] for the U.S. Army, which does an excellent job of describing the steps to be taken in the acquisition of an FMS. The Handbook notes that systematic evaluation procedures will only "indicate" which FMS configuration, if any, should be chosen to produce a given family of parts. The analysis cannot "choose the best configuration." The final decision must be tempered by judgment.

11.10 References

1. Lenz, JE, Talavage, JJ (1977) General Computerized Manufacturing Systems Simulation. Optimal Planning of Computerized Manufacturing Systems, Report No. 7, August 1977. School of Industrial Engineering, Purdue University, West Lafayette, Indiana
2. Solberg, JJ (1980) CAN-Q User's Guide. Optimal Planning of Cumputerized Manufacturing Systems. Report No. 9 (Revised), July 1980. School of Industrial Engineering, Purdue University, West Lafayette, Indiana
3. Runner, JA, Leimkuhler, FF (1978) CAMSAM: A Simulation Analysis Model for Computerized Manufacturing Systems. Optimal Planning of Computerized Manufacturing System, Report No. 13, December 1978. School of Industrial Engineering, Purdue University, West Lafayette, Indiana
4. Prenting, TO, Thomopoulis, NT (1974) Humanism and Technology in Assembly Line System. Spartan Books, Rochelle Park, New Jersey
5. Steffy, WR, Bawol, L. LaChance, Polacsek, D (1967) Numerical Control Justification: A Methodology. Institute of Science and Technology, University of Michigan, Ann Arbor, Michigan
6. Newnan, DG (1980) Engineering Economic Analysis. Engineering Press. San Jose, California (Revised)
7. Skinner, W (1980) The Factory of the Future-Always in the Future? A Managerial Viewpoint. In: Factory of the Future. American Society of Mechanical Engineers, Winter Annual Meeting, Special Volume PED-1, November 1980

8. Moriarty, BJ (1983) Flexible Manufacturing System Handbook. Charles Stark Draper Laboratory, Inc., Cambridge, Massachusetts, February 1983. Prepared for U.S. Army Tank Automotive Command, Warren, Michigan
9. Leimkuhler, FF (1982) Production Economics, Chap. 9.6. In: Salvendy, G (ed) Handbook of Industrial Engineering. John Wiley, New York
10. Rose, LM (1976) Engineering Investment Decisions. Elsevier, Amsterdam

Acknowledgement. Research described in this chapter was supported in part by a National Science Foundation Grant, Number-APR 74-15256, to Purdue University.

11.11 Appendix

CONA-COST Computer Program

CANQ-COST makes cost comparisons between two different manufacturing systems. The program was written by J. T. Hsu at Purdue University and is not intended for commercial use. The author assumes no responsibility for errors or difficulties of use. The program is very similar to Solberg's original CANQ program in reference [2].

Input Procedures

The first card of the input data indicates the output option and the number of cycles desired. The values are read as integers in the FORTRAN format I1 and I4. The output option must be an integer between 1 and 9, inclusive, and must appear in the first column. However, the output option is not important and a fixed integer constant of 3 is recommended for simplification. The number of cycles must be an integer and also no less than 1. It is suggested that the maximum number of cycles be 15.

The second card specifies the number of worksstations, the number of workpieces, number of part types, labor cost per operator per hour, and number of shifts per day. The number of workpieces must be at least 2, but no more than 100. All values must be integer and are read with a format of 5I5.

The third card reads the anticipated rate of return and estimated years of machine depreciation. The rate of return must be in decimal and has a format of F5.3. In general, an estimated before tax rate of return is between 0.2 (20%) and 0.3 (30%). The machine depreciation must be an integer and is given in years. It has a format of I5, to be read after the rate of return.

The fourth card describes the transporter, its cost and its labor requirement. The formats are A5, I5, F10.5, I5, and F5.3. The first five columns can be used to give the name of the transporter, such as TRANS or ROBOT. Then the number of transporters, average transporting time in minutes, the cost of each transporter in thousands of dollars, and the number of laborers are provided.

The next set of cards describes the workstations. Each workstation is represented by one card. The total number of workstation cards must equal the number of workstations given in the second card. The format for each workstation is given as A5, I5, I5, and F5.3. The general description of each workstation is very much like the transporter discussed earlier. The first five columns may be used for machine name, followed by the number of servers which must be one or more. Then, there are the cost of the machines in thousands of dollars and the number of laborers needed for each machine.

The final set of input data involves the part processing information. The set is divided into subsets and each subset represents a part type. The first card of each subset provides the name of the part, the number of operations that are required to produce the part, and what fraction of total output the part will be. An example of such a card would be: PART 1 xxxx 5 xxx 0.2, which means that part PART 1 has five operations and makes up 20% of the total production. It is important to realize that the fractions of total output for all parts must add up to one.

The cards following the first card of a subset describe the operations to be performed on the part. Each operation is described by the number of the workstations performing the operation, which process will be performed and the processing time at that workstation expressed in minutes. The ordering of the operations within each part has no significance on the system, but the number of operations cards must match the number of operations specified on the first card of the subset. An example of this situation would be if a part requires five operations, then it must be followed by five cards to describe those operations.

This program is written in such a way that two sets of input data are required, each one representing a production system. CANQ-COST determines the output rate at various production levels, the cost summaries and the utilization performance measures for both systems. An example with input formulation and output results is illustrated on the following pages.

Example Input Data

The production systems are: (a) a system called LINE, using conventional standalone machines, and (b) a computerized manufacturing system, called FMS, using multi-function DNC machining centers and a fully automated transporter system. The two systems produce the same product.

The operation sequences, operation times, equipment costs and labor requirements for LINE and CMS are shown in Table 11.A1 and 11.A2.

Example Output Data

The first page in the output shows the summary of each system. This summary indicates the composition of various factors and output rates for production. The next two pages of the output pertains top the cost analysis, it specifies various cost elements at each production level. This is followed by the station performance measures which are the machine utilizations of each workstation. Examples of the output are shown below followed by the program listing (Table 11.A3).

Table 11.A1

```
3--15                    (output option, cycle number)
----6---15----1---20----3
                         (number of stations, workpieces,
                         part types, labor cost per
                         hour, number of shifts)
-.200---10               (rate of return, machine
                         depreciation)
TRANS---50---2.0         (transporter)
MACH1----1------------250--1.0
MACH2----1             500 .1.0
MACH3----1             200  1.0
MACH4----1             500  1.0
MACH5----1             200  1.0
MACH6----1             150  1.0

PART1----7---1.0
    1 14.77
    2---9.17
    3---3.22             (part processing information)
    4--23.20
    3---3.13
    5--17.23
    6--15.50
3--15
----2----9----3---20----3
-.200---10
TRANS----1---2.0------------500--1.0
LOAD1    1                    0  1.0
MACH2    1                  500  1.0
PART1    2---.333
    1--6.0
----2--17.99
PART2----2---.333
----1--6.0
    2  35.50
PART3----2---.333
----1--6.0
    2  32.73

Note: Each-represents a blank space.
```

Table 11.A2

STATION PERFORMANCE MEASURE

SYSTEM NUMBER	MACHINE UTILIZATION MACH1	MACH2	MACH3	MACH4	MACH5	MACH6	TRANS
1	.620	.385	.267	.974	.723	.651	.012
2	.799	.496	.343	.627	.932	.838	.015
3	.903	.560	.388	.709	.527	.947	.017
4	.994	.617	.427	.781	.580	.522	.019
5	.630	.782	.542	.989	.735	.661	.024
6	.758	.942	.652	.794	.885	.796	.029
7	.822	.510	.707	.861	.959	.863	.031
8	.890	.553	.765	.932	.692	.934	.034
9	.930	.578	.800	.974	.724	.651	.035
10	.993	.617	.854	.780	.773	.695	.038
11	.757	.705	.976	.891	.883	.794	.043
12	.816	.760	.526	.962	.952	.857	.046
13	.849	.790	.547	.800	.990	.891	.048
14	.928	.864	.598	.875	.812	.974	.053
15	.977	.910	.630	.921	.855	.769	.056

STATION PERFORMANCE MEASURE

SYSTEM NUMBER	MACHINE UTILIZATION LOAD1	MACH2	TRANS
1	.209	1.000	.139
2	.418	1.000	.278
3	.626	.999	.417
4	.824	.987	.550
5	.961	.920	.640
6	.522	1.000	.696
7	.624	.996	.832
8	.708	.969	.944
9	.744	.891	.992
10	.834	.999	.556
11	.930	.990	.620
12	.986	.945	.658
13	.696	1.000	.696
14	.765	1.000	.765
15	.835	.999	.835

SYSTEM NUMBER	WORK STATIONS/MACHINE COST IN THOUSANDS MACH1 $ 250	MACH2 $ 500	MACH3 $ 200	MACH4 $ 500	MACH5 $ 200	MACH6 $ 150	TRANS $ -0	OUTPUT RATE PART / HOUR	CAPITAL COST	# OF MEN	UTILIZATION MEASURE CAPITAL	LABOR
1	1	1	1	1	1	1	50	2.5187	1800	6	.63	.60
2	1	1	1	2	1	1	50	3.2445	2300	7	.63	.67
3	1	1	1	2	2	1	50	3.6673	2500	8	.66	.66
4	1	1	1	2	2	2	50	4.0385	2650	9	.68	.64
5	2	1	1	2	2	2	50	5.1180	2900	10	.79	.74
6	2	1	1	3	2	2	50	6.1619	3400	11	.81	.80
7	2	2	1	3	2	2	50	6.6776	3900	12	.77	.80
8	2	2	1	3	3	2	50	7.2318	4100	13	.79	.80
9	2	2	1	3	3	3	50	7.5587	4250	14	.80	.78
10	2	2	1	4	3	3	50	8.0705	4750	15	.76	.77
11	3	2	1	4	3	3	50	9.2211	5000	16	.83	.83
12	3	2	2	4	3	3	50	9.9495	5200	17	.86	.84
13	3	2	2	5	3	3	50	10.3428	5700	18	.81	.83
14	3	2	2	5	4	3	50	11.3101	5900	19	.86	.86
15	3	2	2	5	4	4	50	11.9087	6050	20	.88	.86

OUTPUT SUMMARY

SYSTEM NUMBER	WORK STATIONS/MACHINE COST IN THOUSANDS LOAD1 $ -0	MACH2 $ 500	TRANS $ 500	OUTPUT RATE PART / HOUR	CAPITAL COST	# OF MEN	UTILIZATION MEASURE CAPITAL	LABOR
1	1	1	1	.6966	1000	3	.57	.20
2	1	2	1	1.3931	1500	3	.76	.40
3	1	3	1	2.0876	2000	3	.85	.60
4	1	4	1	2.7509	2500	3	.90	.79
5	1	5	1	3.2050	3000	4	.87	.69
6	2	5	1	3.4819	3000	5	.95	.60
7	2	6	1	4.1636	3500	5	.97	.71
8	2	7	1	4.7257	4000	5	.97	.81
9	2	8	1	4.9640	4500	5	.90	.85
10	2	8	2	5.5674	5000	6	.91	.80
11	2	9	2	6.2044	5500	7	.92	.76
12	2	10	2	6.5831	6000	7	.90	.81
13	3	10	2	6.9659	6000	8	.95	.75
14	3	11	2	7.6620	6500	8	.96	.82
15	3	12	2	8.3534	7000	8	.98	.90

COST ANALYSIS

Table 11.A2 (continued)

SYSTEM NUMBER	OUTPUT RATE PART/YEAR	MACHINE COST/YEAR	LABOR COST/YEAR	TOTAL COST/YEAR	TOTAL COST/HOUR	MACHINE COST/PART	LABOR COST/PART	AVERAGE COST/PART
1	15112.2	429341.0	720000.0	1149341..	191.6	28.4	47.6	76.05
2	19467.1	548602.3	840000.0	1388602.3	231.4	28.2	43.1	71.33
3	22003.5	596306.9	960000.0	1556306.9	259.4	27.1	43.6	70.73
4	24231.1	632085.3	1080000.0	1712085.3	285.3	26.1	44.6	70.66
5	30707.9	691716.0	1200000.0	1891716.0	315.3	22.5	39.1	61.60
6	36971.2	810977.4	1320000.0	2130977.4	355.2	21.9	35.7	57.64
7	40065.6	930238.8	1440000.0	2370239.8	395.0	23.2	35.9	59.16
8	43390.5	977943.3	1560000.0	2537943.3	423.0	22.5	36.0	58.49
9	45352.1	1013721.7	1680000.0	2693721.7	449.0	22.4	37.0	59.40
10	48423.1	1132983.1	1800000.0	2932983.1	488.8	23.4	37.2	60.57
11	55326.9	1192613.8	1920000.0	3112613.8	518.8	21.6	34.7	56.26
12	59697.2	1240318.3	2040000.0	3280318.3	546.7	20.8	34.2	54.95
13	62056.7	1359579.7	2160000.0	3519579.7	586.6	21.9	34.8	56.72
14	67860.3	1407284.3	2280000.0	3687284.3	614.5	20.7	33.6	54.34
15	71452.4	1443062.7	2400000.0	3843062.7	640.5	20.2	33.6	53.78

COST ANALYSIS

SYSTEM NUMBER	OUTPUT RATE PART/YEAR	MACHINE COST/YEAR	LABOR COST/YEAR	TOTAL COST/YEAR	TOTAL COST/HOUR	MACHINE COST/PART	LABOR COST/PART	AVERAGE COST/PART
1	4179.5	238522.8	270000.0	508522.8	84.8	57.1	64.6	121.67
2	8358.6	357784.1	300000.0	657784.1	109.6	42.8	35.9	78.70
3	12525.6	477045.5	330000.0	807045.5	134.5	38.1	26.3	64.43
4	16505.7	596306.9	360000.0	956306.9	159.4	36.1	21.8	57.94
5	19229.8	715568.3	390000.0	1105568.3	184.3	37.2	20.3	57.49
6	20891.6	715568.3	510000.0	1225568.3	204.3	34.3	24.4	58.66
7	24981.7	834829.6	540000.0	1374829.6	229.1	33.4	21.6	55.03
8	28354.5	954091.0	570000.0	1524091.0	254.0	33.6	20.1	53.75
9	29784.0	1073352.4	600000.0	1673352.4	278.9	36.0	20.1	56.18
10	33404.5	1192613.8	720000.0	1912613.8	318.8	35.7	21.6	57.26
11	37226.5	1311875.2	750000.0	2061875.2	343.6	35.2	20.1	55.39
12	39498.5	1431135.5	780000.0	2211136..	368.5	36.2	19.7	55.98
13	41795.2	1431136.5	900000.0	2331136.5	388.5	34.2	21.5	55.78
14	45972.0	1550397.9	930000.0	2480397.9	413.4	33.7	20.2	53.95
15	50120.3	1669659.3	960000.0	2629659.3	438.3	33.3	19.2	52.47

Table 11.A3

```
      PROGRAM CANQ(INPUT,OUTPUT,TAPE5=INPUT,TAPE7=OUTPUT,TAPE6,TAPE8,
     6TAPE9)
      DIMENSION NAME(50),NS(50),Q(50),PT(50),R(50),C(50)
      DIMENSION F(100),G(100),PROB(100),CPROB(100)
      DIMENSION PIDEN(80),NOPS(80),FRAC(80),APTS(50),ANOPS(80)
      DIMENSION RS(80,50),U(50),VAL(80),DVDP(80),UTILL(80)
      DIMENSION NSTA(50,50),PTOP(50,50),FREQ(50,50)
      DIMENSION Z(50),S(50),X(50),Y(50),MEN(50)
      INTEGER CSUM,COST(50),OPT,DDD,CYCLE
      REAL LA,MA,LEFF,MEFF,MEN,MCY,LCY
C
C           THIS PROGRAM, CALLED CAN-Q, ANALYZES MANUFACTURING
C           SYSTEM PERFORMANCE BY MODELING THE WORKFLOW.  IT
C           USES QUEUEING NETWORK THEORY, DESCRIBED IN A PAPER
C           ENTITLED -A MATHEMATICAL MODEL OF COMPUTERIZED
C           MANUFACTURING SYSTEMS.-  DETAILED INSTRUCTIONS FOR
C           DATA PREPARATION ARE GIVEN IN -A USER=S GUIDE TO
C           CAN-Q (FORTRAN VERSION).-  BOTH PAPERS ARE AVAILABLE
C           FROM THE AUTHOR, PROF. JAMES SOLBERG, SCHOOL OF
C           INDUSTRIAL ENGINEERING, PURDUE UNIVERSITY, WEST
C           LAFAYETTE, INDIANA, 47907.  THIS WORK WAS CARRIED
C           OUT UNDER NATIONAL SCIENCE FOUNDATION GRANT NUMBER
C           APR74 15256.
C
C
C           THIS VERSION WAS LAST REVISED 4/26/79.
C
C
C           READ DESIRED OUTPUT OPTION AND NUMBER OF CYCLES
C           NEEDED (LESS THAN 20)
C                OUTPUT OPTION:
C                1=SYSTEM MEASURES ONLY
C                2=BRIEF STATION SUMMARY
C                3=ABOVE PLUS FUNCTIONS OF N
C                4=FULL STATION SUMMARY
C                5=FULL STATION SUMMARY AND DISTRIBUTIONS
C                6=ALL OF ABOVE PLUS SENSITIVITY
C
C
      SSS=0.
      WHILE(SSS.LE.1) DO
      READ(5,200) OPT,CYCLE
  200 FORMAT(I1,I4)
      DDD=0
C
C
C          READ SIZE OF PROBLEM. M = NUMBER OF WORKSTATIONS
C          (LESS THAN 50)
C          N = NUMBER OF WORKPIECES (AT LEAST 2 BUT NOT MORE
C          THAN 100)
C          NPART = NUMBER OF PART TYPES (LESS THAN 80)
C          LCOST=LABOR COST PER HOUR
C          NSHIFT=NUMBER OF SHIFTS PER DAY (BETWEEN 1 TO 3)
C
C
      READ(5,202)M,N,NPART,LCOST,NSHIFT
  202 FORMAT(5I5)
C          READ THE RATE OF RETURN (RR) AND NUMBER OF YEARS
C          OF MACHINE DEPRECIATION (NYD)
C
C
      READ(5,2002)RR,NYD
 2002 FORMAT(F5.3,I5)
C
C
C          READ NAME OF TRANSPORT SYSTEM, NUMBER OF CARRIERS,
C          AVERAGE TRAVEL TIME IN MINUTES, THE COST OF EACH TRANSPORTER,
C          NUMBER OF LABORER AT EACH STATION.
C
C
```

Table 11.A3 (continued)

```
      MM = M+1
      READ (5,204) NAME(MM),NS(MM),PT(MM),COST(MM),MEN(MM)
  204 FORMAT(A5,I5,1F10.5,I5,F5.3)
C
C
C          FOR EACH STATION, READ NAME,NUMBER OF SERVERS,THE INDIVIDUAL
C          MACHINE COST, AND THE NUMBER OF LABORER REQUIRED.
C
C
      DO 2 I=1,M
      READ (5,206) NAME(I),NS(I),Q(I),PT(I),COST(I),MEN(I)
      Z(I)=PT(I)
      S(I)=Q(I)
  206 FORMAT(A5,I5,2F5.3,I5,F5.3)
      U(I)=0.
      APTS(I)=0.
    2 CONTINUE
      IF(NPART.EQ.0) GO TO 12
      ANOP = 0.
C
C
C          READ, COMPUTE, AND PRINT PART PROCESSING INFORMATION
C
C
      DO 10 K=1, NPART
      READ (5,208) PIDEN(K),NOPS(K),FRAC(K),VAL(K)
  208 FORMAT (A5,I5,2F10.5)
      IF(VAL(K).LE.0.) VAL(K)=1.
      ANOPS(K)=0.
      NEND=NOPS(K)
      DO 11 NO=1,NEND
      READ (5,214)NSTA(NO,K),PTOP(NO,K),FREQ(NO,K)
  214 FORMAT(I5,2F10.5)
   11 CONTINUE
   10 CONTINUE
C
C          DO FOR WRITE STATEMENTS
C
      ISUM=1
  999 DO 130 K=1,NPART
C
C          READ PART IDENTIFIER, NUMBER OF OPERATIONS,
C          DESIRED FRACTION OF PRODUCTION FOR EACH PART TYPE,
C          AND (OPTIONALLY) VALUE OF THE PART
C
      IF(DDD.EQ.0) THEN
      WRITE(6,210) PIDEN(K)
  210 FORMAT(1H1,17HROUTING FOR PART ,A5)
      WRITE (6,400)
      WRITE (6,212)
  212 FORMAT (1X,16HOPERATION NUMBER,5X,14HSTATION NUMBER,5X,15HPROCESSI
     CNG TIME,5X,9HFREQUENCY)
      END IF
      ANOPS(K)=0.
      NEND=NOPS(K)
C
C          READ STATION NUMBER, PROCESSING TIME, AND FREQUENCY
C          (IF NOT 1) FOR EACH OPERATION
C
      DO 13 NO=1,NEND
      IF (FREQ(NO,K).LE.0.)FREQ(NO,K)=1.
      IF(DDD.EQ.0)THEN
      WRITE (6,216)NO,NSTA(NO,K),PTOP(NO,K),FREQ(NO,K)
  216 FORMAT (4X,I5,15X,I5,15X,F10.5,5X,F10.5)
      END IF
      U(NSTA(NO,K))=U(NSTA(NO,K))+FREQ(NO,K)
      APTS(NSTA(NO,K))=APTS(NSTA(NO,K))+PTOP(NO,K)
      ANOPS(K)=ANOPS(K)+FREQ(NO,K)
   13 CONTINUE
```

Table 11.A3 (continued)

```
C
C         PRINT SUMMARY FOR THIS PART TYPE
C
      IF(DDD.EQ.0) THEN
      WRITE(6,400)
      WRITE(6,218)
      WRITE(6,219)
  218 FORMAT(3X,7HSTATION,2X,9HNUMBER OF,2X,7H  VISIT,4X,
     C8H   TOTAL,6X,9H  AVERAGE,5X,8HRELATIVE)
  219 FORMAT(12X,7H VISITS,4X,9HFREQUENCY,2X,12HPROCESS TIME,2X,
     C12HPROCESS TIME,2X,8HWORKLOAD)
      END IF
      ANOP = ANOP + FRAC(K)*ANOPS(K)
      DO 6 J=1,M
      IF(V(J).EQ.0) GO TO 6
      VISFR = V(J)/ANOPS(K)
      IF(V(J).GT.0.) APT = APTS(J)/V(J)
      IF(V(J).LE.0.) APT = 0.
      RELWK = VISFR*APT
      IF(DDD.EQ.0) THEN
      WRITE(6,220) J,NAME(J),V(J),VISFR,APTS(J),APT,RELWK
  220 FORMAT(1X,I3,2X,A5,F8.3,4X,F8.3,3X,F8.3,6X,F8.3,6X,F7.3,
     CF10.5,10X,F10.5)
      END IF
      RS(K,J) = RELWK
      PT(J) = PT(J) + APTS(J)*FRAC(K)
      Q(J) = Q(J) + V(J)*FRAC(K)
      APTS(J) = 0.
      V(J) = 0.
    6 CONTINUE
      ANO=ANOPS(K)
      IF (DDD.EQ.0) THEN
      WRITE(6,400)
      WRITE(6,221) ANO
  221 FORMAT(1X,51HAVERAGE NUMBER OF OPERATIONS TO COMPLETE ONE PART =,
     CF10.5)
      WRITE(6,222) FRAC(K)
  222 FORMAT(1X,32HDESIRED FRACTION OF PRODUCTION =,F10.5)
      WRITE(6,223) VAL(K)
  223 FORMAT(1X,19HVALUE OF ONE UNIT =,F10.2)
      END
  130 CONTINUE
C
C
C
C         COMPUTE RELATIVE WORKLOADS AND IDENTIFY MAX
C
   12 CONTINUE
      Q(MM) = 1./ANOP
      R(MM)=PT(MM)
      RMAX=R(MM)/FLOAT(NS(MM))
      IMAX=MM
      DO 16 I=1,M
      IF(Q(I).EQ.0) GO TO 16
      IF(NPART.GT.0) PT(I) = PT(I)/Q(I)
      IF(NPART.GT.0) Q(I) = Q(I)/ANOP
      R(I) = Q(I)*PT(I)
      TEST=R(I)/FLOAT(NS(I))
      IF (TEST.LE.RMAX) GO TO 14
      RMAX=TEST
      IMAX=I
   14 CONTINUE
   16 CONTINUE
C
C
C         PRINT INPUT PARAMETERS AND RELATIVE WORKLOADS
C
C
```

Table 11.A3 (continued)

```
      IF(DDD.EQ.0) THEN
      WRITE (6,232)
  232 FORMAT(1H1,18HINPUT DATA SUMMARY)
      WRITE (6,410)
  410 FORMAT(///)

      WRITE (6,234)
  234 FORMAT(14X,9HNUMBER OF,5X,5HVISIT,9X,7HAVERAGE,8X,8HRELATIVE,
     C5X,8HWORKLOAD)
      WRITE (6,236)
  236 FORMAT(2X,7HSTATION,6X,7HSERVERS,4X,9HFREQUENCY,3X,
     C15HPROCESSING TIME,4X,8HWORKLOAD,4X,10HPER SERVER)
      END IF
      DO 18 I=1,MM
      IF(DDD.EQ.0) THEN
      WRITE (6,400)
  400 FORMAT(/)
      END IF
      WPS = R(I)/FLOAT(NS(I))
      IF(DDD.EQ.0) THEN
      WRITE (6,238) I,NAME(I),NS(I),Q(I),PT(I),R(I),WPS
  238 FORMAT(1X,I2,3X,A5,6X,I2,4X,F10.5,6X,F10.5,6X,F10.5,2X,F10.5)
      END IF
   18 CONTINUE
      IF(DDD.EQ.0) THEN
      WRITE (6,410)
      WRITE (6,240) N
  240 FORMAT(1X,32HNUMBER OF WORKPIECES IN SYSTEM =,1X,I2)
      WRITE (6,405)
      WRITE (6,242) ANOP
  242 FORMAT(1X,51HMEAN NUMBER OF OPERATIONS TO COMPLETE A WORKPIECE =,
     CF10.5)
      END IF
C
C
C         COMPUTE MATRIX G
C
C
      NN = N+5
      DO 40 J=1,NN
      F(J)=0.
      G(J)=0.
      PROB(J)=0.
      CPROB(J)=0.
   40 CONTINUE
      DO 54 I=1,MM
      IF(R(I).LE.0.) GO TO 53
      TEMP=1.
      IF (NS(I).GT.1) GO TO 44
C
C             SINGLE SERVER CASE
C
      DO 42 J=1,NN
      TEMP = R(I)*TEMP + G(J)
      G(J) = TEMP
   42 CONTINUE
      GO TO 52
C
C             MULTIPLE SERVER CASE
C
   44 CONTINUE
      DO 50 J=1,NN
      JJ = NN-J+1
      DO 46 K=1,JJ
      A = FLOAT(NS(I))
      CHK = FLOAT(JJ-K+1)
      IF(CHK.LT.A) A=CHK
      TEMP = (R(I)*TEMP)/A + G(K)
```

Table 11.A3 (continued)

```
   46 CONTINUE
      G(JJ) = TEMP
      TEMP = 1.
   50 CONTINUE
   52 CONTINUE
   53 CONTINUE
   54 CONTINUE
C
C
C
C          COMPUTE AND PRINT SYSTEM PERFORMANCE MEASURES
C

C
      WRITE (6,260)
  260 FORMAT(1H1,27HSYSTEM PERFORMANCE MEASURES)
      WRITE (6,405)
  405 FORMAT(//)
      P = 60.*Q(MM)*G(N-1)/G(N)
      WRITE (6,262) P
  262 FORMAT(10X,17HPRODUCTION RATE =,F8.4,2X,15HPIECES PER HOUR)
      IF(NPART.EQ.0) GO TO 62
      WRITE(6,400)
      WRITE(6,264)
  264 FORMAT(10X,29HPRODUCTION RATES BY PART TYPE)
      WRITE(6,400)
      WRITE(6,265)
  265 FORMAT(25X,6HNUMBER,6X,5HVALUE)
      TVAL = 0.0
      DO 60 K=1,NPART
      PRPT = P*FRAC(K)
      VALUE = VAL(K)*PRPT
      TVAL = TVAL + VALUE
      WRITE(6,266) PIDEN(K),PRPT,VALUE
  266 FORMAT(15X,A5,3X,F8.4,3X,F8.2)
   60 CONTINUE
      WRITE(6,267) TVAL
  267 FORMAT(20X,13HTOTAL VALUE =,F9.2)
      WRITE(6,405)
   62 CONTINUE
      T = FLOAT(N)*G(N)/(Q(MM)*G(N-1))
      WRITE (6,268) T
  268 FORMAT(10X,24HAVERAGE TIME IN SYSTEM =,F9.2,2X,7HMINUTES)
      WRITE (6,400)
      TP = (G(1)-R(MM))/Q(MM)
      WRITE (6,270) TP
  270 FORMAT (22X,10HPROCESSING,2X,F9.2)
      TT = R(MM)/Q(MM)
      WRITE (6,272) TT
  272 FORMAT(22X,9HTRAVELING,3X,F9.2)
      DT = T-TP-TT
      WRITE (6,274) DT
  274 FORMAT(22X,7HWAITING,5X,F9.2)
      IF (OPT.EQ.1) GO TO 195
      IF (OPT.LE.2) GO TO 68
      WRITE(6,405)
      WRITE(6,276)
  276 FORMAT(10X,36HFUNCTIONS OF N, NUMBER OF WORKPIECES)
      WRITE(6,400)
      WRITE(6,278)
  278 FORMAT(16X,1HN,4X,15HPRODUCTION RATE,5X,22HAVERAGE TIME IN SYSTEM)
      WRITE(6,400)
      PN = 60.*Q(MM)/G(1)
      T = ANOP*G(1)
      WRITE(6,280) PN,T
  280 FORMAT(16X,1H1,9X,F8.4,12X,F9.2)
      LIML = N - 5
      LIMU = NN
      IF(NN.LT.12) LIML=2
      IF(LIML.EQ.2) GO TO 64
      WRITE(6,282)
```

Table 11.A3 (continued)

```
  282 FORMAT(15X,1H.,13X,1H.,22X,1H.)
      WRITE(6,282)
      WRITE(6,282)
   64 CONTINUE
      DO 66 NW=LIML,LIMU
      PN = 60.*Q(MM)*G(NW-1)/G(NW)
      T = FLOAT(NW)*60./PN
      WRITE(6,284) NW,PN,T
  284 FORMAT(14X,I3,9X,F8.4,12X,F9.2)
   66 CONTINUE
      WRITE(6,282)
      WRITE(6,282)
      WRITE(6,282)
      PMAX = 60.*Q(MM)/RMAX
      WRITE(6,286) PMAX
  286 FORMAT(14X,3HINF,9X,F8.4,17X,3HINF)
      WRITE(6,400)
      WRITE(6,288) IMAX
  288 FORMAT(15X,25HTHE BOTTLENECK STATION IS,1X,I2)
C
C
C     PRINT OUTPUT RATE AND NUMBER OF STATIONS
C
C
      IF(DDD.EQ.0) THEN
      WRITE(7,6000)
 6000 FORMAT(1H1,14HOUTPUT SUMMARY)
       WRITE(7,1000)
 1000 FORMAT(/////,3X,6HSYSTEM,7X,
     C39HWORK STATIONS/MACHINE COST IN THOUSANDS)
      WRITE(7,1001) MM,(NAME(I),I=1,MM)
 1001 FORMAT(3X,9HNUMBER    ,*(A5,2X),4X,11HOUTPUT RATE,4X,
     C12HCAPITAL COST,4X,8H# OF MEN,3X,19HUTILIZATION MEASURE)
      WRITE(7,1002)MM,(COST(I),I=1,MM)
 1002 FORMAT(12X,*(1H$,I4,2X),4X,11HPART / HOUR,33X,7HCAPITAL,
     C3X,5HLABOR)
      WRITE(7,1003)
 1003 FORMAT(/)
      WRITE(9,8888)
 8888 FORMAT(1H1,27HSTATION PERFORMANCE MEASURE)
      WRITE(9,889)
  889 FORMAT(/////,3X,6HSYSTEM,7X,19HMACHINE UTILIZATION)
      WRITE(9,890)MM,(NAME(I),I=1,MM)
  890 FORMAT(3X,9HNUMBER    ,*(A5,2X))
      WRITE(9,891)
  891 FORMAT(/)
      END IF
      CSUM=0
      XLABOR=0.
      LABOR=0
      LSUM=0
      LA=0
      MA=0
      DO 111 I=1,MM
      CSUM=CSUM+COST(I)*NS(I)
      XLABOR=XLABOR+MEN(I)*NS(I)
      LABOR=XLABOR+.99999
      LSUM=LSUM+MEN(I)*NS(I)*LCOST
  111 CONTINUE
      PR=P/NPART
      AA=RR*(1+RR)**(NYD)/((1+RR)**(NYD)-1)
      BB=AA/(NSHIFT*2000)
      TC=1/PR*(BB*CSUM*1000+LSUM)
      DO 1111 KK=1,MM
      ANBS=R(KK)*G(N-1)/G(N)
      UTILL(KK)=ANBS/FLOAT(NS(KK))
      MA=MA+UTILL(KK)*NS(KK)*COST(KK)
      LA=LA+UTILL(KK)*MEN(KK)*NS(KK)*LCOST
```

Table 11.A3 (continued)

```
1111 CONTINUE
     MEFF=MA/CSUM
     LEFF=LA/(LABOR*LCOST)
     WRITE(7,1004)ISUM,MM,(NS(I),I=1,MM),PR,CSUM,LABOR,
    CMEFF,LEFF
1004 FORMAT(5X,I2,6X,*(I2,5X),3X,F8.4,5X,I10,8X,I4,9X,
    CF5.2,5X,F5.2)
     WRITE(9,892)ISUM,MM,(UTILL(LL),LL=1,MM)
 892 FORMAT(4X,I2,5X,*(F5.3,2X))
     IF(DDD.EQ.0) THEN
     WRITE(8,1005)
1005 FORMAT(1H1,13HCOST ANALYSIS)
     WRITE(8,1006)
1006 FORMAT(/////,8X,6HSYSTEM,3X,11HOUTPUT RATE,4X,7HMACHINE,7X,
    C5HLABOR,8X,5HTOTAL,9X,5HTOTAL,8X,7HMACHINE,6X,5HLABOR,
    C7X,7HAVERAGE)
     WRITE(8,1007)
1007 FORMAT(8X,6HNUMBER,4X,9HPART/YEAR,4X,9HCOST/YEAR,4X,
    C9HCOST/YEAR,4X,9HCOST/YEAR,5X,9HCOST/HOUR,5X,
    C9HCOST/PART,3X,9HCOST/PART,4X,9HCOST/PART)
     END IF
     AMC=AA*CSUM*1000
     ALAB=LSUM*NSHIFT*2000
     TCY=AMC+ALAB
     PRY=PR*NSHIFT*2000
     MCY=AMC/PRY
     LCY=ALAB/PRY
     TCH=BB*CSUM*1000+LSUM
     WRITE(8,1009)ISUM,PRY,AMC,ALAB,TCY,TCH,MCY,LCY,TC
1009 FORMAT(/,9X,I2,4X,F10.1,5X,F10.1,3X,F10.1,3X,F10.1,
    C1X,F10.1,7X,F7.1,6X,F7.1,4X,F10.2)
     DDD=1
  68 CONTINUE
C
C
C         COMPUTE STATION PERFORMANCE MEASURES
C
C
      DO 175 I=1,MM
      IF (OPT.LE.3) GO TO 140
      ANBS = R(I)*G(N-1)/G(N)
      UTIL = ANBS/FLOAT(NS(I))
      IF (NS(I).GT.1) GO TO 86
C
C             SINGLE SERVER CASE
C
      UTIL = ANBS
      SI = 1. - ANBS
      PB = ANBS
      PROB(1) = (R(I)/G(N))*(G(N-1)-R(I)*G(N-2))
      CPROB(1) = PB - PROB(1)
      QIN = PROB(1)
      QIL = 0.
      IF (N.LE.2) GO TO 82
      IF (N.EQ.3) GO TO 81
      NN=N-2
      DO 80 J=2,NN
      NJ=N-J
      PROB(J) = (R(I)**J/G(N))*(G(NJ)-R(I)*G(NJ-1))
      CPROB(J) = CPROB(J-1) - PROB(J)
      QIN = QIN + FLOAT(J)*PROB(J)
      QIL = QIL + (FLOAT(J)-1.)*PROB(J)
   80 CONTINUE
   81 CONTINUE
      PROB(N-1) = (R(I)**(N-1)/G(N))*(G(1)-R(I))
      CPROB(N-1) = CPROB(N-2) - PROB(N-1)
      QIL = QIL + (FLOAT(N)-2.)*PROB(N-1)
```

Table 11.A3 (continued)

```
   82 CONTINUE
      PROB(N) = R(I)**N/G(N)
      CPROB(N) = CPROB(N-1) - PROB(N)
      QIN = QIN + (FLOAT(N)-1.)*PROB(N-1) + FLOAT(N)*PROB(N)
      QIL = (QIL + (FLOAT(N)-1.)*PROB(N))*G(N)/(R(I)*G(N-1))
      C(I) = QIN - QIL
      GO TO 150
C
C             MULTIPLE SERVER CASE
C
   86 CONTINUE
      DO 92 J=1,N
      TEMP=1.
      IF (J.EQ.1) GO TO 90
      JJ=J-1
      DO 88 K=1,JJ
      A = FLOAT(NS(I))
      CHK = FLOAT(J-K+1)
      IF(CHK.LT.A) A=CHK
      TEMP = F(K) + R(I)*TEMP/A
   88 CONTINUE


   90 CONTINUE
      F(J) = G(J) - R(I)*TEMP
   92 CONTINUE
      SI = F(N)/G(N)
      PB = 1. - SI
      PROB(1) = R(I)*F(N-1)/G(N)
      CPROB(1) = PB - PROB(1)
      QIN = PROB(1)
      CTEMP = N-1
      IF (N.EQ.1) GO TO 150
      NN=N-1
      FI = R(I)
      DO 94 J=2,NN
      A = FLOAT(NS(I))
      CHK = FLOAT(J)
      IF(CHK.LT.A) A=CHK
      FI = (R(I)/A)*FI
      NJ=N-J
      PROB(J) = FI*F(NJ)/G(N)
      CPROB(J) = CPROB(J-1) - PROB(J)
      QIN = QIN + FLOAT(J)*PROB(J)
      A2 = FLOAT(NS(I))
      CHK = FLOAT(N-J+1)
      IF(CHK.LT.A2) A2=CHK
      CTEMP = CTEMP*R(I)/A2 + FLOAT(N-J)*F(J-1)
   94 CONTINUE
      A = FLOAT(NS(I))
      CHK = FLOAT(N)
      IF(CHK.LT.A) A=CHK
      PROB(N) = (R(I)/A)*(PROB(N-1)/F(1))
      CPROB(N) = CPROB(N-1) - PROB(N)
      QIN = QIN + FLOAT(N)*PROB(N)
      QIL = R(I)*CTEMP/G(N-1)
      C(I) = QIN - QIL
      GO TO 150
C
C
C             PRINT STATION PERFORMANCE MEASURES
C

C
  140 CONTINUE
      WRITE (6,300)
  300 FORMAT(1H1,28HSTATION PERFORMANCE MEASURES)
      WRITE (6,405)
      WRITE (6,302)
  302 FORMAT(1X,7HSTATION,5X,7HSTATION,7X,6HSERVER,8X,11HAVE. NO. OF)
      WRITE (6,304)
```

Table 11.A3 (continued)

```
304 FORMAT(2X,6HNUMBER,7X,4HNAME,6X,11HUTILIZATION,5X,12HBUSY SERVERS)
    WRITE (6,400)
    DO 145 II=1,MM
    ANBS = R(II)*G(N-1)/G(N)
    UTIL = ANBS/FLOAT(NS(II))
    WRITE (6,306) II,NAME(II),UTIL,ANBS
306 FORMAT(3X,I2,9X,A5,8X,F6.3,10X,F6.3)
    WRITE (6,400)
145 CONTINUE
    GO TO 195
150 CONTINUE
    ANPW = QIN - ANBS
    WRITE (6,308) I,NAME(I)
308 FORMAT(1H1,26HSUMMARY FOR STATION NUMBER,I3,2H :,2X,A5)
    WRITE (6,410)
    WRITE (6,310)
310 FORMAT(5X,9HNUMBER OF,7X,6HSERVER,8X,11HAVE. NO. OF)
    WRITE (6,311)
311 FORMAT(6X,7HSERVERS,6X,11HUTILIZATION,5X,12HBUSY SERVERS)
    WRITE (6,400)
    WRITE (6,312) NS(I), UTIL,ANBS
312 FORMAT(8X,I2,10X,F6.3,12X,F6.3)
    WRITE (6,405)
    WRITE (6,313)

313 FORMAT(5X,31HSTEADY STATE AVERAGE NUMBER OF:)
    WRITE (6,400)
    WRITE (6,314) QIN
314 FORMAT(20X,16HPARTS AT STATION,5X,F7.3)
    WRITE (6,315) ANBS
315 FORMAT(20X,16HPARTS IN PROCESS,5X,F7.3)
    WRITE (6,316) ANPW
316 FORMAT(20X,13HPARTS WAITING,8X,F7.3)
    WRITE (6,405)
    ANOPI = Q(I)/Q(MM)
    IF (I.EQ.MM) ANOPI=ANOP
    APTW = PT(I)*ANOPI
    ATTO = PT(I)*QIN/ANBS
    ATTW = ATTO*ANOPI
    AWTO = ATTO-PT(I)
    AWTW = ATTW-APTW
    WRITE (6,317)
317 FORMAT(5X,34HAVERAGE TIME SPENT AT THIS STATION,6X,13HPER OPERATIO
   CN,5X,13HPER WORKPIECE)
    WRITE(6,400)
    WRITE(6,318) ATTO,ATTW
318 FORMAT(15X,20HTOTAL TIME (MINUTES),12X,F7.3,12X,F7.3)
    WRITE(6,319) PT(I),APTW
319 FORMAT(18X,10HPROCESSING,19X,F7.3,12X,F7.3)
    WRITE(6,320) AWTO,AWTW
320 FORMAT(18X,7HWAITING,22X,F7.3,12X,F7.3)
    WRITE (6,405)
    IF (OPT.LT.5) GO TO 175
    WRITE (6,410)
    WRITE (6,325)
325 FORMAT(5X,16HFRACTION OF TIME,4X,18HX PARTS AT STATION,7X,
   C16HX PARTS EXCEEDED)
    WRITE (6,400)
    WRITE (6,326) SI,PB
326 FORMAT(10X,6HX =  0,15X,F5.4,20X,F5.4)
    DO 174 JJJ=1,N
    WRITE (6,327) JJJ,PROB(JJJ),CPROB(JJJ)
327 FORMAT(10X,3HX =,I3,15X,F5.4,20X,F5.4)
174 CONTINUE
175 CONTINUE
    IF(OPT.LT.6) GO TO 195
C
C
C         PRINT SENSITIVITY INFORMATION
C
C
```

Table 11.A3 (continued)

```
      WRITE(6,350)
  350 FORMAT(1H1,23HSENSITIVITY INFORMATION)
      WRITE(6,3560)
 3560 FORMAT(//,1X,24HA ONE MINUTE DECREASE IN,15X,17HWILL INCREASE THE)
      WRITE(6,3561)
 3561 FORMAT(1X,26HPROCESSING TIME AT STATION,13X,18HPRODUCTION RATE BY)
      DO 3550 I=1,MM
      PCPT=P*C(I)/PT(I)
      CPT=100.*C(I)/PT(I)
      WRITE(6,3562)I,NAME(I),PCPT,CPT
 3562 FORMAT(/,9X,I2,2X,A5,16X,F6.3,13H UNITS/HOUR (,F7.4,9H PERCENT))
 3550 CONTINUE
      WRITE(6,3563)
 3563 FORMAT(///,1X,25HA ONE PERCENT DECREASE IN,14X,17HWILL INCREASE THE)
     CE)
      WRITE(6,3561)
      DO 3551 I=1,MM
      PC=.01*P*C(I)
      WRITE(6,3562)I,NAME(I),PC,C(I)
 3551 CONTINUE
      WRITE(6,3560)
      WRITE(6,3566)
 3566 FORMAT(1X,31HRELATIVE UTILIZATION AT STATION,8X,18HPRODUCTION RATE B
     C BY)
      DO 3552 I=1,MM
      PCR=P*C(I)/R(I)
      CR=100.*C(I)/R(I)
      WRITE(6,3562)I,NAME(I),PCR,CR


 3552 CONTINUE
      IF(NPART.EQ.0) GO TO 195
      WRITE(6,3567)
 3567 FORMAT(//,1X,24HA DECREASE OF .01 IN THE,15X,17HWILL INCREASE THE)
      WRITE(6,3568)
 3568 FORMAT(1X,23HPROPORTION OF PART TYPE,16X,18HPRODUCTION RATE BY)
      DO 3553 K=1,NPART
      SUM=0.
      DO 3554 J=1,MM
      SUM=SUM+C(J)*(RS(K,J)/R(J))
 3554 CONTINUE
      DPDFP=(ANOPS(K)/ANOP)*SUM
      DPDF=.01*P*DPDFP
      DVDP(K)=.01*(TVAL*DPDFP-P*VAL(K))
      WRITE(6,3569)PIDEN(K),DPDF,DPDFP
 3569 FORMAT(/,9X,A5,15X,F11.3,13H UNITS/HOUR (,F7.4,9H PERCENT))
 3553 CONTINUE
      WRITE(6,3567)
      WRITE(6,4568)
 4568 FORMAT(1X,23HPROPORTION OF PART TYPE,16X,
     C22HVALUE OF PRODUCTION BY)
      DO 4569 K=1,NPART
      DVDPP=100.*DVDP(K)/TVAL
      WRITE(6,3569) PIDEN(K),DVDP(K),DVDPP
 4569 CONTINUE
C
C
  195 CONTINUE
      NS(IMAX)=NS(IMAX)+1
      DO 888 I=1,M
      PT(I)=Z(I)
      D(I)=S(I)
      V(I)=0.
      APTS(I)=0.
```

Table 11.A3 (continued)

```
888 CONTINUE
    ANOP=0.
    ISUM=ISUM+1
    N=N+3
    IF(ISUM.LE.CYCLE) GO TO 999
    SSS=SSS+1.
    END WHILE
    STOP
    END
     3  15
        6    15     1    20     3
      .200   10
     TRANS   50   2.0
     MACH1    1                250   1.0
     MACH2    1                500   1.0
     MACH3    1                200   1.0
     MACH4    1                500   1.0
     MACH5    1                200   1.0
     MACH6    1                150   1.0
     PART1    7    1.0
        1  14.77
        2   9.17
        3   3.22
        4  23.20
        3   3.13
        5  17.23
        6  15.50
     3  15
        2     9     3    20     3
      .200   10
     TRANS    1    2.0         500   1.0
     LOAD1    1                      1.0
     MACH2    1                500   .25
     PART1    2    .333
        1   6.0
        2  17.99
     PART2    2    .333
        1   6.0
        2  35.50
     PART3    2    .333
        1   6.0
        2  32.73
```

Subject Index

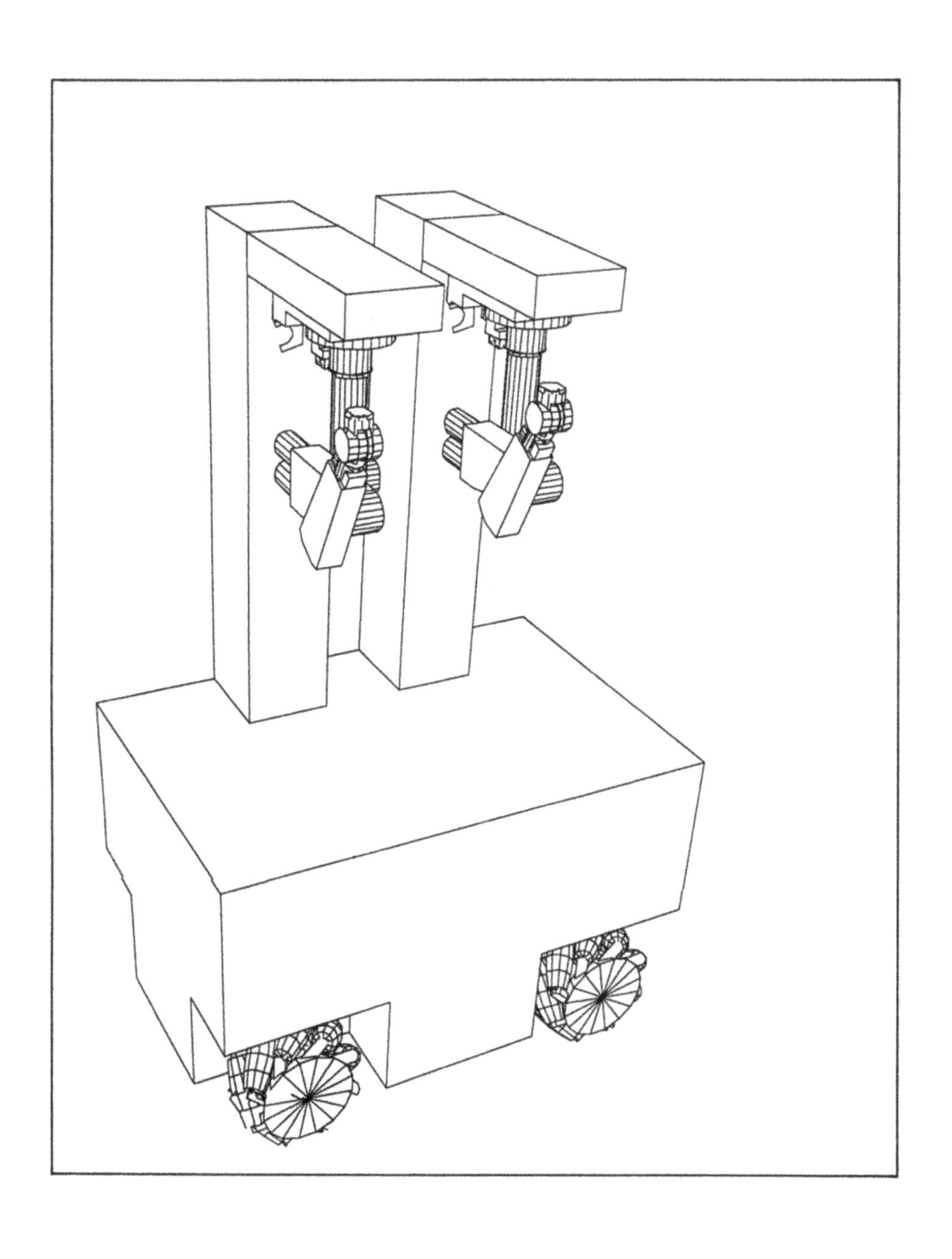

Experimental mobile platform with 2 manipulator arms

Contributors

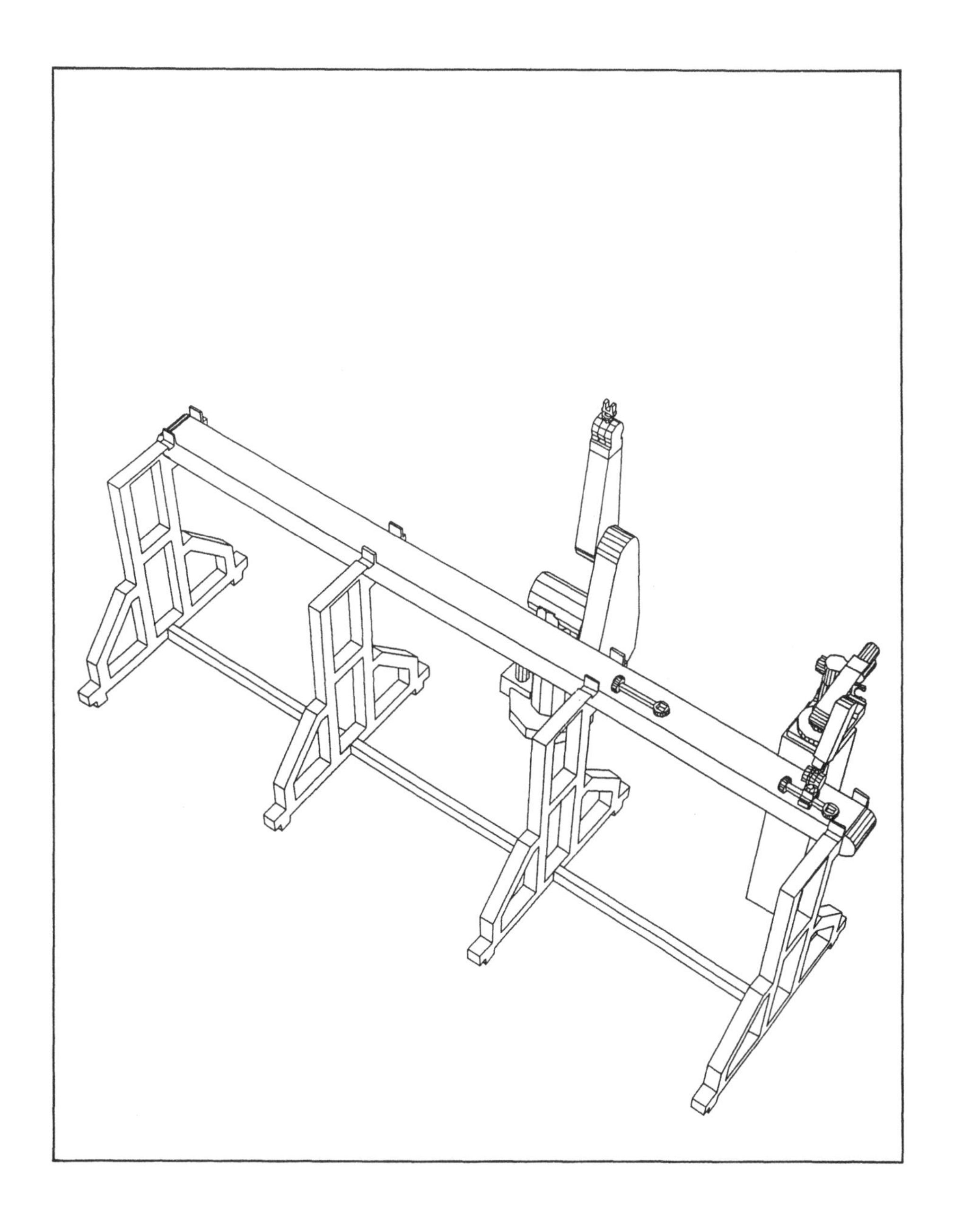

Model of a simple manufacturing cell consisting of two robots and a conveyor

R. Anderl
Institut für Rechneranwendung
in Planung und Konstruktion
University of Karlsruhe
Postfach 69 80
D-7500 Karlsruhe
FRG

H.-M. Anger
V.D.O.
Adolf Schindler AG
Loedenerstraße 9
D-6231 Schwalbach
FRG

C. Berard
Computer Aided Production Group
Laboratoire GRAI
University of Bordeaux I
351, Cours de la Libération
F-33405 Talence Cedex
France

V. Braud
Computer Aided Production Group
Laboratoire GRAI
University of Bordeaux I
351, Cours de la Libération
F-33405 Talence Cedex
France

M. Cividini
Department of Electronics
Politecnico di Milano
Piazza Leonardo da Vinci, 32
I-20133 Milano
Italy

R. Dillmann
Institut für Informatik III
University of Karlsruhe
Postfach 69 80
D-7500 Karlsruhe
FRG

G. Doumeingts
Computer Aided Production Group
Laboratoire GRAI
University of Bordeaux I
351, Cours de la Libération
F-33405 Talence Cedex
France

W. Epple
Swidmutstraße 5
8069 Schweitenkirchen
FRG

G. Gini
Department of Electronics
Politecnico di Milano
Piazza Leonardo da Vinci, 32
I-20133 Milano
Italy

M. Gini
Department of Electronics
Politecnico
Piazza Leonardo da Vinci, 32
I-20133 Milano
Italy

H. Grabowski
Institut für Rechneranwendung
in Planung und Konstruktion
Fakultät für Maschinenbau
University of Karlsruhe
Postfach 69 80
D-7500 Karlsruhe
FRG

W. Grottke
Fraunhofer Institut für
Produktionsanlagen
und Konstruktionstechnik (IPK)
Technische Universität Berlin
Fasanenstr. 12
D-1000 Berlin 12
FRG

F.-L. Krause
Fraunhofer Institut für
Produktionsanlagen
und Konstruktionstechnik (IPK)
Technische Universität Berlin
Pascalstr. 8–9
D-1000 Berlin 10
FRG

F. F. Leimkuhler
School of Industrial Engineering
Purdue University
West Lafayette, Indiana 47907
USA

P. Levi
Forschungszentrum Informatik
an der Universität Karlsruhe
Haid-und-Neu-Str. 10–14
D-7500 Karlsruhe
FRG

C. R. Liu
School of Industrial Engineering
Purdue University
West Lafayette, Indiana 47907
USA

M.-C. Maisonneuve
Computer Aided Production
Group
Laboratoire GRAI
University of Bordeaux I
351, Cours de la Libération
F-33405 Talence Cedex
France

A. H. Redford
Department of Aeronautical and
Mechanical Engineering
University of Salford
Salford M5 4WT
Great Britain

U. Rembold
Institut für Informatik III
University of Karlsruhe
Postfach 69 80
D-7500 Karlsruhe
FRG

A. W. Scheer
Institut für Wirtschaftsinformatik
Universität des Saarlandes
D-6600 Saarbrücken
FRG

L. Sidarous
Norsk Data
Postfach 13 02 20
Solingerstraße 9
D-4330 Mülheim an der Ruhr 13
FRG

G. Spur
Fraunhofer Institut für
Produktionsanlagen
und Konstruktionstechnik (IPK)
Technische Universität
Pascalstr. 8–9
D-1000 Berlin 10
FRG

R. Srinivasan
School of Industrial Engineering
Purdue University
West Lafayette, Indiana 47907
USA

W. Turowski
Fraunhofer Institut für
Produktionsanlagen
und Konstruktionstechnik (IPK)
Technische Universität Berlin
Pascalstr. 8–9
D-1000 Berlin 10
FRG

G. Villa
Department of Electronics
Politecnico di Milano
Piazza Leonardo da Vinci, 32
I-20133 Milano
Italy

Lightning Source UK Ltd.
Milton Keynes UK
UKOW07f2337190116

266715UK00015B/339/P

9 783642 827501